HIGH POLYMERS

A SERIES OF MONOGRAPHS ON THE CHEMISTRY, PHYSICS, AND TECHNOLOGY OF HIGH POLYMERIC SUBSTANCES

VOLUME XVI

PART III

HIGH POLYMERS

Vol. I: **Collected Papers of W. H. Carothers on High Polymeric S**
Edited by H. Mark and G. S. Whitby

Vol. II: **Physical Chemistry of High Polymeric Systems**
Second Edition. By H. Mark and A. V. Tobolsky

Vol. III: **Mechanism of Polymer Reactions**
By G. M. Burnett

Vol. V: **Cellulose and Cellulose Derivatives** (in three parts)
Second Edition. Edited by Emil Ott, H. M. Spurlin, and M. W. Grafflin

Vol. VI: **Mechanical Behavior of High Polymers**
By Turner Alfrey, Jr.

Vol. VII: **Phenoplasts: Their Structure, Properties, and Chemical Technology**
By T. S. Carswell

Vol. IX: **Emulsion Polymerization**
By F. A. Bovey, I. M. Kolthoff, A. I. Medalia, and E. J. Meeh

Vol. X: **Polymer Processes**
Edited by Calvin E. Schildknecht

Vol. XII: **Analytical Chemistry of Polymers** (in three parts)
Edited by Gordon M. Kline

Vol. XIII: **Polyethers** (in three parts)
Edited by Norman G. Gaylord

Vol. XIV: **Polyesters** (in two parts)
Edited by Norman G. Gaylord

Vol. XV: **Radiation Chemistry of Polymeric Systems**
By Adolphe Chapiro

Vol. XVI: **Polyurethanes: Chemistry and Technology** (in two parts)
By J. H. Saunders and K. C. Frisch

Part III: **Analytical Chemistry of the Polyurethanes**
By D. J. David and H. B. Staley

Vol. XVII: **Configurational Statistics of Polymeric Chains**
By M. V. Volkenstein. Translated by Serge N. Timasheff and Marina J. Timasheff.

Vol. XVIII: **Copolymerization**
Edited by George E. Ham

Vol. XIX: **Chemical Reactions of Polymers**
Edited by E. M. Fettes

Vol. XX: **Crystalline Olefin Polymers** (in two parts)
Edited by R. A. V. Raff and K. W. Doak

Vol. XXI: **Macromolecules in Solution**
By Herbert Morawetz

Vol. XXII: **Conformations of Macromolecules**
By T. M. Birshtein and O. B. Ptitsyn. Translated by Serge N. Timasheff and Marina J. Timasheff

Vol. XXIII: **Polymer Chemistry of Synthetic Elastomers** (in two parts)
Edited by Joseph P. Kennedy and Erik G. M. Törnqvist

ANALYTICAL CHEMISTRY OF THE POLYURETHANES

VOLUME XVI

Part III

D. J. DAVID
Group Leader
Analytical Research
Mobay Chemical Company
New Martinsville, West Virginia

H. B. STALEY
Research Librarian
Mobay Chemical Company
New Martinsville, West Virginia

WILEY-INTERSCIENCE

a Division of John Wiley & Sons

NEW YORK · LONDON · SYDNEY · TORONTO

The paper used in this book has pH of 6.5 or higher. It has been used because the best information now available indicates that this will contribute to its longevity.

Library of Congress Catalogue Card Number: 68-8110

SBN 471 39293 6

Printed in the United States of America

10 9 8 7 6 5 4 3 2 1

PREFACE

The two preceding volumes of this series* on polyurethanes by J. H. Saunders and K. C. Frisch thoroughly cover the basic chemistry and chemistry of this important class of high polymers. They describe the preparation of the raw materials, the reactions between the components that yield the polymer, and the practical application of the product.

With the continued expansion of the urethane industry there has been a parallel need for improved analytical techniques. In this volume an attempt has been made to bring together those analytical procedures that have proved useful and applicable to the various phases of urethane chemistry.

The development of an industrial analytical procedure requires some knowledge of the process by which the particular material is prepared. For this reason there is some discussion of the chemistry of the various processes so that the need for the analytical method can be appreciated. We have tried to keep such discussion to a minimum, however, and to confine ourselves to analytical chemistry.

We acknowledge the work done in the Central Analytical Laboratories of Farbenfabriken Bayer A.G., Leverkusen, in the development of many of the basic analytical procedures used in research as well as in quality control of the various processes invovled in the urethane industry. The Mobay Chemical Company, with which we have been associated, was formed in 1954 as a joint venture of Monsanto Chemical Company and Farbenfabriken Bayer. This arrangement provided the new company not only with the technical know-how for the manufacture and application of isocyanates, polyesters, and many related materials based on Bayer's years of experience in this field, but also with many of the analytical methods described here. Although Mobay has pursued an extensive program for development of analytical procedures, we have had access by consultation and correspondence to the developments pertinent to urethane chemistry made in their Central Analytical Laboratories. Inasmuch as it is difficult to trace the original designers of an analytical method, particularly in this case where it is the result of the combined efforts of several investigators, we must make this general acknowledgement of the substantial contributions made by the chemists of our parent company.

We also acknowledge the part played by the entire technical staff of Mobay, including the staff of the Engineering, Quality Control, and Research Departments. Industrial analytical problems usually arise out of a practical, sometimes critical, process problem, and an analytical group can be effective only if it becomes aware of these problems Moreover, its

* High Polymers, Volume XVI, Part I (1962) and Part II (1964).

members must have access to background information about the process in order to arrive at a feasible analytical procedure. Our technical staff have contributed substantially by discussing these problems and making suggestions about suitable analytical approaches.

We extend special appreciation to Dr. J. H. Saunders, Director of Research, and Dr. D. H. Chadwick, Assistant Director of Research, who gave the necessary encouragement and made many helpful comments during the preparation of the manuscript. For the secretarial assistance provided by Mrs. Roberta Amos, Mrs. Lorraine Demkovich, and Mrs. Jo Ann Wade who typed the manuscript although heavily burdened with other assignments we are particularly grateful. We also appreciate the efforts of Mr. Charles Crookshanks who prepared the numerous illustrations and figures.

Finally, we express our appreciation for their approval to prepare this manuscript to Mr. J. D. Mahoney and Dr. A. M. Gemassmer, who were President and Vice President respectively of Mobay Chemical Company when this volume was begun, and to Mr. B. R. Nason and Dr. G. K. Rockstroh, who now hold these positions, for their continued support.

November 1969

D. J. DAVID*
H. B. STALEY

* Presently with the Analytical Instruments Division of Tracor, Inc., Austin, Texas.

CONTENTS

CHAPTER I

GENERAL INTRODUCTION

As few as forty years ago the discipline of analytical chemistry appeared to have reached its zenith. Furthermore, the tools of the analytical chemist were relegated to those which essentially were embodied in the fields of classical gravimetric and volumetric analysis; combustion techniques for the determination of carbon, hydrogen, nitrogen and halogens; and physical property characterization methods such as melting point, boiling point, and refractive index.

Thus, the early organic chemists, such as Wurtz (1), Gautier (2), Hofmann (3), Hentschel (4), and Schroeter (5), among others, who prepared isocyanates and/or simple carbamates had limited analytical tools available to adequately define the product or mixtures of products that were obtained as a result of their experiments. At that time even the techniques that were available, such as those previously mentioned, were crude by comparison to the same types of tools of today.

For many intervening years the workers who prepared compounds which eventually had some bearing on the field of urethane chemistry, as well as others, had to be satisfied with analytical techniques which represented those available for identification and estimation of purity.

After the scientific concept of 'Macromolecular Chemistry' had been developed by H. Staudinger and the invention of super polyamides by Carothers had opened the door to a large scale commercialization of polymers and polycondensates, Professor O. Bayer and co-workers in the Central Research Laboratory of Farbenfabriken Bayer at Leverkusen, Germany discovered in 1937 the diisocyanate polyadditions process as a third important and universal method for building up high molecular weight polymers (6,7).

These workers did not stop at this point but began to investigate the field of polyurethane chemistry from the broad, new concept of the polyaddition principle. Through the broad vision of these workers, the polyurethanes have experienced rapid development and growth into both cellular and noncellular markets, resulting in a myriad of useful, relatively low-cost products with excellent properties.

Because of the market and technical growth of the polyurethanes, the analytical information necessary to establish the quality of the raw materials and intermediates, as well as the diisocyanates and polyols themselves, became more demanding and subtle. Fortuitously or otherwise, it was precisely around this time when rapid and significant advances in the field of analytical chemistry began to occur. From roughly that time to the present, many advances in the field of analytical instrumentation made possible new tools previously unavailable to the analytical chemist. In the early 1940's the first commercially obtainable infrared spectrometers were introduced, and significant improvements to increase the accuracy of pH measurements were achieved. In this same period good ultraviolet-visible spectrophotometers and polarographs also became available.

Then in 1952 James and Martin published the first account of gas–liquid chromatography which opened up a whole new field of analytical chemistry. This technique possessed such amazing separating abilities that both qualitative and quantitative work could be accomplished with a minimum expenditure of time and effort. The spectacular growth of this field, of course, is still going on.

With the advent of the advances in the technology of the electronics field between the 1950's and the present time came the perfection of many existing analytical instruments such as the mass spectrometer, electron microscope, x-ray diffractometer, carbon–hydrogen analyzers, differential thermal analyzers, thermogravimetric analyzers, and completely new instruments embodied by the wedding of theory and electronics that can be seen in instrumental techniques such as neutron activation analysis, NMR, EPR, and atomic absorption spectroscopy. The fields of optics and physics have also contributed in making available higher resolution, greater accuracy, and wavelength coverage spectrophotometers. Currently, the laser-excited Raman spectrometer has appeared and may become as routine an instrument in the laboratory as the present-day infrared.

With these techniques at one's disposal, it is far easier than ever before to look into a material under investigation and see which molecules are doing what. But in reality, it isn't always that easy, since the technology behind the "looking" has become vastly more complicated and subtle. Furthermore, it is difficult, at times, for the analytical chemist to know which tool to choose in order to accomplish his objective. Therefore, the following chapters in this monograph are intended to show the analytical techniques and procedures that have been used for the analysis and characterization of polyurethanes and the chemicals from which they are prepared as well as the pertinent chemistry of the analytical procedures or preparation reactions that may affect or have some bearing upon the analytical methods.

Naturally, the number of types of methods group themselves somewhat according to the length of time the technique has been available since it is logical to assume that more information is available for wet, GLC, and infrared methods than for NMR methods. While this is true and some information has become available after the literature search was closed (January 1, 1966), it is believed that this monograph embodies many techniques and methods that are capable of providing a maximum amount of necessary information. Necessarily, many methods are present because they are employed by the urethane industry and others because they provide definitive information using "common" laboratory equipment. Of course it should be recognized that in many cases the best analytical method to determine some property or component may be a relatively simple titration method and not that method described by a complicated instrumental method or technique. In many cases several methods or bases for methods have been given and the reader is free to exercise his own preference. It is in this spirit, then, that the analytical procedures and techniques of characterization are presented.

It was felt that the complete listing of isocyanates and their physical properties given in Chapter 7 would serve a useful purpose for anyone concerned with identifications, comparisons, or who at some time may require a readily available reference as to which isocyanates have been prepared.

The analytical accuracies and reproducibilities have been given for those methods for which they have been reported or definitely established. In many methods these quantities had to be omitted because they simply have not been definitively ascertained. Furthermore, these quantities will often vary from one laboratory to another depending upon the skill and available time of the chemist to control those parameters affecting these quantities.

References

1. Wurtz, A., *Ann.*, **71**, 326 (1849).
2. Gautier, A. *Ann.*, **149**, 313 (1869).
3. Hofmann, A. W., *Ann.*, **74**, 9 (1850).
4. Hentschel, W., *Ber.*, **17**, 1284 (1884).
5. Schroeter, G., *Ber.*, **42**, 2336, 3356 (1909).
6. Bayer, O., *Angew. Chem.*, **A59**, 275 (1947).
7. Bayer, O., Rinke, H., Siefken, W., Orthner, L., and Schild, H., Ger. Pat. 728,981 (to I. G. Farbenindustrie) (Nov. 12, 1942).

CHAPTER II

RAW MATERIALS FOR ISOCYANATES

I. INTRODUCTION

In order to discuss properly the two major raw materials used for isocyanate preparation, some background regarding the preparation of the intermediate compounds would be desirable. Thus, we will attempt to provide perspective as to the reasons why the amine, phosgene, and/or the final isocyanate are generally analyzed for the presence of certain compounds. Without somewhat of a basic understanding of the origin of certain compounds, possible analytical interferences, as well as the importance of these materials and their effects on the final properties of the isocyanates, the reason for certain analytical methods may not be understood or realized.

Only the compounds of greatest interest for those isocyanates that are commercially important will be discussed. In most cases the basis for these methods can be applied to others of the same class. For example, the gas–liquid chromatography technology for dinitrotoluene is similar to that for the analysis of nitrobenzene except for specific operating conditions.

The analytical methods which apply to the quality control of the raw and intermediate materials used in the commercial preparation of isocyanates will be discussed in this chapter. It is necessary to consider the manufacturing processes by which each of these materials are prepared to provide an understanding of the need for the various analytical methods. Toward this end, we will consider in detail the preparation and analysis of phosgene, the nitrotoluenes, and tolylenediamine.

II. PHOSGENE

A. Preparation

Various methods and experimental conditions have been reported for the preparation of phosgene. Plotnikov (1) prepared phosgene by passing carbon monoxide and chlorine over aluminum chloride at 30–35°C or into chloroform saturated with aluminum chloride.

The reaction of carbon monoxide with metallic chlorides has been studied by Belladen, Noli, and Sommariva (2). The reaction

$$MCl_2 + CO \rightleftharpoons M + COCl_2 \quad (1)$$

was investigated for $PbCl_2$, AgCl, CuCl, and $NiCl_2$. This work appears to be similar to that of Plotnikov. In the presence of metallic chlorides the following reactions have been reported to occur:

$$2CO_2 \rightleftharpoons 2CO + O \quad (2)$$
$$2MCl_2 + O \rightleftharpoons 2MO + Cl_2 \quad (3)$$

At the temperatures utilized, 450–750°C, the yields of phosgene from the reaction between carbon dioxide and the metallic chloride are as great as those with carbon monoxide. Many side reactions occur and this work has shown why the actual yields of phosgene differ so greatly from those calculated from known dissociation data.

A process (3) similar to the one above has been described that is particularly suited to processes that have large amounts of hydrochloric acid as a by-product. The process involves two steps:

(*1*) oxidation–neutralization:

$$2CuCl + 2HCl + \tfrac{1}{2}O_2 \rightarrow 2CuCl_2 + H_2O \quad (4)$$

(*2*) chlorination:

$$2CuCl_2 + CO \rightarrow COCl_2 + 2CuCl \quad (5)$$

Phosgene can also be prepared by the reaction of carbon tetrachloride and oleum (4,5). The reaction is

$$CCl_4 + 2SO_3 \rightarrow S_2O_5Cl_2 + COCl_2 \quad (6)$$

with an additional reaction occurring in the presence of excess hot oleum:

$$CCl_4 + 4SO_3 \rightarrow 2S_2O_5Cl_2 + CO_2 \quad (7)$$

Presently, most commercial processes employ the direct combination of carbon monoxide and chlorine (6–9).

The thermal formation and decomposition of phosgene in glass was investigated by Bodenstein and Plaut (10) over the temperature range 349–450°F. The velocity and equilibrium of the overall reaction

$$2CO + Cl_2 \rightleftharpoons 2COCl_2 \quad (8)$$

may be represented by the following:

$$Cl_2 \rightleftharpoons 2Cl^{\cdot} \quad (9)$$
$$Cl^{\cdot} + Cl_2 \rightleftharpoons Cl_3^{\cdot} \quad (10)$$
$$Cl_3^{\cdot} + CO \rightleftharpoons COCl_2 + Cl^{\cdot} \quad (11)$$

These reactions were found to be in agreement with the observed facts that reaction (10) consumed 51,400 calories and that its state of equilibrium conformed to the Nernst heat principle.

Nitzschmann (11,12) has reported the heat of formation to be 24,600 calories according to the equation $CO + Cl_2 \rightarrow COCl_2$.

If the technique described by Peacock (9) is employed, in which a mixture of air and chlorine is passed over coke at a temperature of 1200–1500°F, many impurities may be introduced which will not all be separated from phosgene by condensation. The use of wet carbon monoxide in the direct reaction with chlorine will also result in similar difficulties according to the reaction

$$H_2O + COCl_2 \rightarrow 2HCl + CO_2 \tag{12}$$

Therefore, the expected impurities in commercial phosgene will be: (*1*) carbon dioxide, (*2*) hydrogen chloride, (*3*) chlorine originating from an excess of chlorine during the reaction or an incomplete one, (*4*) nitrogen and possibly oxygen from the presence of air, and (*5*) carbon monoxide (from an excess of CO during the reaction or an incomplete one). The presence of free chlorine is especially important since this may later result in ring chlorination of the aromatic amine during phosgenation.

B. Analytical Techniques

1. Wet Methods

a. Liquid Phosgene

Liquid phosgene is hazardous to handle because of its extreme toxicity and its low boiling point. The boiling point has been reported as 8.2°C (13) and 280.66°K (14). The safety precautions and equipment required make any analysis of liquid phosgene time-consuming and costly. Therefore, analysis of the gaseous product is to be preferred wherever possible.

Methods have been described by Reeves (15), Pletnev (16), and Rush and Danner (17). In the method of Reeves (15) phosgene is decomposed with 13*N* sodium hydroxide to form sodium carbonate and sodium chloride, and phosgene is determined by means of a double indicator titration or by determination of the chloride content of the solution. The presence of chlorine or hydrogen chloride interferes.

The method of Rush and Danner (17) appears to be the only one applicable to liquid phosgene as well as being the most technically sound.

If cylinder phosgene is to be analyzed for assay, the two previously mentioned methods would inevitably suffer from sampling due to the fact that it is extremely difficult to obtain a representative gaseous sample since any chlorine, nitrogen, and other low-boiling impurities will be concentrated in the initial samples taken, causing the assay to be low.

Liquid Phosgene Assay (17)

Scope

This analysis permits the determination of the purity of liquid phosgene.

Principle

In acetone solution sodium iodide reacts with phosgene to release iodine

$$COCl_2 + 2NaI \rightarrow CO + I_2 + 2NaCl \qquad (13)$$

The released iodine is titrated with sodium thiosulfate

$$I_2 + 2Na_2S_2O_3 \rightarrow 2NaI + Na_2S_4O_6 \qquad (14)$$

When water is present, some hydrolysis of the phosgene occurs:

$$COCl_2 + H_2O \rightarrow CO_2 + 2HCl \qquad (15)$$

However, the decomposition reaction of phosgene with water proceeds slowly (18).

Equipment

(*a*) Heavy-walled glass-stoppered Erlenmeyer flasks, 500 and 750 ml.

(*b*) Sampling apparatus as shown in Figure 2.1.

(*c*) Sampling bulbs, about 12–15 mm in diameter.

(*d*) Anhydrous acetone, prepared by treating 2 liters of reagent grade acetone contained in a brown bottle with 200 g of 20 mesh anhydrous calcium sulfate and allowing to stand at least 24 hr with occasional shaking.

(*e*) Neutral anhydrous sodium iodide, prepared by placing 500 g of reagent grade sodium iodide in a porcelain evaporating dish, adding about 20 ml of 1:2 hydrochloric acid and stirring until sodium iodide is uniformly moist. Place on a hot plate and heat slowly with constant stirring until the crystals are white and dry, but avoid decomposition by too vigorous heating. After grinding to a uniform powder, store at 100–110°C. The prepared sodium iodide is tested for acidity by adding about 5 g to 50 ml of boiled distilled water, adding 2–3 drops of bromothymol blue–phenol red indicator and titrating with 0.1*N* sodium hydroxide. One drop should make the solution basic (yellow-green to violet).

(*f*) Bromothymol blue–phenol red indicator. Mix equal quantities of 0.1% aqueous solutions of the sodium salts of each indicator.

(*g*) Standardized sodium thiosulfate, 0.1*N*.

(*h*) Standardized sodium hydroxide, 0.1*N* (for Acidity Test described below).

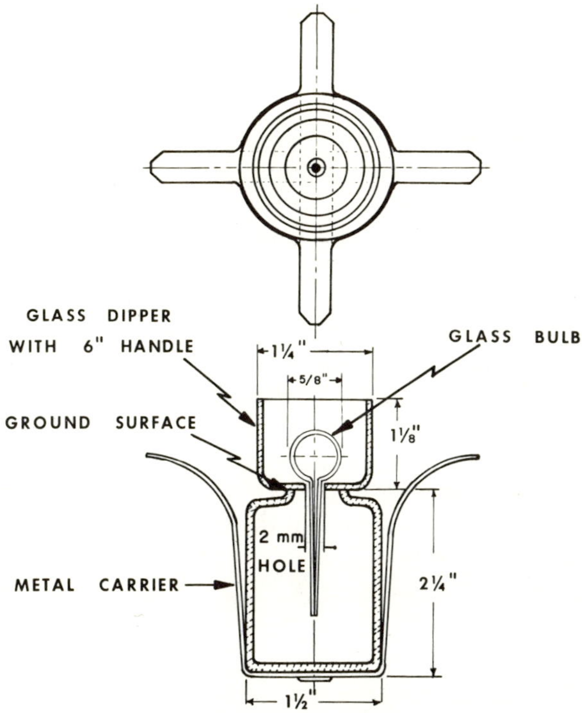

Fig. 2.1. Phosgene sampling apparatus (17).

PROCEDURE

(*1*) Cool the cylinder containing the phosgene in an ice–salt bath for at least 2 hr.

(*2*) Place the previously dried 50 ml sampling bottle in the metal carrier; suspend the assembly in a Dewar previously charged with Dry Ice and isopropanol.

(*3*) Without venting the phosgene cylinder, discharge sufficient phosgene into the sample bottle to fill it to near the shoulder.

(*4*) Place a previously dried and tared glass bulb into the dipper and close the sample bottle with the dipper.

(*5*) Place pulverized Dry Ice around the bulb, leaving sufficient surface around the bulb so as to allow observation.

(*6*) When approximately 0.4 g of sample has been drawn in, remove the dipper, cover the bulb with more solid carbon dioxide, and hold at about a 45° angle with the stem pointing downward.

(*7*) Expel any liquid from the capillary stem by warming and seal quickly in a flame. Dry and reweigh.

(*8*) Transfer, directly from the oven, approximately 5 g of sodium iodide into a dry, warm 500-ml flask.

(*9*) Grease the stopper with pure petroleum jelly, add approximately 30 ml of acetone, place the bulb containing the phosgene in the flask, stopper tightly, and cool to below room temperature.

(*10*) Wrap the flask with a towel and while holding the stopper tightly, shake to break the bulb.

(*11*) Continue to shake for 3 min., while cooling in the ice bath. Remove the stopper and wash down the stopper and neck of the flask with a small amount of boiled distilled water.

(*12*) Without using starch, titrate with the 0.1*N* sodium thiosulfate.

(*13*) Titrate a prepared blank containing the same amount sodium iodide and acetone with 0.1*N* thiosulfate if the solution is colored or with 0.1*N* iodine if the solution is clear. The blank should not exceed 0.1 ml.

(*14*) Reserve the sample solution and the blank solution for the acidity determination described below.

Calculations

$$\frac{(A \pm B) \times N \times 4.946}{W} - (C \times 1.396) - D \times 0.304 = \%\ COCl_2$$

where: A = ml of standard thiosulfate solution used for determination
B = ml of standard thiosulfate solution used for blank
C = % free chlorine
D = % ferric chloride
N = normality of standard solution
W = grams of sample

The correction factor for ferric chloride is generally negligible.

Rush and Danner (17) have found the precision to be 5 parts in 10,000, providing careful attention is given to details.

Acidity (17)

Scope

This analysis permits the determination of the acidity of liquid phosgene.

Principle

The acidity of the liquid phosgene is determined by a standard neutralization reaction. This procedure is a continuation of the assay method in that the solutions remaining from the assay determination are titrated with 0.1*N* sodium hydroxide solution.

PROCEDURE

(*1*) Add 3–4 drops of bromothymol blue–phenol red indicator solution to the flasks containing the sample and the blank from the phosgene determination.

(*2*) Titrate each flask with 0.1*N* sodium hydroxide solution to the same blue-green endpoint with a minimum amount of agitation.

CALCULATIONS

$$\frac{(A - B) \times N \times 3.65}{W} = \%\text{ acidity as HCl}$$

where: A = ml of 0.1*N* NaOH used for the sample
B = ml of 0.1*N* NaOH used for the blank
N = normality of the NaOH solution
W = grams of sample

Free Chlorine (17)

SCOPE

This analysis permits the determination of the free chlorine content of liquid phosgene.

PRINCIPLE

Phosgene is absorbed in aqueous KI and undergoes hydrolysis (eq. 16). The free chlorine reacts with the KI to give free iodine (eq. 17), which is titrated with standard sodium thiosulfate (eq. 18).

$$COCl_2 + H_2O \rightarrow CO_2 + HCl \qquad (16)$$

$$Cl_2 + 2KI \rightarrow 2KCl + I_2 \qquad (17)$$

$$I_2 + 2Na_2S_2O_3 \rightarrow 2NaI + Na_2S_3O_6 \qquad (18)$$

PROCEDURE

(*1*) Introduce approximately 1 g of liquid phosgene into the sample bulb as previously described and place into a 750-ml flask containing 100 ml 3% cold aqueous potassium iodide solution.

(2) Stopper the flask and break the bulb as before.

(3) Cool the flask in an ice bath, remove the stopper, and titrate with 0.05*N* sodium thiosulfate using starch as an indicator.

CALCULATION

$$\frac{A \times N \times 3.55}{W} = \%\text{ free } Cl_2$$

Rush and Danner (17) recommend the use of cold distilled water instead of the 3% aqueous potassium iodide solution for samples containing chlorine in the range of 0.05% and below. This avoids the formation of trace amounts of iodine by the reaction of phosgene with the potassium iodide. Aqueous potassium iodide is necessary for the complete absorption of higher concentrations of chlorine. We have found that 1% potassium iodide solution is sufficient for the complete absorption of chlorine without giving rise to high values due to side reactions.

Nonvolatile Residue (19)

Scope

This analysis permits the determination of nonvolatile residue in liquid phosgene.

Principle

A weighed sample of liquid phosgene is allowed to vaporize and the residue remaining is determined gravimetrically.

Equipment

(*a*) Narrow-mouth bottle with cap, 2 oz.

(*b*) Special Dry Ice–isopropanol bath and container (see Fig. 2.2).

(*c*) Phosgene decomposition apparatus (see Fig. 2.3) according to Farbenfabriken Bayer (20).

(*d*) Three-necked 250-ml flask with 35/25 joints.

(*e*) Reagent grade hydrochloric acid.

Procedure

(*1*) Introduce approximately 100 g of liquid phosgene into a tared 2-oz narrow-mouth bottle precooled with Dry Ice and isopropanol in a special insulated container similar to that shown in Figure 2.2. A tinned can insulated with rigid foam will suffice nicely. The bottle will be approximately ¾ full.

(*2*) Remove the 2-oz bottle from the container and reweigh quickly. Rapidly transfer the contents to a tared 250-ml three-necked flask through the sample port. The flask should be connected to a decomposition tower as shown in Figure 2.3.

(*3*) Adjust the nitrogen flow into the flask so that there is a positive sweep of nitrogen to carry the phosgene vapor into the decomposition column.

(*4*) When the sample is completely vaporized, remove the sample flask and reweigh.

(*5*) Retain the residue in the flask for the determination of iron described below.

Calculation

$$\frac{R \times 100}{W} = \%\ \text{residue}$$

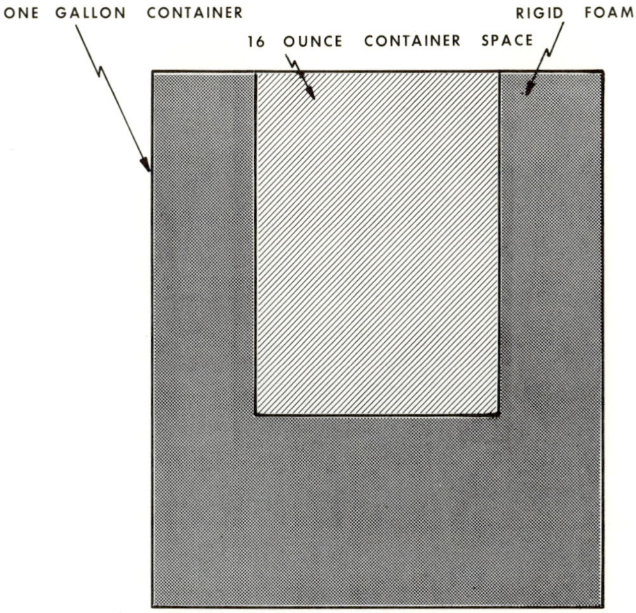

Fig. 2.2. Rigid foam dewar (19).

where R = grams of residue
W = grams of sample

Iron (19)

Scope

This method describes the procedure for determining iron in liquid phosgene. Ferric chloride may be present due to corrosion of equipment.

Principle

This procedure is a continuation of the determination of non-volatile residue in liquid phosgene. The residue from the "Nonvolatile Residue" determination above is treated with hydrochloric acid, and the iron determined by any of several standard methods.

Equipment (Additional)

Suitable reagents and equipment for a standard iron method.

Procedure

(*1*) Add sufficient hydrochloric acid to the residue from the "Nonvolatile Residue" determination to dissolve iron and/or iron salts.

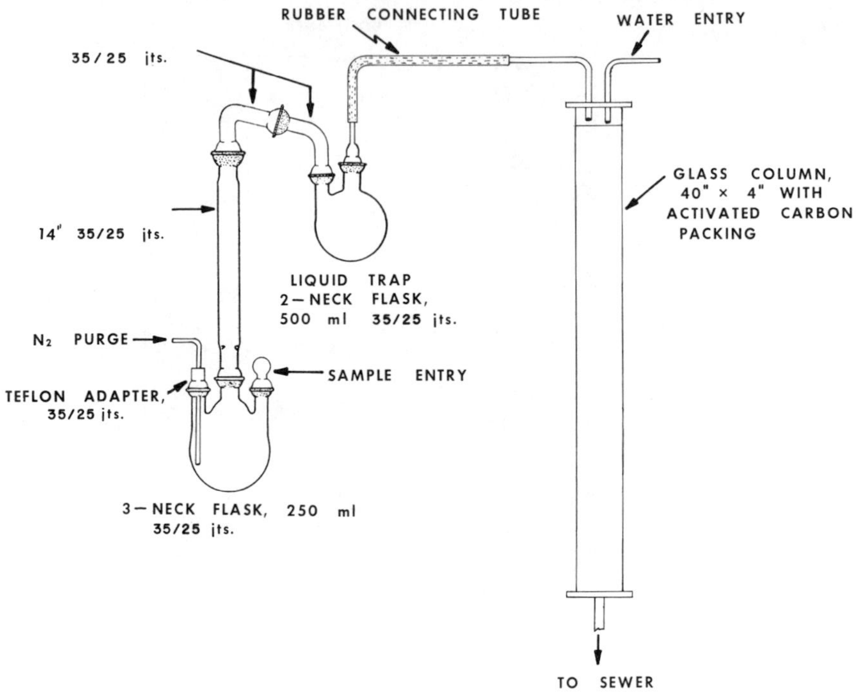

Fig. 2.3. Phosgene decomposition apparatus (20).

(*2*) Dilute with distilled water to an appropriate volume. The iron may be determined by colorimetric procedures, by titration techniques, or by atomic absorption.

CALCULATION

The iron may be expressed as Fe or $FeCl_3$.

The exact procedure employed will be dictated by the previous history of the sample. Typical values have been found to be in the range 20–500 ppm Fe.

Matuszak (21) developed a method which corrects for side reactions that may give low phosgene results and high acidity. It was designed for acid-free phosgene and makes no provision for determining the acidity. In a modified procedure, sodium iodide is used in place of potassium iodide and provision is made for the determination of hydrochloric acid. After the color of the released iodine has been discharged, potassium iodate is added to react with the free acid and the iodine color is discharged again with thiosulfate. This second titration is used in the modified procedure after correction has been made for hydriodic acid resulting from the following side reactions:

$$CH_3COCH_3 \rightleftharpoons CH_2C(OH){:}CH_2 \quad (19)$$
$$CH_2C(OH){:}CH_2 + I_2 \rightleftharpoons CH_3CI(OH)CH_2I \quad (20)$$
$$CH_3CI(OH)CH_2I \rightleftharpoons CH_3COCH_2I + HI \quad (21)$$

The iodoacetone is then reacted with thiosulfate and then back-titrated with iodine

$$CH_3COCH_2I + Na_2S_2O_3 \rightarrow CH_3COCH_2NaS_2O_3 + NaI \quad (22)$$

Rush and Danner (17) have found that side reactions do not occur to any extent if water is absent and the hydrochloric acid content is below 0.50%.

b. Gaseous Phosgene

During the manufacture of phosgene it is often desirable and more convenient to monitor the gas for the presence of chlorine in the gaseous phase rather than in the liquid.

Free Chlorine (22)

SCOPE

This method permits the determination of free chlorine in gaseous phosgene.

PRINCIPLE

See method for the determination of free chlorine in liquid phosgene.

EQUIPMENT

(*a*) Gas-sampling glass bulb, 250 cc.
(*b*) Semimicro buret, 5 ml capacity.
(*c*) 1% KI.
(*d*) Standardized sodium thiosulfate, 0.01*N*.
(*e*) Starch indicator.
(*f*) Erlenmeyer flask, 250 ml.

PROCEDURE

(*1*) Allow the phosgene gas to flush through the sampling bulb for several minutes. After taking the sample, equalize the pressure by opening a stopcock briefly.

(*2*) Attach one end of the sampling bulb to a buret containing 1% potassium iodide by means of a short piece of rubber tubing.

(*3*) By means of gravity flow, allow about 25 ml of the potassium iodide to enter the sampling bulb.

(*4*) After 10 min, shake briefly and transfer the contents of the bulb to a 250-ml Erlenmeyer flask.

(*5*) Rinse the bulb with distilled water and add the rinse water to the flask.

(*6*) Add about 0.5 ml of the starch indicator to the flask and titrate with the 0.01*N* sodium thiosulfate to a colorless end point.

CALCULATIONS

$$\frac{\text{ml } Na_2S_2O_3 \times N \times 35.5 \times 1000}{\text{Volume of sample bulb}} = \text{mg } Cl_2 \text{ per liter}$$

The sensitivity of this method is approximately 0.2 mg/liter or 0.005%. The sensitivity can obviously be improved by utilizing larger sampling bulbs. However, the sensitivity cannot be increased much beyond 0.003% because of the difficulty in washing a bulb much greater than 500 ml in capacity. Using greater amounts of potassium iodide solution results in much larger blanks, and errors due to technique may become appreciable.

Glass sampling containers should be used whenever possible due to the reaction of phosgene with iron and/or iron oxides. Delepine (18) has reported that phosgene containing a little chlorine as an impurity slowly attacks iron, forming iron chloride. In the absence of free chlorine the phosgene will attack iron rust, the iron chloride in commercial phosgene probably being due to this source. This implies that perhaps phosgene without chlorine may not attack iron. On the other hand, Kretov (23) has reported that phosgene does not corrode pure iron, even in the presence of chlorine, but impure iron rapidly forms iron chloride. At any rate, the action of phosgene upon pure iron would seemingly be negligible but small quantities of iron oxides would probably be attacked and cause erroneous analytical results.

Density

The density and thermal expansion of liquid phosgene has been studied by Davies (24) over the range −20°C to 60°C. The density in this temperature range was defined as:

$$\rho = 1.42014 - 0.0023120t - 0.000002872t^2 \quad (23)$$

2. GAS CHROMATOGRAPHIC METHODS

It appears that there are no published methods that are directly applicable to the analysis of phosgene for the impurities previously discussed. However, there are several methods where phosgene, hydrogen chloride, and chlorine can be determined as constituents of other gases by gas–liquid chromatography.

Baker and Hartwell (25) have discussed phosgene in a publication covering mine-gas analysis. Turkel'taub et al. (26) have analyzed boron trichloride for phosgene, hydrogen chloride, and chlorine by GLC. They used Celite 545 and stationary phases of dinonyl phthalate, hexadecane, and polyorganosiloxane VKZ-94. With dinonyl phthalate the boron trichloride was irreversibly retained.

Runge (27) has reported a GLC method for the determination of corrosive gases containing mixtures of HCl, Cl_2, CO, NOCl, ClCN, (CN_2), HCN, and others. He specifies all parts should be made of poly(tetrafluoroethylene) or poly(trifluoromonochloroethylene) using a chlorinated biphenyl or a low polymer of the packing material as the liquid phase.

Carbon monoxide, one of the most important constituents of phosgene to monitor during the manufacturing process, can be determined by vapor phase chromatography.

Carbon Monoxide Content of Gaseous Phosgene (28)

SCOPE

This analysis permits the determination of carbon monoxide content of gaseous phosgene.

PRINCIPLE

The sample is introduced into a helium carrier stream of the chromatograph and is separated by retention depending upon the molecular size of the holes in the particular molecular sieve packing that is utilized. The carbon monoxide content is then obtained by comparing the peak height with the peak height of a standard.

EQUIPMENT

(*a*) Any commercially obtainable gas chromatograph will suffice.
(*b*) 2.5-ml capacity gas-sampling loop.
(*c*) Gas sampling bulb, 250 ml capacity.
(*d*) Standard gas, containing a known amount of carbon monoxide.
(*e*) A 1 mv recorder.

CHROMATOGRAPHIC CONDITIONS

(*a*) Hot wire filament detector.

(*b*) Detector current: 250 mA or sufficient current to provide the desired sensitivity.

(*c*) Detector temperature: 50°C.

(*d*) Column: 8-ft Molecular Sieve 13A.

(*e*) Column temperature: 40°C.
(*f*) Flow: 70 ml/min.
(*g*) Sample size: 2.5 ml.

PROCEDURE

(*1*) Introduce the sample into the gas-sampling loop from the sample bulb by means of a leveling bulb.

(*2*) Exhaust the phosgene from the sample loop into a caustic solution or to the outside atmosphere.

(*3*) Introduce the sample into the chromatograph. After the CO peak has eluted, introduce a standard gas containing a known amount of CO by the same procedure.

(*4*) Calculate the peak heights of the standard and the sample.

CALCULATIONS

$$\frac{(\%\text{CO in std}) \times (\text{Sample peak height})}{\text{Peak height of std.}} = \%\text{CO}$$

Figure 2.4 is a typical chromatogram of gaseous phosgene which shows

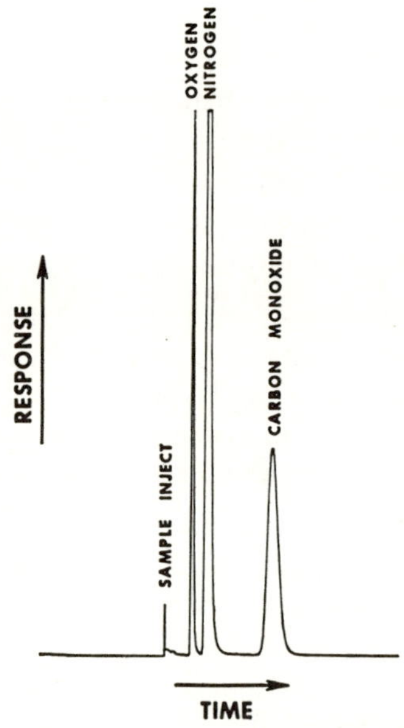

Fig. 2.4. Chromatogram of gaseous phosgene showing response and separation of carbon monoxide (28).

the response and separation of carbon monoxide under the conditions described.

Vapor Phase Chromatographic Analysis of Gaseous Phosgene for Carbon Monoxide, Carbon Dioxide, Nitrogen, Hydrogen Chloride, and Chlorine (29)

SCOPE

This method permits the quantitative determination, after calibration, of carbon monoxide, carbon dioxide, nitrogen, hydrogen chloride, and chlorine in phosgene by gas–liquid and gas–solid chromatography.

PRINCIPLE

The sample is introduced into a helium carrier stream of the chromatograph and is separated by retention of each respective component upon a silicone oil coating. Each separated component is suitably detected and graphically recorded as a function of time and may then be determined quantitatively by either area or peak height measurement, after calibration. In order to determine CO and N_2 separately, two columns and a dual column valve must be employed.

EQUIPMENT

(*a*) Any commercially obtainable gas chromatograph will suffice.
(*b*) Gas-sampling loop, 2.5 ml capacity.
(*c*) Gas-sampling bulb, 250 ml capacity.
(*d*) Standard gas, containing known amounts of impurities.
(*e*) A 1 mv recorder.
(*f*) Dual-column valve.

CHROMATOGRAPHIC CONDITIONS

(*a*) Hot wire filament detector and power supply.
(*b*) Detector current: 250 mA or sufficient current to provide the desired sensitivity.
(*c*) Detector temperature: 55°C.
(*d*) Columns: 30 ft ¼ in. Teflon tubing packed with 15% silicone SF-96 on Gas Pack F; 4-ft Molecular Sieve 5A.
(*e*) Column temperature: 55°C.
(*f*) Flow: 57 ml/min.
(*g*) Sample size: 5 ml.

PROCEDURE

(*1*) Introduce the sample into the gas-sampling loop from the sample bulb by means of a leveling bulb.

(*2*) Exhaust the phosgene from the sample loop into a caustic solution or to the outside atmosphere.

(*3*) Introduce the sample into the chromatograph. If CO and N_2 are not required separately, allow all the components to elute.

(*4*) If CO and N_2 are to be determined separately, turn the dual-column valve so that the CO and N_2 are channeled into the molecular sieve column which is behind the silicone oil column. The exact sequence for this operation should be previously established by measuring the elution time of these components.

(*5*) After the CO and N_2 have been stored in the molecular sieve column, i.e., eluted from the first or silicone oil column, but before they have eluted from the sieve column, turn the dual-column valve so that the remaining components elute from the silicone oil column. If a faster analysis time is desired, the phosgene may be stored in this column and eluted at a later time.

(*6*) After the Cl_2 or $COCl_2$ has eluted, turn the dual-column valve so that the CO and N_2 are eluted from the molecular sieve column.

(*7*) Repeat the same procedure utilizing a standard gas. This may be done using a matrix of helium rather than phosgene.

Calculations

$$\frac{(\%\ \text{Component in std}) \times (\text{Sample peak height})}{\text{Peak height of std}} = \%\ \text{Component}$$

Figure 2.5 is a chromatogram that shows the component separation on the silicone oil only, while Figure 2.6 shows complete separation of all components utilizing the molecular sieve column as well as the silicone oil column in conjunction with a dual-column valve.

Teflon was selected for the column tubing because of its excellent resistance to corrosive substances. Corrosion-resistant ¼ in. tubing or other impervious materials may be used but may offer greater difficulties in fabricating into a suitable shape which will fit conveniently into the column oven.

In the case of the dual column operation (Fig. 2.6), the phosgene was not eluted from the silicone oil column before dual column operation was again utilized to lessen the complete analysis time, since the CO and N_2 concentrations were obtained immediately after the Cl_2 had eluted. The dual-column valve must then be turned to single-column operation once again in order to elute the phosgene from the silicone oil column and prolong the life of the sieve column. The elution of the phosgene is not shown on this chromatogram. The column flows must be precisely matched so that difficulties are not encountered in the switching operation which will

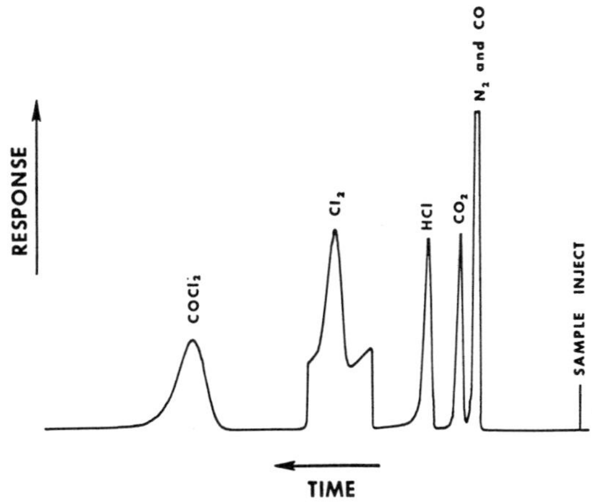

Fig. 2.5. Chromatogram of gaseous phosgene on GasPack F coated with 15% silicone oil SF-96 (29).

interfere with the baseline. This can be accomplished by utilizing a suitable adjustable restrictor to balance single-column flow against dual-column flow.

3. Spectroscopic Studies

One of the earliest spectroscopic studies of the structure of the phosgene molecule was performed by Henri and Howell (30). Further studies were performed by Mulliken (31), Ananthakrishnan (32), and McMurry (33).

Ugryumov (34) reported that the divergence between the experimental and calculated dipole moments of phosgene, 1.18 and 1.13 D, respectively, indicates the presence of electronic resonance. He described the eight possible forms as:

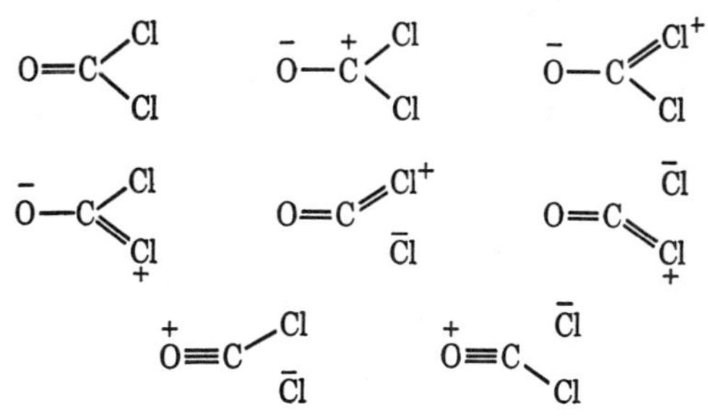

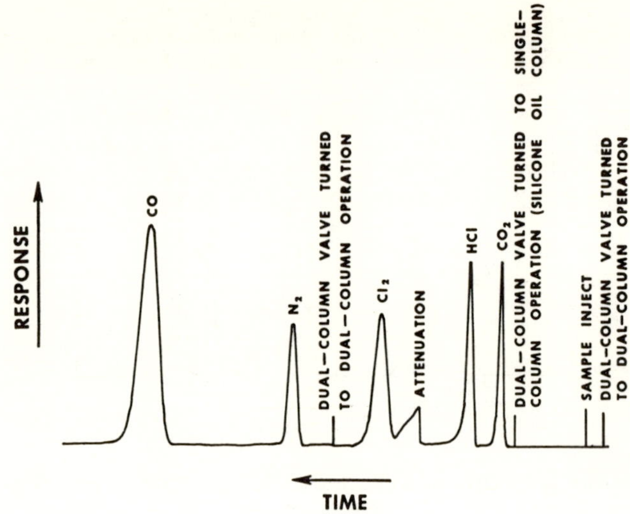

Fig. 2.6. Chromatogram of gaseous phosgene using a 5 Å molecular sieve column and a GasPack F column coated with 15% SF-96 silicone oil (29).

Lovell and Jones (35) calculated bond moments and effective charges from infrared intensities measured by the Wilson-Well's pressure broadening technique for the ν_1, ν_2, and ν_4 bands of phosgene. The permanent moment measured by Smyth (36) and Coop and Sutton (37) is 1.18 D. With the assumption that the oxygen forms the negative end of the dipole, Lovell and Jones have expressed the permanent moment as

$$-1.18 \times 10^{-18} = \mu_{CO} + 1.129\mu_{CCl} \tag{24}$$

The choice of polarity for $COCl_2$ moment is dictated by the electronegativity of oxygen (3.5) as compared to 2.5 for carbon and 3.0 for chlorine; also, by the large dipole moment (~2.3 D) usually found for the C=O as compared to (~1.5 D) usually found for the C—Cl bond.

Later spectroscopic studies of interest are those of Bailey and Hale (38), Nielsen et al. (39), and Catalano and Pitzer (40).

Table 2.1 lists the observed infrared bands of phosgene and their assignments according to Catalano and Pitzer (40).

TABLE 2.1
Assignment and Symmetry of Observed Infrared Bands of Carbonyl Chloride in Gas Phase and in Argon Matrix at 20°K (40)

Assignment symmetry		Obsd. (cm.$^{-1}$)		Calcd. (cm.$^{-1}$)	
		Gas phase	Argon matrix	Gas phase	Argon matrix
ν_1	A_1	(570)[a]	568[b]		
ν_2	A_1	1827[c]	1803		
ν_3	A_1	285	(284)[d]		
ν_4	B_1	849[c]	835		
ν_5	B_1	440[c]	(438)[d]		
ν_6	B_2	(585)[a]	582		
$2\nu_3$	A_1	—[e]	563	570	568
$\nu_3 + \nu_5$	B_1	740		725	
$\nu_1 + \nu_3$	A_1	—[f]	808	855	852
$3\nu_3$	A_1			855	852
$\nu_1 + \nu_5$	B_1	1011[c]	1006	1010	1006
$2\nu_1$	A_1	1159[c]	1160	1140	1136
$2\nu_6$	A_1			1170	1164
$\nu_4 + \nu_5$	A_1	1276[c]		1289	
$\nu_1 + \nu_4$	B_1	1412[c]	1390	1419	1403
$\nu_5 + 2\nu_1$	B_1	1570[c]		1580	
$\nu_5 + 2\nu_6$	B_1			1610	
$2\nu_4$	A_1	1677[c]	1645	1698	1670
$\nu_1 + 2\nu_6$	A_1	—[g]	1763	1740	1732
$\nu_1 + \nu_4 + \nu_5$	A_1	—[g]	1835	1859	1841
$\nu_2 + \nu_3$	A_1	2130[c]		2112	
$\nu_1 + \nu_2$	A_1	2405[c]		2397	
$\nu_2 + 2\nu_5$	A_1	2705[c]		2703	
$\nu_2 + 3\nu_1$	A_1	3527[c]		2537	
$2\nu_2$	A_1	3636[c]		3654	
$\nu_5 + 2\nu_2$	B_1	4073[c]		4094	
?		4180[c]			
$\nu_1 + 2\nu_2$	A_1	4255[c]		4224	
$2\nu_2 + 2\nu_4$	A_1	5335[c]		5352	
?		5408[c]			

[a] Estimate from matrix data (see ref. 40). Center of gas phase band at 575 cm^{-1}.

[b] Average for isotope split band (see ref. 40).

[c] Gas phase work of Nielsen and co-workers (see ref. 39).

[d] Estimate from gas phase data.

[e] This band would be masked by ν_1 and ν_6 in gas phase.

[f] This band would be masked by ν_4 in gas phase.

[g] This band would be masked by ν_2 in gas phase.

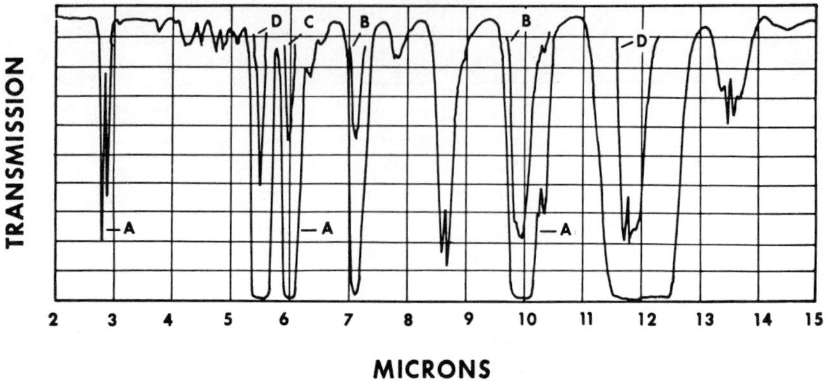

Fig. 2.7. Infrared spectrum of gaseous phosgene (42).

Curve letter	*A*	*B*	*C*	*D*
Pressure, mm	710	80	17	1.5
Resolution	927	927	927	927

Nielsen et al. (39) have previously reported that $COCl_2$ belongs to a class of molecules whose symmetry is that of the point group $C_{2}\nu$. As Table 2.1 shows, there are six nondegenerate fundamental frequencies, the forms of which have been described (41), and whose symmetry species are A_1, for ν_1, ν_2, and ν_3, B_1 for ν_4 and ν_5, and B_2 for ν_6. Thus, if the electric moment oscillates along the C_2 or twofold axis, an A_1 symmetry results; if it oscillates along an axis perpendicular to the C_2 but in the plane of the model, a B_1 symmetry results; and if it oscillates perpendicularly to the plane of the model, a B_2 results.

The ν_1 is essentially the in-plase C—X vibration, ν_2 is the C═O vibration, ν_3 is the bending of the X—C—X angle, ν_4 is the out-of-phase C—X, ν_5 is the rocking of the X—C—X triangle, and ν_6 is the out-of-plane vibration.

The infrared spectrum of phosgene, obtained by Pierson et al. (42), is shown in Figure 2.7. The spectrum was obtained using a Perkin-Elmer Model 21 spectrophotometer using a 10-cm gas cell with sodium chloride optics. The pressures utilized were as shown on Figure 2.7.

In many cases, as will be shown later, phosgene is utilized in an inert solvent, such as benzene, for the phosgenation of amines to the corresponding isocyanate. Thus, the carbonyl stretching vibration at 1827 cm^{-1} or the first overtone at 3636 cm^{-1} can be employed readily as analytical bands for quantitative infrared measurements with little interference.

One of the earliest ultraviolet spectroscopic studies of phosgene was performed by Henri and Howell (43). Henri and Howell reported that the ultraviolet spectrum consists of approximately 270 bands between 3050 and 2380 Å, with absorption becoming more intense toward the ultraviolet region. A regular distribution includes doublets and triplets with a separation of 0.5–1.0 Å and also single bands. The distribution of the bands is given by

$$1/\lambda = 33551 + 422\,p' - 582p^\circ + 254\,q' - 302q^\circ \tag{25}$$

or

$$1/\lambda = 33655 + 422(p' + \tfrac{1}{2}) - 582(p^\circ + \tfrac{1}{2}) + 254(q' + \tfrac{1}{2}) - 302(q^\circ + \tfrac{1}{2}) \tag{26}$$

where p° and q° are the vibration quantum numbers of the normal molecule, and p' and q' are the vibration quantum numbers of the activated molecule.

Additional electronic band spectra of phosgene and structures have been reported (31,33,44,45). The vacuum ultraviolet absorption spectrum of phosgene has also been studied (46).

A recent study of the ultraviolet spectrum of phosgene, in the region of 2100–3150 Å at high dispersion and for gas temperatures from −70 to +250°C, was performed by Giddings and Innes (47). Absorbing paths were 0.005–5 m-atm. The molecule was reported as having a strongly asymmetric top, and resulting complex band structures were further complicated by overlapping lines of corresponding $CO^{35}Cl_2$ and $CO^{35}Cl^{37}Cl$ bands (of relative intensity 3 and 2, respectively). This interpretation was verified by the absence of rotational fine structure even when resolving powers of 150,000–200,000 were employed.

The key to the vibrational assignments was the observation of the effect of temperature on relative band intensities. The best data obtained for phosgene were at −80°C and +90°C between 34,900 and 35,600 cm^{-1} and at 25°C and 170°C between 33,400 and 34,000 cm^{-1}. Only four or five bands of several dozen studied definitely originated in excited vibrational levels. Most of the bands appeared in pairs. Henri and Howell (43) interpreted these doublets as corresponding bands of $CO^{35}Cl_2$ and $CO^{35}Cl^{37}Cl$, respectively. This interpretation has been supported by other evidence (48) as well as this study.

Since most of the phosgene transitions appeared to originate in the ground vibrational state, most observed band intervals represented vibrational differences of the excited electronic state. The bands at 580 and 430 cm^{-1} dominate the strongest bands. The fundamental frequencies found from the $CO^{35}Cl_2$ band heads are $\nu_5' = 430$, $\nu_6' - 581$, $\nu_5'' = 438$, and $\nu_6'' = 587$ cm^{-1}, and the system origin is at 32,730 cm^{-1}.

The 200 strongest sharp heads below 3050 Å were interpreted as Q-branches of A-type asymmetric top bands. Two vibrational frequencies of the excited state were established as ν_5–430 cm^{-1} and ν_6–580 cm^{-1}. The interpretation thus indicated that the electronic transition moment is perpendicular to the CO band and that in the excited state the molecule does not have a plane of symmetry.

The assignments by Giddings and Innes are shown in Table 2.2. A very slow trend to larger separations were observed toward higher energies, but the isotope shifts of the active vibrations are clearly very small. They did not feel assignments could be made easily nor would they be meaningful for the many remaining weak bands in the spectrum.

4. Additional Methods and Techniques

a. Methods

There are a number of additional, important methods that have been described in the literature. Many of these methods do not appear to be amenable to routine operation where ease of handling and time requirements are of significant interest.

Nenitzescu and Pana (49) have reported a method for the analysis of industrial phosgene. The phosgene is passed through a weighed tube containing metallic antimony. The tube is connected to an absorption bulb that contains 20 ml of 10% alcoholic potassium hydroxide and a tube filled with silica gel is used to prevent the escape of vapors from the absorption bulb. After phosgene is drawn through the apparatus for 5 min, it is flushed with air to remove phosgene from the tube containing the antimony. The gain in weight of the antimony represents the weight of chlorine in the mixture. The gain in weight of the absorption and silica gel tubes represent the weight of phosgene in the gas.

Terentév, Buzlanova, and Obtenperanskay (50) have reported a method of analysis for the determination of phosgene in the presence of chlorine and hydrogen chloride. The method is based on the reaction of phosgene with excess hexamethylenimine and indirect titration of the excess after its distillation from the alkaline solution into a measured volume of 0.1N acid. Hydrogen chloride forms a salt with $(CH_2)_6NH$ and chlorine forms chloramine which is destroyed by the addition of sodium thiosulfate after phosgene has reacted with the hexamethylenimine.

Meisel and Mazor have described a method for determination of phosgene based on its dissociation temperatures (51). The decomposition was

TABLE 2.2

Assignments for Strong $CO^{35}Cl_2$ Band Head Occurring in the Ultraviolet Spectrum of Phosgene (47)

(ν_{vac} cm^{-1})[a]

		32,439 (0) *580*	33,019 (3) *581*	33,600 (4)[b]*582*	34,182 (3)		35,378 (1)			
		439	*439*	*437*	*445*		*436*			
		32,878 (3) *580*	33,458 (4)[b]*579*	34,037 (5) *590*	34,627 (4) *593*	35,220 (1) *594*	35,814 (1) *577*	36,391 (–)		37,565 (–)
		438	*440*	*441*	*431*	*434*	*432*	*442*		*439*
	32,730 (1) *586*	33,316 (4) *582*	33,898 (3) *582*	34,480 (5) *578*	35,058 (5)[c]*596*	35,654 (2) *592*	36,246 (–) *587*	36,833 (–) *590*	37,423 (–) *581*	38,004 (–)
	434	*429*	*429*	*427*	*434*	*428*	*433*	*427*	*415*	
	33,164 (2) *581*	33,745 (3)[c]*582*	34,327 (5) *580*	34,907 (5)[c]*585*	35,492 (5) *590*	36,082 (5) *597*	36,679 (6) *581*	37,260 (4) *578*	37,838 (–)	
	424	*421*	*419*	*425*	*423*	*416*	*413*	*405*	*408*	
33,010 (4) *578*	33,588 (1)[c]*578*	34,166 (3) *580*	34,746 (2) *586*	35,332 (4) *583*	35,915 (4) *583*	36,498 (5) *594*	37,092 (1) *573*	37,665 (–) *581*	38,246 (–)	
403	*406*	*410*	*411*	*416*	*419*	*416*	*397*	*415*	*416*	
33,413 (3) *581*	33,994 (4) *582*	34,576 (4) *581*	35,157 (5) *591*	35,748 (4) *586*	36,334 (5) *580*	36,914 (1) *575*	37,489 (–) *591*	38,080 (–) *582*	38,662 (–)	
414	*422*	*421*	*423*	*421*	*419*	*411*	*418*	*396*		
33,827 (2) *589*	34,416 (2) *581*	34,997 (4)[c]*583*	35,580 (2) *589*	36,169 (3) *584*	36,753 (5) *572*	37,325 (–) *582*	37,907 (–) *569*	38,476 (–)		
419	*417*	*417*	*410*	*417*	*412*	*426*	*404*	*410*		
34,246 (2) *587*	34,833 (2) *580*	35,413 (1) *577*	35,990 (4) *596*	36,586 (3) *579*	37,165 (–) *586*	37,751 (–) *560*	38,311 (–) *575*	38,886 (–)		
	426	*420*	*420*	*419*	*414*	*434*	*419*			
	35,259 (2) *574*	35,833 (3) *577*	36,410 (2) *595*	37,005 (4) *574*	37,579 (–) *606*	38,185 (–) *545*	38,730 (–)			
	402	*413*	*412*	*408*	*408*	*395*				
35,071 (5)[b]*588*	35,661 (5) *585*	36,246 (–) *576*	36,822 (5) *591*	37,413 (2) *574*	37,987 (–) *593*	38,580 (–)				
403	*403*	*397*		*397*	*396*	*414*				
35,474 (3) *590*	36,064 (–) *579*	36,643 (4)		37,810 (–) *573*	38,383 (–) *611*	38,994 (–)				
	396	*403*			*396*					
	36,460 (1) *586*	37,046 (1)			38,779 (–)					
	397									
	36,857 (1)									

[a] Bands are diffuse at frequencies higher than about 36,000 cm^{-1}. Henri's measurements are listed for those frequencies higher than 34,500 cm^{-1} (see ref. 30).
[b] Bands which increase in intensity when gaseous phosgene is heated.
[c] Bands which do not increase in intensity when gaseous phosgene is heated.

reported as being

$$COCl_2 \leftrightarrows CO + Cl_2 \quad (27)$$

4.4% at 300°C, 42% at 420°C, 74% at 520°C, 86% at 600°C, 95% at 700°C, and 100% at 800°C. The amounts of decomposition and the temperatures of occurrence appear to agree, in general, with those previously reported (52,53). This is especially true of the 100% decomposition temperature. Stock et al. (52,53) have found that the presence of materials such as aluminum chloride, glass, silica, and carbon, which may act as catalysts, affect the time-temperature decomposition relationships.

Meisel and Mazor (51) passed the phosgene through a tube containing silver which was held at 800°C. The carbon monoxide formed was measured in an azotometer.

b. Phosgene Measurement and Detection

Numerous methods have been reported in the literature for the detection and determination of small amounts of phosgene both in gases and in solution (55–71). In working with phosgene on a laboratory or plant scale, it may be advisable to have certain methods at one's disposal for analytic use as well as to be aware of certain safety considerations in monitoring off-gases. The subsequent section is included with this in mind.

Anger and Wang (54) have reported the following spot test for phosgene.

PROCEDURE

One drop of the solution to be tested is placed in a microcrucible. A small amount of phenylhydrazine cinnamate is added and after 5 min, one drop of 1% copper sulfate solution. Phosgene reacts with the reagent to form diphenylcarbazide which gives the red-violet copper salt.

DETECTION

As little as 0.5 γ of phosgene can be detected.

Pu and Lo (60) have reported that methyl violet, methyl violet B, gentian violet B, and rosaniline are specific for the detection of phosgene in air. Distinct color changes were obtained at gas concentrations of 0.7 mg/liter of air within 1 sec.

PROCEDURE

Impregnate filter paper with a mixture of equal parts of an 8% solution of *p*-dimethylaminobenzaldehyde in 95% ethyl alcohol and an alcoholic solution of an aromatic amine. Dimethylaniline (25% in 95% ethyl alcohol) which gives a blue color is to be preferred because of sensitivity.

An even more sensitive paper can be prepared by using an alcoholic solution saturated with the *p*-dimethylaminobenzaldehyde.

DETECTION

As little as 10 mg of phosgene per cubic meter of air can be detected in the presence of 10–12 times that concentration of chlorine.

Lamouroux (65) has reported that phosgene in air can be determined by passing the air through $(CH_3)_2NCHO$ or diethyl phthalate containing (4-nitrobenzyl)pyridine and comparing the transmission of the solution at 415 mμ to that of known concentrations. The calibration curve was found to be linear up to 5 γ of phosgene/ml.

Liddell (66) has described a spot test for phosgene that provides a color change from white to bright blue that is relatively unaffected by mineral acid vapors. The paper is suitable for use in a pump and the reagent solutions will keep for many months in the dark although the papers should be prepared as required.

PROCEDURE

Dip a filter paper in a solution of 1.68 g of *N*-ethyl-*N*-2-hydroxyethylaniline, 0.75 g of *p*-dimethylaminobenzaldehyde and 2.5 ml of diethyl phthalate in 25 ml of ethanol. The solvent should be allowed to evaporate before use since the wet paper is quite insensitive.

DETECTION

When 500 ml of air containing as little as 1 γ of phosgene/liter are drawn through a circular detector paper of diameter 0.110 in., a blue color is produced.

Dixon and Hands (69) have reported the following field method for the detection of phosgene.

PROCEDURE

A 120 ml sample of air is drawn by means of a rubber-bulb hand aspirator through a 1 cm diameter circle of filter paper at a rate not exceeding 3 ml/sec. The paper is impregnated with a solution containing 2% of 4-(*p*-nitrobenzyl)pyridine and 4% *N*-benzylaniline in benzene. If phosgene is present, a red stain appears. If traces of chlorine are present, a dried paper impregnated with aqueous 4% sodium iodide and 10% sodium thiosulfate placed on the inlet side will remove it. The test paper is not affected by hydrogen chloride up to 500 ppm.

DETECTION

As little as 0.25 ppm of phosgene can be detected. The usable range in which standard stains can be compared is 0.25–10 ppm. The presence of monochlorobenzene, trichloroethylene, in the presence of phosgene offer some interference but the color produced by monochlorobenzene generally fades after 3 min, thus not seriously affecting the test. This is true for concentrations up to 200–300 ppm of impurities.

Grosskopf (71) has described an apparatus to detect the presence of phosgene that utilizes *N*-phenylbenzylamine in alcohol. No detection sensitivity ranges were reported in this patent.

Witten and Prostak (72) have described a means for detecting and semiquantitatively measuring trace quantities of phosgene in air. They have employed detector crayons, which are prepared by mixing suitable reagents with inert fillers and binders (blanc fixe such as barium sulfate) and forming into crayons. They claim a much better stability for this technique due to the smaller surface area of the crayon. The crayons can be used to write on most surfaces where phosgene detection is desired. The sensitivity of the crayon marks is expressed as a Ct value, where C is the concentration of the gas and t is the time necessary for a mark to respond to concentration C.

In similar environments, the color change is, to a first approximation, inversely proportional to the concentration of the reactive gas. Although the detector crayons are primarily qualitative, Ct is approximately a constant and therefore gives a rough estimate of concentration.

Phosgene detector crayons ($Ct = 0.03$ mg min/m^3) will turn red in 1 min when suspended at 90°F in a turbulent atmosphere containing 7.5 ppb of phosgene.

INGREDIENTS

4-(*p*-Nitrobenzyl)pyridine, 2%.
N-Phenylbenzylamine, 5%.
Sodium carbonate, CP, 5%.
Blanc fixe, neutral, amorphous, dry, 88%.

PROCEDURE

Absorb a benzene solution of the nitrobenzylpyridine and phenylbenzylamine in blanc fixe so that the fixe is completely wetted and absorbed. Allow the benzene to evaporate overnight and impregnate the mixture with a water solution containing the sodium carbonate as before. Dry the mixture and press into crayons. A laboratory hydraulic press using approximately 6000 psi will suffice. The light yellow marks turn red in the presence of phosgene. High concentrations of hydrogen chloride interfere because of destruction of the sodium carbonate. The crayons are stable at room temperature for over one year.

Detection

As little as 8 ppb of phosgene in 1 min can be detected. In very high concentrations of phosgene the red color of the crayon mark tends to fade.

It will be noted that this technique is similar to that used by Dixon and Hands except for the incorporation of sodium carbonate and the subsequent physical state.

For high concentrations of phosgene the following method can be used.

Ingredients

2-Amino-4,4′-bis(dimethylamino)benzophenone (2-amino-Michler's ketone), 1 g.
4,4′-Bis(dimethylamino)benzophenone (Michler's ketone), 0.5 g.
N-Phenyl-1-naphthylamine, 5 g.
Acetone, 80 ml.
Molded plain white chalk crayons.

Procedure

Dissolve the three active ingredients in acetone. Completely immerse the crayons for 24 hr, remove, and dry.

Detection

Approximately one-half the sensitivity of the preceding crayons.

In any procedure involving the detection and measurement of one gas in another, standard concentrations and methods of calibration assume paramount interest. Once a method of analysis has been chosen and detection limits, interferences, etc., have been considered, it is often difficult to accurately prepare known concentrations of the component of interest so that the concentration of this component may be determined in the sample gas.

Saltzman (73) has described a means of calibration and measurement for a number of gases, one of which is phosgene. In this technique, the dynamic method is recommended since losses are less serious in flow dilution systems because of less absorption loss and the system is more compact. The physical dimensions of such a system are a compromise between allowable pressure drops (lower losses being favored by larger dimensions) and the desired minimal dead space and rapid flushing time (favored by small dimensions).

Two all-glass flow dilution devices are shown in Figure 2.8. The simplest system, A is fabricated from a standard three-way T stopcock having a solid stopper plug bored with 3 or 4 mm diameter holes and side arms of 6 or 7 mm i.d. The stopcock permits passing the mixture for a

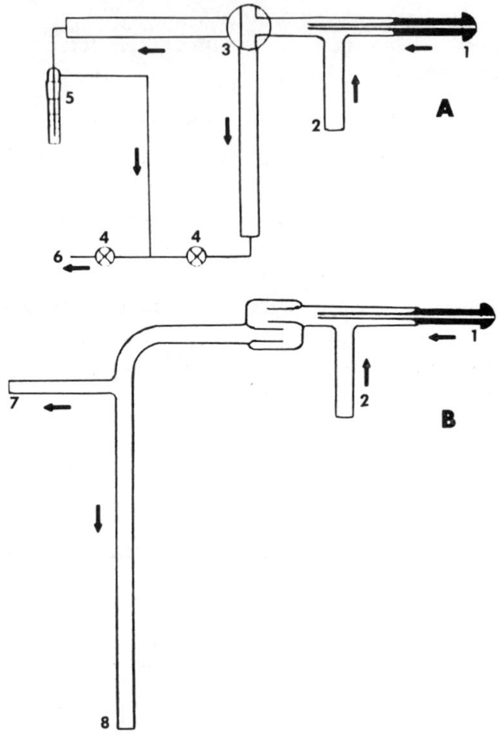

Fig. 2.8. Flow dilution devices (73). (*A*) With three-way stopcock to pass entire flow through either sampling device or bypass of equal resistance. Connections for vacuum system illustrated. (*B*) With provision for splitting mixture stream. (*1*) Inlet for metered test gas. (*2*) Inlet for metered purified air. (*3*) Three-way stopcock shown in bypass position (¼ turn counterclockwise for sampling). (*4*) Needle valve. (*5*) Sampling device. (*6*) Vacuum source (aspirator in hood). (*7*) Sampling connection. (*8*) Waste mixture outlet.

timed interval through a sampling apparatus or at other times through a bypass of equal resistance. This permits adjustments before sampling and subsequent collection without disturbing flow conditions. The test gas inlet should be as short and small a diameter as practicable. With a 0.5 mm i.d. nozzle flows up to 0.3 liter/min may be passed with only a slight pressure drop. Air dilution flows up to 10 liters/min may be metered into the T connection on the side arm with ground joint.

Figure 2.8*B*, shows a somewhat more elaborate system which may be utilized for aliquoting only a portion of the entire stream. Flow reversals in the traplike device provide complete mixing of the gas stream with minimal pressure drops. The portion of the gas mixture to be utilized is sampled from the sidearm (*7*, Fig. 2.8*B*).

Purification of the air used for dilution must always be considered in relation to the component being analyzed and the sum total effects upon the analysis of the mixture.

Figure 2.9*A* shows an asbestos plug flowmeter (74) useful for metering flows from a few hundredths of a milliliter to a few milliliters per minute. It is fabricated from a three-way T stopcock and a graduated cylinder. One leg of the stopcock bore is packed with asbestos fiber of the type used for Gooch crucibles. The gas flow through this plug is directly proportional to the pressure drop across it. The pressure on the upstream face of the plug is maintained constant at the depth of liquid (water, etc.) used in the graduate, over a reasonably wide range of tank needle valve settings.

Saltzman states that a valve setting providing only a few bubbles per minute of waste gas is adequate as long as the flow is not so low that stoppage occurs during a run. Flows are calibrated by attaching a

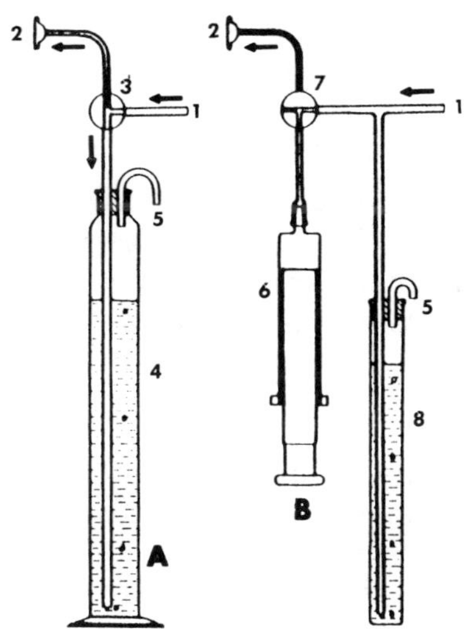

Fig. 2.9. Apparatus for the micrometering of gases (73). (*A*) Asbestos plug flowmeter. (*B*) Motor-driven syringe. (*1*) Test gas inlet. (*2*) Metered gas outlet, 12/2 ball joint. (*3*) Three-way T-shaped stopcock, with asbestos packing in upper leg of stopcock plug bore (stopcock never turned). (*4*) 250-ml. graduated cylinder containing water or oil. (*5*) Waste gas outlet. (*6*) Motor-driven, 50-ml. glass syringe. (*7*) Three-way T-shaped stopcock, shown in syringe filling position (¼ turn clockwise for gas delivery). (*8*) Safety pressure release and flushing vent.

graduated 1 or 0.1 ml pipet to the meter outlet and timing with a stopwatch the movement of a drop of water, oil, etc., past the graduations. A calibration plot of flow vs. water depth (or milliliters) over the outlet of the waste tube may be made.

A 1 mm bore in the stopcock plug is satisfactory for most of the lower flow rates. The asbestos plug is packed by trial and error until the desired flow range is obtained. For accurate work in actual use, the back pressure on the downstream face of the asbestos plug may be measured manometrically and deducted from the water level in the cylinder to give true flow from the calibration chart.

Motor-driven glass syringes provide another means of metering gas as well as liquid flows in the range of fractions of a milliliter to several milliliters per minute. By means of accurate diameter measurements of the syringe and known gear ratios, precise flow rate equivalents may be obtained. A suitable arrangement is shown in Figure 2.9*B*.

Saltzman and Gilbert have evaluated mathematically the factors affecting absorption of gases in low concentrations. In absorption methods the parts per million by volume of test gas may be readily computed by dividing the microliters found chemically by the corrected volume in liters of air sampled. Saltzman's treatment has led to:

$$C_e = C_r + ax \qquad (28)$$

where C_e is the equilibrium concentration or the escaping tendency of the gas from the absorbing reagent since it is reasonable to assume that the absorption rate of the test gas is proportional to a driving force, the difference between the actual and equilibrium concentrations in each bubble at any instant, divided by a resistance which represents the slow diffusion at gas–liquid interfaces and the slow reaction rate, and which is a constant for a fixed temperature and sampling flow rate.

C_r is the blank value for unexposed reagent, a is an empirical constant and must be such that it converts x to gas concentration (ppm), and x is the concentration of the product.

The apparent absorber efficiency is given by

$$K = 1 - C_2/C_1 \qquad (29)$$

and the true absorption efficiency by

$$C_1/C = (1 - C_2/C_1)(1 - C_r/C) \qquad (30)$$

These equations may be applied to any number of identical absorbers in series, if subscripts 1 and 2 denote any pair of consecutive concentrations by chemical analysis, rather than only the first and second, and C is the gas concentration entering the first absorber of the pair.

The value of the empirical constant a (Eq. 28) is obtained from a rearranged form of the definition of the apparent absorption efficiency, K:

$$\frac{1}{K} = \frac{1}{k''} + \frac{aV}{2R} \tag{31}$$

where k'' is the overall rate constant, V is the total volume of the air sample, and R is the gas constant.

A plot of the reciprocal of the apparent absorption efficiencies obtained vs. the variable V/R would have an intercept of $1/k''$ and a slope of $a/2$.

Midget impinging equipment represents a practical way of gas sampling since it is portable, economical, and exact amounts of reagents can be introduced into the impinger and the final color may be developed by adding additional reagents into the impinger before colorimetric analysis is performed.

Three devices are shown in Figure 2.10. The device C is generally utilized when higher absorption efficiencies are required.

If a reasonable pressure drop of no more than 1½ in. of Hg is desired, a 60 μ fritted disk can be employed. For less difficult absorptions a 100 μ pore diameter disk is satisfactory. The pore diameter in (microns) can be determined by just covering the disk with distilled water and measuring the minimum vacuum at which the first bubbles of air can be drawn through. The pore diameter is given by

$$D(\mu) = 30\,S/P \tag{32}$$

where S is the surface tension in dynes/cm at the test temperature and P is the vacuum in mm of Hg.

Saltzman used the iodometric method of Matuszak (21) for phosgene. The sample air stream is bubbled through a 2% sodium iodide solution in anhydrous acetone. Accurate results were obtained by titrating the liberated iodine with standard thiosulfate using starch indicator. Water was added to keep the acetone concentration less than 25% by volume in order to obtain the proper blue end point. The complete procedure recommended by Matuszak was followed.

Crummett (64) determined phosgene concentrations by bubbling the sample gas through 50 ml of aqueous aniline at a rate such that not more than 2 mg of phosgene is passed per minute. Hydrochloric acid was added and the solution diluted with methanol and scanned from 240 to 300 mμ. The net absorbance at 260.5 mμ was multiplied by the absorbance ratio, R, and subtracted from the absorbance at 254.5 mμ which equals, say, A.

When phosgene reacts with aqueous aniline, 1,3-diphenylurea is formed. The absorbance of 1,3-diphenylurea of known concentrations in acidic methanol (HCl and CH_3OH) is read at 254.5 mμ. The concentration is divided by the absorption at this wavelength to calculate a coefficient C.

The aniline interference was determined by reading the absorbance of a known amount of aniline in acidic methanol at 260.5 mμ and 254.5 mμ and obtaining the ratio,

$$R = \left(\frac{\text{Absorbance at 254.5 m}\mu}{\text{Absorbance at 260.5 m}\mu}\right)$$

The baseline method was used.

Then, milligrams of phosgene/volume of solution $= A \times C \times 0.466$. The factor 0.466 represents the molecular weight ratio of phosgene to diphenylurea.

Testing of the procedure was accomplished by breaking sealed capillaries containing known amounts of phosgene in a vaporization chamber under dynamic gas conditions and bubbling the gas through absorption bubblers similar to those shown in Figure 2.10. The method showed good recovery, 95–99% for one bubbler and 97–100% for three bubblers in series. The method is claimed to be specific and eliminate substances which normally interfere with a gravimetric weighing of the 1,3-diphenylurea.

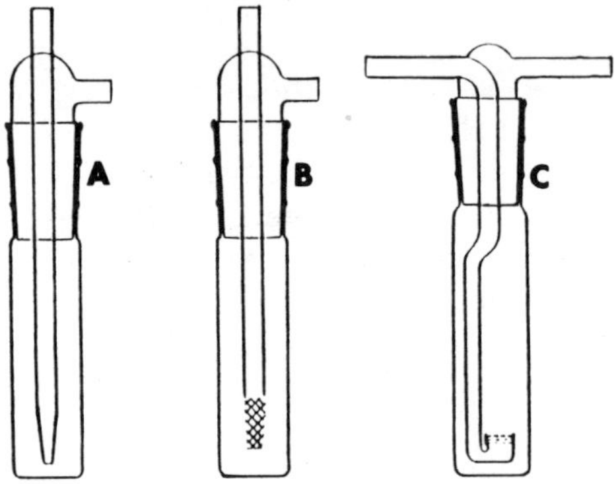

Fig. 2.10. Midget sampling equipment (73). (A) Standard midget impinger. Nozzle about 1-mm. diameter, passes 0.09–0.11 cfm at 12-in. H_2O vacuum. (B) Midget fritted tube end bubbler. (C) Midget fritted disk bubbler. Standard taper 24/40 joints. Capacities of lower (outer) pieces: 46 ml. (A and B), 60 ml. (C), when filled to brim.

Troy (63) has reported on a portable ultraviolet analyzer which is capable of operating at a number of wavelengths, namely 254, 313, 365, 405, and 436 mμ. A suitable calibrating apparatus is also described. Although the method is relatively straightforward, it does require a rather elaborate piece of equipment and appears to suffer from poor minimum detectability of phosgene (5.5 ppm).

III. NITROTOLUENES

A. Preparation

Numerous aromatic and aliphatic nitro compounds have been used as intermediates in the preparation of various isocyanates, but the most important one by far is dinitrotoluene (DNT), used in the preparation of toluenediisocyanate (TDI). The nitration of aromatic compounds will be reviewed (75–94) to serve as a background for the several analytical problems encountered in the manufacture of DNT.

Westheimer and Kharasch (86) have proposed that ionization of nitric acid is involved and the most probable ionizations are:

$$HNO_3 + H_2SO_4 \rightarrow H_2NO_3^+ + HSO_4^- \quad (33)$$

$$HNO_3 + 2H_2SO_4 \rightarrow H_3O^+ + NO_2^+ + 2HSO_4^- \quad (34)$$

Ionization of the type shown in Equation 34 and not that shown in 33 would account for the thousandfold increase in rate of nitration which accompanies an increase in sulfuric acid concentration from 80 to 90%.

The kinetic data on nitration in sulfuric acid indicate that the reaction is of the second order and that the second-order rate constant for the mononitration of toluene is 10,000,000 times that for the nitration of mononitrotoluene.

Gillespie et al. (95) proposed the following mechanisms for nitration:

I. Formation of NO_2^+

$$2HNO_3 \underset{2}{\overset{1}{\rightleftharpoons}} H_2NO_3^+ + NO_3^- \qquad \text{1 and 2 are fast} \qquad (35)$$

$$H_2SO_4 + HNO_3 \underset{4}{\overset{3}{\rightleftharpoons}} H_2NO_3^+ + HSO_4^- \qquad \text{3 and 4 are fast} \qquad (36)$$

$$H_2NO_3^+ \xrightarrow{5} NO_2^+ + H_2O \qquad \text{5 is slow} \qquad (37)$$

II. Entrance of NO_2^+ into Aromatic Nucleus

$$\mathrm{ArH} + \mathrm{NO_2^+} \xrightarrow{6} {}^{+}\mathrm{Ar}\begin{matrix}\diagup \mathrm{H} \\ \diagdown \mathrm{NO_2}\end{matrix} \qquad \text{6 is slow} \tag{38}$$

$${}^{+}\mathrm{Ar}\begin{matrix}\diagup \mathrm{H} \\ \diagdown \mathrm{NO_2}\end{matrix} \xrightarrow{7} \mathrm{ArNO_2} + \mathrm{H^+} \tag{39}$$

Steps 1 and 3 are supported by the fact that both nitrate and bisulfate ions decrease the rate of the overall reaction without affecting the reaction order.

Brennecke and Kobe (92) proposed nitration by the nitronium ion to occur according to the following:

$$\mathrm{HNO_3} + 2\mathrm{H_2SO_4} \rightleftharpoons \mathrm{NO_2^+} + 2\mathrm{HSO_4^-} + \mathrm{H_3O^+} \tag{40}$$

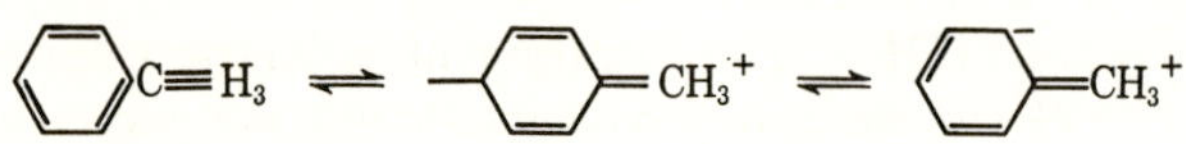

(41)

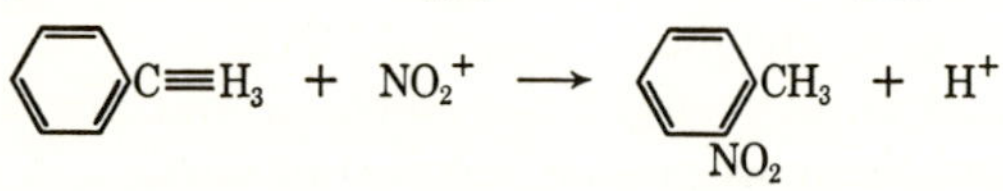

(42)

In a large number of ring substitutions the English workers have shown that orientation can be explained on the basis of probable electron densities around the ring.

Commercial nitration processes quite often consist of continuous operations for both mono- and dinitration steps. The Biazzi process was devised in 1935 by Dr. M. Biazzi (88). Continuous processes using mixed acids (sulfuric and nitric acid) mono- and dinitration of toluene have been similarly described (90,96).

Gibson et al. (76) reported the isomer distribution of mononitrotoluene between 0° and 50°C, prepared under the usual conditions of nitration to be *o*-nitrotoluene 62 $\pm$ 1%, *m*-nitrotoluene 4.2–4.5%, and *p*-nitrotoluene 32–33.5%.

Ingold et al. (78) have reported values for the isomers of mononitrotoluene at 0° and 30°C using acetyl nitrate in acetic anhydride. The average values are 58.4% *o*-nitrotoluene, 4.1% *m*-nitrotoluene, and 37.6% *p*-nitrotoluene.

Holleman (97) reported 58.8% *ortho*, 4.4% *meta*, and 36.8% *para* in the nitration of toluene at 30°. Jones and Russell (87) reported 58.5% *ortho*, 4.4% *meta*, and 37.2% *para* for the nitration of toluene at 30°C.

Knowles et al. (98) have reported isomer distributions similar to those of Gibson (76). The nitrations were carried out at 0°C with nitric acid and acetic anhydride. The isomer distribution, as analyzed by gas chromatography was found to be 61.4% *ortho*, 1.6% *meta*, and 37.0% *para*.

Roberts et al. (99) have shown that the nitration of *ortho*-nitrotoluene results in four isomers; approximately 66.4% 2,4-dinitrotoluene, 33.2% 2,6-dinitrotoluene, and small amounts of 2,4- and 2,5-dinitrotoluenes. The amounts and ratios of 2,4- and 2,6-dinitrotoluene agree well with the results previously reported by Gibson et al. (76).

The nitration of the meta isomer was reported by Gibson et al. (76) to give 25% 2,4-dinitrotoluene, 55% 3,4-dinitrotoluene, 20% 2,5-dinitrotoluene, and a small amount of 3,5-dinitrotoluene. It was also reported that only the 2,4 isomer was obtained as a product of the nitration of *p*-nitrotoluene. This agrees, in essence, with the results of other investigators although approximately 0.2% 3,4-dinitrotoluene appears to be produced along with the 2,4 isomer (100) from the nitration of *p*-nitrotoluene.

Roberts et al. (99) have shown that the proportion of 2,4-dinitrotoluene formed by nitration of *o*-nitrotoluene as determined from isotope dilution analysis is less than 0.4% of the total amount of dinitrotoluenes.

Apparent discrepancies in the literature may be due to slightly different nitration conditions (temperature as well as concentrations) and possible differences in analytical techniques. The newer techniques, such as gas chromatography and infrared, allow more accurate and reproducible results than the classical methods of crystallizing points.

The concentrations of the isomers obtained from both the mono- and dinitration of toluene vary depending upon the temperature of the reaction, concentration of sulfuric acid, nitric acid, and toluene as well as contact effects from agitation. Figure 2.11 shows the average concentrations of isomers normally listed in the literature for the composition of mono- and dinitrotoluene and the average concentrations in which they are formed during both nitration steps. It should be noted, however, that the DNT used ultimately for isocyanate production has approximately 80% of the 2,4 isomer and 20% of the 2,6 isomer.

The concentration of the 2,4 DNT isomer can be changed slightly with subsequent minor changes in the concentrations of some of the other isomers by suitable control of the process variables mentioned above (96).

It can be seen that suitable analytical methods are required to monitor the concentrations of the isomers of mono- and dinitrotoluenes as well as acidity, free nitric acid, mononitrotoluenes in DNT, and other impurities that exist in nitration grade toluene.

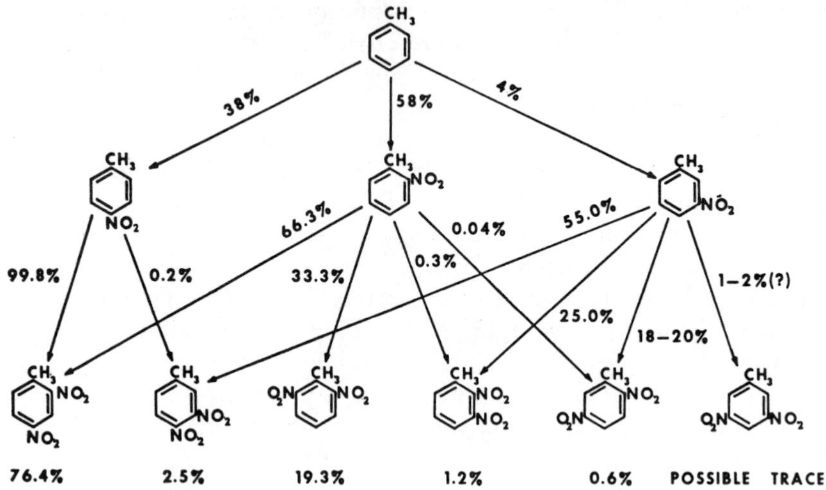

Fig. 2.11. Isomer distribution for the mono- and dinitration of toluene.

B. Analytical Techniques

1. Mononitrotoluenes

The main concern in the analysis of mononitrotoluene is the isomer distribution. Older analytical methods, such as crystallizing points, and less specific methods, such as infrared, will not be considered for mononitrotoluene because of the ease and accuracy of gas chromatographic methods. Gas chromatographic analyses of the mononitroluenes for isomer distributions have been described previously (101,102).

Knowles and Norman (101) used 30% Apiezon L grease coated on Embacel and Parsons et al. (102) used 10% Apiezon L on 60–80 mesh acid-washed Chromosorb W. The following method is very similar to these except for minor variations (103).

Gas Chromatographic Analysis of Mononitrotoluene for Assay and Isomer Ratio (103)

Scope

This method determines the isomer distribution of the mononitrotoluenes and the presence of dinitrotoluenes and other impurities.

Principle

See gas chromatographic analysis of phosgene.

EQUIPMENT

(*a*) Suitable gas chromatograph, preferably one equipped to do relatively high temperature work and fitted with W-2 filaments and katharometer power supply.

(*b*) Apiezon L.

(*c*) 30–60 mesh caustic-washed Kromat CE.

(*d*) A 10 μl syringe.

(*e*) Ethyl acetate.

(*f*) A 1 mv recorder.

CHROMATOGRAPHIC CONDITIONS

(*a*) Hot wire filament detector and power supply.

(*b*) Detector current: 250 mA or sufficient current to provide the desired sensitivity.

(*c*) Detector temperature: 210°C.

(*d*) Column: 4 m ¼ in. column packed with 20% Apiezon L on 30–60 mesh caustic washed Kromat CE.

(*e*) Column temperature: 190°C.

(*f*) Flow: 45 ml/min He.

(*g*) Flash vaporizer, 240°C.

(*h*) Sample size: 5 μl.

PROCEDURE

(*1*) Inject a 5 μl sample into the flash vaporizer.

(*2*) After sample injection, immediately rinse the syringe with ethyl acetate.

(*3*) Identification of components can be made by comparing the elution of the components in the sample to those on the chromatogram shown in Figure 2.12.

CALCULATIONS

Peak heights or area measurements may be utilized for determining the percentage of each component in the sample, after suitable calibration of the response of the system to known quantities of the pure isomers.

The accuracy of the method varies slightly for each isomer but will in general result in an overall accuracy for each component of $\pm 0.5\%$ providing each isomer concentration is normalized. The samples must be accurately measured and introduced with care.

2. DINITROTOLUENE

a. Wet Methods

The concentration of nitric acid in the spent acid is important because of its effect on the reaction with toluene. If not maintained at the proper

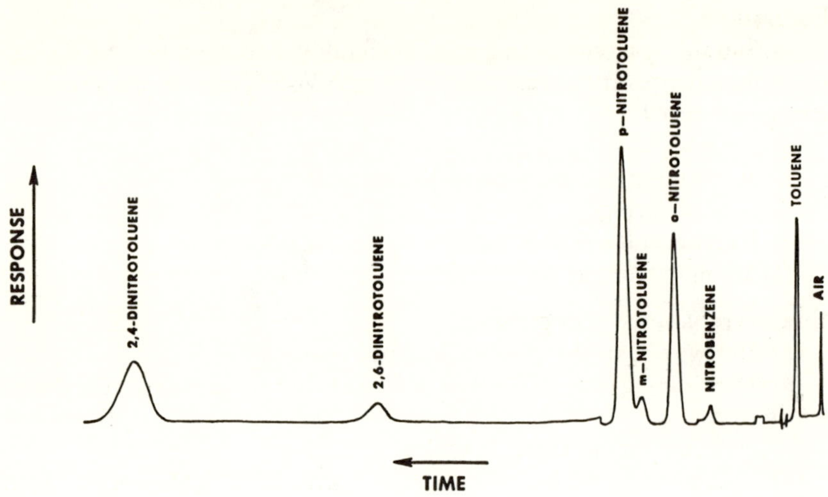

Fig. 2.12. Chromatogram of mononitrotoluene using a 20% Apiezon L column showing isomer distribution (103).

level it will affect the yield of DNT or result in losses of a costly material.

The titration procedure described by Scott (104), and by Kobe (93) may be used to determine nitric acid in the concentration range of 0.02–5.0% found in spent acid, or may be modified by suitable dilution to determine the higher concentration found in mixed acid.

The titration must be carried out in the presence of a large excess of cold concentrated sulfuric acid. Nitrososulfuric acid ($HNOSO_4$), usually present in low concentrations, does not interfere. The oily layer consisting of mononitrotoluene and dinitrotoluene usually found in spent acid does not seriously interfere, except to make the visual end point difficult to detect. This can be remedied by use of a potentiometer and suitable electrode system.

Determination of Free Nitric Acid in Spent Sulfuric Acid (105)

Scope

This method permits the determination of the amount of unreacted nitric acid in the spent sulfuric acid.

Principle

The nitric acid is titrated with a solution of ferrous sulfate to a brown-purple endpoint. The reaction is:

$$4FeSO_4 + 2HNO_3 + 2H_2SO_4 \rightarrow 2Fe_2(SO_4)_3 + NO + NO_2 + 3H_2O$$

When the nitrate is consumed completely, the nitric oxide in solution combines with excess ferrous ion to form a red or brown color.

Equipment

(*a*) Erlenmeyer flask, 500 ml.

(*b*) Ferrous ammonium sulfate solution, 0.1*N*, 40 g of ferrous ammonium sulfate in a cooled mixture of 300 ml of water and 30 ml of sulfuric acid diluted with water to 1 liter (106).

(*c*) Suitable balance.

(*d*) Reagent grade sulfuric acid.

(*e*) 100 ml graduate.

(*f*) Chipped ice.

(*g*) Suitable buret.

Procedure

(*1*) Introduce by means of a graduate 100 ml of concentrated reagent grade sulfuric acid into the Erlenmeyer flask.

(*2*) Cool the acid in an ice bath.

(*3*) Pipet 5–10 ml of sample, depending upon the expected concentration, from a weighed sample container. The weight of the sample is taken as the difference between initial and final weighings.

(*4*) Slowly introduce the sample under the surface of the sulfuric acid near the sides of the flask with a swirling motion to prevent local heating and subsequent loss of nitric acid. Rinse the pipet by drawing the acid into the pipet with a suction bulb.

(*5*) Titrate the solution immediately with the standard ferrous ammonium sulfate solution to the endpoint previously mentioned, maintaining the solution below 40°C at all times. Small amounts of nitric acid will give a pink endpoint.

Calculations

$$\%HNO_3 = \frac{\text{g } HNO_3/\text{ml } FeSO_4 \times 100}{\text{Sample wt.}}$$

Ferrous ammonium sulfate is to be preferred over ferrous sulfate because of better stability over a period of time. Either reagent should be standardized against nitric acid standards, prepared in the same concentration range as expected in the samples, and weighed and diluted with 75% sulfuric acid. The same reagents should be utilized for the determination as are used for the standardization. If ferrous sulfate or ferrous ammonium sulfate greater than 0.3*N* is used, a blank of 0.2 ml must be subtracted from the titration since a definite excess of an indicator blank is then required (93,104).

Acidity of DNT (107)

Scope

It is important to know the total concentration of the acids present in DNT since it will affect the corrosiveness of the material. Excess acidity indicates losses in process and may affect its reactivity in subsequent processes.

Equipment

(*a*) Two 250 ml Erlenmeyer flasks.
(*b*) Beaker, 400 ml.
(*c*) Hot plate.
(*d*) Balance.
(*e*) Bromthymol blue indicator.
(*f*) Distilled water.
(*g*) Graduate, 100 ml.
(*h*) Sodium hydroxide, 0.01*N*
(*i*) Buret, 10 ml.

Procedure

(*1*) Add approximately 300 ml of distilled water to a 400 ml beaker, place on a hot plate and boil for several minutes.

(*2*) Weigh approximately 10 g of sample into a 250 ml Erlenmeyer flask on a suitable balance.

(*3*) Add 100 ml of boiling distilled water to the flask and swirl vigorously for approximately 1 min. then allow the DNT to settle to the bottom of the flask.

(*4*) Decant the water layer into a 250 ml Erlenmeyer flask.

(*5*) Repeat step (*3*) two more times using 50 ml of boiling distilled water each time, and add the washings to the same flask.

(*6*) Cool the washings to room temperature, add 10–15 drops of bromthymol blue indicator, then titrate with $0.01N$ NaOH to the blue-green endpoint.

CALCULATION

$$\%\ \text{Acidity as } H_2SO_4 = \frac{\text{ml} \times N \times 0.049 \times 100}{\text{Sample Weight}}$$

Moisture in Dinitrotoluene (108)

SCOPE

Excessive amounts of water in DNT will create processing difficulties and prevent its recovery. It may also affect the subsequent hydrogenation process.

PRINCIPLE

This method employs the standard Karl Fischer titration. The KF reagent reacts quantitatively with water. Excess reagent is added to a weighed sample of the DNT. The excess reagent is then back-titrated with water–methanol solution. The following reaction occurs (109):

$$C_5H_5N{\cdot}I_2 + C_5H_5N{\cdot}SO_2 + C_5H_5N + H_2O + CH_3OH \rightarrow 2C_5H_5N\begin{smallmatrix}H\\I\end{smallmatrix} + C_5H_5N\begin{smallmatrix}H\\SO_4CH_3\end{smallmatrix} \quad (43)$$

Dry benzene or other suitable dry solvents may be used together with the methanol in order to render some samples soluble.

EQUIPMENT

(*a*) Suitable titrimeter equipped with platinum electrodes.
(*b*) Karl Fischer reagent.
(*c*) Water–methanol reagent.
(*d*) A 500 ml glass-stoppered Erlenmeyer flask.

PROCEDURE

(*1*) Add approximately 200 g of molten DNT to a warm, tared, glass-stoppered Erlenmeyer flask. Weigh the contents to nearest 0.1 g.

(*2*) Add approximately 100 g of dry pyridine and weigh to nearest 0.1 g.

(*3*) Shake vigorously until the contents form a homogeneous mixture or are completely solubilized.

(*4*) Using a tared, preheated eyedropper and vial, transfer approximately the equivalent of a 5 g sample to a preheated vial and reweigh.

(*5*) Titrate the sample according to standard methods for moisture in organics using suitable titrimetry equipment and Karl Fischer reagent.

CALCULATIONS

$$\% \; H_2O = \frac{(\text{ml KF reagent} - \text{ml of W–M}) \; \text{WF} \times 100}{\text{Sample weight of DNT}}$$

where WM is the water–methanol equivalent of KF reagent and WF is water factor expressed in g H_2O/ml KF reagent. A blank should be determined frequently on the pyridine to determine water content.

OTHER METHODS

Additional titration methods for the direct determination of dinitrotoluene, separately or in admixtures, using titanous chloride have been described (110–112).

Sarson (113) has reported methods for the analysis of explosives by nonaqueous titration. Trinitrotoluene, dinitrotoluene, and mononitrotoluene are resolved differently by titrations in methyl isobutyl ketone and dimethylformanide with tetrabutylammonium hydroxide since they behave as acids in these media.

b. Gas Chromatographic Methods

The Analysis of Dinitrotoluene for Impurities and Isomer Ratio by Gas–Liquid Chromatography (114)

SCOPE

The complete analysis employs two different chromatographic columns, silicone grease and Apiezon L grease, each on Kromat CE. From the silicone grease column the 2,5- and 3,4-dinitrotoluenes and the TNT content are calculated. From the Apiezon L grease column, the nitrobenzene, mononitrotoluene, *p*-nitroethylbenzene, nitroxylene, 2,4 dinitrotoluene and 2,4:2,6 DNT isomer ratio are calculated. The 2,4:2,6 DNT isomer ratio is calculated only after the concentration of the 2,4 isomer, from the Apiezon L grease column, has been corrected for interference from the 3,4 isomer. The isomer ratio has an accuracy and precision of $\pm 0.5\%$.

PRINCIPLE

See gas chromatographic analysis of phosgene.

EQUIPMENT

(*a*) Suitable gas chromatograph, preferably one equipped to perform adequately at high temperatures and fitted with W-2 filaments and katharometer power supply.

(*b*) Apiezon L.

(*c*) Silicone grease.

(*d*) A 50 μl syringe.

(*e*) Planimeter.

(*f*) Ethyl acetate.

(*g*) Kromat CE, 30–60 mesh caustic washed.

(*h*) A 1 mv recorder.

(*i*) Sodium sulfate, anhydrous.

CHROMATOGRAPHIC CONDITIONS

(*a*) Detector current: 250 mA or sufficient current to provide the required sensitivity.

(*b*) Detector temperature: 210°C.

(*c*) Columns: Two ¼-in. 4-m columns, one packed with 20% Apiezon L on 30–60 mesh caustic washed Kromat CE, the other column packed with 15% Dow Corning high vacuum grease on 30–60 mesh Kromat CE.

(*d*) Column temperature 190°C.

(*e*) Flow: 120 cc/min for silicone grease column; 70 cc/min for Apiezon L column.

(*f*) Flash vaporizer 260°C.

(*g*) Sample size 5 μl.

PROCEDURE

(*1*) Prepare a 50 wt % solution of DNT in ethyl acetate by weighing 5 g of molten DNT into a vial and adding 5 g of ethyl acetate.

(*2*) Add 1 g of anhydrous sodium sulfate and mix until the solution is clear.

(*3*) Inject five μl samples separately into each of the two columns.

(*4*) Identification of components can be made by comparing the elution of the components in the sample to those on the chromatograms shown in Figures 2.13 and 2.14.

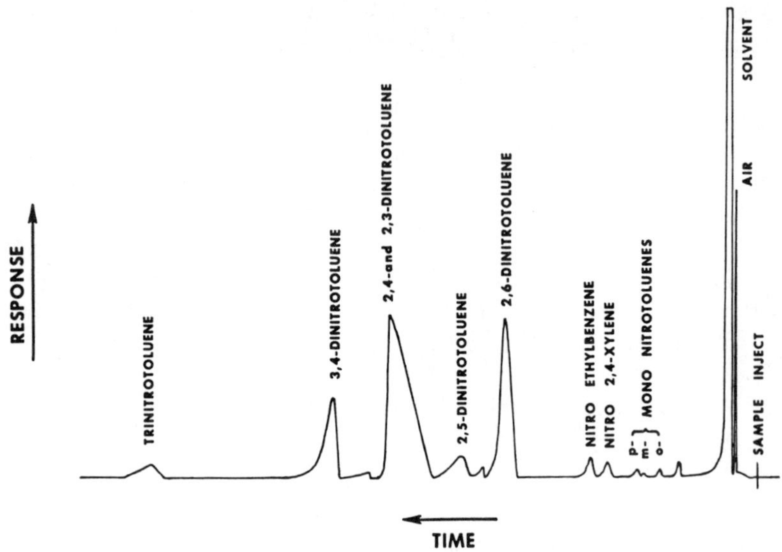

Fig. 2.13. Chromatogram of dinitrotoluene on a 15% silicone grease column showing isomer distribution and impurities (114).

Calculations

From silicone grease column: %3,4-DNT, %2,5-DNT, and %TNT

From Apiezon L column: % nitrobenzene, % MNT, %2,6-DNT, %2,3 + 2,5-DNT, and %2,4 + 3,4-DNT

Area measurements and subsequent area percentages should be utilized for determining the concentration of each component in the sample, after suitable calibration of the response of the system to known quantities of the pure substances. The quantities of the 2,4- and 2,6-dinitrotoluenes are then normalized in order to obtain the isomer ratio.

The temperature of the flash vaporizer must be carefully controlled, since at slightly lower temperatures peak broadening occurs while at higher temperatures (approximately 280°C) decomposition will be apparent. This can normally be observed as an increase in the mononitrotoluene content as a function of temperature. Although this method, especially the portion employing the Apiezon L column is similar to that described by Parsons et al. for DNT (102), and Knowles and Norman for MNT (101), as well as our own MNT method, it does appear to give better results when applied to commercial DNT. Parsons et al. (102) only studied the 2,4 isomer in the range of 5–30%.

c. Spectroscopic Methods and Studies

An excellent infrared study of nitro and other oxidized nitrogen compounds was performed by Brown (115). The fundamental vibrations of the XNO_2 group in nitro compounds and nitrate esters are shown in Figure 2.15 (115). In the study of the simplest nitroalkane, nitromethane, the positions of the C—NO_2 fundamentals were described by Brown to be:

NO_2 *asym* stretching	6.37 μ
NO_2 *sym* stretching	7.25 μ
C—N stretching	10.90 μ
NO_2 bending	15.24 μ
Out-of-plane	16.24 μ
CNO bending	20.75 μ

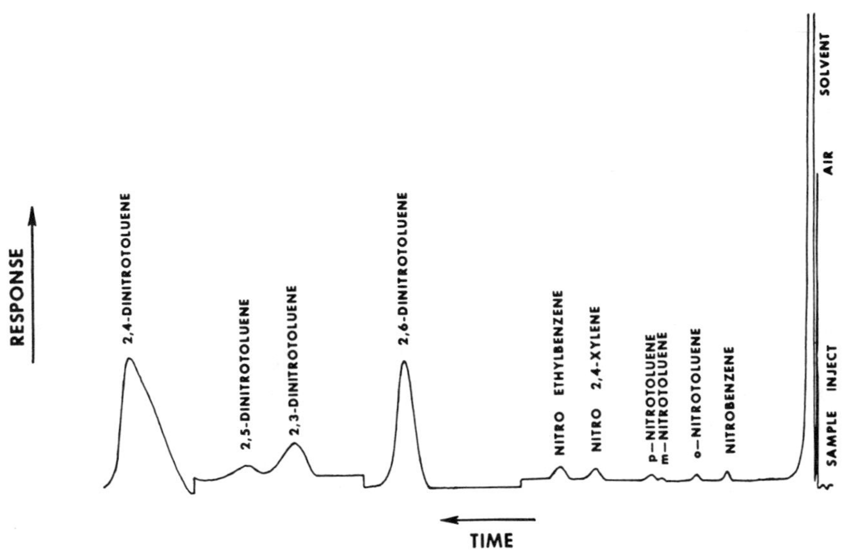

Fig. 2.14. Chromatogram of dinitrotoluene on a 20% Apiezon L grease column showing isomer distribution and impurities (114).

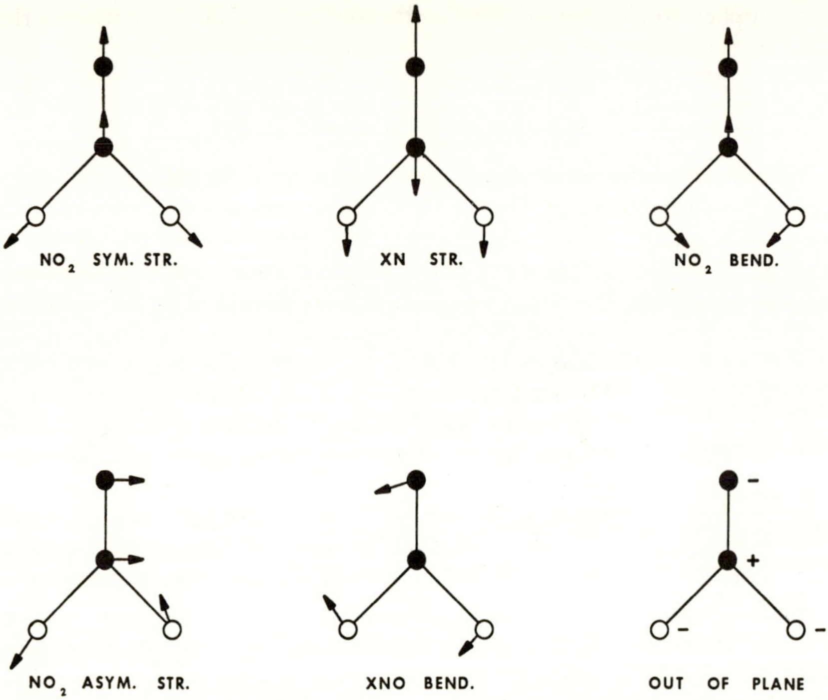

Fig. 2.15. Fundamental vibrations of XNO_2 structures (115).

These assignments and observations are consistent with those of earlier workers and the empirical assignments (116,117).

Randle and Whiffen (118) have reported that aromatic nitro compounds show bands at 6.58 + 0.05 and 7.41 ± 0.05 μ due to NO_2 stretching, and at 11.78 ± 0.12 μ due to C—N stretching. Brown found strong NO stretching bands of variable intensity at 6.46–6.63 and 7.37–7.46 μ for nitrobenzenes in chloroform.

The bands near 6.2 and 6.3 μ, arising from benzene ring vibrations, varied independently of either the orientation or the σ value, while the bands near 6.7 μ were at least roughly related to σ. The NO_2 asymmetrical stretching bands showed a good linear correlation with σ and the wavelengths of the NO_2 symmetrical stretching bands were lengthened by either negative or positive substituents.

The nitro asymmetric stretching frequency is primarily determined by the N—O bond order, and consequently the N—O stretching force constant. Since bond orders are determined by the relative contribution of resonance forms I and II to the overall structure, an increase in the con-

(I) (II)

tribution of form II should result in lower asymmetric stretching frequencies.

Kross and Fassel (119) have offered evidence to show that the C—NO_2 stretching frequency in aromatic nitro compounds is in the 1300 cm^{-1} region, while Brown concluded that the C—N stretching frequencies are less constant and, therefore, still uncertain for diverse aromatic compounds.

Qualitative and quantitative analyses of nitro compounds and mixtures have been described by Pristera (120) and Pristera and Halik (121).

Pristera and Halik (121) developed an infrared method for determining the concentration of *o*-, *m*-, and *p*-mononitrotoluenes and 2,4-dinitrotoluene in mixtures of these components as a result of the nitration of toluene. The method consists in measuring the absorbance of a 20% solution of the sample in cyclohexane at 8.70 μ for *o*-nitrotoluene, 12.49 μ for *m*-nitrotoluene, 8.49 μ for *p*-nitrotoluene, and 9.39 μ for 2,4-dinitrotoluene.

The measured absorbancies were corrected for solvent and cell effects which were determined prior to the analysis. The concentration was approximately 20%, but varied to the extent necessary to maintain the transmission in the 20–65% range. The concentration of each component was then calculated using the above and known absorptivities of the pure compounds. The method of successive approximations was then employed in order to correct for the contribution of each component. It was found that generally three approximations were sufficient to achieve the final results. The error experienced varied from 0.2% to approximately 4% relative, depending upon the concentration of the component being determined. In certain instances, when very low concentrations were determined, an error in excess of 4% was found. The method was reported to be applicable over wide ranges of concentration (0–100%) except in the case of 2,4-DNT.

The spectrograms of *o*-, *m*-, and *p*-mononitrotoluene and 2,4- and 2,6-dinitrotoluene obtained by Pristera and Halik are shown in Figures 2.16–2.20. These spectra were obtained by preparing capillary films of the materials in a sealed cell or heating the solids on a demountable salt plate.

Obviously, simultaneous equations could have been used in place of successive approximations and both techniques would be easier if computerized. This method, as any infrared method of this type, inevitably suffers from the possible interference of other isomers and/or impurities not normally expected. For this reason, as well as subsequent ease in handling and time savings, the gas chromatographic method, previously described is recommended.

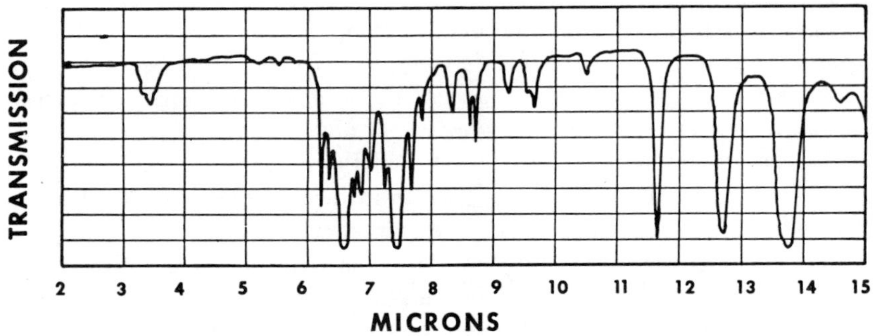

Fig. 2.16. Infrared spectrum of *o*-nitrotoluene (capillary film) (121).

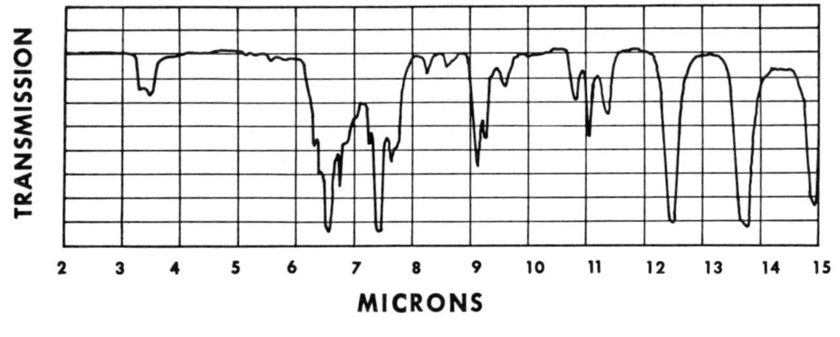

Fig. 2.17. Infrared spectrum of *m*-nitrotoluene (capillary film) (121).

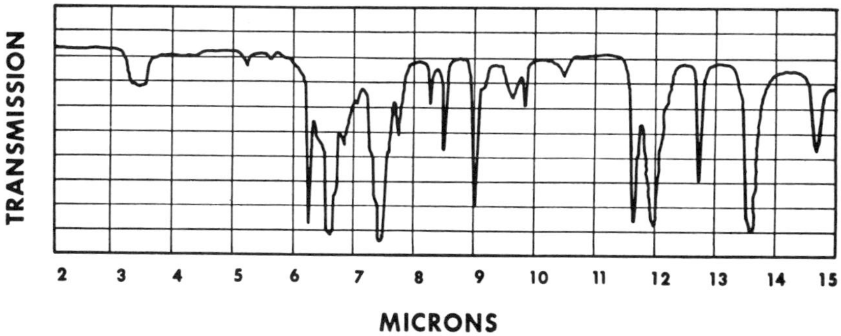

Fig. 2.18. Infrared spectrum of *p*-nitrotoluene (capillary film) (121).

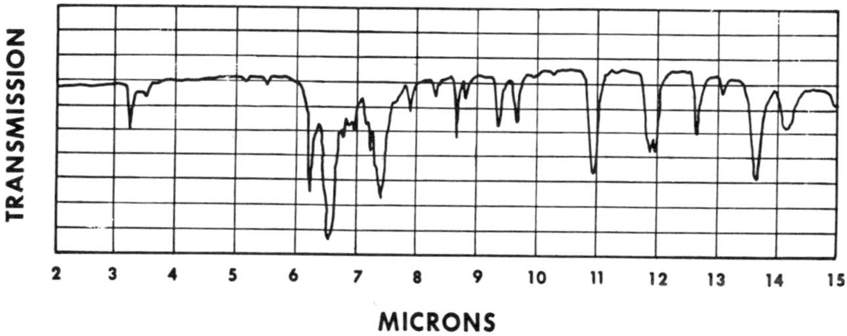

Fig. 2.19. Infrared spectrum of 2,4-dinitrotoluene (capillary film) (121).

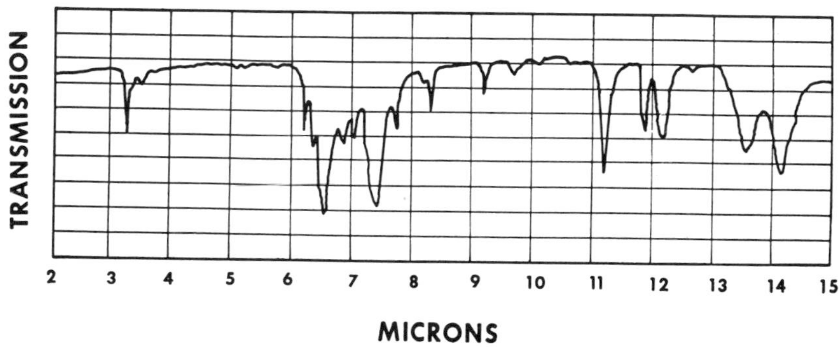

Fig. 2.20. Infrared spectrum of 2,6-dinitrotoluene (capillary film) (121).

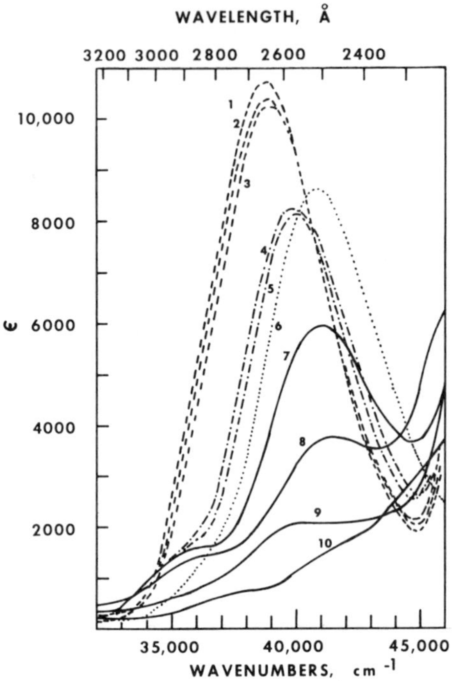

Fig. 2.21. Steric effects of alkyl groups on the nitro group absorption band (122). (*1*) *p*-Nitro-*t*-butylbenzene. (*2*) *p*-Nitrocumene. (*3*) *p*-Nitrotoluene. (*4*) *m*-Nitro-*t*-butylbenzene. (*5*) *m*-Nitrotoluene. (*6*) Nitrobenzene. (*7*) *o*-Nitrotoluene. (*8*) *o*-Nitrocumene. (*9*) Nitromesitylene. (*10*) *o*-Nitro-*t*-butylbenzene.

Brown and Reagan (122) have studied the steric inhibition of resonance caused by alkyl groups in *ortho*-substituted aromatic nitro compounds. The extent of this inhibition was dramatically shown according to the size and position of the substituent group as shown in Figure 2.21.

Doub and Vandenbelt (123) in their study of mono- and *p*-disubstituted benzene derivatives, found that the effect of the nitro group in nitrobenzene was to displace the primary band 65 mμ, from 203.5 mμ for benzene to 268.5 mμ for nitrobenzene. This effect was greatest for those groups capable of greatest electronic interaction across the ring.

Schroeder et al. (124) have studied the ultraviolet spectra of nitro compounds in order to aid in identification of components in smokeless powder and derivatives that are formed.

The spectra of five nitro compounds, of general interest here, are shown in Figure 2.22. It should be noted that the general spectral displacement is consistent with that found by Doub and Vandenbilt (123); it can be explained by the position of the group, and is a consequence of resonance effects.

Brand, Horning, and Thornley (125) have found that the ultraviolet absorption spectrum of aromatic nitro compounds is modified characteristically in highly acid (H_2SO_4 and H_2SO_4—SO_3) solvents by the transfer of a proton to the nitro group, yielding ions of the form Ar·NO·OH$^+$, with alteration of the binding energies of the electrons. They also evaluated ionization constants from the spectra in sulfuric acid media. The characteristic spectral effect of the proton was reported to move the principal absorption band some 6000–9000 cm^{-1} toward lower frequencies and simultaneously to increase ϵ_{max} nearly twofold as shown in Figure 2.23.

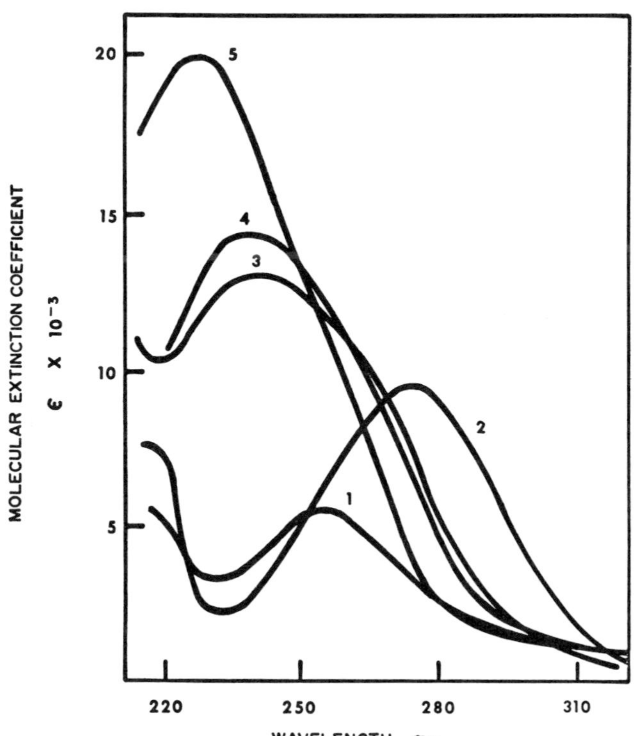

Fig. 2.22. Ultraviolet spectra of nitroalkylbenzenes (124). (*1*) 2-Nitrotoluene. (*2*) 4-Nitrotoluene. (*3*) 2,4-Dinitro-1-ethylbenzene. (*4*) 2,4-Dinitrotoluene. (*5*) 2,4,6-Trinitrotoluene.

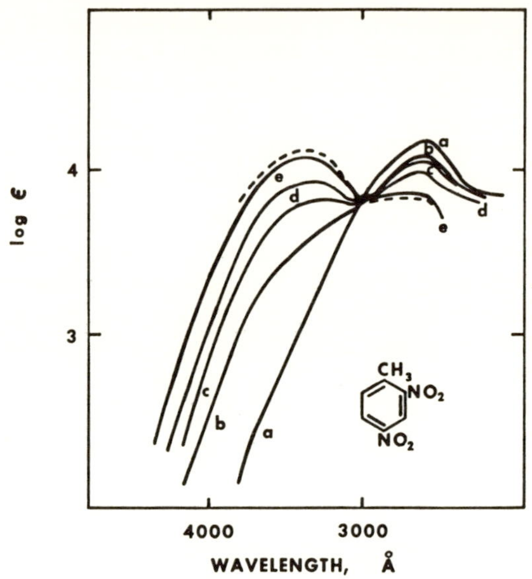

Fig. 2.23. Ultraviolet spectra of 2,4-dinitrotoluene in sulfuric acids (125). (*a*) 99.0% H_2SO_4. (*b*) 102.74% H_2SO_4. (*c*) 105.46% H_2SO_4. (*d*) 107.30% H_2SO_4. (*e*) 114.5% H_2SO_4.

In the spectra of 2,4-dinitrotoluene (Fig. 2.23), two intense bands are present, the first with the frequency and intensity characteristic of the ion of a mononitro compound (3550 Å) and the second characteristic of an un-ionized mononitro compound (2650 Å). Other compounds corroborated this, but the spectra are quite different from that of an ionized mononitro compound (e.g., *m*-nitroaniline). The simplest interpretation is that the 40,000 cm^{-1} transition of nitrobenzene does not involve the benzene orbitals but is localized mainly in the nitro group as Brand et al. (125) have suggested.

The above discussion of spectroscopic methods applicable to nitro compounds will serve as a background for the following infrared method.

Infrared Determination of 2,4- and 2,6-Dinitrotoluene Isomer Ratio and 2,3- and 3,4-Dinitrotoluene Content (124)

Scope

This procedure determines the ratio of the 2,4- and the 2,6-dinitrotoluene and the percent 2,3- and 3,4-dinitrotoluene in a solid mixture consisting of approximately 75% 2,4-, 22% 2,6-, and 3% 2,3- and 3,4-dinitrotoluene isomers combined. The ratio of the 2,4- to the 2,6-dinitrotoluene must be determined first in order that the concentration of the 2,3- and 3,4-dinitrotoluene can be determined by differential analysis.

PRINCIPLE

The absorption of each isomer, at a characteristic wavelength, is measured. This absorption of energy is caused by the C—H out-of-plane deformation of the benzene ring and is, therefore, characteristic of the nitro group substitution on the ring. It suffers from unknown interferences as well as slight contributions of each isomer at more than one wavelength. The characteristic wavelengths are as follows:

2,4-DNT	11.00 μ
2,6-DNT	11.22 μ
2,3-DNT	12.42 μ
3,4-DNT	12.50 μ

EQUIPMENT

(*a*) Suitable infrared spectrophotometer to provide necessary resolution which should be determined on an individual basis.

(*b*) Sealed liquid sodium chloride cells of 0.1 mm for 2,4- and 2,6-DNT and 0.5 mm for 2,3- and 3,4-DNT path length (calibrated before use).

(*c*) Appropriate glassware.

INFRARED CONDITIONS

Should be determined upon the individual instrument to provide optimum resolution and accuracy.

CALIBRATION (RATIO OF 2,4- TO 2,6-DNT)

(*1*) Weigh into a serum vial to 0.1 mg. approximately the following:

(*a*) 1.6 g 2,4-DNT; 0.4 g 2,6-DNT
(*b*) 1.8 g 2,4-DNT; 0.2 g 2,6-DNT
(*c*) 1.4 g 2,4-DNT; 0.6 g 2,6-DNT
(*d*) 1.2 g 2,4-DNT; 0.8 g 2,6-DNT

(*2*) Add the weights of each of the four standards and from the total weight of each mixture calculate the amount of benzene to add to make a concentration of 2.0000 gm. total DNT per 10.00 ml. of benzene. For example, suppose 1.8100 gm. of one isomer and 0.1850 gm. of the other isomer were actually weighed. The total sample weight is 1.9950 g. Then,

$$\frac{1.9950}{2.0000} \times 10.00 \text{ ml} = 9.97 \text{ ml of benzene to add to the particular mix}$$

(*3*) Add the calculated amount of benzene to each vial.

PROCEDURE (RATIO OF 2,4- TO 2,6-DNT)

(*1*) Fill the cell with pure benzene and set the pen at an absorbance of 0.1 at 10.5 μ.

(*2*) Record the spectrum to 11.5 μ.

(*3*) Without making any instrument changes, empty the cell and fill with sample solution and record the spectrum as before, superimposing it upon the benzene spectrum.

(*4*) For the 2,4-DNT, measure the absorbance difference between the benzene spectrum and the sample spectrum at 11.00 μ.

(*5*) For the 2,6-DNT, measure the absorbance difference between the benzene spectrum and sample spectrum at 11.22 μ.

(*6*) Divide absorbance at 11.00 μ by absorbance at 11.22 μ to obtain R.

(*7*) Plot the value of R versus the concentration of 2,4-DNT on two-cycle semilog paper.

CALIBRATION (PERCENT 2,3- AND 3,4-DNT)

(*1*) Prepare a dry mix of approximately 50 g of 2,4- and 2,6-DNT from purified materials in an accurately known weight ratio close to an 80:20 ratio. Melt the entire mix and stir thoroughly.

(*2*) Add approximately 2.0 g of the molten mixture into each of five tared serum vials, and reweigh after cooling to room temperature.

(*3*) Prepare four standards as follows and reserve the fifth one as the reference standard. Add the following amounts of 2,3- and 3,4-DNT to the vials:

(*a*) 0.01 g 2,3-DNT; 0.01 g 3,4-DNT
(*b*) 0.02 g 2,3-DNT; 0.02 g 3,4-DNT
(*c*) 0.04 g 2,3-DNT; 0.04 g 3,4-DNT
(*d*) 0.10 g 2,3-DNT; 0.10 g 3,4-DNT

(*4*) Using the total weight of all the isomers added, calculate the amount of benzene to add as previously described. Use these standards as samples.

CALIBRATION (SPECIAL STANDARDS)

(*1*) Weigh 20.000 g of 2,4-DNT into a clean dry 100 ml glass-stoppered volumetric flask and dilute to volume with benzene.

(*2*) Stopper and mix and label as "Stock A."

(*3*) Prepare a 2,6-DNT standard in the same manner; label as "Stock B."

PROCEDURE (2,3- AND 3,4-DNT CONCENTRATIONS)

(*1*) Calculate the ratio of 2,4- to 2,6-DNT of the 2,3- and 3,4-DNT standards (sample).

(*2*) Prepare a reference standard of pure 2,4-DNT (Stock A) and 2,6-DNT (Stock B) in the same ratio as the sample and in such a way that the total weight is equivalent to 2.0000 g/10.00 ml of benzene. For example, if the ratio found in the sample is 81:19, use 8.10 ml of Stock A and 1.90 ml of Stock B.

(*3*) Fill both cells with the reference standard from Step 2 and record the spectrum from 12.2 to 13.0 μ.

(*4*) Empty and clean the sample cell and fill with sample. Without changing the instrument conditions, record the sample spectrum by superimposing it over the reference spectrum from 12.2 to 13.0 μ.

(*5*) Measure the absorbance difference between the reference versus reference scan and reference versus sample scan at 12.42 and 12.50 μ.

(*6*) Using the absorbance differences at 12.42 μ for 2,3-DNT and 12.50 μ for 3,4-DNT, construct a curve of absorbance versus concentration.

SAMPLE PROCEDURE

(*1*) Place the DNT sample on a steam bath or in an oven at 70°C until it has completely melted. If small amounts of water are known to be present, dry the molten sample with calcium chloride.

(*2*) Weigh 2.0000 g of the liquid into a wide-mouth serum vial. Stopper the vial with a neoprene serum cap.

(*3*) Add clean dry benzene to dissolve the sample and make to a concentration of 2.0000 g of sample/10.00 ml of benzene.

(*4*) Calculate the weight of benzene to add as previously described.

(*5*) Record the spectrum as previously described.

(*6*) Measure the absorbancies as previously described.

CALCULATIONS

(*a*) *Ratio of 2,4- to 2,6-DNT*

1. Divide the absorbance at 11.00 μ by absorbance at 11.22 $\mu(R)$.
2. Using R, read the concentration of 2,4-DNT from the calibration curve.
3. Subtract the 2,4 isomer content from 100 tc find the 2,6 isomer.

(*b*) *2,3- and 3,4-DNT concentrations*

1. Using the absorbance measured at 12.42 μ, read the concentration of 2,3-DNT from the calibration curve.
2. Using the absorbance measured at 12.50 μ, read the concentration of 3,4-DNT from the calibration curve.

PRECISION AND ACCURACY

The precision of the method is $\pm 0.4\%$ standard deviation based on the 2,4 isomer.

d. Additional Methods and Techniques

The color reaction produced by certain dinitro aromatic compounds when treated in acetone solution with alcoholic sodium hydroxide was first described by Rudolph (127) and has been used qualitatively to analyze for the presence of dinitro compounds. English (128) has used the basis of this reaction for the quantitative determination of certain dinitro compounds formed in small amounts during the mononitration step. The preparation of the sodium hydroxide reagent has been described as being critical and in order to obtain quantitative results, sodium monoxide must be used in place of sodium hydroxide. Rudolf (127) suggests the following procedure:

Add 2 g sodium monoxide to 10 ml acetone and 10 ml of 95% ethanol, swirl gently for 2 min, cool to room temperature, and bring the volume to 200 ml with acetone.

Introduce approximately 10 g of sample into a 100 ml volumetric flask. Dilute to 100 ml with acetone. Pipet 25 ml of the sodium hydroxide reagent into a 50 ml beaker. Pipet an aliquot of sample solution equivalent to 1 g of sample into the reagent solution, stir, and read the color on a suitable colorimeter or spectrophotometer at 610 mμ. The absorbance reading should be compared against known amounts of DNT in MNT treated in a similar manner.

English (128) has also described a procedure for the determination of dinitrocresol by reacting 10 ml of the sample with 75 ml of 0.5% aqueous sodium hydroxide, diluting to 100 ml and reading the color intensity of the aqueous layer at 410 mμ.

Nelson and Lakowski (129) have shown the applicability of lithium aluminum hydride as a qualitative test reagent for aromatic nitro compounds. It is relatively easy to conduct the test; only the reagent and sample dissolved in anhydrous ethyl ether are required. The test is subject to interferences which must be checked out on an individual basis as well as different color hues depending upon the variety of aromatic nitro compounds present.

Schonberg et al. have reported color formation of benzene solutions of aromatic nitro compounds with activated alumina (130). The color formed with 2,4 DNT and activated alumina was violet-brown and was reported to be associated with the change from the benzenoid to the quininoid system according to Eq. 44.

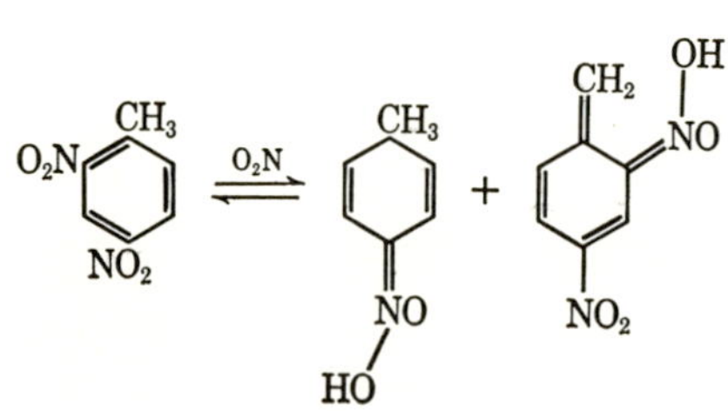

(44)

Porter (131) extended the color reaction of *m*-dinitro and trinitro compounds with acetone or alcohol and alkali (128,132–134) to mononitro aromatic compounds. The reaction was carried out by adding 0.1 ml of a 10% aqueous solution of tetraethylammonium hydroxide (Teah) to a solution of 5–100 γ of nitro compound in 9.9 ml of dimethylformamide (DMFA). For the Janovsky reaction, 10% sodium hydroxide and acetone were used in place of the above reagents. The color reactions and results are shown in Table 2.3.

The reaction of di- and trinitro compounds has been reported to proceed according to Eq. 45 (135,136). Porter (131) has suggested a similar reac-

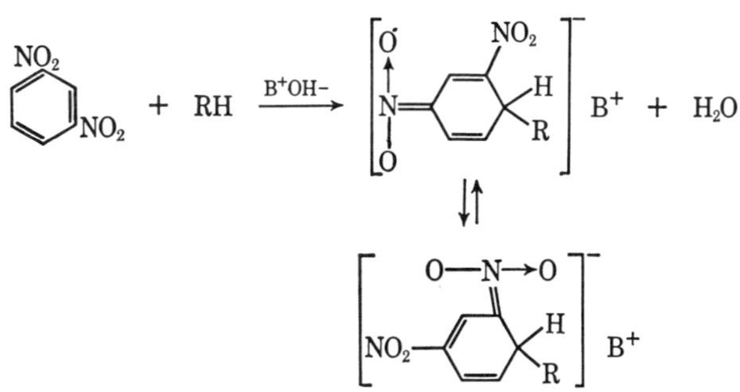

(45)

tion for mono-nitro compounds (Eq. 46).

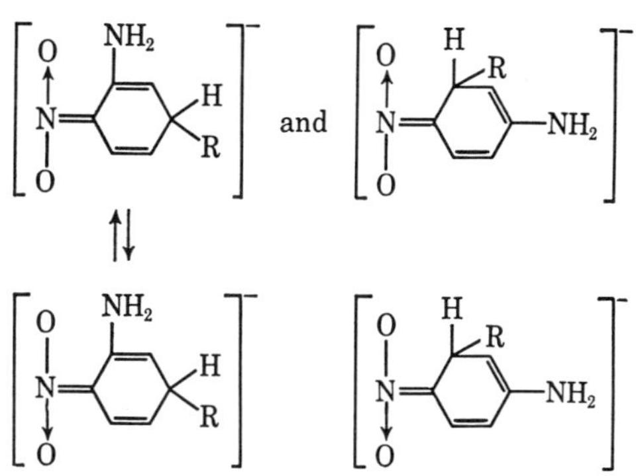

(46)

IV. TOLYLENEDIAMINE

A. Preparation

There are numerous processes by which an aromatic nitro compound may be converted to the corresponding amine (137–149). Commercially

TABLE 2.3
Color Reactions of Various Aromatic Nitro Compounds (131)

No.	Dinitrobenzene derivatives	Teah-DMFA[f] Wave length of max. abs., mμ	Teah-DMFA[f] Color	Teah-DMFA[f] Absorbance[a] at λ of max. abs.	Janovsky reaction[f] Wave length of max. abs., mμ	Janovsky reaction[f] Color	Janovsky reaction[f] Absorbance[a] at λ of max. abs.
1	2,4-Dinitrotoluene	660 (410)	Blue	0.80	580 (670)	Blue	0.33 (0.23)
2	3,5-Dinitrobenzoic acid	545 (370)	Purple	1.23 (0.52)	570 (380)	Purple	0.27 (0.16)
3	*m*-Dinitrobenzene	524	Red	1.48	570	Purple	1.58
4	2,4-Dinitrophenylhydrazine	525	Red	1.09	410 (570)	Green	0.6 (0.3)
5	2,4-Dinitroanisole	490	Orange	0.74	578	Purple	0.42
6	3,5-Dinitrosalicylic acid	455	Yellow	0.68	398	Lt. yellow	0.40
7	2,4-Dinitro-1-naphthol	446	Yellow	0.76	445	Yellow	0.73
8	2,4-Dinitrophenol	430	Yellow	0.76	422	Yellow	0.73
9	2,5-Dinitrobenzoic acid	430	Yellow	0.09	385	Lt. yellow	0.03
10	*p*-Dinitrobenzene	430	Yellow	0.77	554 (400)	Red → yellow	0.27 (0.18)
11	2,4-Dinitrochlorobenzene	428 (520)	Yellow	0.59 (0.14)	550 (660)	Purple	0.58 (0.28)
Mononitro Compounds							
12	2,Nitrodiphenylamine	550	Purple	0.51	430	Yellow	0.32
13	*p*-Nitrotoluene	520	Purple[b]	1.32	—	None	—
14	2-Nitro-4-chloroaniline	510	Red	0.57	417	Yellow	0.32
15	2-Nitro-4-chlorophenylurea	510	Red	0.53	416	Yellow	0.23
16	*o*-Nitroaniline	504	Red	0.68	407	Lt. yellow	0.38
17	4-Nitro-*o*-phenylenediamine	503	Red	1.97	408	Lt. yellow	0.69
18	*o*-Nitrophenylurea	502	Red	0.57	383	Lt. yellow	0.17
19	*o*-Nitroxylidine	500	Red orange	0.69	417	Yellow	0.34
20	*p*-Nitrophenylbiguanide	495	Orange	1.94	395	Lt. yellow	0.58
21	4-Nitrocarbanilide	480	Deep yellow	1.76	360	Lt. yellow	0.19
22	4-Nitro-4 -chlorocarbanilide	480	Deep yellow	1.45	460	Yellow	0.80
23	5-Nitro-2-aminoanisole	478	Deep yellow	1.93	395	Lt. yellow	1.11
24	*m*-Nitrophenol	478	Yellow	0.17	400	Lt. yellow	0.08
25	1-Benzoyl-3-(*p*-nitrophenyl)urea	475	Yellow	0.98	462	Yellow	0.90
26	4-Nitro-4′-cyanocarbanilide	473	Yellow	1.54	450	Yellow	0.58
27	4-Nitro-2-chloroaniline	470	Yellow	1.53	465	Yellow	0.51
28	2,6-Dichloro-4-nitroaniline	470	Yellow	1.41	465	Yellow	1.42
29	*m*-Nitrophenylurea	470	Lt. yellow	0.06	370	None	0.03
30	*p*-Nitrophenylurea	466	Deep yellow	1.04	446	Yellow	0.70
31	*m*-Nitrobenzaldehyde	465	Yellow	0.35	—	None	—
32	*p*-Nitroaniline	465	Deep yellow	1.06	386	None	0.51
33	1-Nitronaphthalene	462	Yellow[c]	0.60	—	None	—
34	8-Nitroquinoline	460	Yellow[d]	0.21	—	None	—
35	*o*-Nitrobromobenzene	460	Yellow[e]	0.13	375	None	0.04

TABLE 2.3 (*Continued*)

No.	Dinitrobenzene derivatives	Teah-DMFA[f] Wave length of max. abs., mμ	Teah-DMFA[f] Color	Teah-DMFA[f] Absorbance[a] at λ of max. abs.	Janovsky reaction[f] Wave length of max. abs., mμ	Janovsky reaction[f] Color	Janovsky reaction[f] Absorbance[a] at λ of max. abs.
37	*o*-Nitrophenol	452	Yellow	0.52	442	Yellow	0.50
38	*o*-Nitrochlorobenzene	450	Yellow[e]	0.17	—	None	—
39	4-Nitrophenylurethane	450	Yellow	0.99	436	Yellow	0.90
40	6-Nitroquinoline	440	Yellow	0.79	415	Yellow[c]	0.45
41	*p*-Nitrophenol	430	Yellow	0.93	424	Yellow	0.90
42	*p*-Nitrochlorobenzene	430	Yellow	0.75	—	None	—
43	4-Nitro-2-amino toluene	420	Lt. yellow[e]	0.50	390	Lt. yellow	0.13
44	*o*-Nitro-toluene	410	Lt. yellow	0.08	—	None	—
45	6-Nitrobenzimidazole	404	Lt. yellow[e]	0.71	392	Lt. yellow	0.40
46	2-Nitroresorcinol	400	Yellow[e]	0.22	488	Lt. orange	0.17
47	*m*-Nitroaniline	400	Lt. yellow	0.11	385	Lt. yellow	0.11
48	*o*-Nitrobenzaldehyde	400	None	0.04	—	None	—
49	*p*-Nitrobenzamide	380	None	0.12	360	None	0.02
50	5-Nitrobenzotriazole	386	Lt. yellow	0.32	382	Lt. yellow	0.26
51	5-Nitrouracil	385	Lt. yellow	0.50	370	None	0.20
52	*p*-Nitrobenzaldehyde	360	None	0.07	—	None	—
53	*p*-Nitrobenzoic acid	360	None	0.04	—	None	—
54	*p*-Nitroanisole	360	None	0.05	—	None	—
55	*m*-Nitroanisole	360	None	0.04	—	None	—
56	*o*-Nitrobenzoic acid	—	None	—	—	None	—
57	*m*-Nitrotoluene	—	None	—	—	None	—
58	Nitrobenzene	—	None	—	—	None	—
Compounds Containing Two Nitro Groups on Separate Aromatic Rings							
59	4,4′-Dinitrodiphenyl-methane	722 (460)	Blue-green	1.04	750 (490)	Blue	0.69
60	4,4′-Dinitrostilbene	596 (375)	Blue	0.53 (0.52)	—	None	—
61	4,4′-Dinitrocarbanilide	556	Purple	1.9	460	Yellow	0.48
62	4-Nitrobenzoyl-4′-nitro-phenylhydrazine	522 (660)	Purple	0.8 (0.67)	547	Purple	1.19
64	4′,3-Dinitrobenzanilide	450	Yellow	0.87	440	Yellow	0.83
65	3′,4-Dinitrobenzanilide	415	Yellow	0.3	400	Yellow	0.25
66	3,3′-Dinitrobenzanilide	375	Yellow	0.12	375	Lt. yellow	0.11
67	2,4′-Dinitrobiphenyl	—	None	—	—	None	—

[a] 10γ compound/ml. solution. Light path, 1.3 cm.
[b] Required 0.15 rather than 0.1 ml. Teah for stable color.
[c] After standing 1 hour.
[d] After standing 0.5 hour.
[e] Increases with time.
[f] See text.

dinitrotoluene is reduced to tolylanediamine by means of hydrogen in the presence of a suitable catalyst.

Albert and Ritchie (139), studied the reduction of several nitro compounds by hydrogen and Raney nickel at atmospheric temperature and pressure, and reported the reaction to follow the route

$$RNO_2 \rightarrow RNO \quad (47)$$

$$RNO \rightarrow RNHOH \quad (48)$$

$$RNHOH \rightarrow RNH_2 \quad (49)$$

Hydrogenation can be carried out by (*1*) dispersing the nitro compound in water with palladium-carbon catalyst and hydrogenating at 70–75°C (142), (*2*) dissolving the nitro compound in cyclohexane and tetrahydrofuran and hydrogenating in the presence of palladium black catalyst (143), and (*3*) dissolving dinitrotoluene in methanol and treating it with hydrogen in the presence of Raney nickel at 80–120°C and 200 atmospheres (146). The distribution of the several isomers in the tolylenediamine will be the same as in the dinitrotoluene, excepting for minor losses of one or two isomers.

The following analytical determinations are the most useful for characterizing the tolylenediamine for isocyanate production: (*1*) assay, (*2*) water content, and (*3*) residue content.

B. Analytical Techniques

1. Wet Methods

Determination of Water in Tolylenediamine (150)

Scope

This method allows the determination of water in tolylenediamine by the Karl Fischer technique. The level of moisture is important since the presence of water contributes to side reactions resulting in the formation of ureas and acid during the phosgenation step.

Principle

The sample of tolylenediamine is dissolved in dry xylene and sufficient xylene distilled over to quantitatively remove (azeotrope) the water. Methanol is added to the water–xylene distillate as a mutual solvent. The mixture is then titrated with Karl Fischer reagent using a direct dead-stop technique.

Equipment

(*a*) Heating mantle, 500 ml.
(*b*) Single-neck flask with 35/25 joint, 500 ml.

(*c*) Distillate column (see Fig. 2.24).
(*d*) Water cooled condenser.
(*e*) Three-neck flask with 35/25 joints, 500 ml.
(*f*) Buret, back-fill, amber, 50 ml.
(*g*) Electrode assembly (see Fig. 2.24).
(*h*) Magnetic stirrer (see Fig. 2.24).
(*i*) Drying tube.
(*j*) Methanol.
(*k*) Karl Fischer reagent (standardized to obtain water equivalent).
(*l*) 50 ml graduate.
(*m*) Suitable electrical circuit such as that shown in Figure 2.24.

PROCEDURE

(*1*) Assemble the equipment as shown in Figure 2.24. All equipment must be thoroughly dried.

(*2*) Add 100 ml of dry xylene to the distillation flask. Weigh by difference, to the nearest 0.1 g, about 200 g TDA into the distillation flask.

(*3*) Apply heat and distil approximately 50 ml into the receiver. Discontinue the distillation and rinse the condenser with exactly 50 ml of dry methanol. Stopper the condenser.

(*4*) The buret should be full of standardized Karl Fischer reagent. The electrodes must be immersed and not touching each other. Start the stirrer and begin adding reagent. The voltmeter will deflect from 0 to about 0.85 V with the first few milliliters of reagent and then become steady as additional reagent is added. With experience, the reagent can be added rapidly until the meter begins to oscillate unsteadily (about 0.1 V) just before the endpoint is reached. Proceed dropwise until at the endpoint the meter deflects sharply through about 0.5 V (to about 1.4–1.5 V). This high reading should be steady for 20–25 sec without further addition of reagent.

(*5*) A blank is determined by repeating the above procedure, omitting the TDA. The voltage range may be slightly lower but the change at the endpoint will be about 0.5 V.

CALCULATIONS

$$\%H_2O = \frac{(\text{ml of KF} - \text{blank}) \times \text{KF Factor}}{\text{Sample weight} \times 10}$$

$$\text{KF factor} = (\text{mg of } H_2O)/(\text{ml of KF})$$

The above method uses the Dean-Stark distillation procedure (151), except for minor details. The standardization to obtain the KF factor

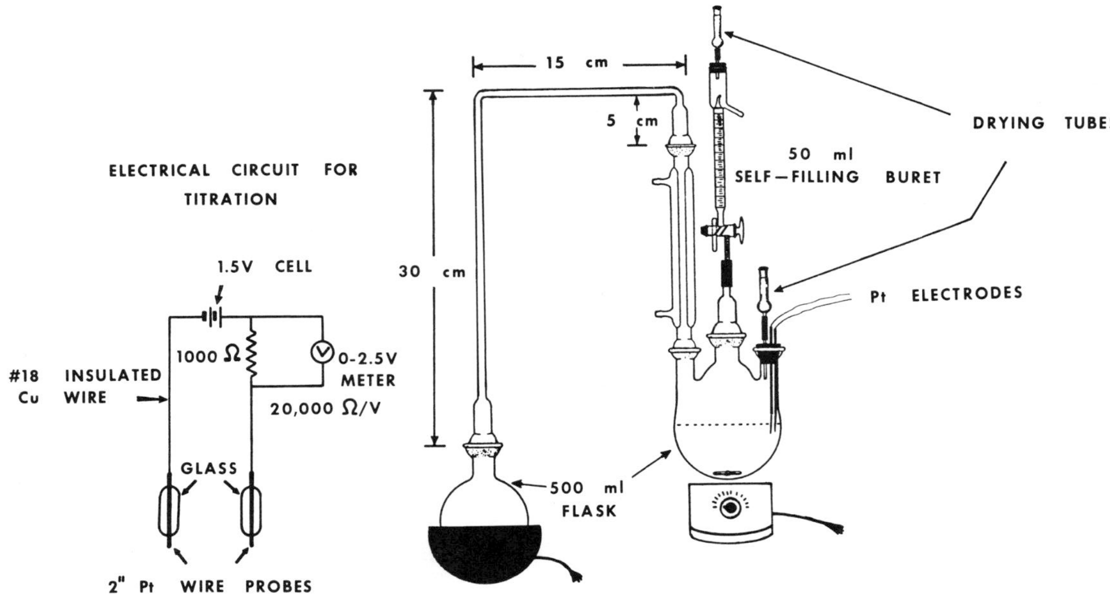

Fig. 2.24. Apparatus for the determination of water in tolylenediamine by Karl Fischer titration (150).

should be carried out using 50 ml of xylene and 50 ml of methanol which are blanked out before adding known amounts of water and titrating to the endpoint as previously described.

Care must be taken to ensure that the equipment is properly dried prior to use, since significant error may be introduced using large flasks, which are employed in this method.

The sample size may have to be adjusted depending upon the water content of the sample to be analyzed. Prior to use, a voltage drop of at least 1.4 V should be registered by measuring the voltage drop across the resistor while the resistor is shorted directly across the terminals of the battery.

Determination of Purity of Tolylenediamine (152)

Scope

This method can be used to conveniently define the purity of the tolylenediamine. However, this procedure will be in slight error if polyamines are present since the amino groups will also be titrated.

Principle

The sample is titrated as a diacid base in a nonaqueous medium using 0.1*N* $HClO_4$ in acetic acid as the titrant.

Equipment

(*a*) Fisher titrimeter or equivalent.
(*b*) Glass electrode.
(*c*) Calomel electrode.
(*d*) Berzelius beakers, 200 ml.
(*e*) Glacial acetic acid.
(*f*) 0.1*N* perchloric acid in acetic acid. Take 8.5 ml of 70% perchloric acid and dilute to 1 liter with glacial acetic acid. Mix well and allow to stand for 24 hr before standardization. Standardize the reagent with potassium acid phthalate.

Procedure

(*1*) On an analytical balance, weigh 0.2–0.25 g of pulverized sample and transfer into a 200 ml Berzelius beaker.

(*2*) With agitation, dissolve the sample in 80 ml of glacial acetic acid.

(*3*) Titrate potentiometrically with the perchloric acid solution, titrating in small increments of about 0.05 ml when nearing the equivalence point.

(*4*) Construct a curve by plotting volume vs. pH and determine the volume of titrant used at the point of inflection.

Calculation

$$\%\text{TDA} = \frac{\text{ml HClO}_4 \times N \times 6.109}{\text{Sample weight}}$$

Determination of Residue in Tolylenediamine (153)

Scope

This method provides a measure of the amount of polyamines that have been formed during, and as a result of, the reduction step. The values derived from this method are not strictly comparable with those of the purity (titration method) since the polyamines are also titratable. The results of both methods, considered in total, will provide an adequate picture of the quality of the tolylenediamine.

Principle

The tolylenediamine is distilled at high vacuum and temperature until all of the material that distils without decomposition has been removed.

Equipment

(*a*) Flask: 500 ml two-neck, 35/25 and 28/15 (obtain tare weight).
(*b*) Thermometer: −10°C to +360°C.
(*c*) Waage oil bath.
(*d*) Pyrex glass distillation delivery tube (60° bend).
(*e*) Manometer (0–30 cm. range).
(*f*) Flask, 35/25, 500 ml (obtain tare weight).
(*g*) Vacuum adapter.
(*h*) Vacuum vapor traps.
(*i*) Thermos for cooling traps.
(*j*) Air condenser.
(*k*) Lab-Jack.
(*l*) Boiling sticks.

Procedure

(*1*) Weigh by differences approximately 200 g of sample into the tared 500 ml boiling flask and add two boiling sticks.

(*2*) Assemble the apparatus as shown in Figure 2.25.

(*3*) Lower the boiling flask into the oil bath which has been heated to about 100°C. Pull a vacuum to 0.5 mm or less. The cooling bath and vapor trap both should be in position.

(*4*) Heat the sample flask to approximately 165°C and slowly increase the temperature until distillation begins. Maintain a distillation rate of 2–3 drops per second by gradually increasing the temperature to 240°C and full vacuum for 10 min after the distillation stops.

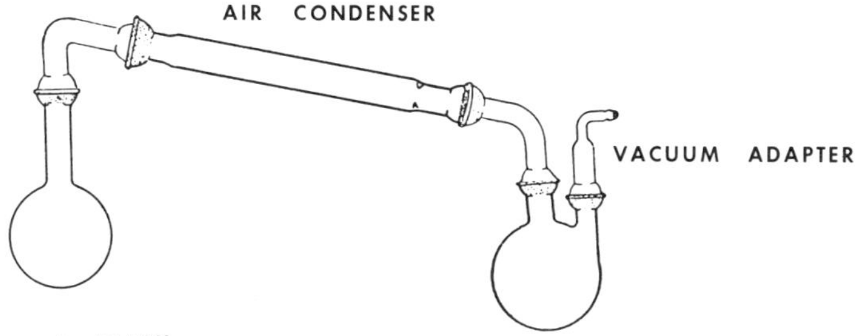

Fig. 2.25. Apparatus used for the determination of residue in tolylenediamine (153).

(*5*) Flame neck of flask, delivery tube, and condenser cautiously to distill over any TDA remaining in the apparatus.

(*6*) Weigh the distillation flask and receiver, and determine weight of residue, and distillate.

CALCULATIONS

$$\%\text{Residue} = \frac{\text{Weight of Residue} \times 100}{\text{Sample weight}}$$

$$\%\text{TDA} = \frac{\text{Weight of distillate} \times 100}{\text{Sample weight}}$$

Determination of Dinitrotoluene in Tolylenediamine (154)

SCOPE

As an additional contaminant in tolylenediamine, there is a possibility that traces of DNT may be present due to incomplete reduction of the nitro compounds. This method serves to further define the quality of the diamine.

PRINCIPLE

Titanium trichloride reduces nitro groups to amino groups according to the following reaction:

$$12TiCl_3 + R(NO_2)_2 + 12HCl \xrightarrow{HF} 12TiCl_4 + R(NH_2)_2 + 4H_2O \qquad (50)$$

Therefore, if a compound containing nitro groups is treated with an excess of titanium trichloride and the excess determined by titration with ferric chloride, the concentration of nitro groups can be calculated. Ammonium or potassium thiocyanate is used as the indicator for the titration since excess iron will react to form the reddish-pink colored complex

$$Fe^{3+} + SCN^- \rightarrow Fe(SCN)^{2-} \qquad (51)$$

EQUIPMENT

(*a*) Titanous chloride solution

Preparation. Add 100 ml. of concentrated reagent grade HCl to a 2-liter flask and dilute to approximately 1 liter with boiled distilled water. Add 560 ml of 20% $TiCl_3$ and dilute to volume. Let stand for 2 days and decant the titanic acid. Store the reagent in a Machlette buret under nitrogen or hydrogen.

(*b*) Nitrogen or hydrogen, prepurified.

(*c*) Ferric chloride solution, 0.1*N*.

Preparation. Dissolve 27 g of $FeCl_3$ in 50 ml of 30% HCl and dilute to volume in a 1-liter flask with previously boiled distilled water. From a buret, introduce 45 ml of the $FeCl_3$ solution into an iodine flask. Add 10 ml of 20% KI, stopper, shake, and allow to stand 5 min with occasional shaking. Titrate with standard 0.1*N* $Na_2S_2O_3$ to a pale straw yellow, then add 1 ml of starch indicator and titrate to a colorless endpoint.

$$N \text{ of } FeCl_3 = \frac{N \text{ of } Na_2S_2O_3 \times \text{ml of } Na_2S_2O_3}{\text{ml of } FeCl_3}$$

(*d*) Methanol, reagent grade.

(*e*) Hydrofluoric acid, 48%.

(*f*) Ammonium thiocyanate, 10%.

(*g*) Machlette buret.

(*h*) Hydrochloric acid, 30%.

(*i*) Round-bottom, two-necked flask, 35/25 center joint, 500 ml. 28/15 vertical neck side joint.

(*j*) Condenser.

(*k*) Magnetic stirrer.

(*l*) Teflon stirring bar.

PROCEDURE

(*1*) By difference, weigh 5 g of TDA into the 500 ml flask. Add 100 ml of methanol and sweep the flask out with nitrogen.

(*2*) Add 25 ml. of 30% HCl, approximately 2–4 ml of HF, mix and add from the Machlette buret exactly 100 ml of $TiCl_3$ and boil for 5 min under reflux in a nitrogen atmosphere.

(*3*) Cool, rinse down the condenser with 25 ml of methanol, and titrate with ferric chloride with continuous stirring using 10 ml of 10% ammonium thiocyanate solution as an indicator to a reddish-pink endpoint.

CALCULATIONS

$$\%\text{DNT} = \frac{(\text{ml blank} - \text{ml sample}) \times 182 \times N \text{ of } FeCl_3}{120 \times \text{Sample weight}}$$

2. Gas Chromatographic Methods

The current literature does not show any specific methods nor give any clues as to chromatographic techniques that might be applied directly to tolylenediamine for isomer separation or impurity examination. However, some techniques generally applicable to high boiling aromatic amines might be useful here. Carbowax (polyethylene glycol) on glass beads has been a column packing that has been utilized successfully for amine compounds.

3. Spectroscopic Methods

Mesorana and Poggio (155) studied the ultraviolet absorption spectra of 15 nitro, halogen, and amino derivatives of toluene. It was found that derivatives in the 2 position, with respect to the methyl group, tend to displace the absorption spectrum farther into the ultraviolet region, while all derivatives in position 4 tend to displace it toward longer wavelengths.

The presence of the amino group was found to displace the absorption bands toward the visible region of the spectrum, the amino group in the *para* position having the largest effect.

Allan and Muzik (156) have described a method for identifying primary aromatic amines. This is accomplished by diazotization followed by coupling with R-salt (or resorcinol) and measurement of the absorption spectra of the resulting azo dyes. The diazotization and coupling conditions and the absorption maxima of approximately 200 azo dyes were reported. Only a small portion of these are directly applicable to the types of compounds under consideration here. This method appears to have little application to multicomponent mixtures.

Finkelshtein et al. (157) have described spectroscopic methods for the analysis of mixtures. One mixture analyzed consisted of 2,4- and 2,6-tolylenediamine for which the absorption bands at 12.55 and 11.82 μ were employed for the 2,4 isomer and 12.85 μ for the 2,6 isomer.

Finkelshtein and Boïtsov (158) have reported a method for the analysis of isomers in mixtures of tolylenediamines and of tolylenediisocyanates. The 2,4 isomer of tolylenediamine is determined by measuring the absorption at 11.86 and 12.57 μ and the 2,6-isomer at 12.87 μ.

The concentrations of the isomers are found from the following equations:

$$D_{\lambda_1} = K'_{\lambda_1}C' + K''_{\lambda_1}C'' \tag{52}$$

$$D_{\lambda_2} = K'_{\lambda_2}C' + K''_{\lambda_2}C'' \tag{53}$$

where D_{λ_1}, D_{λ_2} are the optical densities and K'_{λ_1}, K''_{λ_1}, K'_{λ_2}, and K''_{λ_2} are absorption coefficients of the two components at λ_1 and λ_2.

For the 2,4 and 2,6 isomers of tolylenediamine, where three measurements are made, two pairs of equations are compiled and arithmetic averages are taken from the results. The reported error is ≦2%. The possible interferences, due to the presence of other isomers of tolylenediamine such as the 2,4, 2,5, and 3,4 are not given. It should be noted that the specific method for isomer analysis of tolylenediamine of Finkelshtein and Boĭtsov (158) employs slightly different wavelengths for absorption measurement from the previous method of Finkelshtein et al. (157). Presumably, the latter method (158) represents an improvement over the previous method.

4. Additional Methods and Techniques

Bertetti (159) studied the paper chromatographic behavior of 45 aromatic amines and reported their R_f values. The amines were developed with *n*-butanol saturated with 25% acetic acid. The stains were treated with a solution of potassium dichromate acidified with sulfuric acid which permitted identification of mixtures of aromatic amines with very close R_f values.

Bertetti (160) has also described a systematic method for the identification of 45 aromatic amines using five reagents successively.

Ettel et al. (161) have described paper chromatographic techniques for the separation of aromatic nitro compounds. Compounds that differ slightly from other compounds were separated on paper impregnated with

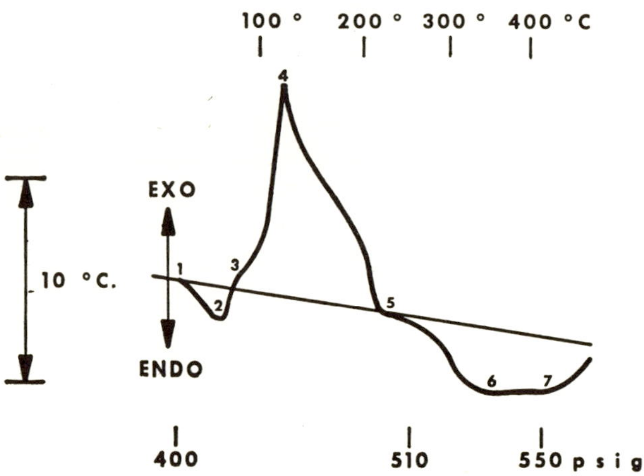

Fig. 2.26. Thermogram of reduction of 0.0753 g dinitrotoluene with 0.2333 g Raney nickel and water at 400 psig N_2 (163). (*1*) Start of dinitrotoluene melting. (*2*) Peak melting endotherm. (*3*) Hydrogenation starting. (*4*) Peak exotherm. (*5*) H_2O boiling. (*6*) Peak of H_2O boiling (244°C bp of H_2O at 510 psig). (*7*) Probable tolylene diamine sublimation.

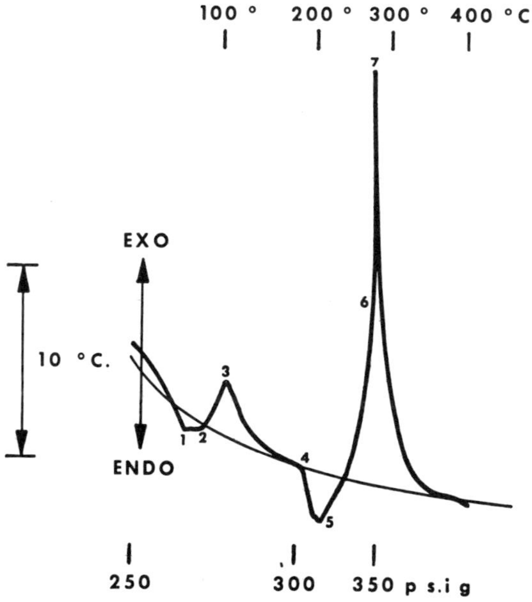

Fig. 2.27. Thermogram showing incomplete reduction of 0.099 g dinitrotoluene with 0.0996 g Raney nickel in 0.0348 g methanol at 250 psig H_2 (163). (*1*) Dinitrotoluene melting endotherm. (*2*) Hydrogenation starting. (*3*) Peak exotherm of reduction. (*4*) H_2O and/or CH_3OH boiling. (*5*) Completion of boiling and/or initiation of reaction. (*6*) Initial condensation. (*7*) High exotherm and pressure surge due to violent decomposition.

$HCONH_2$ by eluting gradually with solvents of increasing polarity. Compounds with close R_f values can be differentiated chromatographically after total reduction to amines.

Genchev and Atanasova (162) have described a new color reaction for the detection of aromatic amines which is sensitive in the range of 1 γ or less depending upon the specific amine. For this reaction the amine is oxidized with cobalt hydroxide. For detection, about 0.25 g. of aromatic amine was suspended in water, 1 ml of 1% cobalt nitrate, 1 ml of 1% sodium hydroxide, and 2 ml of 25% acetic acid was added. The mixture was stirred for a few minutes and then 0.5 ml of 0.1% H_2O_2 was added as a catalyst. The reaction is usually quick and highly colored.

For quantitative determination, after suitable calibration, to 1 ml of aromatic amine was added 0.25–1.00 ml of water or dilute alcohol, 0.1 ml of 1% cobalt nitrate, 0.1 ml 1% sodium hydroxide, 0.15 ml of 0.1% hydrogen peroxide, and 1 ml of 25% acetic acid. The resulting colored solutions can be compared visually or the absorbance determined by means of a colorimeter or spectrophotometer.

A differential thermal analysis study was performed by David (163) on the thermal stability of dinitrotoluene, the reduction temperature of dinitrotoluene with Raney nickel, decomposition reactions of dinitrotoluene with tolylenediamine, and parameters affecting the violent decomposition of reduction mixtures. This study served to point out the serious safety hazards that may exist in macro reductions of dinitrotoluene. If dinitrotoluene is not almost completely reduced, the tolylenediamine formed is capable of a decomposition reaction with the remaining nitro compound.

Figure 2.26 shows a complete reduction of dinitrotoluene when sufficient catalyst was present. This was confirmed by analyzing the resulting products by gas–liquid chromatography.

Figure 2.27 shows a condition in which the dinitrotoluene is only partially reduced, providing a mixture of amine and nitro compound and a subsequent violent reaction between the remaining tolylenediamine and dinitrotoluene as the high exotherm indicates. Other experiments showed that the presence of Raney nickel catalyst tended to intensify the decomposition.

References

1. Plotnikov, V. A., *J. Russ. Phys.-Chem. Soc.*, **48,** 457 (1916); *Chem. Abstr.* **11**:764[6].
2. Belladen, L., Noli, M., and Sommariva, A., *Gazz. Chim. Ital.*, **58,** 443 (1928); *Chem. Abstr.* **23**:787[4].
3. Gorin, E., and Miles, C. B., Socony-Vacuum Oil Co., Inc., U.S. Pat. 2,444,289 (June 29, 1948).
4. Paterno, E., and Mazzucchelli, A., *Gazz. Chim. Ital.*, **50,** 30 (1920); *Chem. Abstr.* **14**:2921[5].
5. Murphy, R. K., and Reuter, F. H., *Australian Chem. Inst. J. and Proc.* **15,** 144 (1948); *Chem. Abstr.* **43**:6930i.
6. Whitehouse, N. L. G., U.S. Pat. 1,231,226 (June 26, 1917).
7. Bradner, D. R., U.S. Pat. 1,457,493 (June 5, 1923).
8. Brundrit, Dennis, I.C.I. Ltd. Brit. Pat. 583,477 (Dec. 19, 1946).
9. Peacock, S., U.S. Pat. 1,360,312 (Nov. 30, 1921).
10. Bodenstein, Max, and Plaut, Heinrich, *Z. Physik. Chem.*, **110,** 399 (1924); *Chem. Abstr.* **18**:2848[9].
11. Nitzschmann, R., *Metallborse*, **17,** 1601 (1927); *Chem. Abstr.* **22**:4330[9].
12. Nitzschmann, R., *Chem. Zentr.*, **II,** 1513 (1927); *Chem. Abstr.* **22**:4330[9].
13. Beckmann, Ernst, *Z. Anorg. Chem.*, **55,** 371 (1907); *Chem. Abstr.* **2**:501[7].
14. Giauque, W. F., and Jones, W. M., *J. Am. Chem. Soc.*, **70,** 120 (1948).
15. Reeves, H. G., *J. Soc. Chem. Ind.*, **43,** 279 (1924).
16. Pletnev, S., *Lab. Praktika*, **4,** No. 2 (1928); *Chem. Abstr.* **23**:4162[9].
17. Rush, Cecil H., and Danner, Charles E., *Anal. Chem.*, **20,** 644 (1948).
18. Delepine, Marcel, *Bull. Soc. Chim. France*, **27,** 283 (1920).
19. David, D. J., and Heinz, E. H., "Mobay Method for The Determination of Nonvolatile Residue and Iron in Liquid Phosgene," (1965) unpublished, Mobay Chemical Co.

20. Farbenfabriken Bayer, A. G., Brit. Pat. 752,460 (July 11, 1956).
21. Matuszak, Maryan, P., *Ind. Eng. Chem., Anal. Ed.*, **6,** 457 (1934).
22. Siok, J., and David, D. J., "Mobay Method for the Determination of Free Chlorine in Gaseous Phosgene" (1965), unpublished, Mobay Chemical Co.
23. Kretov, A. E., *Zh. Prikl. Khim.*, **2,** 483 (1929); *Chem. Abstr.* **24**:313.
24. Davies, C. N., *J. Chem. Phys.*, **14,** 48 (1946).
25. Baker, A. R., and Hartwell, F. J., *Min. Power (Gt. Brit.), Safety Mines Res. Estab.*, **189,** (1960); *Chem. Abstr.* **55**:1283f.
26. Turkel'taub, N. N., Ainshtein, S. A., and Syavtsillo, S. V., *Zavodsk. Lab.*, **28,** 141 (1962); *Chem. Abstr.* **57**:18e.
27. Runge, Henner, *Z. Anal. Chem.*, **189,** 111 (1962); *Chem. Abstr.* **57**:13167i.
28. Castilow, A., Mars, J. J., "Mobay Method for the Determination of Carbon Monoxide in Gaseous Phosgene," (1963), unpublished, Mobay Chemical Co.
29. David, D. J., Bonar, F. M., "Mobay Method for the Analysis of Gaseous Phosgene by VPC," (1964), unpublished, Mobay Chemical Co.
30. Henri, Victor, and Howell, Owen R., *Proc. Roy. Soc. (London), Ser. A*, **128,** 178 (1930); *Chem. Abstr.* **24**:5623[6].
31. Mulliken, Robert S., *J. Chem. Phys.*, **3,** 564 (1935).
32. Ananthakrishnan, R., *Proc. Indian Acad. Sci.*, **5A**, 285 (1937); *Chem. Abstr.* **31**:4596[6].
33. McMurry, Henry L., *J. Chem. Phys.*, **9,** 231 (1941).
34. Ugryumov, P. S., *J. Gen. Chem. (USSR)*, **17,** 2024 (1947); *Chem. Abstr.* **42**:4808a.
35. Lovell, R. J., and Jones, E. A., *J. Mol. Spectry.*, **4,** 173 (1960); *Chem. Abstr.* **55**:13054.
36. Smyth, C. P., *J. Am. Chem. Soc.*, **60,** 183 (1938).
37. Coop, I. E., and Sutton, L. E., *Trans. Faraday Soc.*, **35,** 505 (1939); *Chem. Abstr.* **33**:4839[8].
38. Bailey, C. R., and Hale, J. B., *Phil. Mag.*, **25,** 98 (1938); *Chem. Abstr.* **32**:4075[8].
39. Nielsen, A. H., Burke, T. G., Woltz, P. J. H., and Jones, E. A., *J. Chem. Phys.*, **20,** 596 (1952).
40. Catalano, E., and Pitzer, K., *J. Am. Chem. Soc.*, **80,** 1054 (1958).
41. Herzberg, G., *Infrared and Raman Spectra*, Van Nostrand, New York, 1945.
42. Pierson, R. H., Fletcher, A. N., and Gantz, E. St. C., *Anal. Chem.*, **28,** 1218 (1956).
43. Henri, Victor, and Howell, Owen R., *Proc. Roy. Soc. (London), Ser. A*, **128,** 192 (1930); *Chem. Abstr.* **24**:5623[6].
44. Imanishi, Sunao, and Kanda, Yoshiya, *J. Sci. Res. Inst. (Tokyo)*, **43,** 1–16 (199–214) (1949); *Chem. Abstr.* **43**:8885c.
45. Duchesne, Jules, *J. Chem. Phys.*, **22,** 1464 (1954).
46. La Paglia, S. R., and Duncan, A. B. F., *J. Chem. Phys.*, **34,** 125 (1961).
47. Giddings, Jr., L. E., and Innes, Keith K., *J. Mol. Spectry.*, **8,** 328 (1962); *Chem. Abstr.* **57**:2988i.
48. Kuhn, W., and Martin, H., *Z. Physik. Chem. (Leipzig)*, **B21,** 93 (1933); *Chem. Abstr.* **27**:3138[2].
49. Nenitzescu, Costin, D., and Pana, Constantin, *Bul. Soc. chim. Romania*, **15,** 45 (1933); *Chem. Abstr.* **28**:73[2].
50. Terentév, A. P., Buzlanova, M. M., and Obtenperanskaya, S. I., *Zhur. Anal. Khim.*, **16,** 743 (1961); *Chem. Abstr.* **56**:9419a.
51. Meisel, Tibor, and Mazor, Laszlo, *Magy. Kem. Lapja*, **17,** 421 (1962); *Chem. Abstr.* **58**:2847e.
52. Stock, Alfred, and Wustrow, Werner, *Z. anorg. allgem. Chem.*, **147,** 245 (1925); *Chem. Abstr.* **19**:3228[6].
53. Stock, Alfred, Wustrow, Werner, Lux, Hermann, and Ramser, Hans, *Z. anorg. allgem. Chem.*, **195,** 140 (1931); *Chem. Abstr.* **25**:2935[2].

54. Anger, V., and Wang, S., *Mikrochim. Acta*, **3**, 24 (1938); *Chem. Abstr.* **32**:3295[6].
55. Vallender, R. B., *Chem. Ind. (London)*, **1939**, 330.
56. Kling, Andre, and Schmutz, Rene, *Compt. Rend.*, **168**, 891 (1919); *Chem. Abstr.* **13**:2650.
57. Cox, H. E., *Analyst*, **64**, 807 (1939); *Chem. Abstr.* **34**:541[9].
58. Noriega del Aguila, M., *Prim. congr. peruano quim.* (*Actas y. trav.*) (1938) 1001; *Chem. Abstr.* **34**:4179[3].
59. Hoogeveen, A. P. J., *Chem. Ind. (London)*, **1940**, 550.
60. Pu, Tung-Lieh, and Lo, Chien-Tao, *J. Chinese Chem. Soc.*, **8**, 140 (1941); *Chem. Abstr.* **37**:6053.
61. Studinger, J., *Mitt. Lebensm. Hyg.*, **27**, 8 (1936); *Chem. Abstr.* **31**:6367[9].
62. Moureu, Henri, Chovin, Paul, and Truffert, Louis, *Arch. maladies profess. med. travail et securiti Sociale*, **11**, 445 (1950).
63. Troy, D. J., *Anal. Chem.*, **27**, 1217 (1955).
64. Crummett, W. B., *Anal. Chem.*, **28**, 410 (1956).
65. Lamouroux, A., *Mém. poudres*, **38**, 383 (1956); *Chem. Abstr.* **51**:11174d.
66. Liddell, H. F., *Analyst*, **82**, 375 (1957); *Chem. Abstr.* **51**:11930d.
67. Kinnear, A. M., *Chem. Ind. (London)*, **1959**, 361.
68. Neale, E., and Perry, B. J., *Analyst*, **84**, 226 (1959); *Chem. Abstr.* **53**:18338b.
69. Dixon, B. E., and Hands, G. C., *Analyst*, **84**, 463 (1959); *Chem. Abstr.* **54**:163h.
70. Forostyan, Yu N., and Lazurevskil, G. V., Russ. Pat. 122,634 (Sept. 18, 1959); *Chem. Abstr.* **54**:7439e.
71. Grosskopf, C., U.S. Pat. 3,033,655 (May 8, 1962).
72. Witten, B., and Prostak, A., *Anal. Chem.*, **29**, 885 (1957).
73. Saltzman, B. E., *Anal. Chem.*, **33**, 1100 (1961).
74. Saltzman, B. E., and Gilbert, N., *Am. Ind. Hyg. Assoc. J.*, **20**, 379 (1959).
75. Hough, Arthur, Savage, Wallace, and Van Marle, D. J., *Chem. and Met. Eng.*, **23**, 666 (1920).
76. Gibson, W. H., Duckham, R., and Fairbairn, R., *J. Chem. Soc.*, **121**, 270 (1922).
77. Ingold, C. K., Shaw, F. R., *J. Chem. Soc.*, **1927**, 2918.
78. Ingold, C. K., Lapworth, A., Rothstein, E., and Ward, D., *J. Chem. Soc.*, **1931**, 1959.
79. Hetherington, J. H., and Masson, I., *J. Chem. Soc.*, **1933**, 105.
80. Pounder, F. E., and Masson, I., *J. Chem. Soc.*, **1934**, 1352.
81. Bennett, G. M., and Youle, Philip V., *J. Chem. Soc.*, **1938**, 1816.
82. Bullock, J. L., and Mitchell, E. T., *J. Am. Chem. Soc.*, **63**, 3230 (1941).
83. Othmer, D. F., Jacobs, J. J., Jr., and Levy, J. F., *Ind. Eng. Chem.*, **34**, 286 (1942).
84. Othmer, D. F., Kleinhaus, H. L. Jr., *Ind. Eng. Chem.*, **36**, 447 (1944).
85. McKinley, C., and White, R. R., *Trans. Amer. Inst. Chem. Eng.*, **40**, 143 (1944).
86. Westheimer, F. H., and Kharasch, M. S., *J. Am. Chem. Soc.*, **68**, 1871 (1946).
87. Jones, W. W., and Russell, M., *J. Chem. Soc.*, **1947**, 921.
88. Smith, J. C., *Chemical Industries*, **62**, 929 (1948); *Chem. Abstr.* **43**:1711[2].
89. Kobe, K. A., and Levin, H., *Ind. Eng. Chem.*, **42**, 352 (1950).
90. Meissner, F., Wannschaff, G., and Othmer, D. F., *Ind. Eng. Chem.*, **46**, 718 (1954).
91. Kobe, K. A., Skinner, C. G., and Prindle, H. B., *Ind. Eng. Chem.*, **47**, 785 (1955).
92. Brennecke, H. M., and Kobe, K. A., *Ind. Eng. Chem.*, **48**, 1298 (1956).
93. Barduhn, A. J., and Kobe, K. A., *Ind. Eng. Chem.*, **48**, 1305 (1956).
94. Kobe, K. A., and Lakemeyer, J. L., *Ind. Eng. Chem.*, **50**, 1691 (1958).
95. Gillespie, R. J., Hughes, E. D., Ingold, C. K., Millen, D. J., and Reed, R. I., *Nature*, **163**, 599 (1949); *Chem. Abstr.* **43**:5754g.
96. Adams, R. K., U.S. Pat. 2,947,791 (Aug. 2, 1960).

97. Holleman, A. F., Vermeulen, J., and DeMoody, W. J., *Rec. Trav. Chim. France*, **33,** 1–34 (1914); *Chem. Abstr.* **8:**2372.
98. Knowles, J. R., Norman, R. O. C., and Radda, G. K., *J. Chem. Soc.*, **1960,** 4885.
99. Roberts, R. M., Watkins, J. D., and Kobe, K. A., *J. Am. Chem. Soc.*, **81,** 1167 (1959).
100. Roberts, R. M., Browder, H. P. Jr., and Kobe, K. A., *J. Am. Chem. Soc.*, **81,** 1165 (1959).
101. Knowles, J. R., and Norman, R. O. C., *J. Chem. Soc.*, **1961,** 2938.
102. Parsons, J. S., Tsang, S. M., Di Giaimo, M. P., Feinland, R., and Paylor, R. A. L., *Anal. Chem.*, **33,** 1858 (1961).
103. David, D. J., and Bonar, F. M., "Mobay Method for Assay and Isomer Ratio of Mononitrotoluene," (1960), unpublished, Mobay Chemical Co.
104. Scott, W. W., in *Standard Methods of Chemical Analysis*, Vol. 2, N. H. Furman, Ed., Van Nostrand, Princeton, N. J., 5th Edition, 1939, pp. 2216–2217.
105. Siok, J. F., and Malec, E. J., "Mobay Method for the Determination of Free Nitric Acid in Spent Acid," (1964), unpublished, Mobay Chemical Co.
106. Rosin, J., *Reagent Chemicals and Standards*, Van Nostrand, Princeton, N.J., Fourth Edition, 1961, p. 513.
107. Malec, E. J., "Mobay Method for the Determination of Acidity of DNT," (1964), unpublished, Mobay Chemical Co.
108. Malec, E. J., "Mobay Method for the Determination of Moisture in Dinitrotoluene," (1963), unpublished, Mobay Chemical Co.
109. Mitchell, John, Jr., and Smith, D. M., "Aquametry," Interscience, New York-London, (1948).
110. Butts, P. G., Meikle, W. J., Shovers, John, Kouba, D. L., and Becker, W. W., *Anal. Chem.*, **20,** 947 (1948).
111. Kouba, D. L., Kicklighter, R. C., and Becker, W. W., *Anal. Chem.*, **20,** 948 (1948).
112. Becker, W. W., *Ind. and Eng. Chem.*, Anal. Ed., 5, 152 (1933).
113. Sarson, Roy D., *Anal. Chem.*, **30,** 932 (1958).
114. David, D. J., Baumann, G. F., and Steingiser, Samuel, "Mobay Method for Analysis of Dinitrotoluene for Impurities and Isomer Ratio by Gas Liquid Chromatography," (1960), unpublished, Mobay Chemical Co.
115. Brown, John F., Jr., *J. Am. Chem. Soc.*, **77,** 6341 (1955).
116. Smith, D. C., Pan, C. Y., and Nielsen, J. R., *J. Chem. Phys.*, **18,** 706 (1950).
117. Mathieu, J. P., and Massignon, D., *Compt. rend.*, **212,** 1084 (1941); *Chem. Abstr.* **37:**2658².
118. Randle, R. R., and Whiffen, D. H., *J. Chem. Soc.*, 4153 (1952).
119. Kross, Robert D., and Fassel, Velmer A., *J. Am. Chem. Soc.*, **78,** 4225 (1956).
120. Pristera, Frank, *Anal. Chem.*, **25,** 844 (1953).
121. Pristera, Frank, and Halik, Michael, *Anal. Chem.*, **27,** 217 (1955).
122. Brown, Weldon G., and Reagan, Harriet, *J. Am. Chem. Soc.*, **69,** 1032 (1947).
123. Doub, Leonard, and Vandenbelt, J. M., *J. Am. Chem. Soc.*, **69,** 2714 (1947).
124. Schroeder, W. A., Wilcox, Philip E., Trueblood, Kenneth N., and Dekker, Albert O., *Anal. Chem.*, **23,** 1740 (1951).
125. Brand, J. C. D., Horning, W. C., and Thornley, M. B., *J. Chem. Soc.*, **1952,** 1374
126. Orchem Method 20-1-1-1A, unpublished, Dupont Co.
127. Rudolph, O., *Z. Anal. Chem.*, **60,** 239 (1921); *Chem. Abstr.* **15:**3798.
128. English, F. L., *Anal. Chem.*, **20,** 745 (1948).
129. Nelson, Lloyd S., and Lakowski, Donald E., *Anal. Chem.*, **23,** 1495 (1951).
130. Schönberg, Alexander, Mustafa, Ahmed, and Asker, Wafia, *J. Am. Chem. Soc.*, **74,** 5640 (1952).
131. Porter, Curt C., *Anal. Chem.*, **27,** 805 (1955).

132. Bost, R. W., and Nicholson, F., *Anal. Chem.*, **7,** 190 (1935).
133. Janovsky, J. V., *Ber.*, **24,** 971 (1891).
134. Parks, G. D., and Farthing, A. C., *J. Chem. Soc.*, **1948,** 1275.
135. Meisenheimer, J., *Ann.*, **323,** 205 (1902).
136. Zimmerman, W., *Z. physiol. Chem.*, **245,** 49 (1937); *Chem. Abstr.* **31**:2626[4].
137. Mahood, S. A., and Schaffner, P. V. L., "Organic Synthesis," Vol. XI, 32–5, (1931). *Chem. Abstr.* **25**:2125 and Coll. Vol. II, p. 160.
138. Yoshikawa, Kiyoshi, Yamanaka, Tatsuo, and Kubota, Bennosuke, *Bull. Inst. Phys.-Chem. Research (Tokyo)*, **14,** 409 (1935); *Chem. Abstr.* **30**:1754[3].
139. Albert, Adrien, and Ritchie, Bruce, *J. Proc. Roy. Soc. N. S. Wales*, **74,** 74 (1940); *Chem. Abstr.* **34**:7286[1].
140. Whitmore, W. F., and Revukas, A. J., *J. Am. Chem. Soc.*, **62,** 1687 (1940).
141. Munday, John C., U.S. Pat. 2,620,356, December 2, 1952.
142. Benner, Roland G., and Stevenson, Arthur C., U.S. Pat. 2,619,503, Nov. 25, 1952.
143. Linstead, Reginald P., and Braude, Ernest A., Brit. Pat. 705,919, March 24, 1954.
144. Jadot, J., and Braine, R., *Bull Soc. Roy. Sci. Liege*, **25,** 62 (1956); *Chem. Abstr.* **50**:16651h.
145. Jadot, J., Braine, R., and Roynet, H., *Bull. Soc. Roy. Sci. Liege*, **25,** 79 (1956); *Chem. Abstr.* **50**:16653i.
146. Farbenfabriken Bayer A. G., Brit. Pat. 768,111 (Feb. 13, 1957); *Chem. Abstr.* **51**:16534e.
147. Issoire, J., and Long, Co-Van, *Mem. Poudres*, **39,** 447 (1957); *Chem. Abstr.* **53**:1196d.
148. Dierichs, H., and Holzricher, H., Ger. Pat. 948,784 (Sept. 6, 1956); *Chem. Abstr.* **53**:4209A.
149. Graham, D. P., and Spiegler, L., U.S. Pat. 2,823,235 (Feb. 11, 1958).
150. Castilow, A. R., and Malec, E. J., "Mobay Method for the Determination of Water in Tolylenediamine," (1963), unpublished, Mobay Chemical Co.
151. Dean, E. W., and Stark, D. D., *Ind. Eng. Chem.*, **12,** 486 (1920).
152. Sherwood, N., Malec, E., and Christman, R., "Mobay Method for the Determination of Purity of Tolylenediamine," (1958), unpublished, Mobay Chemical Co.
153. McKay, O., and Malec, E. J., "Mobay Method for the Determination of Residue in Tolylenediamine," (1963), unpublished, Mobay Chemical Co.
154. Gorrel, R., and Malec, E. J., "Mobay Method for the Determination of Dinitrotoluene in Tolylenediamine," (1963), unpublished, Mobay Chemical Co.
155. Poggio, Mesorana, F., and Marquet, F., *Anales Real Soc. Espan. Fis. Quim. (Madrid)*, **49B,** 5 (1953); *Chem. Abstr.* **47**:11000f.
156. Allan, Z. J., and Muzik, F., *Chem. Listy*, **47,** 380 (1953); *Chem. Abstr.* **49**:207c.
157. Finkelshtein, A. I., Roginskaya, Ts. N., Balabanova, P. N., Malachevskaya, F. L., Fisher, A. M., and Machin, G. P., *Zavodsk. Lab.*, **25,** 932 (1959).
158. Finkelshtein, A. I., and Boĭtsov, E. N., *Zavodsk. Lab.*, **26,** 959 (1960).
159. Bertetti, J., *Ann. Chim. (Rome)*, **44,** 495 (1954); *Chem. Abstr.* **49**:6783i.
160. Bertetti, J., *Ann. Chim. (Rome)*, **44,** 313 (1954); *Chem. Abstr.* **49**:13023h.
161. Ettel, V., Pospisil, J., and Deyl, Z., *Chem. Listy*, **52,** 623 (1958); *Chem. Abstr.* **52**:12676f.
162. Genchev, Ml., and Atanasova, B., *Nauch. Tr. Visshiya Med. Inst. Sofiya*, **6** (No. 1), 25 (1959); *Chem. Abstr.* **55**:1484b.
163. David, D. J., *Anal. Chem.*, **37,** 82 (1965).

CHAPTER III

ISOCYANATES

I. INTRODUCTION

The isocyanate group (—NCO) reacts with a wide variety of compounds, as well as with itself to form dimers, trimers, and carbodiimides. These reactions, together with some of those used to prepare isocyanates, will be considered in this chapter. The analytical aspects of these reactions will be emphasized to provide a better understanding of the problems involved in controlling the quality of commercial products. The detailed procedures given are generally applicable to all isocyanates with minor modifications.

II. ISOCYANATE PREPARATION

A. General

Isocyanates are esters of isocyanic acid, which in turn is the imide of carbonic acid (1).

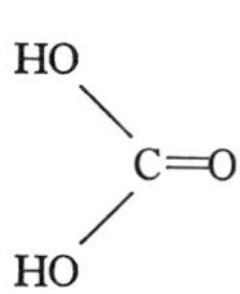

Carbonic acid

HN=C=O

Isocyanic acid

R—N=C=O

Isocyanic acid esters

Isocyanic acid may be regarded as a tautometric form of cyanic acid (Eq. 1).

$$\underset{\text{Isocyanic acid}}{\text{H—N=C=O}} \rightleftharpoons \underset{\text{Cyanic acid}}{\text{HO—C}\equiv\text{N}} \qquad (1)$$

B. Preparation

Saunders and Slocombe (2) classified and reviewed the various methods of isocyanate preparation. More recently, Saunders and Frisch extensively described commercial and laboratory methods of preparation of isocyanates in the first volume of the present work (3).

A portion of these preparative methods will be included here, with the hope that this brief description may serve as background for material presented throughout this volume.

1. Curtius, Hofmann, and Lossen Rearrangements

A series of rearrangements characterized by a common intermediate has been utilized for the laboratory preparation of isocyanates. A similar intermediate is involved which rearranges to the isocyanate in (*1*) the Curtius rearrangement of acid azides, (*2*) the Hofmann rearrangement of acid amides, and (*3*) the Lossen rearrangement of hydroxamic acids. This can be seen from Eq. 2. The Curtius rearrangement of an acid azide in a

Curtius:

Hofmann:

Lossen:

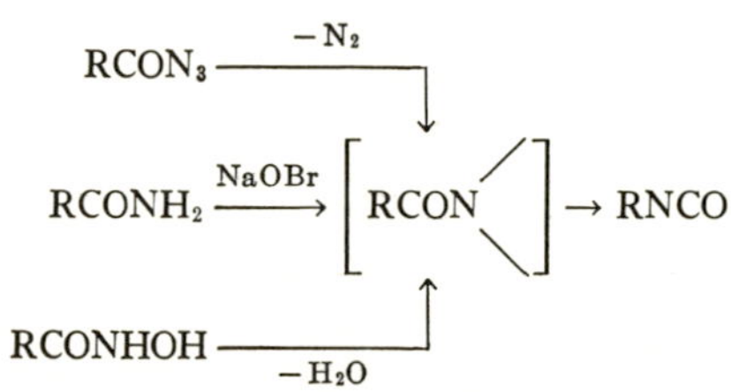

(2)

neutral solvent (4,5) is used primarily for the preparation of short-chain aliphatic diisocyanates and unsaturated isocyanates.

Blomquist and Stevenson (6) prepared a number of substituted furyl isocyanates by means of the Curtius degradation or by Schroeter's (7) modification of the Curtius reaction (Eq. 3). The Hofmann rearrange-

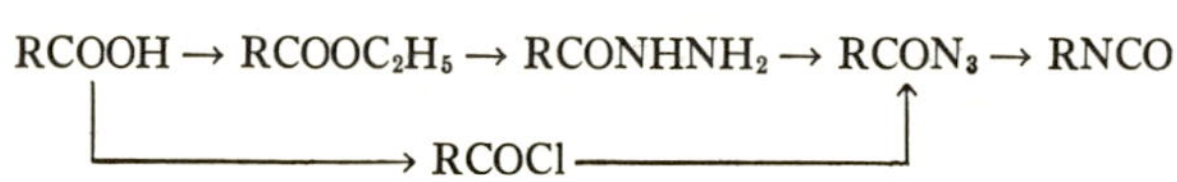

(3)

ment of amides is useful only for those isocyanates which do not react readily with water, since an aqueous medium is required. Pyman (8) has discussed this reaction in detail and cited numerous examples.

Analytically, most of these compounds were identified from boiling point, melting point, derivative preparation, and elemental analysis. Little or no mention was given to the possible presence of trace impurities such as acid chlorides, carbamyl chlorides, and urea.

2. Double Decomposition Reactions

The reaction between organic halides or sulfates and salts of cyanic acid was used by Wurtz (9) in 1849 for the preparation of isocyanates.

$$R_2SO_4 + KCNO \rightarrow RNCO \tag{4}$$

A number of isocyanates that have been prepared by this method include: ethyl, methyl, and propyl isocyanates (10); acyl isocyanates using acid chlorides (11); and inorganic isocyanates (12–14). With regard to much of the work, the previous comments as to the analysis apply here.

3. Preparation from Isocyanate Derivatives

The addition reactions between isocyanates and active hydrogen compounds to form urethanes and ureas can be reversed under sufficiently strenuous conditions. A number of isocyanates have been prepared in the laboratory by utilizing this mechanism at temperatures from 135 to 500°C (15) or by distilling the carbamate from phosphorus pentoxide (16,17). Carbamates, ureas, etc., will be more fully discussed in Chapter 6.

$$\mathrm{RNH\overset{\overset{O}{\|}}{C}OR'} \xrightarrow{P_2O_5} \mathrm{RNCO} \tag{5}$$

Hoshino et al. (18) studied the thermal dissociation and kinetics of the decomposition of diphenylurea in fatty acids. They reported the following reactions, for dissociation in acetic acid:

$$C_6H_5NHCONHC_6H_5 \underset{k_1}{\overset{k_{-1}}{\rightleftharpoons}} C_6H_5NCO + C_6H_5NH_2 \tag{6}$$

$$C_6H_5NCO + CH_3COOH \xrightarrow{k_2} C_6H_5NH\overset{\overset{O}{\|}}{C}\text{—}O\overset{\overset{O}{\|}}{C}\text{—}CH_3 \tag{7}$$

$$C_6H_5NH\overset{\overset{O}{\|}}{C}\text{—}O\text{—}\overset{\overset{O}{\|}}{C}\text{—}CH_3 \xrightarrow{k_3} C_6H_5NHCOCH_3 + CO_2 \tag{8}$$

$$C_6H_5NH_2 + CH_3COOH \rightleftharpoons C_6H_5NHCOCH_3 + H_2O \tag{9}$$

In one experiment, the isocyanate dissociated during the reaction was caught, and in a separate experiment, the isocyanate was allowed to react with acetic acid and the carbon dioxide evolution was measured.

The dissociation of diphenylurea was explained by Eqs. 10 and 11. The

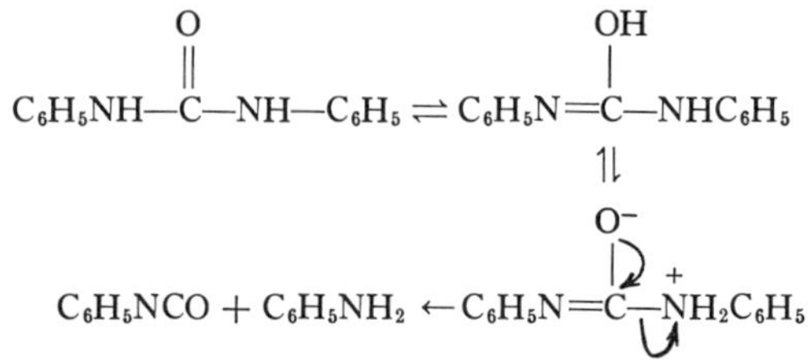

(10)

(11)

effect of solvents on these reactions have also been studied (19).

The thermal dissociation of urethanes and solvent effect has also been studied (20) by measuring the carbon dioxide evolution.

$$RNHCOOR \rightleftharpoons RNCO + ROH \quad (12)$$
$$RNCO + HOOCR \rightleftharpoons RNHCOOCOR \quad (13)$$
$$RNHCOOCOR \rightarrow RNHCOR + CO_2 \quad (14)$$

A rapid and convenient method has been described for the preparation of secondary and tertiary alkyl isocyanates by utilizing the base-catalyzed pyrolysis of a series of *N*-alkyl carbamates at 170–230°C (21).

Bennet, Saunders, and Hardy (22) have prepared isocyanates in good yield by the vaporization (decomposition at 350–370°C) of *sym*-disubstituted urea, followed by cooling in the presence of hydrogen chloride. The isocyanate content of the solvent was determined by the addition of aniline and weighing the *N*-phenylurea formed. The purity of the isocyanate and assay by the di-*n*-butylamine method were not determined because of the limited quantities of isocyanates. The di-*n*-butylamine method is discussed later in this chapter.

Monosubstituted ureas and HCl were also used to prepare isocyanates in low yield (22,23). These preparations apparently involve the dissociation of a substituted urea, with the acid serving to prevent recombination (2), thus

$$RNH_2 + HNCO \rightleftharpoons RNHCONH_2 \quad (15)$$
$$RNHCONH_2 + HCl \rightleftharpoons RNCO + NH_4Cl \quad (16)$$

4. Additional Preparations

Additional methods that are usually less practical are included here.

Diazonium chlorides, when treated with potassium cyanate in the presence of copper, are converted to isocyanates (24,25)

$$ArN_2Cl + KNCO \xrightarrow{Cu} ArNCO + N_2 + KCl \quad (17)$$

Isocyanides (26) or isothiocyanates (27) may be oxidized with mercuric oxide to give the isocyanate:

$$\mathrm{RNC + HgO \rightarrow RNCO} \tag{18}$$

$$\mathrm{RNCS + HgO \rightarrow RNCO} \tag{19}$$

Mukaiyama et al. (28) have described a new method for the preparation of isocyanates by the reaction of various hydroxamic acids with ketene dimer, to give an acetoacetyl hydroxamate adduct which decomposes when pyrolized (350–400°C) to give carbon dioxide, acetone, and the corresponding isocyanate. The isocyanate does not react with the acetone and this permits recovery. The overall reaction may be represented by Eq. 20.

$$\underset{\text{Hydroxamic acid}}{\mathrm{R{-}C({=}O){-}N(H){-}OH}} + \underset{\text{Ketene dimer}}{\mathrm{CH_2{=}C{-}O,\ H_2C{-}C{=}O}} \rightarrow \underset{\text{Acetoacetyl hydroxamate}}{\mathrm{R{-}CO{-}NH{-}O{-}CO{-}CH_2{-}CO{-}CH_3}} \xrightarrow{\Delta}$$

$$\mathrm{RNCO + CO_2 + CH_3{-}C({=}O){-}CH_3} \tag{20}$$

5. Phosgenation Methods

Phosgenation of an amine is by far the most important commercial method for the preparation of isocyanates on an industrial scale. The reactions involved are discussed in the following paragraphs, to show the origin of some of the impurities which give rise to analytical problems.

In 1884, Hentschel (29) showed that an isocyanate could be obtained from the reaction between phosgene and the salt of a primary amine. Gattermann and Schmidt (30) modified this reaction and reported an almost quantitative yield of methyl isocyanate. They treated molten methylamine hydrochloride with phosgene at 250°C and obtained methylcarbamoyl chloride, which was decomposed by heating with lime:

$$\mathrm{CH_3NH_2{\cdot}HCl + COCl_2 \rightarrow CH_3NHCOCl + 2HCl} \tag{21}$$

$$\mathrm{2CH_3NHCOCl + 2CaO \rightarrow 2CH_3NCO + CaCl_2 + Ca(OH)_2} \tag{22}$$

The reaction of phosgene with primary amines can be represented (3):

$$\mathrm{RNH_2 + COCl_2 \rightarrow \left[R{-}N(H)(H){:} \rightarrow C(Cl)(Cl){=}O \right] \rightarrow RNCO + 2HCl} \tag{23}$$

A number of side reactions are possible, such as the formation of amine hydrochlorides, carbamoyl chlorides, and substituted ureas:

$$\mathrm{RNH_2 + HCl \rightleftharpoons RNH_2{\cdot}HCl} \quad (24)$$

$$\mathrm{RNCO + RNH_2 — RNHCONHR} \quad (25)$$

$$\mathrm{2RNH_2 + COCl_2 — RNHCONHR + 2HCl} \quad (26)$$

$$\mathrm{RNCO + HCl \rightleftharpoons RNHCOCl} \quad (27)$$

$$\mathrm{RNCO + Cl_2 \rightleftharpoons R\underset{\displaystyle Cl}{\underset{|}{N}}\overset{\displaystyle O}{\overset{\|}{C}}Cl} \quad (28)$$

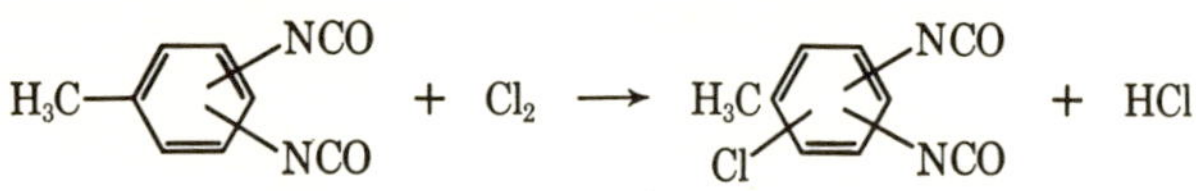
(29)

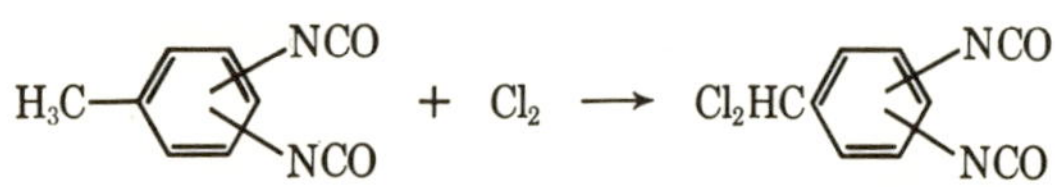
(30)

A number of aliphatic or cycloaliphatic isocyanates were prepared by reacting phosgene with the halogen salts of primary aliphatic or cycloaliphatic amines and liberating the halogen acid (31).

Numerous isocyanates, both aromatic and aliphatic, have been prepared by methods which are now commercial (32–46).

Siefken (47) has reported experimental details of the liquid-phase phosgenation of many amines and amine salts giving their physical properties. Additional techniques involving vapor-phase reactions have also been described (48).

In many patents describing the commercial preparation of mono-, di-, and polyisocyanates, a two-stage reaction is carried out, the first stage of which consists in phosgenating the amine in an inert organic solvent at low temperatures (−20°C to 60°C), the second stage of which consists in transferring the mixture to another zone of higher temperature (80–320°C), further phosgenating followed by degassing (stripping and/or heating) to remove the unreacted phosgene and hydrogen chloride (40,44,46,49–51) before final distillation of the isocyanate.

The advantages of various solvents have been frequently discussed. o-dichlorobenzene (39,40,50–52), and monochlorobenzene (45,53) are generally used, although ethyl acetate (44) is sometimes recommended. Effective removal of hydrogen chloride from the solvent is oftentimes a factor in its choice.

For a more comprehensive treatment of the preparation of isocyanates the following review articles should be consulted: Siefken (47), Peterson and Piepenbrink (54), Saunders and Slocombe (3), and Arnold, Nelson, and Verbanc (55).

A comprehensive listing of isocyanates and their physical properties is given in Chapter 7.

III. ANALYTICAL METHODS

A. General

Methods are described in the following sections for determining the percent NCO, acidity, hydrolyzable chloride, and total chloride content of isocyanate materials by "wet methods." Residual solvents and other impurities may be detected by the gas chromatographic method. These methods are generally applicable to any kind of isocyanate if provision is made for a suitable solvent and the proper size of sample is taken.

Inasmuch as most of the isocyanates are prepared by the phosgenation process, free phosgene may be present if it has not been removed in the final purification step. This will be included in the hydrolyzable chloride test, and partly included in the acidity test, if hydrolysis takes place.

The phosgenation reactions (23–30) discussed in Section II-5 give rise to free HCl and to carbamyl chloride type of (RNHCOCl) intermediates, which decompose into free HCl at elevated temperature (48). This free acidity may act as a mild catalyst, or it may neutralize a basic catalyst used for the isocyanate–polyol reaction. In any case, experience has shown that free acid markedly affects the reactivity of the isocyanate. Consequently the determination of the acidity and hydrolyzable chloride content are important tests for controlling the quality of the material. The total

TABLE 3.1
Tentative ASTM Specifications for Commerical Tolylene Diisocyanates (ASTM D-1786-60T)

Specifications	80/20 Isomers	65/35 Isomers	2,4 Isomer
2,4-Isomer, %	80 ± 2	65 ± 2	97.5 min
2,6-Isomer, %	20 ± 2	36 ± 2	
Purity, min, %	99.5	99.5	99.5
Hydrolyzable chloride, max, %	0.01	0.01	0.01
APHA color, max	50	50	50
Total Chlorine, max, %	0.2	0.2	0.2

chloride determination may show the presence of other chlorine-containing compounds resulting from the phosgenation.

It has been demonstrated that the isomer ratio in commercial toluene diisocyanate (that is, the proportion of the 2,4 and 2,6 isomers) affects its reactivity. This ratio may be determined by the infrared method described under Spectroscopic Methods. Tentative specifications have been established by ASTM for the various grades of toluene diisocyanate as shown in Table 3.1.

B. Wet Methods

The Determination of Isocyanates—Review of Method Development

Primary and secondary amines react with isocyanate to yield the corresponding urea $RNH_2 + R'NCO \rightarrow RNHCONHR'$. This reaction is quantitative in the presence of an excess of the amine and is the basic reaction for most of the methods by which the isocyanate content of a sample is determined. After completion of the reaction the excess amine is determined by back-titration with an acid. Since this involves acid–base reactions, the results must be corrected for any free acid in the sample. Several different amines have been proposed and reported in the literature, all of which are satisfactory.

The choice of a particular amine and solvent system will be governed largely by the nature of the isocyanate material. The solvent must be inert toward the isocyanate and the reagents. In the case of some isocyanates, it is desirable to carry out the reaction at elevated temperatures to obtain complete reaction. For this reason, a relatively high boiling solvent and amine have certain advantages. Toluene or monochlorobenzene are frequently used as suitable solvents, and di-*n*-butylamine is recommended in many of the standard procedures.

The method used at I. G. Farben-Leverkusen early in 1945, as disclosed in an interview with Professor Otto Bayer and reported in *B.I.O.S. Final Report* 719 (68), used 2*N* diisobutylamine in chlorobenzene. An excess of this reagent was added to the sample dissolved in chlorobenzene and the excess amine back-titrated with 1*N* HCl using bromphenol blue indicator. The same procedure was described by Bohme and Roos and reported in *FIAT Final Report* 712 (139).

Stagg (60) described a volumetric method in 1946 in which an excess of piperidine was added to the sample dissolved in acetone, and after reaction the excess piperidine was removed with an excess of 0.5*N* HCl, the excess of which was determined by titration with 0.5*N* NaOH using methyl red–

methylene blue indicator. This method was compared with a gravimetric procedure in which the urea formed from aniline and the isocyanate was filtered, dried, and weighed. Stagg considered the gravimetric method to be more reproducible. Sometimes the volumetric procedure gave higher values than the gravimetric, in which case he estimated the amount of carbamoyl chloride from the results of the two procedures.

Baker and Holdworth (107) modified the Stagg procedure in 1947 for semimicro determinations, using piperidine and dibutyl ether as the solvent. They used this method in their extensive studies of the reaction between isocyanates and alcohols.

Siggia and Hanna (62) compared several amines (1948) including *n*-amyl, *n*-propyl, and *n*-butyl amines, and concluded that the higher boiling *n*-butyl amine was most satisfactory. They preferred dioxane as the solvent and stated that the average precision was better than 0.4%.

Siefkin (47) indicated in 1949 that di-*n*-butyl amine or diisobutylamine reagents in chlorobenzene were used by Spielberger for the analysis of the large number of isocyanates described in his paper. This appears to be the first reference in the literature to the use of di-*n*-butylamine for this reaction.

Williamson (140) found (1952) that the presence of methanol and water were desirable when the final solution is titrated using pH electrodes instead of the usual bromphenol blue indicator used in the previous methods.

Scope (59)

This method determines the NCO function in an isocyanate compound. By proper choice of sample and solvent it can be applied to most isocyanate materials.

Principle (59)

The reaction between dibutylamine and an isocyanate may be written:

$$RNCO + (C_4H_9)_2NH \rightarrow RNHCON(C_4H_9)_2 \qquad (31)$$

This reaction requires an excess of amine to make it proceed quantitatively, the excess of which is titrated with an acid.

The results of this determination may be expressed as "percent NCO," as amine equivalent (AE), or as an isocyanate number. The amine equivalent is defined as the weight of sample which will combine with one equivalent of an amine. Suitable corrections for the acidity or hydrolyzable chloride content of the sample must be made using consistent terms. The amine equivalent or isocyanate number is sometimes more convenient for calculating formulas involving the isocyanate and a polyol, whereas the percent NCO expression permits a clear statement of purity or assay.

EQUIPMENT

(*a*) Isopropyl alcohol, A.R. grade.

(*b*) Dry methanol, A.R. grade.

(*c*) Dry toluene, A.R. grade.

(*d*) Dry dioxane, A.R. grade.

(*e*) Dibutylamine reagent in toluene, $2N$ (DBA reagent). Dilute 338 ml of di-*n*-butylamine (Sharples No. 107) to 1 liter with dry reagent grade toluene.

(*f*) Bromphenol blue indicator prepared by mixing 1.0 g of bromphenol blue with 1.5 ml of $1N$ NaOH and adding 1000 ml of distilled water and mixing well.

(*g*) Bromcresol green indicator prepared by mixing 0.100 g of bromcresol green with 1.5 ml $0.1N$ NaOH, titrating until complete solution is obtained and diluting to 100 ml.

(*h*) Standard $1N$ HCl.

(*i*) Carborundum boiling chips.

PROCEDURE

(*1*) Weigh, by difference, on an analytical balance, 6.7 ± 0.2 g of TDI into a 500 ml Erlenmeyer flask containing two or three Carborundum boiling chips.

(*2*) Carefully pipet 50 ml of DBA reagent into the flask.

(*3*) Using a dispensing buret or pipet, add 40 ml of toluene, carefully washing all sample from the sides of the flask.

(*4*) Stopper loosely and allow to stand 15 min at room temperature.

(*5*) Add 225 ml of isopropyl alcohol from a 250 ml graduate and 0.5 ml of 0.1% bromcresol green indicator.

(*6*) Titrate with standardized $1.0N$ HCl solution while swirling the flask to effect rapid mixing. Near the endpoint add the HCl dropwise and shake the solution vigorously. Consider the endpoint reached with the disappearance of the blue color and the appearance of a yellow color that persists for at least 15 sec.

(*7*) Always run a blank at the same time as the samples, following exactly the same procedure with the blank in place of the sample.

CALCULATIONS

The following equations may be used to calculate the results from the above procedure. The symbols used have the following definitions.

B = milliliters of HCl required for titration of the blank

S = milliliters of HCl required for titration of the sample

N = normality of the HCl

E = equivalent weight of the toluene-diisocyanate, in grams (87.08)
W = grams of sample used

$$\text{Amine equivalent (uncorrected)} = \frac{1000\,W}{(B - S)N}$$

$$\text{Percent NCO (uncorrected)} = \frac{(0.042)(100)N(B - S)}{W}$$

$$\text{Assay (\% TDI, uncorrected)} = \frac{(B - S)NE}{1000\,W} \times 100$$

The results should be rounded to 0.10 unit.

The relationship between AE and percent NCO is

$$\text{AE} = 4200/\%\ \text{NCO}$$

Since the hydrolyzable chloride value is reported as % HCl, it must be converted to % NCO, for making corrections to the above. Letting HC represent the % HCl found in the hydrolyzable chloride test, then

$$\text{HC} \times \frac{42.00}{36.46} = \text{hydrolyzable chloride as \% NCO}$$

and

$$\%\ \text{NCO (corrected)} = \%\ \text{NCO (uncorrected)} - \left(\text{HC} \times \frac{42.00}{36.46}\right)$$

Many isocyanates require heat to drive the DBA reaction to quantitative completion, in which case the hydrolyzable chloride correction may be of even greater interest. The flask should be stoppered while cooling in order to prevent condensation of moisture which may react with the last traces of isocyanate. The addition of the isopropanol provides a suitable medium for the titration.

Obviously, the absolute level of the hydrolyzable chlorides will dictate its importance as a correction factor. For refined isocyanates, such as 80:20 TDI where the level of HC is generally in the range of 0.001–0.01%, it will not account for any differences in assay since the DBA method is accurate only to ±0.05 units. However, in the case of undistilled isocyanates where the HC level is in the range from 0.1–0.5%, the HC correction can have a major effect upon assay.

For most isocyanates, toluene is employed as the solvent. However, for undistilled or modified isocyanates where thermal treatment is employed, we have found that better reproducibility can be attained using dry trichlorobenzene.

In the case of undistilled or modified isocyanates, such as Mondur MR, the following changes in procedure are recommended: (*1*) 2.5 g of sample, (*2*) 20 ml of dibutylamine reagent, (*3*) 25 ml of trichlorobenzene, (*4*) the flask should be heated on a hot plate with stirring until a condensate ring rises ½ in. up the side of the flask, (*5*) stopper the flask and allow the contents to stand for 20 min, and (*6*) add 100 ml of dry methanol before the titration.

This procedure is different from the SPI procedure used for NCO or AE content of undistilled or modified isocyanates (63). The major differences in the SPI method are: (*1*) toluene is employed as the solvent, (*2*) the sample mixture is heated to 95–100°C in approximately 4 min (the temperature is monitored by means of a suitable thermometer), (*3*) the mixture is allowed to stand for 30 min, and (*4*) 225 ml of dry isopropanol is added before the titration.

The Determination of Acidity (64)

SCOPE

This method is used to determine the acidity of refined TDI of various isomer concentrations and can be specifically used for 2,4/2,6; 80/20; 65/35; and 99/1 contents.

PRINCIPLE

The sample is reacted with excess *n*-propyl alcohol to form urethanes. The acidic components released are absorbed in the reaction mixture and titrated potentiometrically with methanolic KOH.

EQUIPMENT

(*a*) Suitable pH meter or automatic titrator.
(*b*) Glass and calomel electrodes.
(*c*) Buret, 10 ml.
(*d*) Beakers, 400 ml.
(*e*) Magnetic stirrer and stirring bar.
(*f*) Methanolic KOH, 0.01*N*.
(*g*) HCl, 0.01*N*.
(*h*) Alcohol, *n*-propyl.
(*i*) Suitable stirrer.
(*j*) Dispensing buret, 1000 ml.

PROCEDURE

(*1*) Place a clean, dry, 400-ml beaker containing a stirring bar on a magnetic stirrer, add by dispensing buret or graduate 100 ml of *n*-propanol, and turn on stirrer.

(2) Add 12 ± 1 g of sample from a weighing vial (weighed by difference). Cover the beaker with a watch glass and stir for exactly 10 min. If the acidity content is expected to be less than 0.005%, use 200 ml of *n*-propanol and 24 ± 1 g of sample and a 5 ml buret for titration.

(3) Titrate with 0.01*N* methanolic KOH with constant stirring. When near the endpoint, titrate in increments of 0.10 ml.

(4) Construct a curve by plotting volume vs. pH and determine the volume of titrant at the point of inflection.

(5) Run a blank omitting the sample.

Calculations

$$\% \text{ HCl} = \frac{(A + B) \times N \text{ KOH} \times 3.65}{\text{Sample weight}}$$

where A = volume of KOH to titrate the sample, and B = volume of blank titration.

The blank titration must be added to or subtracted from the sample titration depending upon the acidity or alkalinity of the blank, titrated to the apparent pH of the endpoint observed with the sample present.

The acidity procedure for crude or modified isocyanates differs from the above on the following points: (1) 200 ml of acid number solvent are used which consists of 333 ml of reagent grade benzene, 335 ml of isopropanol, and these solvents are diluted to 1 liter with reagent grade methanol, (2) approximately 1 g of sample is added to the 200 ml of solvent mixture and stirred for 25 min, and (3) 15 ml of boiled distilled water are added before titration with 0.1*N* methanolic KOH. The solvent system was changed so as to completely solubilize the isocyanate before and during the carbamate reaction to insure complete release of all acidic materials. This is important since much difficulty will be encountered without the modified solvent system.

Obviously, the sample size must be changed in the case of undistilled or modified isocyanates, since they will have higher values.

Distilled water is added before titration to provide a sharper endpoint.

This method differs from the SPI acidity method (65) which uses 50 ml of toluene and 50 ml of anhydrous methanol as the solvent reaction mixture. After the isocyanate is introduced, the mixture is heated with stirring to 60–65°C, and is not allowed to remain on the hot plate for more than 1 min after it reaches 60°C, nor to exceed 65°C. Twenty-five additional milliliters of anhydrous methanol are used to wash down the glassware before titration with 0.02*N* methanolic KOH.

Both the Mobay method and the SPI method appear to give satisfactory results, although the latter appears to be more cumbersome, and requires stricter control. This always presents the possibility of greater error.

It would appear to be an advantage to use $0.01N$ reagent rather than the $0.02N$ since the lower normality will generally place the final volume of titrant in the 3–5 ml range, a usable range when a 10 ml buret is employed.

Determination of Hydrolyzable Chlorides (66)

SCOPE

This method is used for the determination of the hydrolyzable chloride content of TDI of various isomer concentrations and is specifically applicable to materials having 2,4/2,6 ratio of 80/20, 65/35, and 99/1.

PRINCIPLE

The sample is reacted with methanol to form the carbamate, liberating the chloride which forms HCl. Many of the components previously determined from the acidity method will be determined here (assuming halogen acidic components are present) in addition to the main source of hydrolyzable chloride, carbamyl chloride. The reaction mixture is then titrated with $AgNO_3$.

$$COCl_2 + H_2O \longrightarrow 2HCl + CO_2 \tag{32}$$

$$RNHCOCl \xrightarrow[\Delta]{90\text{–}110^\circ C} RNCO + HCl \tag{33}$$

$$RNCO + H_2O \longrightarrow RNHCOOH \xrightarrow{\Delta} RNH_2 + CO_2 \tag{34}$$

$$RNCO + CH_3OH \longrightarrow RNHCOOCH_3 \tag{35}$$

$$RNCO + RNH_2 \longrightarrow RNH\overset{\overset{\displaystyle O}{\|}}{C}NHR \tag{36}$$

$$HCl + AgNO_3 \longrightarrow AgCl + HNO_3 \tag{37}$$

EQUIPMENT

(*a*) Beakers, 400 ml.
(*b*) Heater, stirrer unit.
(*c*) Suitable pH meter or titrator.
(*d*) Silver–silver chloride electrode.
(*e*) Silver electrode.
(*f*) Standard $0.01N$ silver nitrate.
(*g*) Standard $0.005N$ silver nitrate.

PROCEDURE

(*1*) Weigh, by difference, approximately 10 g of sample into a clean, dry 400 ml beaker. If the HC content is expected to be less than 0.01%, use approximately 20 g of sample.

(*2*) Add, by graduate, very slowly and cautiously, 50 ml of methanol with a slow, even rate of stirring. This step is critical and must be carried out properly because a finely divided precipitate must be obtained. If large globules of precipitate form, some HC material will be occluded and the results will be inaccurate and nonreproducible.

(*3*) Continue stirring until the reaction begins as is evidenced by a slight exotherm and crystal separation on the sides of the beaker. Add sufficient distilled water to fill the beaker half full, with continuous stirring. The water must be added quickly to prevent hard solidification of the reactants and minimize any loss of HCl.

(*4*) After the addition of water, boil gently for 30 min with continuous agitation, wash the stirrer and sides of the beaker with distilled water, and place in cooling bath.

(*5*) When cool, add 10 drops of nitric acid and titrate potentiometrically with $0.1N$ $AgNO_3$ or $0.005N$ $AgNO_3$ if the HC is expected to be less than 0.01%.

(*6*) Construct a curve of volume vs. pH and determine the volume of titrant at the inflection point.

CALCULATIONS

$$\% \text{ Hydrolyzable chloride (as HCl)} = \frac{\text{ml } AgNO_3 \times N\ AgNO_3 \times 3.646}{\text{Sample weight}}$$

Duplicate analyses should agree to $\pm 0.0005\%$.

This procedure is essentially the same as the ASTM method for urethane raw materials (61).

The hydrolyzable chloride procedure for undistilled or modified isocyanates utilizes 1 g of sample which is placed in a 400-ml beaker. Benzene, 100 ml, is added before 100 ml of methanol in order to solubilize the entire sample, including any polyisocyanates that are present.

The reaction mixture is stirred, brought to boiling, and allowed to boil for 15 min. The beaker and contents are cooled to room temperature and then potentiometrically titrated.

Determination of Total Chloride (67)

SCOPE

This method is used to determine the total chloride content of TDI of various isomer concentrations and is specifically applicable to materials having 2,4/2,6 ratios of 80/20, 65/35, and 99/1.

PRINCIPLE

Tolylene diisocyanate, as well as other organic materials, is oxidized by combustion in a bomb containing oxygen under pressure. The resultant chloride ion is absorbed in a 1% sodium carbonate solution and the concentration of chloride present is determined potentiometrically.

EQUIPMENT

(*a*) Parr oxygen bomb, double valve, self-sealing, 360 ml. capacity.
(*b*) Combustion capsule, stainless steel.
(*c*) Fuse wire, 10 cm.
(*d*) Oxygen bomb filling connection.
(*e*) Sodium carbonate, 1%.
(*f*) Silver nitrate, 0.01*N*.
(*g*) Nitric acid, A.R.
(*h*) Oxygen, free of combustible material and halogen compounds.
(*i*) Bubble counter: A 100-ml graduate and a delivery tube. The graduate is filled to the 50 ml mark with distilled water to which 3 ml of 0.1*N* $AgNO_3$ and 1 drop of concentrated HNO_3 have been added. Any turbidity that develops indicates that hydrogen chloride is being lost when venting the bomb.
(*j*) Silver–silver chloride electrode.
(*k*) Silver electrode.
(*l*) Suitable pH meter or titrator.

PROCEDURE

(*1*) Check all equipment to make sure all oxygen equipment and fittings are oil free.

(*2*) Weigh by difference on an analytical balance 0.9500 ± 0.05 g into the combustion capsule. Do not use more than a 1 g sample, since a severe safety hazard would be present.

(*3*) Fit a 10 cm length of No. 34 B and S gauge Fe-Ni-Cr alloy fuse wire onto the two electrodes. (An excellent visual demonstration of how to fasten this wire can be seen on p. 38 of Parr *Manual* No. 120.) Place the combustion capsule in place on the loop electrode.

(*4*) Adjust the fuse wire so that it is under the surface of the sample in the capsule but does not touch the capsule at any point and wash down the walls of the bomb cylinder with ca. 10 ml of 1% Na_2CO_3 solution.

(*5*) Put the bomb head in the bomb cylinder, the contact ring on top of the bomb head, and screw the cap down finger tight.

(*6*) Close the outlet valve securely with the special wrench provided for this purpose and then open the main oxygen cylinder valve slightly.

(*7*) Place the bomb in the mounted holder and secure.

(*8*) Attach the union on the oxygen filling connection to the inlet valve of the bomb. Normally, finger tight is sufficient to give a satisfactory seal if the joint is held together while tightening the union.

(*9*) Admit oxygen slowly to prevent blowing the sample from the cup until 20–25 atm of oxygen is admitted. Close the operating valve of the oxygen cylinder and observe the pressure on the gauge. If the pressure drops more than 1 atm, refill and check again. Continual loss indicates a leak. Check all connections to insure a tight seal. Do not continue until the bomb holds pressure satisfactorily. This is necessary to have sufficient oxygen for complete combustion.

(*10*) Release the pressure from the oxygen-filling system and disconnect the bomb from the oxygen-filling assembly. Place the valve thumb nut on the oxygen inlet valve and tighten finger tight. **Caution:** Continual care must be exercised from this point on until the bomb has been fired, cooled, and bled free of oxygen. In addition, a serious shock hazard exists around the bomb ignition receptacle should the ignition unit be shorted. Always pull the electric plug before touching this receptacle.

(*11*) Disconnect plug to bomb ignition unit, turn on cooling water valve, and fill the bomb ignition receptacle $\frac{3}{4}$ full with water.

(*12*) Submerge the bomb in the center of the ignition receptacle and visually inspect for oxygen leaks. If the needle valve is not gas tight, tighten the packing gland slightly. In the event this procedure does not remedy the leaks, the bombs should not be fired until they are corrected.

(*13*) Allow cooling water to circulate around the bomb the entire time the bomb is in the receptacle. Connect the terminal found at the top of the ignition receptacle to the terminal on top of the bomb.

(*14*) Connect plugs to cooling receptacle and insert plug to ignition unit. Fire the ignition unit by pushing the button. The red indicator light should flash on and then off, indicating that the bomb fired properly. Pull electrical plug. If the bomb does not fire as indicated by light and if the bomb is cold to the touch, repeat the above steps.

(*15*) Allow bomb to cool at least 20 min, remove from receptacle, and connect to a bubble counter at the outlet valve. Release the pressure at a slow uniform rate over a period of at least 2 min.

(*16*) After all the gas has been released, open the bomb and examine for incomplete combustion, as evidenced by traces of unburned sample or sooty material, in which case the combustion must be repeated.

(*17*) Carefully wash the bomb head, electrodes, capsule, and cylinder walls with distilled water and transfer to a 250 ml beaker. Remove the needle valve and rinse with ca. 5 ml of distilled water.

(*18*) Carefully transfer the combustion capsule and the valve needle into the bomb. Rinse the walls of the cylinder down with ca. 25 ml of distilled water and fill the capsule. Scrub the interior of the bomb and the inner surface of the bomb cover with a rubber policeman.

(*19*) Place the bomb head in the bomb cylinder obliquely so as not to touch the seat. Support the bomb with wire mesh screen on a tripod and boil. After boiling transfer the washings quantitatively into a 250 ml beaker and repeat the washing procedure twice.

(*20*) Concentrate the washings until the volume is approximately 25 ml. Cool the solution to approximately 10°C and acidify with concentrated nitric acid. Add 60–100 ml of denatured ethanol (Formula 2B will suffice) and cool to ca. 10°C.

(*21*) Titrate potentiometrically with $AgNO_3$ as previously described in the hydrolyzable chloride method. Use 0.01*N* $AgNO_3$ unless the total chloride is expected to be less than 0.1%, in which case 0.005*N* $AgNO_3$ should be employed.

CALCULATIONS

$$\%\ \text{Total chloride as Cl}^- = \frac{\text{ml AgNO}_3 \times N\text{AgNO}_3 \times 3.546}{\text{Sample weight}}$$

The reproducibility of this method is ±0.02% of the actual value in the range of 0.0–0.2% total chloride. For values greater than this, the reproducibility is approximately ±5% of the total chloride obtained.

This method is essentially identical to the ASTM procedure (61). Other methods have been used for the determination of total chloride in other organic materials as well as TDI, such as the Schöniger method (61) or conversion to sodium chloride by reaction with excess alcohol and subsequent refluxing with metallic sodium (68). The above procedure can be used for virtually any isocyanate.

Additional methods for the determination of specific gravity, color, and freezing point of isocyanates can be found in many analytical texts. The exact procedures for isocyanates have been described by ASTM in D-1638-61T (61). While these tests may aid in establishing the quality of the isocyanate, they appear to be less definitive in characterizing the isocyanate. Once a "norm" for these tests has been established, variation will not provide an answer as to the possible causes. Therefore, it is advisable to utilize the methods given previously as well as the methods in the following sections on gas chromatography and spectroscopy in order to detect and define any variations or abnormalities in quality. In many cases trace quantities of impurities present will not be detected by methods such as specific gravity and refractive index due to their lack of sensitivity, whereas gas chromatographic methods and spectroscopic methods will detect many of the impurities normally encountered.

C. Gas Chromatographic Methods

The literature concerning gas–liquid chromatographic methods for the analysis of isocyanates is sparse.

Felton (69) has described high temperature gas chromatographic techniques in which Dow Corning high vacuum grease (20%) on Celite (2 m) is used at 250°C for separation of mono- and triisocyanates from the diisocyanate. The *o*- and *p*-tolylisocyanates and the 2,4,6-triisocyanato toluene were separated from the 2,4- and 2,6-tolylenediisocyanate. The isomers of tolylenediisocyanate were not separated and the authors know of no current gas–liquid chromatographic method that describes their separation.

Kjaer and Jart (70) have described a method for separating isothiocyanates. A 2-m column packed with Dixon rings coated with silicone grease was used to separate methyl, ethyl, isopropyl, *n*-butyl, and isoamyl thiocyanates.

Vagelos et al. (71) separated isocyanates using silicone gum SE 30. They acetylated hydroxamic acids and converted the acetate to the isocyanates in the flash heater and subsequently separated the isocyanates formed by the decomposition.

Jart (72) studied the behavior of 32 isothiocyanates. The liquid phases utilized were Apiezon L, silicone elastomer E-301, squalene, dinonyl phthalate, and tritolyl phosphate.

In our own laboratory we have investigated many possible liquid phases for use in the separation and analysis of isocyanates. The silicones have proved to be superior in all respects and some form of silicone seems to be the substrate generally employed. Silicones appear to possess the necessary inertness or resistance to chemical attack by the isocyanate as well as good thermal stability. The polar Si—O—Si bond undoubtedly has a decided beneficial effect on the —NCO group in terms of separation ability. However, the silicone columns must be checked periodically as they do have a limited life. The first evidence of deterioration will be evidenced by tailing, followed by loss of resolution. When this occurs, poor quantitative results will be obtained even though the column is calibrated frequently.

Determination of Impurities in TDI by Gas–Liquid Chromatography (73)

Scope

This vapor phase chromatographic method offers a rapid means of detecting and determining impurities in tolylenediisocyanate.

Principle

See Chapter 2.

Equipment

(*a*) Commercially obtainable gas chromatograph capable of operating at high temperatures.

(*b*) One 2.5-m conditioned silicone grease column.

(*c*) Suitable syringe, capacity of which will depend somewhat on chromatographic sensitivity.

(*d*) Ethyl acetate.

Chromatographic Conditions

(*a*) Hot wire filament detector.

(*b*) Detector current: 250 mA or sufficient current to provide desired sensitivity.

(*c*) Flash vaporizer temperature: 225°C.

(*d*) Column temperature: 195°C.

(*e*) Detector temperature: 250°C.

(*f*) Column: 2.5 m of 10% Dow Corning high vacuum silicone grease; 30–60 mesh firebrick.

(*g*) Flow: 40 ml/min of helium.

(*h*) Sample size: approximately 5 μl.

(*i*) Recorder: standard 1 MV.

(*j*) Chart speed: ½ in./min.

Procedure

(*1*) Inject the TDI sample rapidly into the chromatograph and attenuate when required.

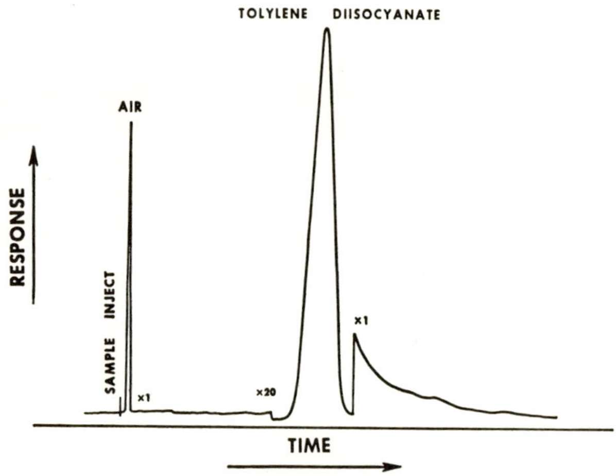

Fig. 3.1. Chromatogram of tolylene diisocyanate on 2.5 m of 10% silicone grease on 30–60 mesh firebrick.

CALCULATIONS

The chromatogram should be examined for the possible presence of trace impurities such as solvents and monoisocyanates. An example of a chromatogram is shown in Figure 3.1.

Many other silicones and solid supports have proven useful for the analysis of a great variety of isocyanates.

D. Spectroscopic Methods

1. Ultraviolet Region

Dixon (74) studied the absorption of the free —NCO radical in the visible and ultraviolet regions of the spectrum. Two systems of absorption bands were observed during the flash photolysis of several organic cyanates and were photographed under high resolution with long absorbing path lengths.

Extensive vibrational and rotational analyses were made for the bands of one system and showed that the spectrum was due to an electronic transition $A(^2\Sigma^+) \leftarrow X(^2\Pi_6)$ of the free —NCO radical, which is linear in both states. The electronic states of NCO were correlated with those of its dissociation products which showed that the bond dissociation energy of the CO bond was slightly greater than that of the CN bond in the three known states.

Mangini and Modena (75) have studied the near ultraviolet spectra of a number of aromatic isocyanates and isothiocyanates. They reported that the —NCO group does not absorb in the ultraviolet even if attached to a $CH_2{:}CH$ group and the benzene spectrum is only modified slightly which may be caused by a moderate interaction with the π electrons.

The —NCS group attached to an aliphatic chain showed a medium absorption and the benzene spectrum was modified considerably.

Holland et al. (76) studied the absorption and emission spectra of C_2H_5NCO, CH_3NCS, CH_3SCN, and HNCO. The spectra were assigned to the NCO and NCS free radicals.

2. Near-Infrared Region

A study of the near-infrared spectra of a number of aliphatic and aromatic isocyanates was performed by David (77). It was reported that for both aromatic and aliphatic isocyanates, bands which give information about the chain length and branching can be obtained by examining the spectra in the region 2.27–2.34 μ. No exact correlation of absorptivities or wavelengths of bands in this region with specific structures or molecules were found.

TABLE 3.2
Near-Infrared Absorption Data Used to Study Effects of Substitution on Isocyanate Structure (77)

Scan No.	Isocyanate	Molar absorptivity	σ^a	Wave number cm^{-1}
18	4,4′-Diphenylmethane diisocyanate	2.88	—	4062
21	*o*-Biphenyl isocyanate	2.56	—	4060
14	*p*-Chlorophenyl isocyanate	1.87	+0.227	4077
33	*p*-Bromophenyl isocyanate	1.75	+0.232	4072
11	*p*-Tolyl isocyanate	1.51	−0.170	4062
9	Phenyl isocyanate	1.51	0.000	4070
32	*p*-Ethoxyphenyl isocyanate	1.50	−0.250	4066
31	*p*-Methoxyphenyl isocyanate	1.37	−0.268	4067
20	*p*-Dodecylphenyl isocyanate	1.36	—	4067
13	*m*-Chlorophenyl isocyanate	1.31	+0.373	4077
19	*m*-Phenylene diisocyanate	1.26	+0.299	4072
27	*o*-Nitrophenyl isocyanate	1.23	—	4153
29	*p*-Nitrophenyl isocyanate	1.19	+0.778	4088
12	*m*-Tolyl isocyanate	1.07	−0.069	4055
25	*m*-Xylylene diisocyanate	0.95	—	4049
23	4,4′,4″-Triisocyanatotriphenylmethane	0.92	—	4065
28	*m*-Nitrophenyl isocyanate	0.78	+0.710	4095
24	1,3-Diisopropylphenylene-4,6-diisocyanate	0.71	—	4065
4	*n*-Amyl isocyanate	0.62	—	4078
5	Isopropyl isocyanate	0.56	—	4078
2	*n*-Butyl isocyanate	0.54	—	4075
30	*o*-Methoxyphenyl isocyanate	0.43	—	4110
1	Ethyl isocyanate	0.43	—	4070
15	2,4-Tolylene diisocyanate	0.43	—	4085
7	*n*-Dodecyl isocyanate	0.37	—	4067
16	2,6-Tolylene diisocyanate	0.22	—	4085
26	2-Isocyanatobicyclohexyl	0.24	—	4057
10	*o*-Tolyl isocyanate	0.24	—	4082
22	*p*-Bisphenyl isocyanate	Not determined	+0.009	4067
17	2,5-Tolylene diisocyanate	Not determined	—	—
6	*n*-Octyl isocyanate	—	—	—
8	*n*-Tetradecyl isocyanate	—	—	—

[a] These Hammett substituent constants were correlated with wavelength. Half intensity band width for these bands is 13–14 mμ.

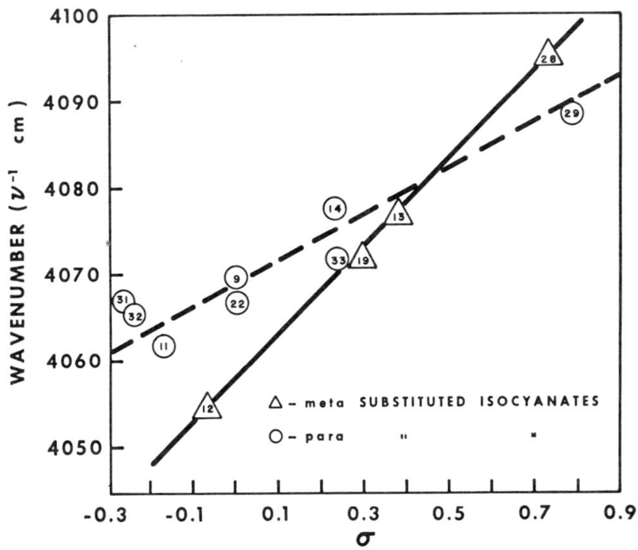

Fig. 3.2. Wavenumber vs. σ for near-infrared band near 2.48 μ (77).

Aromatic 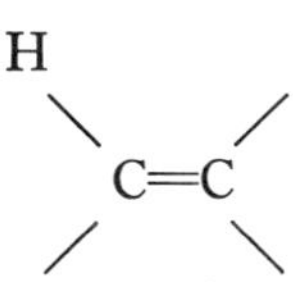 could readily be detected at 1.66 and 2.15 μ.

Resonance properties of the aromatic isocyanate were obtained by an examination of the molar absorptivities and wavelength of the band near 2.45 μ, since the wavelength of this band was correlated with Hammett's σ values.

Table 3.2 shows the wavelength and absorptivity of the band in the 2.45 μ region. It should be noted that the aliphatic isocyanates grouped themselves since they exhibited the lowest molar absorptivities. Table 3.2 also gives the σ values used to obtain the wavelength correlation shown in Figure 3.2.

The wavelength of the first overtone of the —NCO stretching vibration was reported to occur near 2.642 μ and could be employed for quantitative work.

3. Infrared Region

Eyster and Gillette (78) studied the vibration spectra of hydrazoic acid, methyl azide, and methyl isocyanate. This was an excellent study and

TABLE 3.3
Principle Absorption Wavelength of the Isocyanate Group[a] (80)

	Wavelength, μ
Aliphatic and cycloaliphatic isocyanates	
n-Butyl isocyanate	4.400
sec-Butyl isocyanate	4.421–4.422
Isobutyl isocyanate	4.419
n-Amyl isocyanate	4.398
2-Methyl-3-isocyanato-butane	4.408
3-Isocyanato-*n*-pentane	4.416–4.418
2,2-Dimethyl-3-isocyanato-butane	4.408
Stearylisocyanate	4.409
1-Chlor-6-isocyanato-*n*-hexane	4.411–4.414
1-Cyan-3-isocyanato-propane	4.399
1-Cyan-5-isocyanato-*n*-pentane	4.397–4.400
Cyclohexyl isocyanate	4.434
1-Methyl-2,4-diisocyanato-cyclohexane[b]	4.434–4.439
4,4′-Diisocyanato-dicyclohexylmethane[b]	4.434
3-Isocyanato-tetramethylensulfone	4.414
Aromatic isocyanates	
Phenyl isocyanate	4.424
3-Chlorophenyl isocyanate	4.416
4-Chlorophenyl isocyanate	4.414
2,5-Dichlorophenyl isocyanate	4.422
4-Nitrophenyl isocyanate	4.422
3-Nitro-4-chlorophenyl isocyanate	4.418
o-Tolyl isocyanate	4.399
m-Tolyl isocyanate	4.414
p-Tolyl isocyanate	4.418
Bis(4-isocyanato-phenyl)-methane	4.406
2,2-Bis(4-isocyanato-phenyl)-propane	4.409
1,1-Bis(4-isocyanato-phenyl)-cyclohexane	4.419
1,4-Diisocyanato-benzol	4.429
1-Methyl-2,4-diisocyanato-benzol	4.416
1-Methyl-2,5-diisocyanato-benzol	4.434–4.435
1-Methyl-2,6-diisocyanato-benzol	4.423
1-Methyl-3,5-diisocyanato-benzol	4.437
1-Methyl-3,5-diethyl-2,4-diisocyanato-benzol	4.401–4.402
1-Nitro-3-methyl-4,6-diisocyanato-benzol	4.423
1-Nitro-2-methyl-4,6-diethyl-3,5-diisocyanato-benzol	4.401
4,4′,4″-Triisocyanato-triphenylmethane	4.425
3-Methoxyphenyl isocyanate	4.412
4-Ethoxyphenyl isocyanate	4.398
3,3′-Dimethoxy-4,4′-diisocyanato-diphenyl	4.460
3,3′-Diethoxy-4,4′-diisocyanato-diphenyl	4.464

TABLE 3.3 (*Continued*)

	Wavelength, μ
2,5-Dimethoxyphenyl isocyanate	4.446–4.447
1-Naphthyl isocyanate	4.411
2-Isocyanato-chrysen	4.397
1-Isocyanato-2-ethoxy-5,6,7,8-tetrahydronaphthalin	4.447
Benzyl isocyanate	4.414–4.416
1,3-Bis(isocyanato-methyl)-benzol	4.421
Acylisocyanate	
Acetyl isocyanate	4.453

[a] Wavelengths of the isocyanate group were deteımined in carbon tetrachloride solutions.

[b] *Cis–trans* mixtures.

quite definitive considering the instrumental limitations at that time. From the similarity of the vibration spectra of methyl azide, they concluded the NCO group was linear. From the increase in the triatomic stretching frequencies on going from the azide to the isocyanate and other data, Eyster and Gillette predicted this might be explained if the C—O bond were somewhat stronger than the N—C bond. This seems to have been confirmed by Dixon (74). The following resonance structures were postulated:

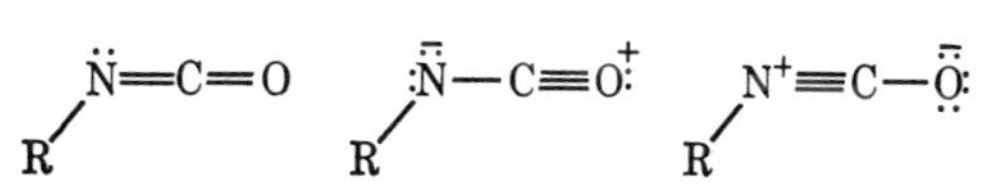

(38)

Davison (79) studied the infrared absorption of the isocyanate group of ten isocyanates. The results showed that ν_{as}(NCO) exhibited an extremely intense absorption at 2269 $\pm$ 6 cm^{-1}. For three of the isocyanates weak absorptions at 1380–1395 cm^{-1} were observed and tentatively assigned to the ν_s(NCO) which agrees with the assignment of ν_s(NCO) at 1412 cm^{-1} in methyl isocyanate by Eyster and Gillette (78).

Hoyer (80) also studied the infrared absorption of the isocyanate group. A substantial number of aliphatic and aromatic isocyanates were examined and the ν_{as}(NCO) stretching vibration of the —NCO group was found to absorb near 4.4 μ. Hoyer's data are shown in Table 3.3.

Recently, Hirschmann et al. (81) studied the infrared spectra of alkyl isocyanates. The frequency ranges for the characteristic vibrations of the isocyanate group were summarized as shown in Table 3.4.

The characteristic vibrations found by investigators (78,80,82) in both the infrared and Raman spectra confirm Thomas's (83) suggestion that the

TABLE 3.4
Infrared Frequency Ranges for the Characteristic Vibrations of the Isocyanate Group (81)

Vibration	Frequency[a]	Intensity
ν_{as}(NCO)	2300–2250 (l)	Very strong
ν_s(NCO)	1450–1400 (l)	Very weak
$\delta_\perp$(NCO)	650–600 (l)	Poorly defined
$\delta_\parallel$(NCO)	600–590 (s)	Strong

[a] l = Liquid state; s = Solid state.

TABLE 3.5
Assignments for the Fundamental Modes of Methyl Isocyanate (81)

Vibration	Frequency (cm^{-1})[a]
$\nu_{as}(CH_3)$	3000
$\nu_s(CH_3)$	2960
ν_{as}(NCO)	2288
$\delta_{as}(CH_3)$	1453
ν_s(NCO)	1437[b]
$\delta_s(CH_3)$	1416
$\gamma_\parallel(CH_3)$	1132
$\gamma_\perp(CH_3)$	1023
$\nu(C_\alpha$—N)	852
$\delta_\perp$(NCO)	613[b]
$\delta_\parallel$(NCO)	596
δ(CNC)	353[c]
Γ_{CH_3}	96[d]

[a] All frequencies are from the liquid state unless otherwise indicated.
[b] Solid state frequency.
[c] Frequency (82).
[d] Calculated value from observed frequency at 143 cm^{-1}.

stretching vibrations are coupled as ν_{as}(NCO) and ν_s(NCO) rather than ν(N═C) and ν(C═O).

The assignments for the fundamental modes of methyl isocyanate by Hirschmann et al. (81) are given in Table 3.5. The vapor and liquid spectra are shown in Figure 3.3. The assignments for the combination bands for three of the isocyanates studied are shown in Table 3.6.

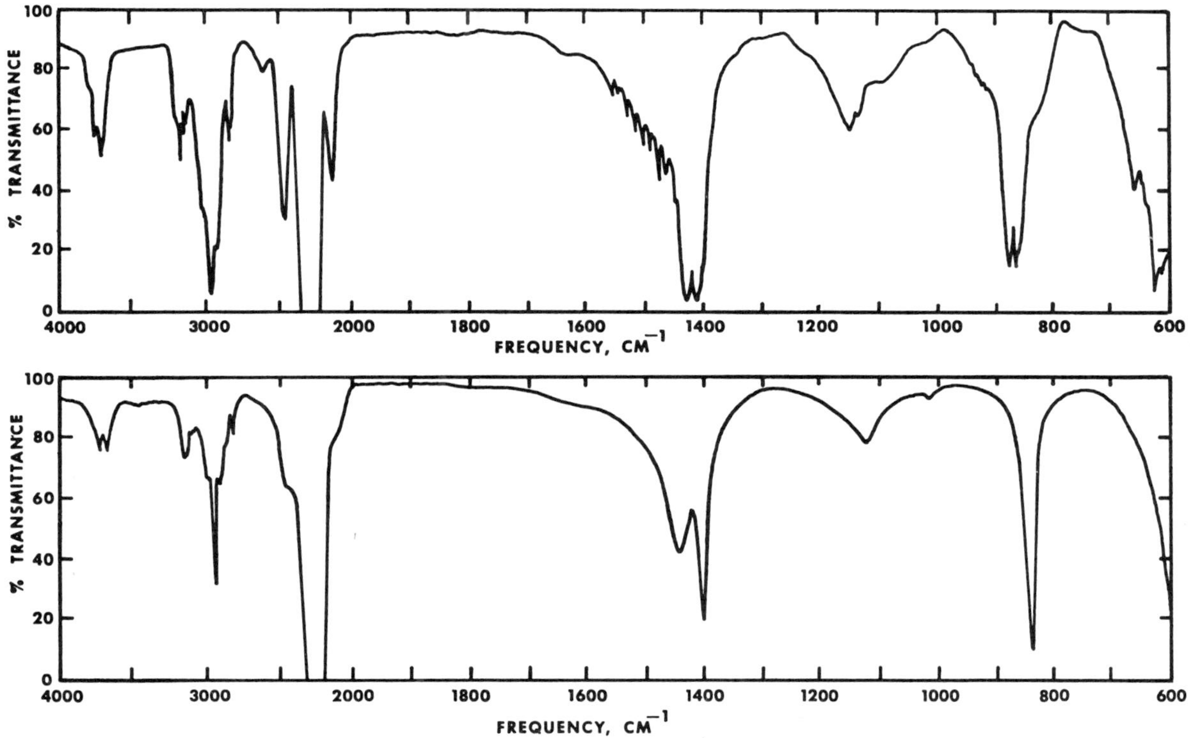

Fig. 3.3. Infrared spectra of the vapor (upper) and liquid (lower) states of methyl isocyanate (81).

TABLE 3.6
Assignments for the Combination Bands in Isocyanates (81)

Frequency (cm^{-1})[a]			
Methyl isocyanate	Ethyl isocyanate	Isopropyl isocyanate	Assignment
3726	—	—	$\nu_{as}(NCO) + \delta_{as}(CH_3)$
—	3698	3674	$\nu_{as}(NCO) + \nu_s(NCO)$
3680	—	—	$\nu_{as}(NCO) + \delta_s(CH_3)$
—	3633	—	$2218 + \nu_s(NCO)$
—	3263	—	$\nu_{as}(NCO) + \nu(CC)$
3160	—	—	$2318 + \nu(C_\alpha—N)$
3135	3075	—	$\nu_{as}(NCO + \nu(C_\alpha—N)$
3089	—	—	$2254 + \nu(C_\alpha—N)$
2912	2901	—	$2 \times \delta_{as}(CH_3)$
2871[b]	—	—	$2 \times \nu_s(NCO)$
2813	—	—	$2 \times \delta_s(CH_3)$
2570	—	—	$\delta_{as}(CH_3) + \gamma_{\parallel}(CH_3)$
2460	—	—	$\delta_{as}(CH_3) + \gamma_{\perp}(CH_3)$
—	2419	—	$\nu_s(NCO) + \nu(CC)$
—	—	2393	$\delta(CH) + \nu(CC)$
2318	2218	2172	$\nu_s(NCO) + \nu(C_\alpha—N)$
2254	—	—	$\delta_s(CH_3) + \nu(C_\alpha—N)$
—	2140	—	$CH_2Wag + \nu(C_\alpha—N)$
2131	—	—	$\nu_{as}(CH_3)—\nu(C_\alpha—N)$
—	2090	—	CH_2 bend $+ \delta_{\parallel}(NCO)$
—	1399	—	$\nu(CC) + \delta_{\perp}(CCN)$
—	1375[b]	—	$\nu(CC) + \delta_{\parallel}(CCN)$
—	1262	1242	$2 \times \delta(NCO)$
—	—	1221	$\delta(NCO) + \delta(NCO)$

[a] Frequencies from liquid state spectra unless otherwise indicated.
[b] Solid state frequencies
— No band observed.

The assignment tables and spectra of the relatively simple isocyanate, methyl isocyanate, were included here because these data lay a foundation for a basic understanding of the spectra of other isocyanates as well as implementing the practical interpretation of a large number of isocyanate spectra.

Determination of Isomer Ratio of TDI by Infrared Spectroscopy (84)

SCOPE

This method permits the determination of the isomer ratio by suitable infrared absorption measurements on several commercially obtainable tolylene diisocyanates.

Principle

This method is based on the quantitative measurement of absorption bands due to out-of-plane C—H deformation vibrations of the benzene ring. The absorption bands occur at different wavelengths because of differences in the position of ring substituents.

Equipment

(*a*) Volumetric flasks, 25 ml.
(*b*) Sealed liquid cell, 0.1 mm.
(*c*) Cyclohexane (Spectro grade).
(*d*) Infrared spectrophotometer (Beckman IR-4 or equivalent).

Procedure

Part A

This procedure is used for TDI having a 2,4/2,6 isomer ratio of 65/35 or 80/20.

(*1*) A 0.12 mm cell is used for this analysis; however, if this cell size is not available, the same results may be produced by using a different known path length cell and adjusting the sample weight.

$$\text{Sample weight} = \frac{0.24}{\text{Cell size, mm}}$$

(*2*) Weigh the sample on an analytical balance (to the nearest 0.01 g) into a 25 ml volumetric flask and dilute to volume with pure cyclohexane.

(*3*) Fill the cell with pure cyclohexane, after flushing the cell three times and adjust the instrument to give a scan of medium high resolution.

(*4*) Place the cell in the instrument and scan the spectrum from 13.0 to 12.0 μ three times, manually or automatically, moving the recording paper each time.

(*5*) Remove and drain the cell, flush the cell three times with sample solution, then fill the cell.

(*6*) Scan the spectrum three times from 13.0 to 12.0 μ superimposing on the spectra of the solvent alone.

(*7*) Remove the sample from the cell and flush with cyclohexane as before, then with dry acetone, and dry the cell with nitrogen.

Calculations

(*1*) Read or calculate the absorption from the cyclohexane baseline at 12.35 and 12.80 μ.

(*2*) $$\text{Conc. of 2,4 isomer} = \frac{\text{Abs. at } 12.35 - 0.226}{0.00372} + 55.0$$

$$\text{Conc. of 2,6 isomer} = \frac{\text{Abs. at } 12.80 - 0.150}{0.0120} + 11.6$$

(*3*) Normalize the concentrations to 100% as follows:

$$\frac{\text{Concentration of 2,4 isomer}}{\text{Conc. of 2,4} + \text{conc. of 2,6}} = \text{True \% 2,4 isomer}$$

$$\frac{\text{Concentration of 2,6 isomer}}{\text{Conc. of 2,4} + \text{conc. of 2,6}} = \text{True \% 2,6 isomer}$$

All results should be reported to the nearest 0.1%.

CALIBRATION

(*1*) Prepare known weight concentrations of the isomers from pure 2,4 and 2,6 TDI in the analytical ranges desired.

(*2*) Begin with step No. 2, Part A, and continue through step *7*.

(*3*) Plot the absorbance for each isomer vs. concentration.

(*4*) Determine the x-axis intercept (% 2,4 or % 2,6 isomer) and the y-axis intercept (absorbance) that corresponds to the x-axis intercept.

(*5*) The concentration of each isomer is given by: Absorbance at 12.35 or 12.80 μ minus absorbance that corresponds to % isomer intercept on the x-axis divided by the units of absorbance per unit of isomer (slope) plus the % isomer intercept on the x-axis. In the present case, the equations obtained are shown in step *2* of calculations.

Part B

This procedure is used for samples that contain more than 90% of the 2,4 isomer of TDI.

(*1*) Use a cell of known path length between 0.04 and 0.11 mm.

(*2*) Fill the cell with the undiluted sample after flushing the cell three times with sample and adjust the instrument to give a scan of medium high resolution.

(*3*) Place the cell in the instrument and scan the spectrum three times from 13.5 to 12.0 μ.

(*4*) Remove the sample from the cell and flush with cyclohexane, as before, then with dry acetone, and dry the cell with dry nitrogen.

CALCULATIONS

(*1*) Using the standard base line technique, calculate the absorbance at 12.70 μ.

(*2*) Calculate the normalized absorbance by dividing the absorbance of the sample by the cell thickness in mm.

$$\frac{\text{Absorbance}}{\text{Cell thickness (mm)}} = \text{Normalized absorbance}$$

(*3*) % 2,4 isomer = 100 − (normalized absorbance × 0.75570) − 0.0372. The factor 0.75570 is the slope of the line found by plotting the known concentration of 2,6 isomer on the y axis and normalized absorbance on the x axis from calibration using the procedure outlined above. The slope should be determined using the standard least squares calculation.

The factor 0.0372 is the intercept of the curve found according to the formula $b = \bar{y} - m\bar{x}$, where b = intercept, $\bar{y}$ = average concentration of 2,6 isomer, and $\bar{x}$ = average normalized absorbance. All results should be reported to the nearest 0.1%.

The precision of the above methods is ±0.2%.

Possible changes in the cell path length should be checked daily by the interference fringe method of cell calibration. This will allow the absorbances to be normalized, which must be done in order to obtain more accurate and reproducible results.

Two methods for normalizing the absorbance are given above, to show how these corrections can be made. In Part *A* the changing cell path length is taken into account by adjusting the sample size (step *1*). In Part *B* the absorbance is normalized before the equation of the curve is obtained. This same technique could have been used in Part A with equally good results but either preferred technique can be used.

This procedure is essentially the same as the SPI procedure (61) and uses the same wavelengths for absorption measurements. However, the two methods are different in many details.

We feel that the correction factors, as given by ASTM to correct for absorption interferences of the 2,4 isomer at the primary 2,6 isomer absorption wavelength and vice versa, are in error. Obviously, this would cause some analytical inaccuracies. We suspect this may be due to the fact that extremely pure isomers were not used for the calibration causing greater than the theoretical interferences, which results in overcorrections to be applied to the samples.

Over a period of years we have found this method to be precise to ±0.2%. The ASTM procedure gives the precision as being ±0.8%. It is our opinion that the method is capable of giving far better precision than the ±0.8%.

Both the Mobay procedure and the ASTM procedure are based on the work and method as originally developed by Lord (85). The typical spectra of the 2,4 isomer, which absorbs at 12.35 μ and the 2,6 isomer, which absorbs at 12.80 μ, are shown in Figure 3.4.

Lord (85) reported that the 2,5 isomer interferes slightly with the 2,4 isomer determination but will not cause serious error since it is present normally in low amounts in the refined isocyanates.

Interference from the 2,3 and 3,4 isomers should be negligible since it has been reported that the *ortho* diamines do not phosgenate (47); therefore, we could expect only trace amounts, at best, to be present in the refined diisocyanate.

Since the 2,4 and 2,5 isomers both represent 1,2,4-trisubstituted benzenes, we would expect overlapping absorbancies and mutual interference as shown in Figure 3.5. Lord (85) found that the interference with the 2,4

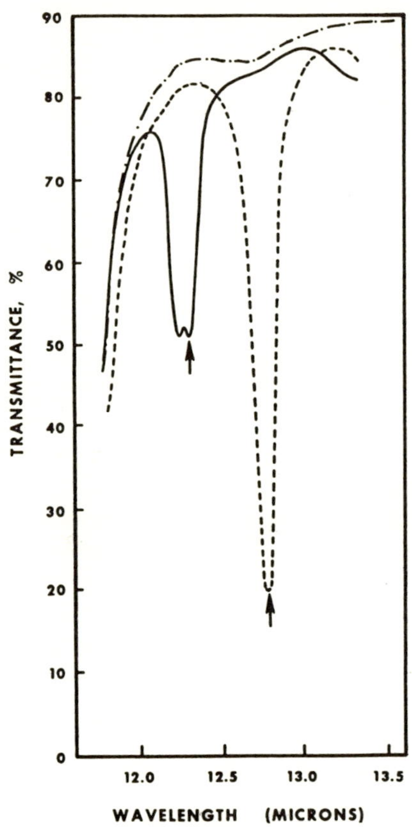

Fig. 3.4. Infrared analytical absorption bands used for the determination of 2,4- and 2,6-tolylene diisocyanate isomer ratio (85). (- · -) Cyclohexane, 0.208-mm cell. (——) Tolylene 2,4-diisocyanate, 0.0203 g/ml in cyclohexane, 0.208-mm cell. (- - -) Tolylene 2,6-diisocyanate, 0.0200 g/ml in cyclohexane, 0.208-mm cell.

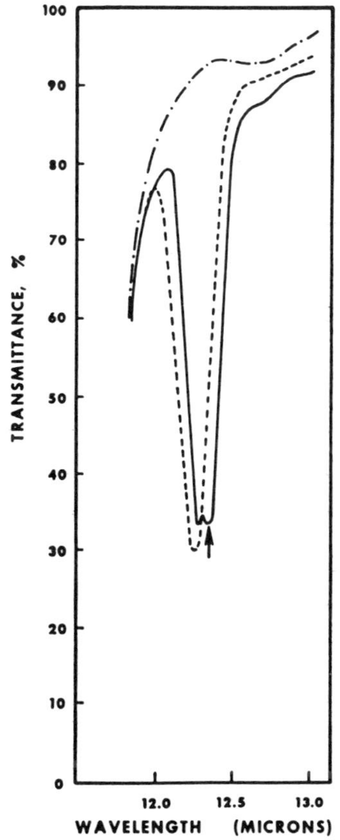

Fig. 3.5. Infrared absorption spectra showing interference of the 2,5 isomer and 2,4-tolylene diisocyanate (85). (- · -) Cyclohexane, 0.208-mm cell. (——) Tolylene 2,4-diisocyanate, 0.0400 g/ml in cyclohexane, 0.208-mm cell. (- - -) Tolylene 2,5-diisocyanate, 0.0400 g/ml in cyclohexane, 0.208-mm cell.

isomer determination was equivalent to approximately 50% of the actual amount of 2,5 isomer present and does not interfere with the 2,6 isomer.

E. Miscellaneous Methods

Another method for the determination of the isomer ratio of tolylene diisocyanate has been described by Steingiser, Darr, and Hardy (86). This method is based on the principle of dielectric constant measurements and is easily applicable to continuous monitoring of the isomer ratio during production.

This method suffers because (*1*) it has a lack of specificity, (*2*) it requires a large sample, (*3*) variable ionic impurities in process material changes the dielectric losses, affecting the dielectric constant, and (*4*) it requires precise temperature control.

A Foxboro Capacitance Dynalog analyzer was employed with a capacity span of 20 $\mu\mu$F and a frequency of 1.6 Mc/sec.

The dielectric constants of TDI at 23.1°C were found to be:

2,4-TDI	8.27
2,6-TDI	5.21

which indicated sufficient separation to allow quantitative analysis after calibration with known mixtures. The reported results from these measurements, in general, appeared to give lower results than the infrared isomer ratio method in the normal (80% and 65% 2,4 isomer) range and higher results in the higher (90% 2,4) range.

IV. REACTIONS OF ISOCYANATES WITH THEMSELVES

A. Dimers (Uretedinediones)

Aryl isocyanates readily polymerize in the presence of catalysts into dimers containing 2 moles of the monomeric isocyanate. The reaction and structure is generally considered to be:

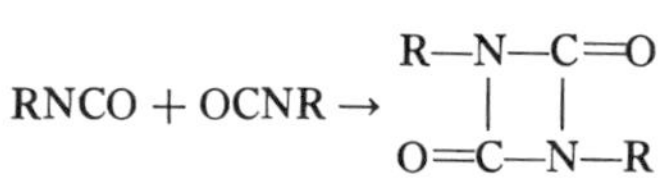

These compounds are listed currently in *Chemical Abstracts* as uretidinediones, although the term dimer is frequently applied.

Hofmann (87) prepared the dimer of phenyl isocyanate using triethyl phosphine as a catalyst as early as 1870 and Dennstadt (88) prepared the analogous compound from bromophenyl isocyanate employing the same catalyst. Snape, in 1886 (89), found that pyridine catalyzed the dimerization of phenyl isocyanate.

The papers of Hale (90) and of Hale and Lange (91) on the four-membered cyclic ureas are of historical interest pertaining to the nomenclature of this class of compounds. They proposed the term *uretedone* for the four-member heterocyclic ring containing two nitrogen and two carbon atoms in alternate positions. They adopted the "ur" from urea to show that these compounds are derivatives of urea. The "ete" portion was derived from earlier names such as "keturet" and "alduret" attributed to Fromm for similar compounds. The "one" suffix indicates the ketone group.

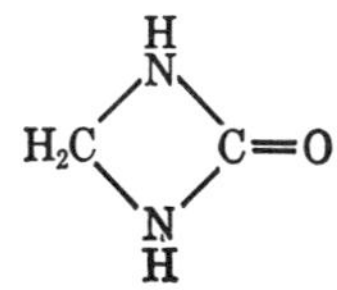

Blair and Smith (92) discuss the "tautomerism of cyanourea with guanyl isocyanate" and conclude that cyanourea exists as a ring structure tautomeric with a chain structure. They prepared diphenyl diisocyanate (dimer) by the method of Snape (89) using pyridine, and studied its properties to confirm their conclusions pertaining to cyanourea.

Raiford and Shelton (93) mentioned the compound prepared by Warren and Wilson (94) in 1935 as "diphenyluretidone." In 1939 (55) (95) Chemical Abstracts indexed isocyanate dimers as uretediones. Subsequently this was changed to uretidinediones, the name now in use. They compared the structure of a complex dimer containing bromine, hydroxyl, methyl, and tolyl sulfonyl groups with the dimer of phenyl isocyanate. Subsequently, Raiford and Freyermuth (96) prepared a series of some 13 dimers from phenyl and substituted phenyl isocyanates and described them as derivatives of 1,3-diphenyluretedione, evidently inserting the "di" to indicate the two ketone groups. The physical properties of selected uretediones are given in Table 3.7.

The 1,3 positions refer to substituents on the nitrogen atoms of the ring.

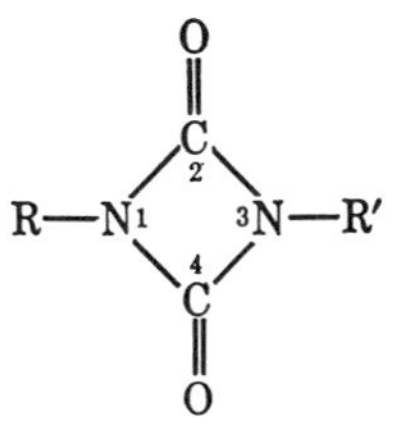

Numerous biurets were prepared from these dimers, as will be discussed in Chapter 4.

Raiford and Freyermuth (97) found that thionyl chloride reacted with phenylurethane (i.e., the *N*-substituted aryl carbamate, ethyl phenyl carbamate C_6H_5—NH—CO—O—C_2H_5) to give 1,3-diphenyluretedione. They found that the esters of carbamic acid, e.g., ethyl carbamate,

NH_2—CO—O—Et, and butyl carbamate, NH_2—CO—O—Bu, yielded allophanates when treated with thionyl chloride.

Lyons and Thomson (98) prepared the dimer of 1-naphthylisocyanate in their study of the formation of urethanes from this isocyanate and polyhydric phenols. Triethylamine was used as a catalyst. They use the name 1,3-di-1′-naphthyluretidione and gave the alternate designation of 1,3-di-1′-naphthyl-1,3-diazacyclobuta-2,4-dione.

Saunders and Hardy (99) showed that the dimer of 2,4-tolylene diisocyanate, which has two free NCO groups, reacts with dibutylamine as though it contains more than two equivalents of NCO, even at a temperature as low as 50°C. At higher temperatures (170–180°) in the presence

TABLE 3.7
Physical Properties of Selected Dimers

1. Phenyl isocyanate dimer
2. 4-Aminophenyl dimer
3. 4-Amino-3-methylphenyl dimer
4. 3-Amino-4-methoxyphenyl dimer
5. 3-Amino-5-chlorophenyl dimer
6. 4(4′Aminophenyl)phenyl dimer
7. Monobromophenyl dimer
8. 4-Bromophenyl dimer
9. 3-Bromo-2-hydroxy-5-methylphenyl dimer
10. 2-Chlorophenyl dimer
11. 3-Chlorophenyl dimer
12. 4-Chlorophenyl dimer
13. 2,5-Dichlorophenyl dimer
14. 3,4-Dichlorophenyl dimer
15. *para*-Dodecylphenyl dimer
16. 4-Ethoxyphenyl dimer
17. 3-Methylphenyl (*m*-tolyl isocyanate) dimer
18. 4-Methylphenyl (*p*-tolyl isocyanate) dimer
19. 3-Nitrophenyl dimer
20. 4-Nitrophenyl dimer
21. 4-Phenylphenyl dimer
22. 4-Phenyl azo phenyl dimer
23. 1-Naphthyl dimer
24. 2-Naphthyl dimer
25. 4-Aminonaphthyl dimer
26. 2,4-Tolylene diisocyanate dimer
27. 4,4′-Biphenylene diisocyanate dimer
28. 4-Chloro-*m*-phenylene diisocyanate dimer
29. 4-Methoxy-*m*-phenylene diisocyanate dimer
30. 4-Phenoxy-*m*-phenylene diisocyanate dimer
31. *p*-(2-Isocyanato-isobutyl)phenyl isocyanate dimer
32. α(1-Isocyanato-cyclohexyl)-*p*-tolylisocyanate dimer

TABLE 3.7 (*Continued*)

No.	Name and structure	Analysis: Elem.	Analysis: Calc.	Analysis: Fd.	Infrared data: Solvent	Infrared data: Wavelength	Ref.
1.	Phenyl isocyanate dimer 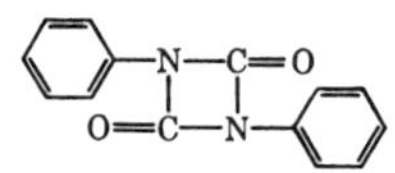		11.78	11.90			55, 104, 87
							89, 96
							100
	Formula $C_{14}H_{10}O_2N_2$ Mol. wt. 238.24	M.P. 175°C					
	Solvents for recrystallization. *N*-Butyl ether, benzene, or CCl_4	B.P.					
	Appearance						
2.	4-Aminophenyl dimer 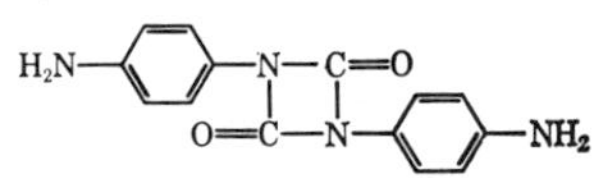						55
							106
	Formula $C_{14}H_{12}O_2N_4$ Mol. Wt. 268.27	M.P. Not given					
	Solvents for recrystallization	B.P.					
	Appearance						
3.	4-Amino-3-methylphenyl dimer 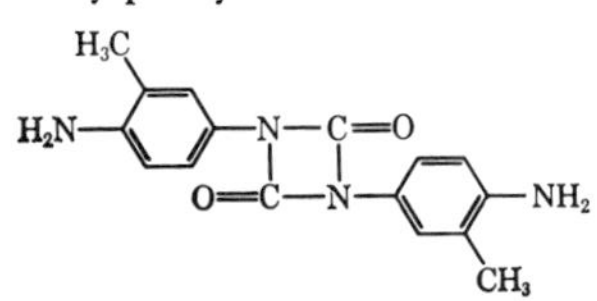						55
							106

(*continued*)

TABLE 3.7 (*Continued*)

No.	Name and structure	Analysis: Elem.	Analysis: Calc.	Analysis: Fd.	Infrared data: Solvent	Infrared data: Wavelength	Ref.
	Formula $C_{16}H_{16}O_2N_4$ Mol wt. 296.32	M.P. Not given					
	Solvents for recrystallization	B.P.					
	Appearance						
4.	3-Amino-4-methoxyphenyl dimer 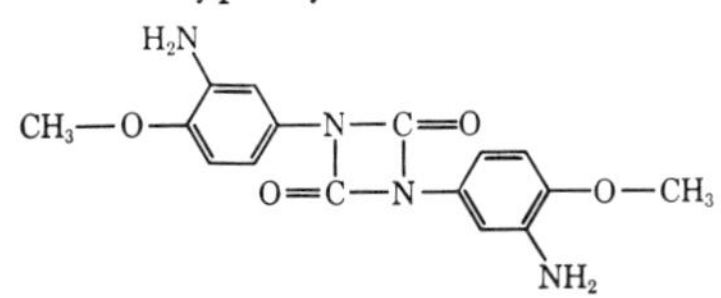						55, 106
	Formula $C_{16}H_{16}O_4N_4$ Mol. wt. 328.32	M.P. Not given					
	Solvents for recrystallization	B.P.					
	Appearance						
5.	3-Amino-5-chlorophenyl dimer 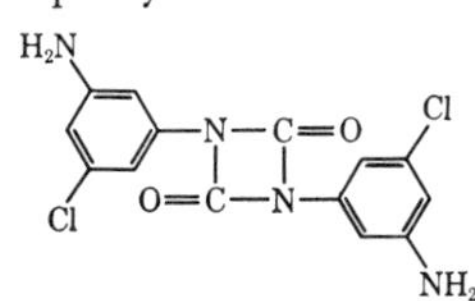						55, 106

Compound						Ref.
Formula $C_{14}H_{10}O_2N_4Cl_2$ Mol. wt. 337.17	M.P. Not given					
Solvents for recrystallization	B.P.					
Appearance						
6. 4(4′Aminophenyl)phenyl dimer 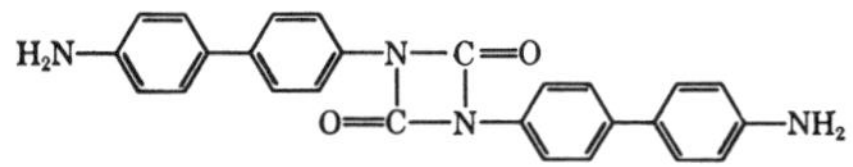						55, 106
Formula $C_{26}H_{20}O_2N_4$ Mol. wt. 420.45	M.P. Not given					
Solvents for recrystallization	B.P.					
Appearance						
7. Monobromophenyl dimer 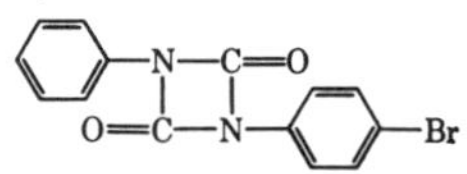	Br	25.1	25.1	Nujol mull	5.65–5.70 μ	88, 105
Formula $C_{14}H_9O_2N_2Br$ Mol. wt. 317.14	M.P. 190–191°C					
Solvents for recrystallization Methanol	B.P.					
Appearance						
8. 4-Bromophenyl dimer 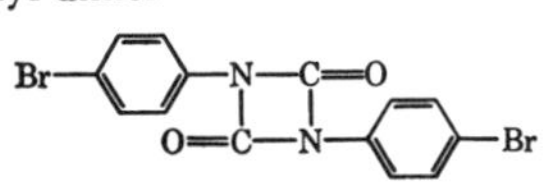	Br	40.40	40.40			96, 104
						55, 99
						88

(continued)

TABLE 3.7 (*Continued*)

No.	Name and structure	Analysis			Infrared data		Ref.
		Elem.	Calc.	Fd.	Solvent	Wavelength	
	Formula $C_{14}H_8O_2N_2Br_2$ Mol. wt. 396.05	M.P. 203–204°C					55, 96
		187–192°C					100
	Solvents for recrystallization *N*-Butyl ether, benzene	199°C					88
	Appearance						
9.	2-hydroxy-3-Bromo-5-methylphenyl dimer						55
	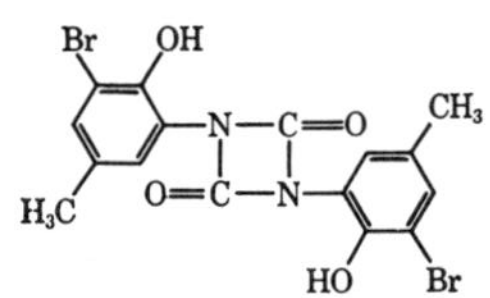	Br	35.08	35.83			93
		N	6.14	6.15			
	Formula $C_{16}H_{12}O_2N_4Br_2$ Mol. wt. 456.10	M.P. 170°C					
	Solvents for recrystallization Dilute alcohol	B.P.					
	Appearance						
10.	2-Chlorophenyl dimer						
	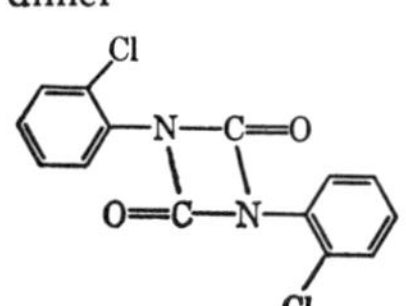	Cl	23.12	23.12			104, 96
							55, 100
		N	9.12	9.20			

Compound						Ref.
Formula $C_{14}H_8O_2N_2Cl_2$ Mol. wt. 307.13	M.P. 234–235°C					55, 96
	218–219°C					100
Solvents for recrystallization Acetic acid	B.P.					
Appearance Colorless plates						
11. 3-Chlorophenyl dimer 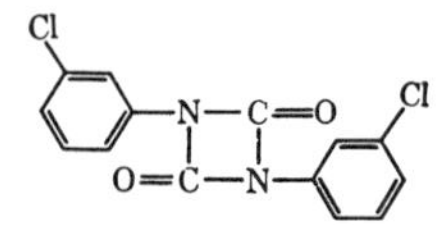						96, 104
						55, 100
Formula $C_{14}H_8O_2N_2Cl_2$ Mol. wt. 307.13 Fd 314	M.P. 153–154°C					55, 96
	171–172°C					100
Solvents for recrystallization *N*-Butyl ether, benzene	B.P.					
Appearance Colorless Needles						
12. 4-Chlorophenyl dimer 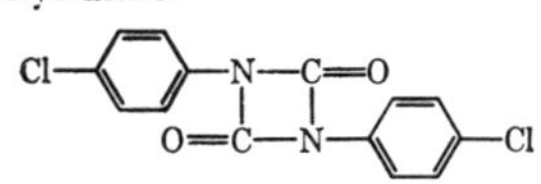	Cl	23.12	22.91			96, 104
	N	9.12	9.12			55, 102
						100, 105
Formula $C_{14}H_8O_2N_2Cl_2$ Mol. wt. 307.13	M.P. 172–174°C					100
	155–156°C					55, 96
Solvents for recrystallization *N*-Butyl ether	170–171°C					105
Appearance						
13. 2,5-Dichlorophenyl dimer 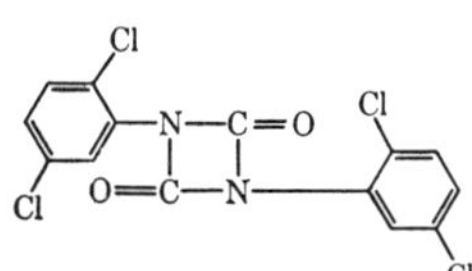						100

(continued)

TABLE 3.7 (*Continued*)

No.	Name and structure	Analysis: Elem.	Calc.	Fd.	Infrared data: Solvent	Wavelength	Ref.
	Formula $C_{14}H_6Cl_4N_2O_2$ Mol. wt. 376.03	M.P. 233–240°C					
	Solvents for recrystallization Benzene	B.P.					
	Appearance						
14.	3,4-Dichlorophenyl dimer 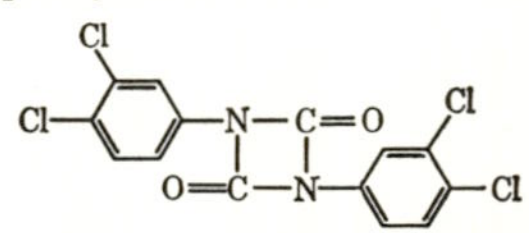						107
	Formula $C_{14}H_6Cl_4N_2O_2$ Mol. wt. 376.03	M.P.					
	Solvents for recrystallization	B.P.					
	Appearance						
15.	*para*-Dodecylphenyl dimer 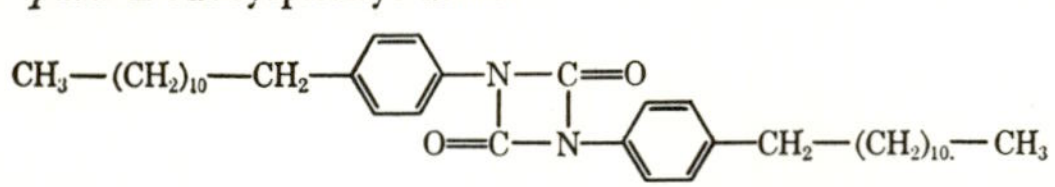						100
	Formula $C_{38}H_{58}O_2N_2$ Mol. wt. 574.86	M.P. 117°C					
	Solvents for recrystallization Pyridine	B.P.					
	Appearance						

Compound	Analysis	Calcd.	Found			Ref.
16. 4-Ethoxyphenyl dimer 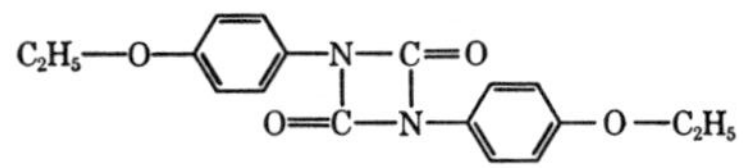	N	8.58	8.54			96, 104
						55, 100
Formula $C_{18}H_{18}O_4N_2$ Mol. wt. 326.34	M.P. 181–182°C					55, 96
	171–175°C					100
Solvents for recrystallization *N*-Butyl ether, Benzene	B.P.					
Appearance						
17. 3-Methylphenyl (*m*-tolyl isocyanate) dimer 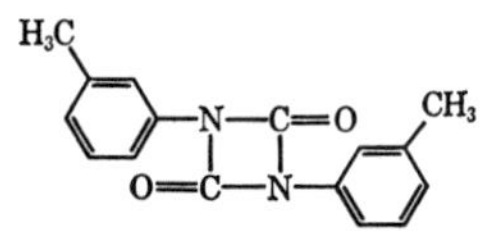						
	N	10.52	10.57			
						55, 96, 104
Formula $C_{16}H_{14}O_2N_2$ Mol. wt. 266.29	M.P. 159–160°C					
Solvents for recrystallization *N*-Butyl ether	B.P.					
Appearance Colorless leaflets						
18. 4-Methylphenyl (*p*-tolyl isocyanate) dimer 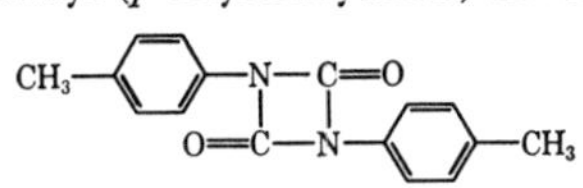						
						55, 96, 104
Formula $C_{16}H_{14}N_2O_2$ Mol. wt. 266.29	M.P. 185°C					
Solvents for recrystallization	B.P.					
Appearance						

(continued)

TABLE 3.7 (*Continued*)

No.	Name and structure	Analysis			Infrared data		Ref.
		Elem.	Calc.	Fd.	Solvent	Wavelength	
19.	3-Nitrophenyl dimer 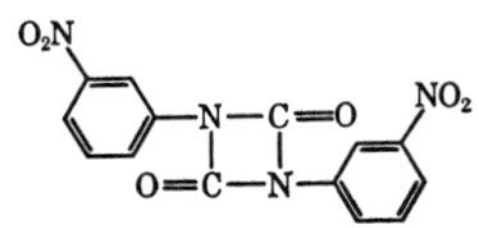						100
	Formula $C_{14}H_8N_4O_6$ Mol. wt. 328.24	M.P. 183–184°C					
	Solvents for recrystallization Benzene	B.P.					
	Appearance						
20.	4-Nitrophenyl dimer 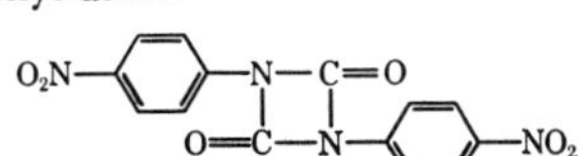	N	17.07	16.78		5.65 μ	96, 100
							105
	Formula $C_{14}H_8N_4O_6$ Mol. wt. 328.24	M.P. 300°C subl.					96
		208–210°C					100, 105
	Solvents for recrystallization Nitrobenzene, CCl_4	B.P.					
	Appearance Yellow needles or long white needles						
21.	4-Phenylphenyl dimer 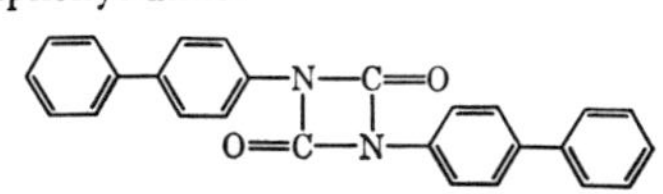	N	7.18	7.24			96, 104

Compound	Analysis					References
Formula $C_{26}H_{18}N_2O_2$ Mol. wt. 390.42	M.P. 270°C					
Solvents for recrystallization Nitrobenzene	B.P.					
Appearance						
22. 4-Phenylazophenyl dimer 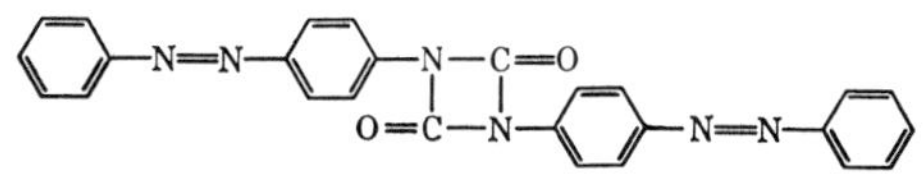						96, 104
	N	18.83	18.95			55
Formula $C_{26}H_{18}N_6O_2$ Mol. wt. 446.45	M.P. 281–282°C					
Solvents for recrystallization Nitrobenzene	B.P.					
Appearance						
23. 1-Naphthyl dimer 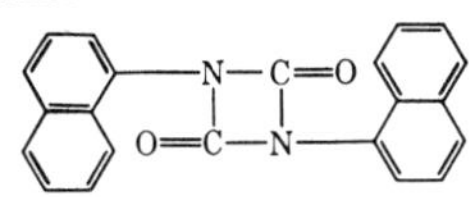	N	8.28	8.49			101, 96
						55, 98
Formula $C_{22}H_{14}N_2O_2$ Mol. wt. 338.35	M.P. 296°C					96, 104
	295°C					98
Solvents for recrystallization Nitrobenzene	B.P.					
Appearance						
24. 2-Naphthyl dimer 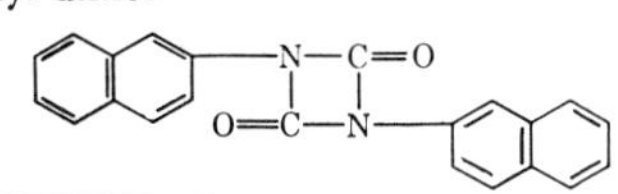	N	8.23	8.41			105
						55, 96

(continued)

TABLE 3.7 (*Continued*)

No.	Name and structure	Analysis			Infrared data		Ref.
		Elem.	Calc.	Fd.	Solvent	Wavelength	
	Formula $C_{22}H_{14}N_2O_2$ Mol. wt. 338.35	M.P. 196–197°C					
	Solvents for recrystallization	B.P.					
	Appearance						
25.	4-Aminonaphthyl dimer 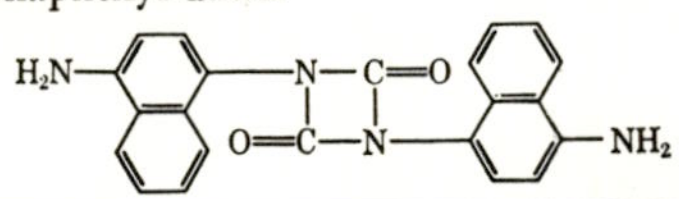						55, 103 106
	Formula $C_{22}H_{16}N_4O_2$ Mol. wt. 368.38	M.P. Not given					
	Solvents for recrystallization	B.P.					
	Appearance						
26.	2,4-Tolylene diisocyanate dimer 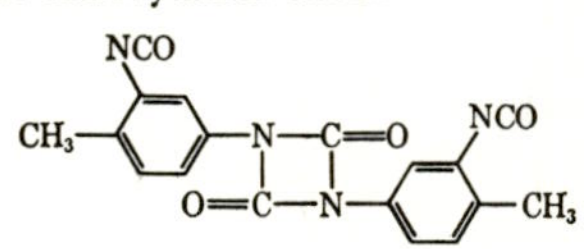						105 55 100

Compound	Properties	Ref.
Formula $C_{18}H_{12}N_4O_4$ Mol. wt. 348.31	M.P. 156–157°C	55, 105
	158°C	100
Solvents for recrystallization Ethyl acetate; Soluble 1 pt in 20 pt chlorobenzene	B.P.	
Appearance		
27. 4,4′-Biphenylene diisocyanate dimer 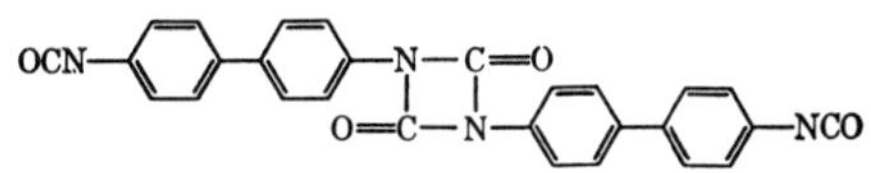		100
Formula $C_{28}H_{16}N_4O_4$ Mol. wt. 472.44	M.P. 325°C	
Solvents for recrystallization Slurried in hexane	B.P.	
Appearance Pale yellow, amorphous		
28. 4-Chloro-*m*-phenylene diisocyanate dimer 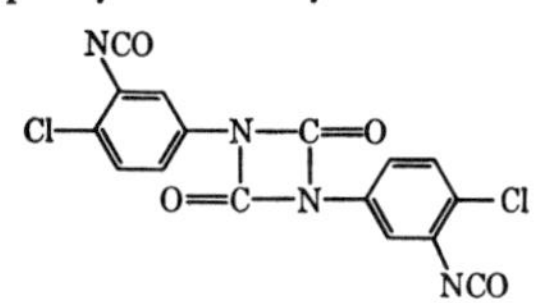		55
Formula $C_{16}H_6N_4O_4Cl_2$ Mol. wt. 389.15	M.P. 172–174°C	
Solvents for recrystallization	B.P.	
Appearance		

(continued)

TABLE 3.7 (*Continued*)

No.	Name and structure	Analysis			Infrared data		Ref.
		Elem.	Calc.	Fd.	Solvent	Wavelength	
29.	4-Methoxy-*m*-phenylene diisocyanate dimer 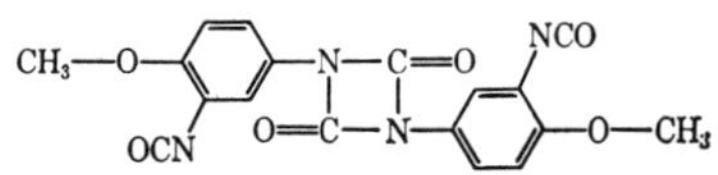				Nujol mull 0.025 cm	5.60 μ	104
	Formula $C_{18}H_{12}N_4O_6$ Mol. wt. 380.31	M.P. 197°C					92, 104
	Solvents for recrystallization	B.P.					
	Appearance						
30.	4 Phenoxy-*m*-phenylene diisocyanate dimer 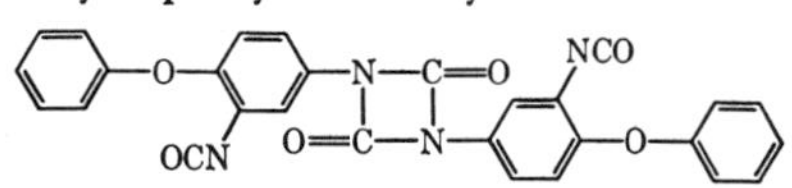				Nujol mull 0.025 cm	5.60 μ	
	Formula $C_{28}H_{18}N_4O_6$ Mol. wt. 504.44	M.P. 179–181°C					92, 104
	Solvents for recrystallization	B.P.					
	Appearance						

31. *p*-(2-Isocyanato-isobutyl)phenyl isocyanate dimer

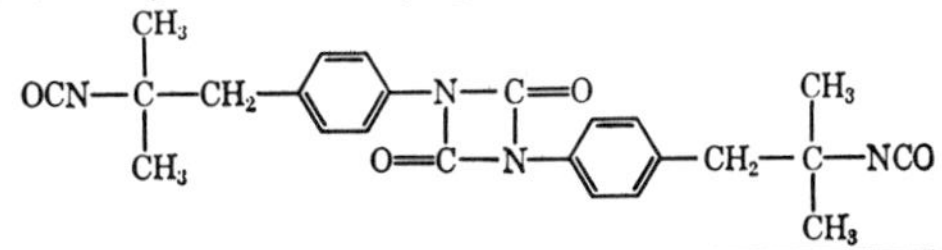

			Nujol mull	$5.60\ \mu$	104
			0.025 cm		

Formula $C_{24}H_{24}N_4O_4$ Mol. wt. 432.46

Solvents for recrystallization

Appearance

M.P. 168–169°C	92, 104
B.P.	

32. α(1-Isocyanato-cyclohexyl)-*p*-tolylisocyanate dimer

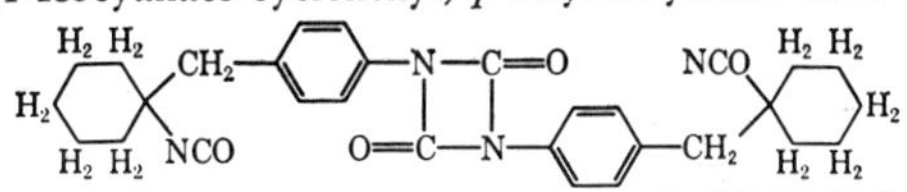

			Nujol mull	$5.60\ \mu$	92, 104
			0.025 cm		

Formula $C_{30}H_{32}N_4O_4$ Mol. wt. 512.59

Solvents for recrystallization

Appearance

M.P. 185–186°C	
B.P.	

of excess amine (100%) the reaction is rapid and four equivalents of amine are consumed per mole of dimer. This demonstrates that the uretedione ring reacts in a stepwise fashion to quantitatively consume 2 moles of di-*n*-butylamine at 160°C or higher.

Stallmann (100) patented the use of a variety of mixed aromatic–aliphatic tertiary phosphines. His remarks concerning catalysts in general are quoted as follows:

> A variety of catalysts has been suggested for the preparation of dimers, including trialkylphosphines, trialkylarsines, trialkylamines, oxalic esters and pyridine. Of these only triethyl phosphine and pyridine have been used with some success. Triethyl phosphine has the disadvantage of being very sensitive to air oxidation, so that it may be used only under a blanket of nitrogen . . . It is a highly active catalyst making the reaction difficult to control. If the temperature is allowed to rise without restraint, there are formed undesired higher reaction products, particularly isocyanuric esters. Pyridine is a very much milder catalyst and large amounts of it, up to from 1 to 5 times the weight of isocyanate, are ordinarily required to catalyze the reaction.

Stallmann's invention describes the use of mixed aromatic–aliphatic tertiary phosphine catalysts at temperatures of 0–60°C. Typical catalysts, ordinarily used in concentrations of 0.1–0.5% of the weight of isocyanate without solvent in the following examples, and yields of phenyl isocyanate dimer are tabulated as follows:

	Yield of Phenyl Dimer
Phenyl dimethyl phosphine, $C_6H_5—P—(CH_3)_2$	90–100% (?)
Phenyl di(*n*-butyl) phosphine, $C_6H_5—P—(C_4H_9)_2$	85%
Diphenyl ethyl phosphine, $(C_6H_5)_2—P—C_2H_5$	62%
4-Tolyldimethyl phosphine, $CH_3—C_6H_4—P—(CH_3)_2$	92%
2,5-Xylyldimethyl phosphine, $(CH_3)_2—C_6H_3—P—(CH_3)_2$	84%
4-Anisyldimethyl phosphine, $CH_3O—C_6H_4—P—(CH_3)_2$	96%

A number of the dimers described in the patent are included in Table 3.7 (on physical properties of dimers). Yields using specific catalysts varied depending upon the reactant (isocyanate) chosen.

Balon and Stallmann (101) patented inhibitors to control the polymerization of isocyanates, that is, the dimerization induced by tertiary phosphine catalysts, including those described by Stallmann (100). These inhibitors are alkyl halides, dialkyl sulfates, alkyl sulfates, alkyl esters of aryl and cycloaryl sulfonic acids, and aralkyl halides such as benzyl chloride. Methyl cyclohexane sulfonate and other similar compounds were preferred. These inhibitors apply only to phosphine-catalyzed polymerization and not to self-polymerization of isocyanates that apparently occur in many isocyanates on aging.

It is interesting to observe that this patent (101) states that "excessive polymerization of aliphatic and cycloaliphatic isocyanates in the presence of tertiary phosphine catalysts may be inhibited by the process of this invention." This may be interpreted to imply that aliphatic isocyanates yield dimers, which they do not, since dimerization is the sole polymerization process discussed in this patent.

Brown (102) determined the crystal structure of phenyl isocyanate dimer, prepared with pyridine and recrystallized from benzene; the abstract is quoted as follows:

> The crystal structure of the phenyl isocyanate dimer has been determined from X-ray diffraction data. A complete three-dimensional analysis has been carried out and from the resulting co-ordinates the geometry of the molecule has been thoroughly investigated. The dimer is found to be isostructural with *p*-terphenyl, $C_6H_5—C_6H_5—C_6H_5$.

The interatomic bond lengths and bond angles as well as cell dimensions were reported. These data seem to confirm the usual structure of the dimer as

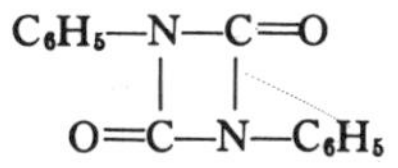

rather than that proposed by Gaylord and Crowdle (103). The latter obviously were not aware of this X-ray diffraction analysis at the time of their report. (Their manuscript was dated 12/20/54, whereas that of Brown was dated 3/24/55.)

Gaylord and Crowdle (103) proposed the following structure for phenyl isocyanate dimer:

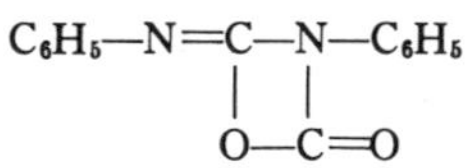

as representing some of the reactions of the dimer, particularly with Grignard reagents and with lithium aluminum hydride which yields carbanilide. Arnold et al. (55) suggest that rearrangement in solution from the original structure to that proposed above may be due to catalytic effects of the reagents.

Arnold, Nelson, and Verbanc (55), in their review of isocyanate chemistry, list 17 dimers of monoisocyanates and five dimers of diisocyanates, all of which have appeared in other publications. Their properties are included in Table 3.7. They review the reactions of dimers as disclosed in previous publications, as well as that which appeared subsequently as presented by Balon (104) and by Barthel and coauthors (105).

Balon and coauthors (104) presented a paper in which the catalytic effect of the mixed aryl–aliphatic tertiary phosphine compounds patented by Stallmann (100) is reviewed. Four new diisocyanate dimers are described (compound numbers 29–32, Table 3.6). They discuss the inhibitors of phosphine catalysts described in Balon's patent (101). They point out that

> . . . the scope of the dimerization reaction is limited to aromatic isocyanates. Aryl monoisocyanates free of *ortho* substituents yield dimers and a large number have been reported in the literature. Dimer formation is inhibited by *ortho* substitution. Thus, tolylene 2,4-diisocyanate with a methyl substituent *ortho* to one isocyanate group readily forms a simple dimer, whereas meta-phenylene diisocyanate yields a high melting amorphous polymer.

This paper also describes the preparation of carbanilides and trimers which are discussed in Section IV-B.

The dissociation of phenyl isocyanate dimer and the bis(methyl urethane) from 2,4-TDI dimer was studied by Barthel, et al. (105). They found that phenyl isocyanate dimer dissociates 89% in 6 hr at 140° in anisole solvent.

Alcohols react with dimers to give allophanates when reacted in dioxane at 80°C with triethylamine catalyst. The order of decreasing velocity was found to be methyl > isopropyl > lauryl > water. *tert*-Butyl alcohol does not react under these conditions (105). Phenol and benzyl alcohol under the same conditions give carbanilates rather than allophanates.

Phenyl isocyanate dimer reacts with aniline at 135°C to give the urea, but at room temperature it will react almost quantitatively in the presence of glacial acetic acid to give 1,3,5-triphenyl biuret. This reaction is suggested as a useful analytical reaction. Phenyl isocyanate dimer reacts with Grignard reagents and with lithium aluminum hydride to give ureas, according to Raiford and Freyermuth (96). It does not react with dry HCl, nor with hydrogen in the presence of Raney nickel or copper chromite catalysts, nor with cold alkaline or acidic $KMnO_4$. It reacts with hot aqueous $KMnO_4$ to give the urea. It can be chlorinated, brominated, and nitrated, yielding the corresponding halogenated derivatives.

Phenyl isocyanate dimer reacts with 100% sulfuric acid to give sulfanilic acid.

It would appear that the dimer of *p*-chloro phenyl isocyanate (PCPI) dissociates within 15 min at 200°C in nitrobenzene solvent according to data given in this paper.

The spectra of various uretedione compounds have been studied by Barthel et al. (105), among others. These workers have studied, mainly, the absorption frequency of the carbonyl group of the uretedione ring.

Table 3.7 lists the reported properties of 32 selected isocyanate dimers and the measured carbonyl absorption frequencies. The carbonyl frequency is reported from 5.60 to 5.65 μ. Our own work and observations indicate that the frequency range is perhaps narrower, from 5.61 to 5.63 μ.

We would expect the frequency of the uretedione carbonyl to occur at lower frequencies than other carbonyls, such as urethane, urea, isocyanurate, due to ring strain. This is in fact the case as Table 4.12, given in Chapter 4, shows.

One must exercise care in the qualitative analysis for the presence of the uretedione ring by infrared since the carbamyl chloride absorbs at the same frequency. However, known molar concentrations of the carbamyl chloride may be prepared and, after calibration, the amount of absorbance due to the carbamyl chloride found from the hydrolyzable chloride determination may be subtracted in order to determine the dimer present.

Blair and Smith (92) had difficulty in determining nitrogen in dimer compounds by the Kjeldahl method because of the apparent complete insolubility of the materials in concentrated sulfuric acid. This caused the compounds to distil or sublime from the flask, preventing digestion and causing low results. This difficulty was overcome by using dilute acid and fitting the flask with a reflux condenser and hydrolyzing the material for 72 hr, after which the condenser was removed and the analysis was continued in the usual manner.

Blair and Smith (92) had difficulty in finding a suitable solvent for the determination of the molecular weight of diphenyl diisocyanate dimer. Benzene and carbon tetrachloride were finally chosen but in both cases the solubility at low temperatures was too small to permit the use of the freezing point method. Therefore, the boiling point method was employed but the concentration was restricted to about 2.5% because of solubility. This corresponds to a boiling point elevation of less than 0.3°C, which limits the accuracy of the results.

These same problems also often occur for trimers and carbodiimides, making molecular weight determination difficult.

B. Trimers (Isocyanurates)

Isocyanates, under certain conditions, will trimerize into a ring structure consisting of alternating nitrogen and carbon atoms. This structure is

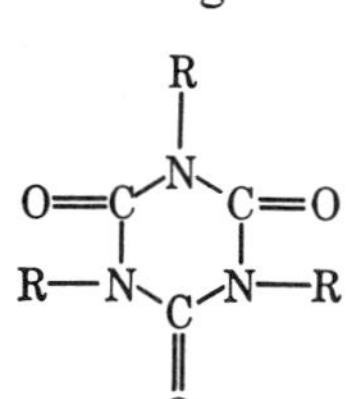

(1)

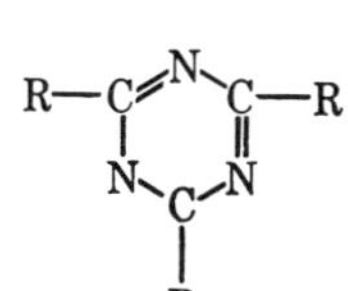

(2)

called the 1,3,5-triazine or isocyanurate ring. Two general types of derivatives based on this ring are recognized (Ref. 1, p. 344), those having substituents on the nitrogen atoms, as in 1, and those having substituents on the carbon atom (2). These two structures may be considered as trimerization products of the tautomeric forms of cyanic acid (Eq. 39).

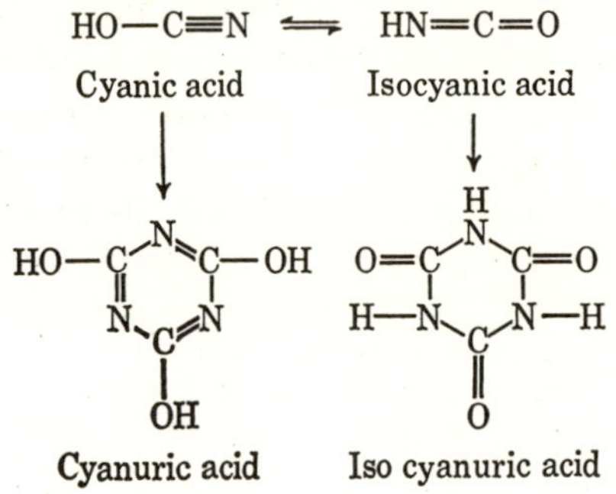

(39)

Esters derived from cyanuric acid are described by Sidgwick (1) as *O*-esters, and those from isocyanuric acid as *N*-esters. The latter compounds are derived from the isocyanates.

It is interesting to compare a few analogous structures on the two series of compounds (Eq. 40). Sidgwick (1) states that "tri-*O*-methyl ester melts

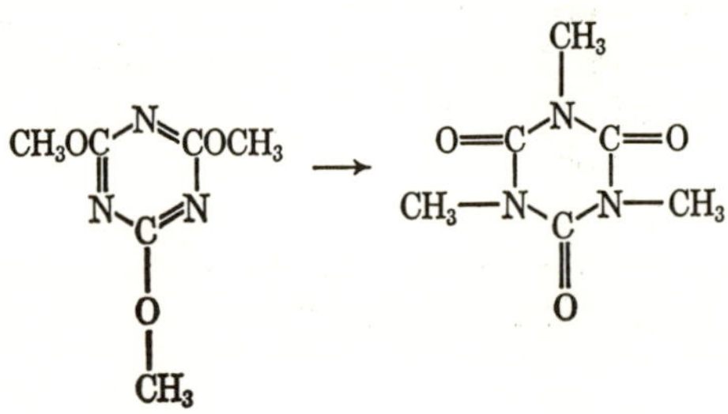

(40)

at 135° and boils at 265°, at which temperature it is slowly converted into the tri-*N*-methyl ester, a behavior common to all the *O*-esters."

The triaryl *s*-triazines are known as "kyaphenina" (sometimes spelled cyaphenin). The properties of the *N*-esters are given in Table 3.8. Cook and Jones (108) describe the preparation of a number of these kyaphenine compounds. Spencer and Beggs (109) in a recent paper describe the preparation, chromatographic separation, and infrared analytical procedures for a "monochloro," "dichloro," and "trichloro" (*para* position) and triphenyl-*s*-triazines.

Graymore (110) prepared several saturated triazine compounds, described as trialkyl trimethylene-triamines having the general formula $(\cdot CH_2 \cdot NR)_3$

by the reaction of a primary amine, (anhydrous or aqueous) with aqueous formaldehyde. Methyl, ethyl, and propyl compounds are described, together with derivatives of each, all of which are inadequately characterized.

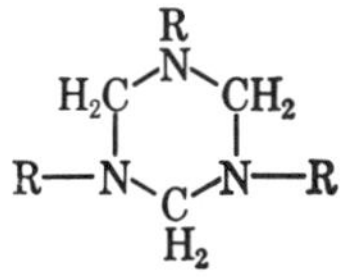

(3)

Burkus (111) utilized the triazine derivatives as prepared by Graymore (110) as catalysts for the trimerization of isocyanates particularly of phenyl isocyanate. He demonstrates their usefulness for catalysis of polyester-TDI prepolymers into stable foams.

Frazier, Little, and Lloyd (112) investigated the preparation of isocyanurate derivatives of triazine (2,4,6-(1H,3H,5H)Trione) directly from the cyanuric acid parent compound. They pointed out that the three general

TABLE 3.8

Physical Properties of Selected Trimers

1.	Phenyl isocyanuric acid
2.	1,3,5-Triphenyl isocyanurate
3.	1,3,5-Tri-*o*-tolyl isocyanurate
4.	1,3,5-Tri-*p*-tolyl isocyanurate
5.	1,3,5-Tri-*o*-methoxyphenyl isocyanurate
6.	1,3,5-Tri-*p*-methoxyphenyl isocyanurate
7.	1,3,5-Tri-*o*-nitrophenyl isocyanurate
8.	1,3,5-Tri-*p*-nitrophenyl isocyanurate
9.	1,3,5-Tri-*o*-chlorophenyl isocyanurate
10.	1,3,5-Tri-*m*-chlorophenyl isocyanurate
11.	1,3,5-Tri-*p*-chlorophenyl isocyanurate
12.	Tri-cyclohexyl isocyanurate
13.	1,3,5-Trimethyl isocyanurate
14.	1,3,5-Triisopropyl isocyanurate
15.	Trimer of 2,4-tolyene diisocyanate
16.	Trimer of 4-*sec*-butyl-*m*-phenylene diisocyanate
17.	Trimer of 2,4-cumene diisocyanate
18.	Trimer of 4-*tert*-butyl-*m*-phenylene diisocyanate
19.	Tris-(2-hydroxyethyl) isocyanurate
20.	Triallyl isocyanurate
21.	Diallyl isocyanurate
22.	Tris(carbethoxymethyl) isocyanurate
23.	Tris(2-cyanoethyl) isocyanurate
24.	Tris(2-carboxyethyl) isocyanurate

TABLE 3.8 (*Continued*)

No.	Name and structure	Analysis			Infrared data		Ref.
		Elem.	Calc.	Fd.	Solvent	Wavelength	
1.	Phenyl isocyanuric acid 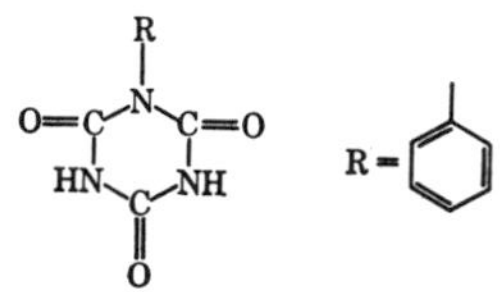						126, 127
	Formula $C_9H_7N_3O_3$ Mol. wt. 205.17	M.P. 305°C					
	Solvents for recrystallization	B.P.					
	Appearance						
2.	1,3,5-Triphenyl isocyanurate 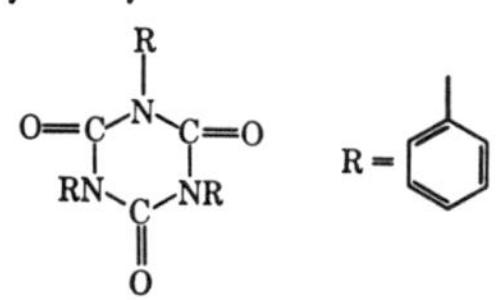	N	11.7	11.6	Nujol mull 0.025 mm	5.90 μ	111, 121, 123
	Formula $C_{21}H_{15}O_3N_3$ Mol. wt. 357.35	M.P. 280–281°C 268–270°C					
	Solvents for recrystallization Ethyl alcohol	B.P.					
	Appearance						

Compound	Analysis	Calcd.	Found	IR medium	IR bands	Ref.
3. 1,3,5-Tri-*o*-tolyl isocyanurate 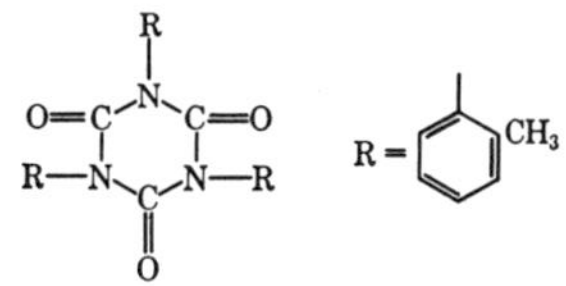	N	9.4	9.5			121
Formula $C_{24}H_{21}O_3N_3$ Mol. wt. 399.43	M.P. 180°C					
Solvents for recrystallization Ethyl alcohol	B.P.					
Appearance						
4. 1,3,5-Tri-*p*-tolyl isocyanurate 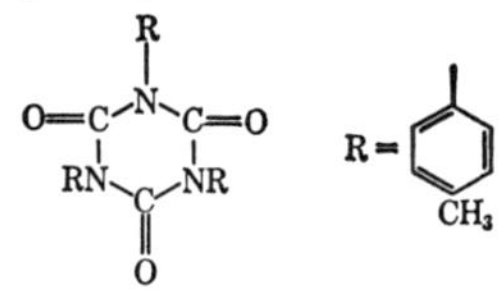	C	72.1	72.2			121
	H	5.2	5.3	Nujol mull	5.68, 5.86, and 5.94 μ	
	N	10.5	10.4			
Formula $C_{24}H_{21}O_3N_3$ Mol. wt. 399.43	M.P. 264–265°C					
Solvents for recrystallization Ethyl alcohol	B.P.					
Appearance						

(*continued*)

TABLE 3.8 (*Continued*)

No.	Name and structure	Analysis			Infrared data		Ref.
		Elem.	Calc.	Fd.	Solvent	Wavelength	
5.	1,3,5-Tri-*o*-methoxyphenyl isocyanurate	N	9.4	9.2			121
	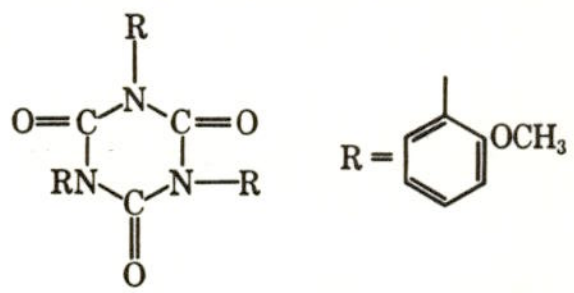				Nujol mull	5.88 μ	
	Formula $C_{24}H_{21}O_6N_3$ Mol. wt. 447.43	M.P. 261°C					
	Solvents for recrystallization Ethyl alcohol	B.P.					
	Appearance						
6.	1,3,5-Tri-*p*-methoxyphenyl isocyanurate	C	64.3	64.0			
	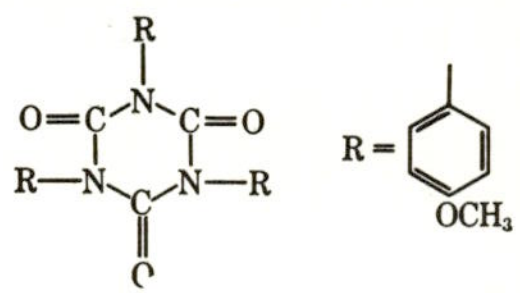	H	4.7	4.8	Nujol mull	5.70, 5.83	121
		N	9.4	9.4		5.95 μ	
	Formula $C_{24}H_{21}O_6N_3$ Mol. wt. 447.43	M.P. 259–260°C					
	Solvents for recrystallization	B.P.					
	Appearance						

7. 1,3,5-Tri-*o*-nitrophenyl isocyanurate

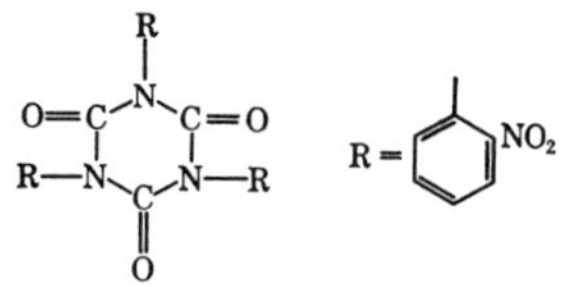

N	17.1	17.8			121

Formula $C_{21}H_{12}O_9N_6$ Mol. wt. 492.35

Solvents for recrystallization

Appearance

M.P. 260°C

B.P.

8. 1,3,5-Tri-*p*-nitrophenyl isocyanurate

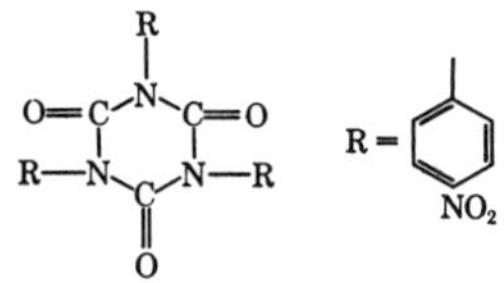

C	57.2	57.3			121
H	2.4	2.4	Nujol mull	5.67, 5.83	
N	17.1	17.2		5.94 μ	

Formula $C_{21}H_{12}O_9N_6$ Mol. wt. 492.35

Solvents for recrystallization Ethyl Alcohol

Appearance

M.P. 350°C

B.P.

9. 1,3,5-Tri-*o*-chlorophenyl isocyanurate

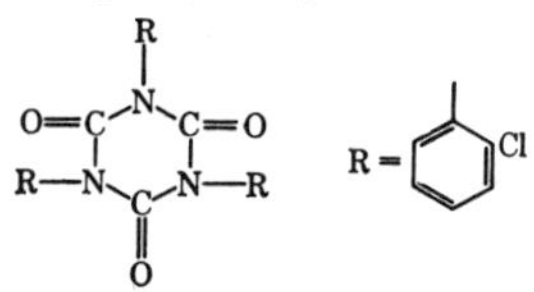

C	54.7	54.2			121
H	2.6	3.1	Nujol mull	5.88 μ	
N	9.1	9.2			

(continued)

TABLE 3.8 (*Continued*)

No.	Name and structure	Analysis			Infrared data		Ref.
		Elem.	Calc.	Fd.	Solvent	Wavelength	
	Formula $C_{21}H_{12}O_3N_3Cl_3$ Mol. wt. 354.33	M.P. 291°C					
	Solvents for recrystallization Ethyl alcohol	B.P.					
	Appearance						
10.	1,3,5-Tri-*m*-chlorophenyl isocyanurate 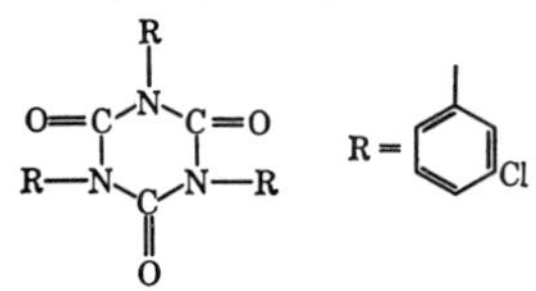	N	9.2	9.7			121
	Formula $C_{21}H_{12}O_3N_3Cl_3$ Mol. wt. 354.33	M.P. 218°C (decomposed)					
	Solvents for recrystallization Ethyl alcohol	B.P.					
	Appearance						
11.	1,3,5-Tri-*p*-chlorophenyl isocyanurate 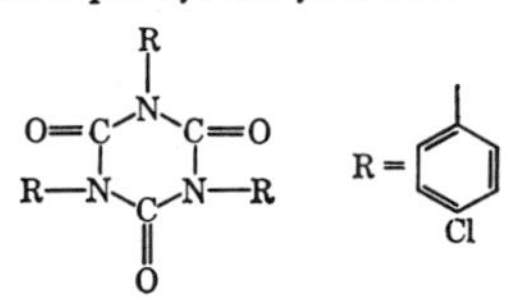	N	9.2	9.1			121
		Cl	23.1	23.1			

	Compound						Ref.
	Formula $C_{21}H_{12}O_3N_3Cl_3$ Mol. wt. 354.33	M.P. 315–316°C					
	Solvents for recrystallization Ethyl alcohol	B.P.					
	Appearance						
12.	Tricyclohexyl isocyanurate 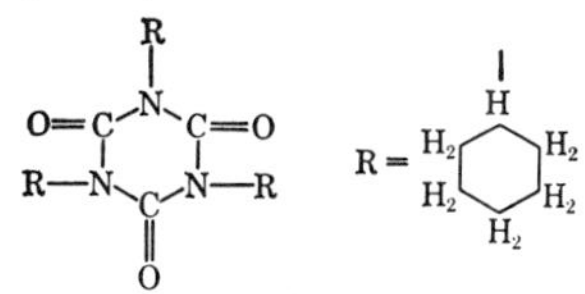	N	11.2	11.6			21
	Formula $C_{21}H_{33}O_3N_3$ Mol. wt. 375.50	M.P. 230–232°C (Impure)					
	Solvents for recrystallization	B.P.					
	Appearance						
13.	1,3,5-Trimethyl isocyanurate 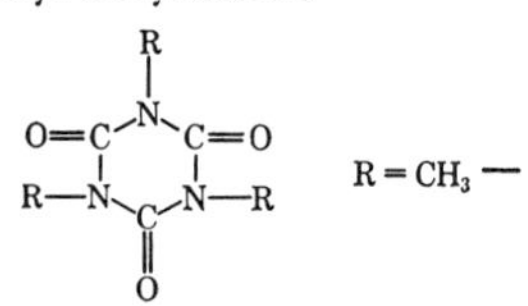						21
	Formula $C_6H_9O_3N_3$ Mol. wt. 171.16	M.P. 173–176°C (impure)					
	Solvents for recrystallization	B.P.					
	Appearance						

(continued)

TABLE 3.8 (*Continued*)

No.	Name and structure	Analysis			Infrared data		Ref.
		Elem.	Calc.	Fd.	Solvent	Wavelength	
14.	1,3,5-Triisopropyl isocyanurate 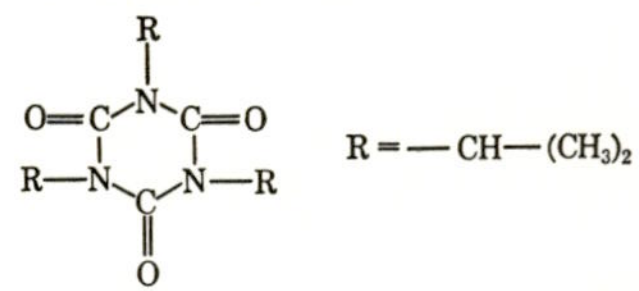	N	16.5	17.7			21
	Formula $C_{12}H_{21}O_3N_3$ Mol. wt. 255.31	M.P. 160–165°C (impure)					
	Solvents for recrystallization Ether	B.P.					
	Appearance						
15.	Trimer of 2,4-tolylene diisocyanate 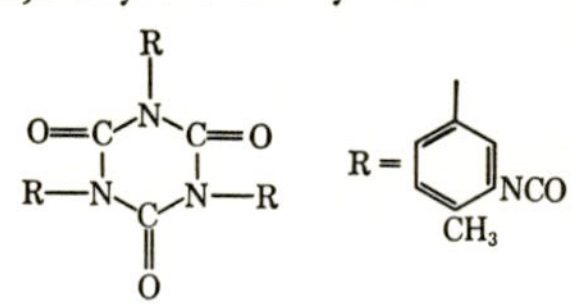	NCO	24.1	23.0		5.85 μ	122
	Formula $C_{27}H_{18}O_6N_6$ Mol. wt. 522.46	M.P. Not given					
	Solvents for recrystallization Ether	B.P.					
	Appearance						

Compound						Ref.
16. Trimer of 4-*sec*-butyl-*m*-phenylene diisocyanate 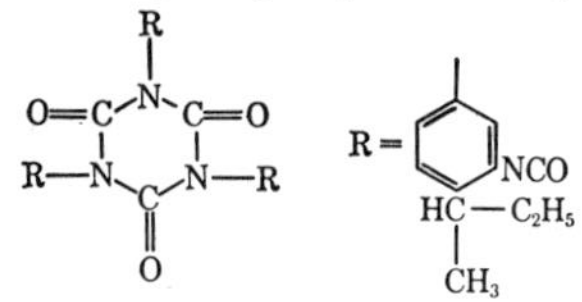						122
Formula $C_{36}H_{36}O_6N_6$ Mol. wt. 648.80	M.P. 364–367°C					
Solvents for recrystallization Ethyl acetate, carbon tetrachloride, benzene, tetrahydrofuran	B.P.					
Appearance						
17. Trimer of 2,4-cumene diisocyanate 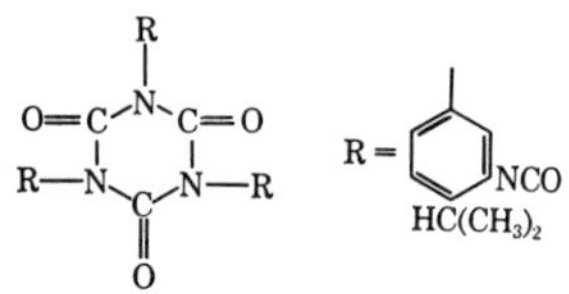						122
Formula $C_{33}H_{30}O_6N_6$ Mol. wt. 606.62	M.P. 346–350°C					
Solvents for recrystallization Ethyl acetate, carbon tetrachloride, benzene, tetrahydrofuran	B.P.					
Appearance White powder						
18. Trimer of 4-*tert*-butyl-*m*-phenylene diisocyanate 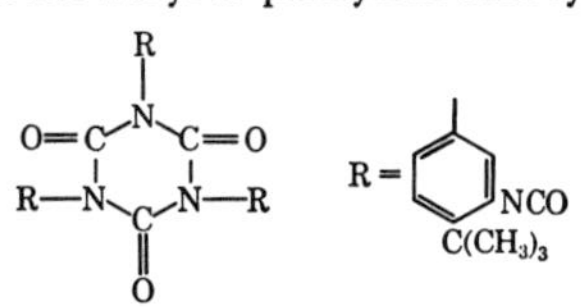	NCO	19.45	19.1			122

(*continued*)

TABLE 3.8 (*Continued*)

No.	Name and structure	Analysis: Elem.	Analysis: Calc.	Analysis: Fd.	Infrared data: Solvent	Infrared data: Wavelength	Ref.
	Formula $C_{36}H_{36}O_6N_6$ Mol. wt. 648.80	M.P. 310–311°C					
	Solvents for recrystallization Ethyl acetate, tetrahydrofuran, benzene, carbon tetrachloride, methanol	B.P.					
	Appearance Colorless needles						
19.	Tris-(2-hydroxyethyl) isocyanurate	C	41.38	41.40			112
	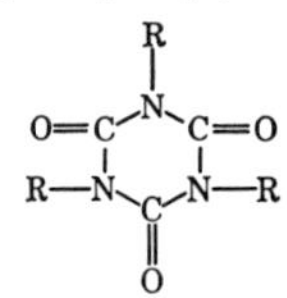 R = — CH_2CH_2OH	H	6.00	5.90			
		N	16.09	16.08			
	Formula $C_9H_{15}N_3O_6$ Mol. wt. 261.23	M.P. 134–136°C					
	Solvents for recrystallization Methanol	B.P.					
	Appearance						
20.	Triallyl isocyanurate	C	57.84	57.85			112
	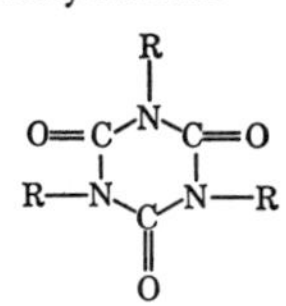 R = —$CH_2CH{=}CH_2$	H	6.06	6.40			
		N	16.86	16.86			

Formula $C_{12}H_{15}N_3O_3$ Mol. wt. 249.26	M.P. 19–22°C					
Solvents for recrystallization Benzene	B.P. 100–110°C at 0.2–0.5 mm					
Appearance						

Compound						Ref.
21. Diallyl isocyanurate	C	51.65	51.87			112
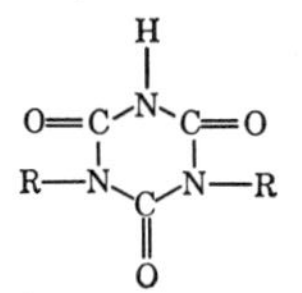 $R = —CH_2—CH{=}CH_2$	H	5.27	5.39			
	N	20.10	19.90			

Formula $C_9H_{11}N_3O_3$ Mol. wt. 209.20	M.P. 143–145°C	
Solvents for recrystallization	B.P.	
Appearance		

Compound						Ref.
22. Tris(carbethoxy methyl) isocyanurate	C	46.51	46.65			112
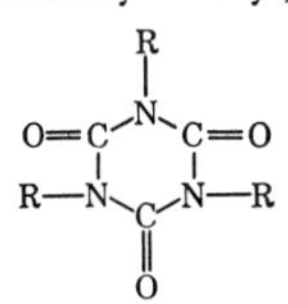 $R = —CH_2COOC_2H_5$	H	5.4	5.89			
	N	10.85	10.72			

Formula $C_{15}H_{21}N_3O_9$ Mol. wt. 387.34	M.P. 71–78°C	
Solvents for recrystallization Ethanol	B.P. 210–220°C at 0.45 mm	
Appearance		

(contnued)

TABLE 3.8 (*Continued*)

No.	Name and structure	Analysis			Infrared data		Ref.
		Elem.	Calc.	Fd.	Solvent	Wavelength	
23.	Tris(2-cyanoethyl) isocyanurate	C	50.00	50.01			112
	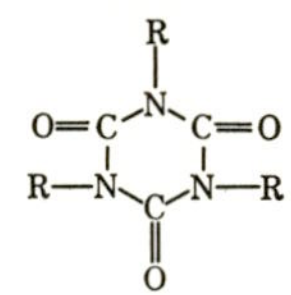 R = —CH_2CH_2CN	H	4.17	4.26			
		N	29.18	29.10			
	Formula $C_{12}H_{12}N_6O_3$ Mol. wt. 288.26	M.P. 228–230°C					
	Solvents for recrystallization Dimethylformamide	B.P.					
	Appearance						
24.	Tris(2-carboxyethyl) isocyanurate	C	41.75	41.87			112
	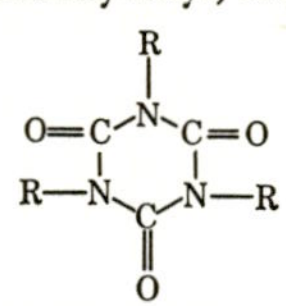 R = —CH_2CH_2COOH	H	4.35	4.39			
		N	12.17	12.10			
	Formula $C_{12}H_{15}N_3O_9$ Mol. wt. 345.26	M.P. 228–229°C					
	Solvents for recrystallization Water	B.P.					
	Appearance						

methods for the preparation of isocyanurates are: (*1*) thermal rearrangement of cyanurates (113,114), (*2*) trimerization of certain alkyl or aryl isocyanates (115,116), and (*3*) condensation of alkyl halides with certain salts of cyanuric acid (117–119).

Frazier, Little, and Lloyd (112) condensed ethylene oxide, allyl chloride ethyl chloroacetate, and acrylonitrile with cyanuric acid under alkaline conditions, to obtain the corresponding isocyanurates.

Bortnick (21) has described the preparation of trimethyl isocyanurate from methyl amine and dimethyl carbonate, and partially characterized isopropyl and cyclohexyl triisocyanurates. These were obtained as residues in their base-catalyzed pyrolysis of *N*-alkyl carbamates to give the corresponding isocyanates.

Kogan (120) states that the metal carboxylates catalyze the formation of trimers from phenyl isocyanate at room temperature, although no data are given to support this conclusion. He shows that the same carboxylates catalyze the reaction of phenyl isocyanate and ethyl carbanilate to form allophanates at room temperature. The isocyanurate of phenyl isocyanate was prepared using the alkaline catalyst *N*-methyl morpholine (121).

Kogon (121) studied the infrared absorption of a number of substituted aryl isocyanurates and concluded that the tri-*ortho*-substituted compounds have a single carbonyl absorption band at 5.87–5.90 μ, whereas the tri-*para*-substituted isocyanurates have three main absorption peaks at 5.66–5.70, 5.83–5.86, and 5.94–5.95 μ.

Balon (122) patented the preparation and gives the properties of the trimers of 4-*tert*-butyl-*meta*-phenylene diisocyanate, 4-*sec*-butyl-*meta*-phenylene diisocyanate, 2,4-toluene diisocyanate, and cumene-2,4-diisocyanate, using tri-*n*-butyl phosphine catalyst. Sodium benzoate in dimethyl formamide was also used as a catalyst.

Burkus (123,124) in his study of the catalysis of the reaction of phenyl isocyanate with butanol-1 by pyridine *N*-oxide, states that this catalyst in combination with propylene oxide also catalyzes the trimerization of phenyl isocyanate. This is in accord with a previous report by Jones and Savill (125) that the combination of a tertiary amine and an alkylene oxide catalyzes the trimerization.

The isocyanurates are normally called trimers, which may be misleading at times, since under proper conditions and catalyst a polymer reaction may occur due to the coupling of many isocyanurate rings. These compounds exhibit the characteristic spectra of the isocyanurate carbonyl group in the infrared, the main band of which occurs at 5.85–5.92 μ.

Table 3.8 lists the reported properties of 24 selected trimers and the measured carbonyl frequencies, when reported.

C. Carbodiimides

These compounds are derivatives of the hypothetical diimide of carbonic acid. The structural relationship between carbonic acid, urea, and carbodiimides are shown. The diimide is isomeric with cyanamide.

Sidgwick's *Organic Chemistry of Nitrogen* (1) states that "two methods of

Carbonic acid — Diamide of carbonic acid (urea) — Diimide of carbonic acid (carbodiimides) — Cyanamide

preparation are known," (*1*) by the action of mercuric oxide on *sym*-disubstituted thioureas in dry benzene and (*2*) by the action of ethyl isothiocyanate, ethyl isocyanate, or CO_2 on triethyl phosphene ethylimine.

Carbodiimides enter easily into addition reactions. Thus, substituted ureas are formed by the addition of water (in the presence of HCl) to the carbodiimides.

An extensive review of the literature is available in "The Chemistry of Carbodiimides" by H. G. Khorana (128). References are made to some 80 carbodiimides prepared by the desulfurization of thioureas. The chemical properties and uses are discussed at length. The carbodiimides have strong absorption bands at 4.75–4.77 μ, according to Khorana.

Stolle (129) prepared diphenylcarbodiimide by heating phenyl isocyanate in a sealed tube at 180°C for 24 hr and stated, "Although interesting, the reaction obviously is of no preparative value." Contrary to this statement, the revived interest in carbodiimides, as evidenced by the several references described below, is due to the ease with which they are prepared from isocyanates.

Other carbodiimides are reported in the early literature. Thus, Huhn (130) prepared the diimides of phenyl, *p*-tolyl, and phenyl-*o*-tolyl using the corresponding thioureas and mercuric oxide. These were converted to ureas and guanidine derivatives.

Dyer and Newborn (131) included in their study of the thermal degradation of carbamates a pyrolysis study of 4,4′-diisocyanato-diphenylmethane (MDI). They report that MDI, heated at 300°C under dry nitrogen for 45 min, yielded 92% carbon dioxide, and a nonvolatile residue which was polymeric, insoluble in common solvents, and infusible. It contained the infrared spectral lines characteristic of the carbodiimides (4.7–4.8 μ).

Smeltz (132) patented the process for making the carbodiimide of

2,4-TDI using 2-ethyl-3-methyl-phospholene-1 oxide catalyst. He describes previous methods for making carbodiimides by treating disubstituted thiourea with a metallic oxide, such as mercuric oxide, or with an aqueous alkaline solution of a hypohalite. These products tend to be contaminated with ureas from the reaction of the carbodiimide with water, that is:

$$C_6H_5{-}N{=}C{=}N{-}C_6H_5 + H_2O \rightarrow C_6H_5{-}NH{-}\overset{\overset{\displaystyle O}{\|}}{C}{-}NH{-}C_6H_5 \qquad (41)$$

The carbodiimide of 2,4-TDI, di(3-isocyanato-4-methylphenyl) carbodiimide, contains free isocyanate groups at the 2-position. The solvent used must be one which will dissolve the isocyanate but not the carbodiimide, for example petroleum ether, and must not react with the free isocyanate group. This carbodiimide will react with ethanol at 78° to give a diurethane which has characteristic infrared bands. It will also react with polyethylene glycol (MW 550) to give a diurethane.

Campbell and Verbanc (133) and Balon (137) describe the preparation of carbodiimides from monoisocyanates, diisocyanates, and mixed isocyanates to yield polymeric materials having a carbodiimide linkage using catalysts such as:

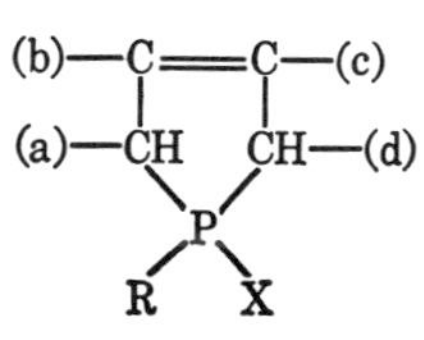

Phospholenes

where a, b, c, and d are radicals selected from H, halogen, lower alkyl, lower alkenyl phenyl, cyclohexyl, and polymethylene groups, R = lower alkyl or phenyl group, and X = chalcogen of atomic weight 15–33.
or phospholidines, such as:

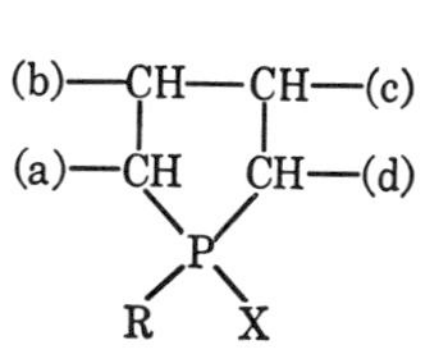

where a, b, c, d, R, and X are defined above.

where a, b, c, d, R, and X are defined above.

Isocyanate terminated polymers may be reacted further to incorporate a carbodiimide linkage by means of these catalysts. Campbell and Verbanc (133) give as an example the polymer prepared from polytetramethylene ether glycol and 2,4,6-triisocyanato toluene. This is reacted further in the presence of the catalyst to yield carbodiimide polymers having cellular (foam) properties.

Dyer and Read (134) studied the decomposition of 1-naphthyl isocyanate. In the early stages, when heated at 255–275°C, a high content of carbodiimide was observed by infrared absorption. This catalyzed the decomposition and an overall conversion to carbon dioxide of 90% or better was observed. Dinaphthylurea likewise catalyzed the reaction. The residue (after 90% conversion to CO_2) contained a mixture of yellow substances, the major component of which was a compound melting at 167–169° which appeared to be an unstable dimer of the carbodiimide.

Samples (5–20 g) were heated in a tube with N_2 flushing. The evolved gases were cooled to −78° to remove organic vapors and CO_2 was absorbed in Ascarite. Solid residues were separated by solubility differences and chromatography on alumina. Liquid residues were separated by distillation or vapor-phase chromatography.

Dyer and Read (135) show that isocyanates (1-naphthyl and MDI) can be converted to carbodiimides at relatively low temperatures (135° boiling xylene) by use of aluminum isopropoxide, or preferably by manganese naphthanate as catalyst. The naphthyl carbodiimide is a well-defined compound, whereas that from MDI is a polymeric mixture.

The mechanism of the formation of carbodiimides using metal alcoholates is postulated to be the formation of metal complexes with isocyanates. Campbell and Monagle (136) briefly describe the preparation of carbodiimides by the use of a unique catalyst, 1-ethyl-3-methyl-3-phospholene-1-oxide (**4**), which is not only effective to produce monocarbodiimides, but

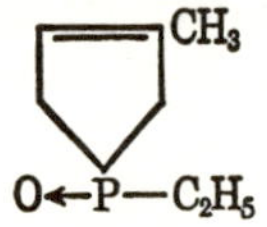

(**4**)

gives a tough high molecular weight condensation polymer from diisocyanates. They describe the carbodiimides of *p*-nitrophenyl isocyanate and the polymeric material from MDI.

The physical properties of a selected group of 19 carbodiimides are listed in Table 3.9.

It should be noted that the carbodiimides absorb between 4.72 and 4.77 μ, which is characteristic of the —N=C=N— group.

TABLE 3.9
Physical Properties of Selected Carbodiimides

1.	Diallyl carbodiimide
2.	Diamyl carbodiimide
3.	Dibutyl carbodiimide
4.	Diethyl carbodiimide
5.	1,3-Di-*tert*-octyl carbodiimide
6.	Dipropyl carbodiimide
7.	Diphenyl carbodiimide
8.	Di-*p*-chlorophenyl carbodiimide
9.	Di-*p*-methoxyphenyl carbodiimide
10.	Di-*p*-nitrophenyl carbodiimide
11.	Di-*p*-tolylphenyl carbodiimide
12.	Phenyl-*o*-tolyl carbodiimide
13.	Phenyl-*p*-tolyl carbodiimide
14.	Di-α-naphthyl carbodiimide
15.	Di-β-naphthyl carbodiimide
16.	Poly(hexamethylene diisocyanate) carbodiimide
17.	Di(3-isocyanato-4-methylphenyl) carbodiimide (2,4-TDI carbodiimide)
18.	Poly methylene bis(4-phenyl) carbodiimide
19.	Dicyclohexyl carbodiimide

TABLE 3.9 (*Continued*)

No.	Name and structure	Analysis: Elem.	Analysis: Calc.	Analysis: Fd.	Infrared data: Solvent	Infrared data: Wavelength	Ref.
1.	Diallyl carbodiimide						130, 137
	$CH_2{=}CH{-}CH_2{-}N{=}C{=}N{-}CH_2{-}CH{=}CH_2$						
	Formula $C_7H_{10}N_2$ Mol. wt. 222.17	M.P.					
	Solvents for recrystallization	B.P. 58–59°C at 10 mm					
	Appearance Colorless liquid						
2.	Diamyl carbodiimide						135, 138
	$CH_3{-}(CH_2)_4{-}N{=}C{=}N{-}(CH_2)_4{-}CH_3$						
	Formula $C_{11}H_{22}N_2$ Mol. wt. 182.30	M.P.					
	Solvents for recrystallization	B.P. Not given					
	Appearance						
3.	Dibutyl carbodiimide						130, 137
	$C_4H_9{-}N{=}C{=}N{-}C_4H_9$						

Compound						Ref.
Formula $C_9H_{18}N_2$ Mol. wt. 154.25	M.P.					
Solvents for recrystallization	B.P. 83–85°C at 10 mm					
Appearance Colorless liquid						
4. Diethyl carbodiimide						135, 138
$C_2H_5—N=C=N—C_2H_5$					4.72 μ	
Formula $C_5H_{10}N_2$ Mol. wt. 98.15	M.P.					
Solvents for recrystallization	B.P. Not given					
Appearance						
5. 1,3-Di-*tert*-octyl carbodiimide	C	76.6	76.9			136
	N	12.9	12.8			
$(CH_3)_3CCH_2C(CH_3)_2—N=C=N—C(CH_3)_2CH_2C(CH_3)_3$	N	10.5	10.5			
Formula $C_{17}H_{34}N_2$ Mol. wt. 266.46	M.P.					
Solvents for recrystallization	B.P. 110–112°C at 3 mm					
Appearance						
6. Dipropyl carbodiimide						130, 137
$C_3H_7—N=C=N—C_3H_7$						

(continued)

TABLE 3.9 (*Continued*)

No.	Name and structure	Analysis: Elem.	Analysis: Calc.	Analysis: Fd.	Infrared data: Solvent	Infrared data: Wavelength	Ref.
	Formula $C_7H_{14}N_2$ Mol. wt. 126.20	M.P.					
	Solvents for recrystallization	B.P. 56–58°C at 11 mm					
	Appearance Nearly colorless liquid						
7.	Diphenyl carbodiimide 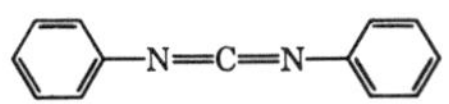						135, 138
	Formula $C_{13}H_{10}N_2$ Mol. wt. 194.23	M.P.					
	Solvents for recrystallization	B.P. 107–121° at 0.3 mm					
		110–120° at 0.5 mm					
	Appearance Pale green, slightly viscous liquid	171–175° at 12 mm					
8.	Di-*p*-chlorophenyl carbodiimide 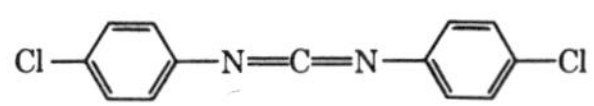						135, 138
	Formula $C_{13}H_8N_2Cl_2$ Mol. wt. 253.12	M.P. not given					130, 137
	Solvents for recrystallization	B.P.					
	Appearance						

Compound		Calc.	Found			Ref.
9. Di-*p*-methoxyphenyl carbodiimide	C	54.9	54.8			
	H	2.81	2.88			
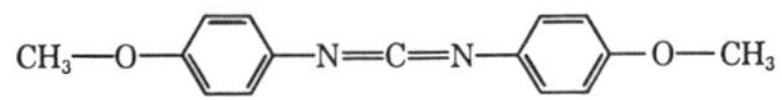	N	19.7	19.9		4.76 μ	
Formula $C_{15}H_{14}O_2N_2$ Mol. wt. 254.28	M.P. 52–53°C					
Solvents for recrystallization Petroleum ether	B.P. 192–194°C at 1.3 mm					135, 138
Appearance Pale yellow distillate						
10. Di-*p*-nitrophenyl carbodiimide						135, 138
						133,
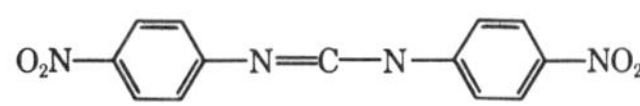						
Formula $C_{13}H_8O_4N_4$ Mol. wt. 284.23	M.P. 164–166°C					
Solvents for recrystallization Petroleum ether	B.P. Not given					
Appearance						
11. Di-*p*-tolylphenyl carbodiimide					4.7 μ	130, 137
	C	81.1	80.9			135, 138
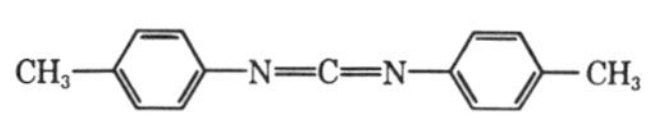	H	6.3	6.2			
	N	12.6	12.4			
Formula $C_{15}H_{14}N_2$ Mol. wt. 222.28	M.P. 56.5–57.5°					135, 138
Solvents for recrystallization Petroleum ether	B.P. 165°C at 2.4 mm					130, 137
Appearance White crystals						130, 137

(continued)

TABLE 3.9 (*Continued*)

No.	Name and structure	Analysis			Infrared data		Ref.
		Elem.	Calc.	Fd.	Solvent	Wavelength	
12.	Phenyl-*o*-tolyl carbodiimide 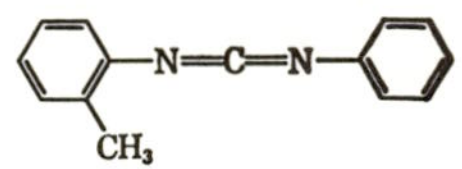	C	80.77	80.61			131
		H	5.77	5.65			
	Formula $C_{14}H_{12}N_2$ Mol. wt. 208.25	M.P. 320–325°C					
	Solvents for recrystallization	B.P.					
	Appearance						
13.	Phenyl-*p*-tolyl carbodiimide 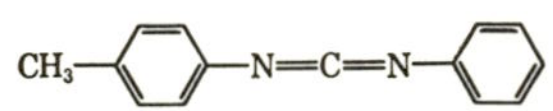	C	80.77	80.41			131
		H	5.77	6.03			
	Formula $C_{14}H_{12}N_2$ Mol. wt. 208.25	M.P. 325–330°C					
	Solvents for recrystallization	B.P.					
	Appearance						
14.	Di-α-naphthyl carbodiimide 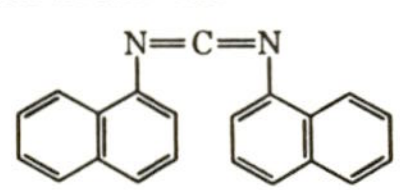	C	85.71	85.75			131
		H	4.76	4.86			
						4.75 μ	134

Compound						Ref.
Formula $C_{21}H_{14}N_2$ Mol. wt.	M.P. 93–94°C					131
	89–90°C					134
Solvents for recrystallization Petroleum ether and ether	B.P.					
Appearance						
15. Di-β-naphthyl carbodiimide 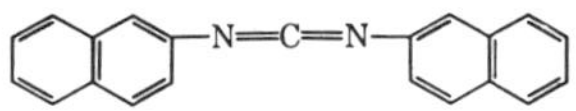	C	85.71	85.18			
	H	4.76	5.24			
Formula $C_{21}H_{14}N_2$ Mol. wt.	M.P. 145–146°C					131
Solvents for recrystallization Benzene and ether	B.P.					
Appearance						
16. Poly(hexamethylene diisocyanate) carbodiimide $[OCN—(CH_2)_6—N=C=N—(CH_2)_6—NCO]_n$						138
Formula Mol. wt.	M.P. Not given					
Solvents for recrystallization	B.P.					
Appearance						
17. Di(3-isocyanato-4-methylphenyl) carbodiimide (2,4-TDI carbodiimide) 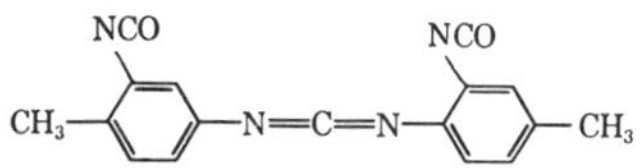						134, 135

(continued)

TABLE 3.9 (*Continued*)

No.	Name and structure	Analysis: Elem.	Calc.	Fd.	Infrared data: Solvent	Wavelength	Ref.
	Formula $C_{17}H_{12}O_2N_4$ Mol. wt. 304.30	M.P. 113–115°C					
	Solvents for recrystallization Hexane, benzene, chloroform, pet. ether	B.P.					
	Appearance						
18.	Polymethylene bis(4-phenyl) carbodiimide 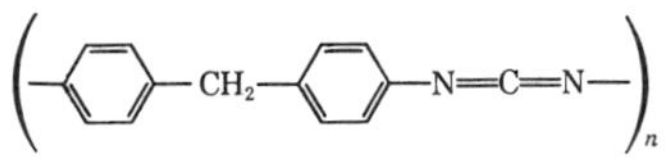						131, 133
	Formula Mol. wt.	M.P. Not given					
	Solvents for recrystallization	B.P.					
	Appearance						
19.	Dicyclohexyl carbodiimide 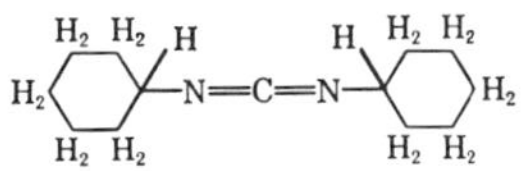					4.75 μ	130, 137
	Formula $C_{13}H_{22}N_2$ Mol. wt. 206.32	M.P.					130, 137
	Solvents for recrystallization	B.P. 150–152°C					
	Appearance Colorless liquid						

References

1. Sidgwick, N. V., *Organic Chemistry of Nitrogen*, Oxford at the Clarendon Press, London, England, 1945, Chap. IX.
2. Saunders, J. H., and Slocombe, R. J., *Chem. Rev.*, **43,** 203 (1948).
3. Saunders, J. H., and Frisch, K. C., *Polyurethanes, Chemistry and Technology*, Part I, Interscience, New York, 1962.
4. Curtius, T., Brüning, G. v., and Derlon, H., *J. Prakt. Chem.*, **125,** 63 (1930); *Chem. Abstr.* **24**:3230 (1930).
5. Curtius, T., and Hechtenberg, W., *J. Prakt. Chem.*, **105,** 289 (1923); *Chem. Abstr.* **17**:2868[6] (1923).
6. Blomquist, A. T., and Stevenson, H. B., *J. Am. Chem. Soc.*, **56,** 146 (1934).
7. Schroeter, G., *Ber.*, **42,** 3356 (1909); *Chem. Abstr.* **4**:48 (1910).
8. Pyman, F. L., *J. Chem. Soc.*, **103,** 852 (1913).
9. Wurtz, A., *Ann.*, **71,** 326 (1849).
10. Slotta, K. H., and Lorenz, L., *Ber.*, **58B,** 1320 (1925); *Chem. Abstr.* **19**:3083 (1925).
11. Hill, A. J., and Degnan, W. M., U.S. Pat. 2,379,486 (to American Cyanamid) (July 3, 1945).
12. Anderson, H. H., *J. Am. Chem. Soc.*, **72,** 193 (1950).
13. Anderson, H. H., *J. Am. Chem. Soc.*, **72,** 196 (1950).
14. Forbes, G. S., and Anderson, H. H., *J. Am. Chem. Soc.*, **70,** 1043, 1222 (1948).
15. Schweitzer, C. E., U.S. Pat. 2,409,712 (to du Pont) (Oct. 22, 1946).
16. Hofmann, A. W., *Ber.*, **3,** 653 (1870).
17. Wenker, H., *J. Am. Chem. Soc.*, **58,** 2608 (1936).
18. Hoshino, T., Mukaiyama, T., and Hoshino, H., *J. Am. Chem. Soc.*, **74,** 3097 (1952).
19. Mukaiyama, T., and Matsunaga, T., *J. Am. Chem. Soc.*, **75,** 6209 (1953).
20. Mukaiyama, T., and Hoshino, Y., *J. Am. Chem. Soc.*, **78,** 1946 (1956).
21. Bortnick, N., Luskin, L. S., Hurwitz, M. D., and Rytina, A. W., *J. Am. Chem. Soc.*, **78,** 4358 (1956).
22. Bennet, W. B., Saunders, J. H., and Hardy, E. E., *J. Am. Chem. Soc.*, **75,** 2101 (1953).
23. Slocombe, R. J., and Saunders, J. H., U.S. Pat. 2,773,086 (to Monsanto) (Dec. 4, 1956).
24. Gattermann, L., *Ber.*, **23,** 1218 (1890).
25. Gattermann, L., and Cantzler, A., *Ber.*, **25,** 1086 (1892).
26. Gautier, A., *Ann.*, **149,** 311 (1869).
27. Kühn, B., and Liebert, M., *Ber.*, **23,** 1536 (1890).
28. Mukaiyama, T., Tokizawa, M., Nohira, H., and Takei, H., *J. Org. Chem.*, **26,** 4381 (1961).
29. Hentschel, W., *Ber.*, **17,** 1284 (1884).
30. Gattermann, L., and Schmidt, G., *Ann.*, **244,** 30 (1888).
31. I. G. Farbenindustrie A.G., Brit. Pat. 462,182 (Mar. 3, 1937).
32. Greenhalgh, R., Piggott, H. A., Brit. Pat. 483,308 (to Imperial Chemical Industries, Ltd.), (Apr. 14, 1938).
33. Kropa, E. L., and Nyquist, A. S., U.S. Pat. 2,468,713 (to American Cyanamid Co.) (Apr. 26, 1949).
34. Slocombe, R. J., Hardy, E. E., and Saunders, J. H., U.S. Pat. 2,480,089, (to Monsanto Chemical Co.) (Aug. 23, 1949).
35. Hardy, E., U.S. Pat. 2,480,088 (to Monsanto Chemical Co.) (Aug. 23, 1949).
36. Hrubesch, A., Ger. Pat. 801,993 (to I. G. Farbenindustrie Akt.-Ges. "In Auflosung) (Feb. 1, 1951).
37. Allen, T. C., and Chadwick, D. H., U.S. Pat. 2,733,254 (to Monsanto Chemical Co.) (Jan. 31, 1956).

38. Goodyear Tire and Rubber Co., Brit. Pat. 749,496 (May 23, 1956).
39. Gemassmer, A., Ger. Pat. 926,845 (to Farbenfabriken Bayer A.G.) (Apr. 25, 1955).
40. Böhme, O., Mott, F., Pfirschke, J., and Wollthan, H., Ger. Pat. 949,228 (to Farbenfabriken Bayer Akt-Ges.) (Sept. 13, 1956).
41. Wollthan, H., Ger. Pat. 952,086 (to Farbenfabriken Bayer Akt.-Ges) (Nov. 8, 1956).
42. Pfirschke, J., Altner, W., and Roos, H., Ger. Pat. 955,860 (to Farbenfabriken Bayer Akt.-Ges.) (Jan. 10, 1957).
43. Tazuma, J. J., U.S. Pat. 2,884,435 (to Food Machinery and Chemical Corp.) (Apr. 28, 1959).
44. Skiles, B. F., U.S. Pat. 2,908,704, (to E. I. du Pont de Nemours and Co.), (Oct. 13, 1959).
45. Latourette, H. K., and Johnson, O. H., U.S. Pat. 2,908,703 (Oct. 13, 1959).
46. Farbenfabriken Bayer A.-G., Brit. Pat. 901,377 (July 18, 1962).
47. Siefken, W., *Ann.*, **562,** 75 (1949). page 99.
48. Slocombe, R. J., Hardy, E. E., Saunders, J. H., and Jenkins, R. L., *J. Am. Chem. Soc.*, **72,** 1888 (1950).
49. Slocombe, R. J., Flores, H. W., and Cleveland, T. H., U.S. Pat. 2,680,127 (to Monsanto Chemical Co.) (June 1, 1954).
50. Vaganay, J., and Wevert, Simone, Fr. Pat. 1,071, 629 (to Etat Francais) (Sept. 2, 1954).
51. Kantyka, T. A., and Toyne, C., Brit. Pat. 806,903 (to Imperial Chemical Industries Ltd.) (Jan. 7, 1959).
52. Irwin, C. F., U.S. Pat. 2,683,160 (to E. I. du Pont de Nemours and Co.) (July 6, 1954).
53. Wingfoot Corp., Ger. Pat. 946, 138 (July 26, 1956).
54. Petersen, S., and Piepenbrink, H. F., in *Houben-Weyl, Methoden der organischen chemie*, Vol. 8, E. Müller, Ed., Thieme-Verlag, Stuttgart, 1952, p. 75.
55. Arnold, R. G., Nelson, J. A., and Verbanc, J. J., *Chem. Rev.*, **57,** 47 (1957).
56. Heiss, H. L., Combs, F. P., Gemeinhardt, P. G., Saunders, J. H., and Hardy, E. E., *Ind. Eng. Chem.*, **51,** 929 (1959).
57. Saunders, J. H., and Frisch, K. C., *Polyurethanes, Chemistry and Technology*, Part II, *Technology*, Interscience, New York, 1964.
58. Kane, R. P., Kogan, I. C., and Bradley, A. W., Paper presented before the Division of Rubber Chemistry, American Chemical Society, Cincinnati, Ohio, May, 1958.
59. Malec, E. J., and David, D. J., Mobay Method for the Determination of %NCO and Amine Equivalent, March 8, 1965, Mobay Chemical Co., unpublished.
60. Stagg, H. E., *Analyst*, **71,** 557 (1946).
61. American Society for Testing Materials, "Testing Urethane Foam Raw Materials," ASTM-D1638-61T.
62. Siggia, S., and Hanna, J. G., *Anal. Chem.*, **20,** 1084 (1948).
63. Society of the Plastics Industry, Raw Materials Subcommittee, "Proposed Method for the Determination of Amine Equivalent or NCO Content of Crude or Modified Isocyanates," August 1965.
64. Sherwood, N., Malec, E. J., and Christman, R., "Mobay Method for the Determination of Acidity," September 28, 1959, Mobay Chemical Co., unpublished.
65. Society of the Plastics Industry, Raw Materials Subcommittee, "Proposed Method for the Determination of Acidity in Crude or Modified Isocyanates," November, 1965.
66. Sherwood, N., Malec, E. J., and Christman, R., "Mobay Method for the Determination of Hydrolyzable Chlorides," March 9, 1960, Mobay Chemical Co., unpublished.
67. Malec, E. J., "Mobay Method for the Determination of Total Chlorides," August 7, 1963. Mobay Chemical Co., unpublished.
68. *BIOS Final Report* No. 719 (PB 45246).

69. Felton, H. R., in *Gas Chromatography*, (Instr. Soc. Am. Symposium, Aug., 1957) V. J. Coates, H. J. Noebels, and I. S. Fagerson, Eds., Academic Press, New York, 1958, p. 141.
70. Kjaer, A., and Jart, A., *Acta Chem. Scand.*, **11**, 1423 (1957); *Chem. Abstr.*, **52**, 9959e (1958).
71. Vagelos, P. R., Vanden Heuvel, W. J. A., and Horning, M. G., *Anal. Biochem.*, **2**, 50 (1961); *Chem. Abstr.*, **55**, 13523b (1961).
72. Jart, A., *Acta Chem. Scand.*, **15**, 1223 (1961); *Chem. Abstr.*, **56**, 10880a (1961).
73. Siok, J., and David, D. J., "Mobay Method for the Determination of Impurities in TDI by Gas Liquid Chromatography," January 8, 1960, Mobay Chemical Co., unpublished.
74. Dixon, R. N., *Phil. Trans. Roy. Soc. London, Ser. A*, **252**, 165 (1960).
75. Mangini, A., and Modena, G., *Boll. Sci. Fac. Chim. Ind. Bologna*, **16**, 18 (1958); *Chem. Abstr.*, **53**, 14682c (1959).
76. Holland, R., Style, D. W. G., Dixon, R. N., and Ramsay, D. A., *Nature*, **182**, 336 (1958).
77. David, D. J., *Anal. Chem.*, **35**, 37 (1963).
78. Eyster, E. H., and Gillette, R. H., *J. Chem. Phys.*, **8**, 369 (1940).
79. Davison, W. H. T., *J. Chem. Soc.*, **1953**, 3712.
80. Hoyer, H., *Chem. Ber.*, **89**, 2677 (1956).
81. Hirschmann, R. P., Kniseley, R. N., and Fassel, V. A., *Spectrochim. Acta*, **21**, (12), 2125 (1965). *Chem. Abstr.* **64**:4447e (1966).
82. Kopper, H., and Pongratz, A., *Monatsh. Chem.*, **62**, 78 (1933). *Chem. Abstr.* **27**:2382[1] (1933).
83. Thomas, W. J. O., *Chem. Ind. (London)*, **1953**, 567.
84. McKay, O. E., "Mobay Method for the Determination of Isomer Ratio in TDI," June 4, 1963, Mobay Chemical Co., unpublished.
85. Lord, S. S., Jr., *Anal. Chem.*, **29**, 497 (1957).
86. Steingiser, S., Darr, W. C., and Hardy, E. E., *Anal. Chem.*, **31**, 1261 (1959).
87. Hofmann, A. W., *Ber.*, **3**, 765 (1870).
88. Dennstedt, M., *Ber.*, **13**, 228 (1880).
89. Snape, H. L., *J. Chem. Soc.*, **49**, 254 (1886).
90. Hale, W. J., *J. Am. Chem. Soc.*, **41**, 370 (1919).
91. Hale, W. J., and Lange, N. A., *J. Am. Chem. Soc.*, **41**, 378 (1919).
92. Blair, J. S., and Smith, G. E. P., Jr., *J. Am. Chem. Soc.*, **56**, 907 (1934).
93. Raiford, C. L., and Shelton, J. R., *J. Org. Chem.*, **4**, 204 (1939).
94. Warren, W. H., and Wilson, F. E., *Ber.*, **68**, 957 (1935).
95. *Chemcial Abstracts*, Vol. 33, Index page 11101 (1939).
96. Raiford, C. L., and Freyermuth, H. B., *J. Org. Chem.*, **8**, 230 (1943).
97. Raiford, C. L., and Freyermuth, H. B., *J. Org. Chem.*, **8**, 174 (1943).
98. Lyons, J. M., and Thomson, R. H., *J. Chem. Soc.*, **1950**, 1970.
99. Saunders, J. H., and Hardy, E. E., *J. Am. Chem. Soc.*, **75**, 5439 (1953).
100. Stallmann, O., U.S. Pat. 2,671,082 (to E. I. du Pont de Nemours and Co.), (Mar. 2, 1954).
101. Balon, W. J., and Stallmann, O., U.S. Pat. 2,683,144 (to E. I. du Pont de Nemours and Co.) (July 6, 1954).
102. Brown, C. J., *J. Chem. Soc.*, **1955**, 2931.
103. Gaylord, N. G., and Crowdle, J. H., *Chem. Ind. (London)*, **1955**, 145.
104. Balon, W. J., Langerak, E. O., Simmons, D. M., and Stallmann, O., ACS Meeting, September 1956, Paper No. 40, Part 1, Div. of Paint, Plastics and Printing Ink Chemistry.
105. Barthel, E., Kehr, C. L., Langerak, E. O., Pelley, R. L., and Smeltz, K. C., ACS Meeting, September 1956, Paper No. 40, Part 2, Division of Paint, Plastics and Printing Ink Chemistry.

106. Stallmann, O., and Langerak, E. O., U.S. Pat. 2,643,250 (to E. I. du Pont de Nemours and Co.) (June 23, 1953).
107. Baker, J. W., and Holdsworth, J. B., *J. Chem. Soc.*, **1947**, 724.
108. Cook, A. H., and Jones, D. G., *J. Chem. Soc.*, **1941**, 278.
109. Spencer, R. D., and Beggs, B. H., *Anal. Chem.*, **35**, 1633 (1963).
110. Graymore, J., *J. Chem. Scc.*, **1931**, 1490.
111. Burkus, J., U.S. Pat. 2,993,870 (July 25, 1961).
112. Frazier, T. C., Little, E. D., and Lloyd, B. E., *J. Org. Chem.*, **25**, 1944 (1960).
113. Biillmann, E., and Bjerrum, J., *Ber.*, **50**, 503 (1917).
114. Hofmann, A. W., *Ber.*, **19**, 2061 (1886).
115. Gal, F., *Compt. Rend.*, **61**, 527 (1865).
116. Hofmann, A. W., *Ber.*, **18**, 765 (1885).
117. Hofmann, A. W., *Ber.*, **18**, 2796 (1885).
118. Ponomarew, J., *Ber.*, **18**, 3271 (1885).
119. Fischer, E., and Frank, F., *Ber.*, **30**, 2616 (1897).
120. Kogon, I. C., *J. Org. Chem.*, **26**, 3004 (1961).
121. Kogon, I. C., *J. Am. Chem. Soc.*, **78**, 4911 (1956).
122. Balon, W. J., U.S. Pat. 2,801,244 (to E. I. du Pont de Nemours and Co.) (July 30, 1957).
123. Burkus, J., *J. Org. Chem.*, **27**, 474 (1962).
124. Burkus, J., U.S. Pat. 2,979,485 (Apr. 11, 1961).
125. Jones, J. I., and Savill, N. G., *J. Chem. Soc.*, **1957**, 4392.
126. Blair, J. S., *J. Am. Chem. Soc.*, **56**, 904 (1934).
127. Dains, F. B., Greider, H. W. and Kidwell, C. H., *J. Am. Chem. Soc.*, **41**, 1004 (1919).
128. Khorana, H. G., *Chem. Rev.*, **53**, 145 (1953).
129. Stolle, H., *Ber.*, **41**, 1125 (1908).
130. Huhn, A., *Ber.*, **19**, 2405 (1886).
131. Dyer, E., and Newborn, E. G., *J. Am. Chem. Soc.*, **80**, 5495 (1958).
132. Smeltz, K. C., U.S. Pat. 2,840,589 (to E. I. du Pont de Nemours and Co.) (June 24, 1958).
133. Campbell, T. W., and Verbanc, J., U.S. Pat. 2,853,473, (to E. I. du Pont de Nemours and Co.), (Sept. 23, 1958).
134. Dyer, E., and Read, R. E., *J. Org. Chem.*, **26**, 4388 (1961).
135. Dyer, E., and Read, R. E., *J. Org. Chem.*, **26**, 4677 (1961).
136. Campbell, T. W., and Monagle, J. J., *J. Am. Chem. Soc.*, **84**, 1493 (1962).
137. Balon, W. J., U.S. Pat. 2,853,518 (to E. I. du Pont de Nemours and Co.) (Sept. 23, 1958).
138. Bortnick, N., Luskin, L. S., Hurwitz, M. D., and Rytina, A. W., *J. Am. Chem. Soc.*, **78**, 4358 (1956).
139. *FIAT Final Report* No. 712.
140. Williamson, A. G., *Analyst*, **77**, 372 (1952).

CHAPTER IV

ISOCYANATE DERIVATIVES

I. INTRODUCTION

This chapter deals with several reactions in which the isocyanate group (—NCO) reacts with other "active" compounds yielding well-defined and easily characterized derivatives known as carbamates (or urethanes), allophanates, substituted ureas and biurets. These derivatives are of interest not only for identification purposes, but the reactions which produce them occur in the polymerization process yielding polyurethanes. Spectroscopic methods applicable to these compounds and their characteristic functional groups are dealt with in considerable detail. The physical properties of a selected number of compounds in each class are presented which may be useful for identification by conventional methods.

II. URETHANES

A. Structural Relationships

Urethanes are esters of carbamic acid, which in turn may be regarded as the monoamide of carbonic acid. The ethyl ester is referred to frequently as "urethane." Carbamic acid, of course, does not exist as a separate entity.

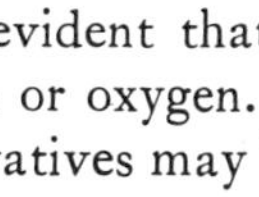

Carbonic acid

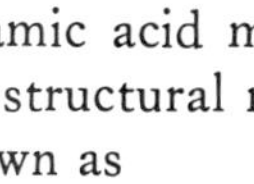

Carbamic acid monamide

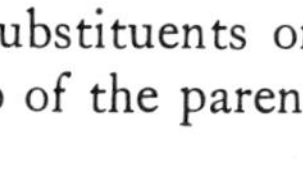

Ethyl ester of carbamic acid or urethane (Ethyl carbamate)

It is evident that carbamic acid may have substituents on either the nitrogen or oxygen. The structural relationship of the parent group and its derivatives may be shown as

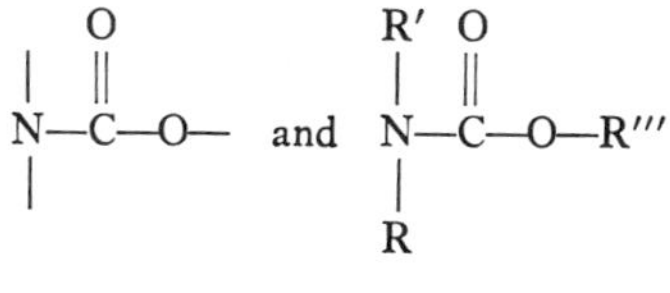

The names applied to this series of compounds can be confusing because of the two positions at which substitution may occur. Some of the structures and their alternate names are given here as **1–5.**

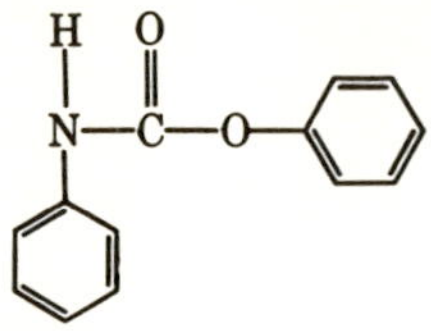

(**1**) Phenyl carbanilate

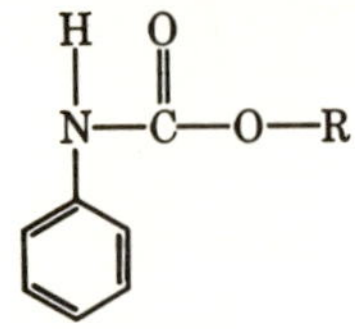

(**2a**) Carbanilate
(**2b**) *N*-Phenyl carbamate

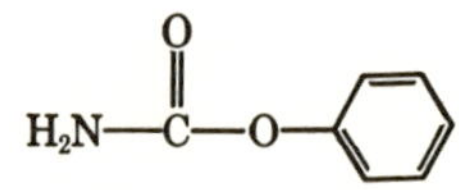

(**3a**) Phenyl carbamate

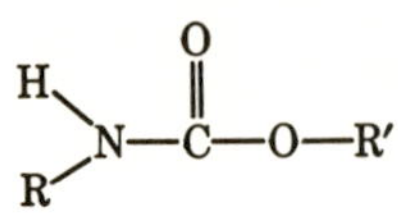

(**4a**) Carbamates
(**4b**) *N*-substituted carbamates

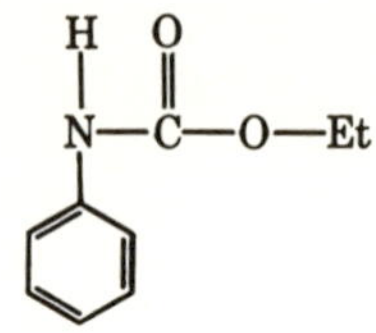

(**5a**) *N*-Phenyl urethane
(**5b**) Ethyl phenyl carbamate

Structure **6** and its name should not be confused with the urethanes. It

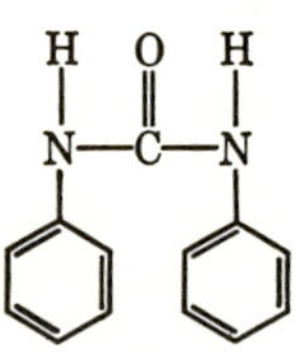

(**6**) Carbanilide

is a substituted urea and is not a urethane, in spite of the similarity in name to that in (1) and in (2a).

B. Preparation

Urethanes are prepared by the general reaction of an isocyanate with an alcohol:

$$\mathrm{RNCO + R'OH \rightarrow R{-}\overset{H}{\overset{|}{N}}{-}\overset{O}{\overset{\|}{C}}{-}R'} \qquad (1)$$

Many of these compounds are well defined and easily characterized. The urethanes of phenyl isocyanate and α-naphthyl isocyanate have become classical derivatives for the identification of various alcohols.

Bickel and French (1) have described the urethanes of some 33 alcohols.

The derivatives of other alcohols have been reported by French and Wirtel (2); McKinley, Nickels, and Sidhu (3); Tarbel, Mallatt, and Wilson (4); and Lyons and Thomson (5).

Kogon (6,7) studied the reaction between phenyl isocyanate and ethyl alcohol and showed that the urethane results from the reaction at room temperature but α,γ-diphenyl allophanate is the main product along with the dimer of phenyl isocyanate when the reaction is carried out at 125°C.

$$C_6H_5NCO + C_2H_5OH \xrightarrow{RT} C_6H_5NHCOOC_2H_5$$

$$C_6H_5NCO + C_2H_5OH \xrightarrow[125°C]{} C_6H_5NHCON(C_6H_5)COOC_2H_5 + (C_6H_5NCO)_2 \qquad (2)$$

When *N*-methylmorpholine was used as a catalyst for reaction 2, triphenylisocyanurate was obtained in excellent yield.

Baker and Holdsworth (8) described the preparation of methyl phenyl carbamate from methyl alcohol and phenyl isocyanate. Baker and Gaunt (9) describe the preparation of isopropyl phenyl carbamate and borneol phenyl urethane. These investigators also studied the mechanisms and kinetics of these reactions. Dennstedt (10) described the ethyl and methyl bromphenyl urethanes.

Bloomquist and Stevenson (11) prepared several methyl-substituted furyl carbamates from the corresponding azides.

Dunlop and Peters (12) described the urethane derivatives of 3,4-furandiamines and gave numerous references to the literature on these less well-known compounds (13–23).

While primary and secondary alcohols give the stable urethane when reacted with isocyanate, the reaction with tertiary alcohols is usually accompanied by olefin formation (24,25).

$$2C_6H_6NCO + (CH_3)_3COH \rightarrow C_6H_6NHCONHC_6H_5 + CO_2 + CH_3\underset{\displaystyle CH_3}{\underset{|}{C}}{=}CH_2 \qquad (3)$$

Other reactions may occur in addition to the urethane reaction when alcohols of more complex structures are used, such as those with α-hydroxy esters. In this case, ring closure results between the ester and urethane if a sodium catalyst is present (26).

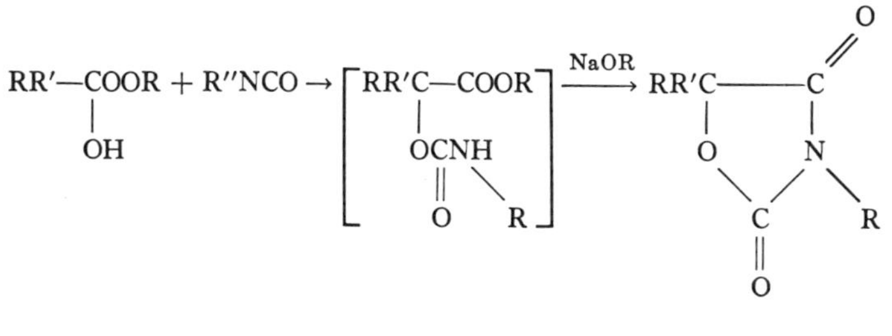

(4)

Additional examples are the urethane from glycidol and phenyl isocyanate (reaction 5) (27).

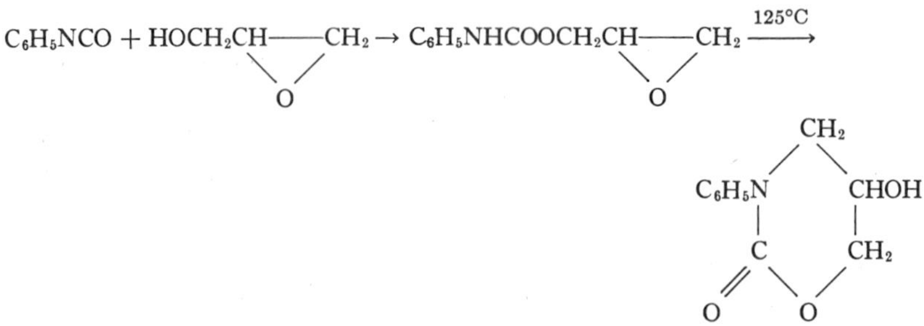

(5)

Under mild conditions 1-alkenyl isocyanates react with alcohols to give the normal urethane but at slightly elevated temperatures a variety of products can be obtained (28).

C. Catalysis and Kinetics

This subject has been treated extensively by Saunders and Frisch (29). It will be mentioned only briefly here since it may prove valuable as background information in studying the formation of urethanes by techniques such as infrared.

Saunders and Frisch (29) have classified the general scheme for the reactions of isocyanates with alcohols as being

$$\text{RNCO} + \text{catalyst} \rightarrow \text{complex} \xrightarrow[\text{compound}]{\text{active hydrogen}} \text{product} + \text{catalyst} \tag{6}$$

This is based on the original proposals of Baker and co-workers (8,9), who studied the alcohol isocyanate reaction

$$\text{R}\cdot\text{N}{:}\text{C}{:}\text{O} + \text{ROH} \underset{k_2}{\overset{k_1}{\rightleftharpoons}} \text{R}\cdot\text{N}{:}\underset{\underset{+}{\text{ROH}}}{\underset{|}{\text{C}}}\text{—O}^- \text{ (complex)} \tag{7}$$

$$\text{Complex} + \text{ROH} \xrightarrow{k_3} \text{R}\cdot\text{NH}\cdot\text{CO}_2\text{R} + \text{ROH} \tag{8}$$

$$\text{R}\cdot\text{N}{:}\text{C}{:}\text{O} + \text{NR}_3 \underset{k_2}{\overset{k_1}{\rightleftharpoons}} \text{R}\cdot\text{N}{:}\underset{\underset{+}{\text{NR}_3}}{\underset{|}{\text{C}}}\text{—O}^- \tag{9}$$

$$\text{R}\cdot\text{N}{:}\underset{\underset{+}{\text{NR}_3}}{\underset{|}{\text{C}}}\text{—O}^- + \text{ROH} \xrightarrow{k_3} \text{R}\cdot\text{NH}\cdot\text{CO}_2\text{R} + \text{NR}_3 \tag{10}$$

Saunders and Frisch (29) have proposed a similar mechanism for the participation of an acid catalyst

$$\mathrm{RNCO + HA \rightleftharpoons [R{-}N{=}C{-}\ddot{\underset{..}{O}}: \rightarrow H \cdots A] \xrightarrow{ROH} RNHCOOR + HA}$$

Baker and co-workers (8,9) followed these reactions experimentally, by the method of Stagg (30), using inert solvents and being careful to exclude water. This method is similar to the di-*n*-butylamine method which was given in Chapter 3 except that Stagg employed acetone as the solvent and piperidine as the reactant. Under stationary state conditions, Baker and Holdworth (8) expressed the reaction rate as:

$$k_{bi} = \frac{k_1k_3[\mathrm{NR_3}]}{k_2 + k_3[\mathrm{ROH}]}$$ for a base-catalyzed reaction as in reactions 9 and 10.

and

$$k_0 = \frac{k_1k_3[\mathrm{ROH}]}{k_2 + k_3[\mathrm{ROH}]}$$ for an alcohol-catalyzed reaction as in reactions 7 and 8.

where k_0 is similar to k_{bi}. These workers found the reaction to follow second-order kinetics, and in the absence of a base to be catalyzed by the weakly basic urethane product. The experimental velocities were R = p-$NO_2C_6H_4 \gg C_6H_5 > p$-Me·$C_6H_4 > p$-MeO·$C_6H_4 \gg$ cyclo-C_6H_{11}. This is in agreement with the view that the reaction velocity is increased by increased conjugation of the unshared electron pair on the nitrogen in the —NCO group with the R group. For example, an electron-withdrawing substituent on the isocyanate molecule should increase the partial positive charge on the —NCO carbon and move the negative charge further from the site of reaction, thus making attack on the carbon by an electron donor easier (reaction 11).

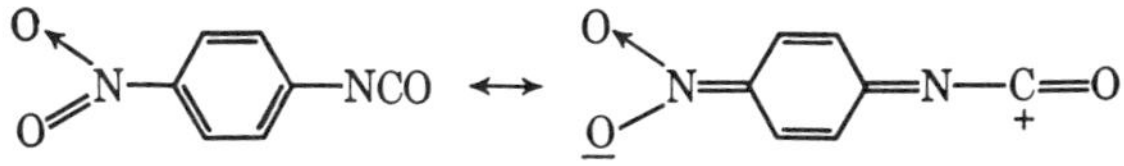
(11)

One must be cautious in investigating catalysts, mechanisms, and rates of isocyanate reactions for the following reasons:

(*1*) When a reactant molecule serves as the catalyst, the reaction may not be truly second order since the rate constant will change as the concentration of the reactant species is varied.

(*2*) The acidity of the isocyanate should be accurately known and reported since this may have a major effect on the reaction rate and render comparisons between investigators invalid.

(*3*) The experimental studies generally represent idealized systems and are not strictly comparable from one solvent system to another and with the actual urethane formation as in foams and elastomers.

(*4*) By monitoring only the disappearance of the —NCO, false conclusions concerning the reaction could be reached. It would be better to employ infrared to monitor disappearance of the NCO and OH, and appearance of urethane, as well as any unexpected side reaction products.

In addition to the electronegative or electropositive effects on ring substituents, steric factors and bulkiness of the groups play a major part in reactivity as well as influencing the effectiveness of the catalyst. This is certainly true of the most important commercial isocyanate, TDI. The most reactive group is in the 4-position which is activated by the 2-position isocyanate group. The 2-position group is similarly activated by the 4-position group but it tends to be deactivated by the 1-position methyl group. After the 4-position group has reacted with a hydroxyl compound, the 2-position group is even less reactive than initially because of the methyl group *ortho* to it, in spite of the slight activation from the urethane in the 4-position.

Bailey et al. (31) studied the reaction rates of several diisocyanates with a large excess of 2-ethyl hexanol and a diethylene glycol adipate of molecular weight 1900. The reactions were carried out in benzene and the rate was followed by monitoring the disappearance of —NCO at 4.4 μ.

A sharp decrease in the rate of reaction was observed for 80:20 TDI when approximately 50% of the reaction has occurred. Other diisocyanates showed only a slight decrease in rate constant as the reaction proceeded. This is apparently due to the completion of reaction of the 4-position —NCO group. Simmons and Arnold have also demonstrated the greater reactivity of the *p*-NCO group in 2,4-TDI (32).

Morton et al. (33,34) studied the reactions of 2,4-tolylene diisocyanate and 4,4′-diphenylmethane diisocyanate with *n*-butanol in xylene. Increasing the alcohol/isocyanate ratio increased the apparent reaction rate constants as in the case of monoisocyanates, and a plot of [ROH]/k against [ROH] gave a straight line with both isocyanates.

Burkus (35) studied the rate of reaction between phenyl isocyanate and butanol-1 catalyzed by pyridine *N*-oxide as compared with the rate when catalyzed by triethylamine and pyridine.

A more recent study on the kinetics of alcohol and polyether polyol–isocyanate reactions using metal catalysts was performed by Frisch et al. (36). The reaction rates of ether alcohols and primary and secondary methoxy-propanols with phenyl isocyanate in the presence of lead naphthenate or dibutyltin dilaurate were found to be faster in toluene than in

Cello-solve acetate. These reactions were investigated both by infrared and NMR. A catalyst–alcohol complex was not detected by infrared at 10–20% catalyst concentrations; some evidence was found for complex formation at 1:1 catalyst/alcohol ratio. NMR spectra indicated the existence of a 1:1 complex in the system dibutyltin dilaurate–1-methoxy-2-propanol.

Baker et al. (37) have studied the effects of various concentrations of triethylamine on the infrared absorption spectra, in the 3 μ region, of methyl and ethyl alcohols in di-*n*-butyl ether and in benzene solutions. The addition of triethylamine gives rise to a new absorption near 3.1 μ due to the OH group association (hydrogen bonding) of an amine alcohol complex. In the absence of amine, there is no absorption at 2.73 μ due to unassociated hydroxyl groups, which shows that OH groups present are solvent-associated and/or self-associated. It appears that this is good evidence for catalyst–alcohol complex formation.

Robertson and Stutchbury (38) studied the reaction of phenyl isocyanate with ethanol, polymethylene glycols, and polyethylene glycols in a number of solvents. These authors found no deviations from second-order kinetics, but they found deviations from Baker and Gaunt's proposed mechanisms (9) in polar solvents which are not understood fully. Effects due to hydrogen bonding were observed but in the case of ethylene and diethylene glycols, plots of $[ROH]/k_{obs}$ vs. [ROH] gave curves of negative slopes, whereas Baker and Gaunt's mechanism predicts a positive slope ($1/k_1$) and a positive intercept (k_2/k_1k_3). Other effects of reactions of phenyl isocyanate with hydroxy compounds in dioxane were noted. For example, triethylene and tetraethylene glycols gave positive slopes while trimethylene, tetramethylene, and hexamethylene glycols gave essentially zero slopes.

Many of these discrepancies could be accounted for solely on the basis of the solvents used and the hydroxy compound's ability to hydrogen bond with a polar solvent, since Baker and Gaunt used di-*n*-butyl ether and benzene as solvents in various studies that they conducted. However, this again points out the difficulty in comparing results in dissimilar systems.

D. Reactions

Since the urethane group is the basic unit in any polyurethane system, and the reactions it undergoes may serve as guide to analytical examination of either pure urethanes (carbamates) or polyurethanes, some of the more important reactions will be presented briefly here.

Of major interest is the thermal stability of the urethane. Urethanes from tertiary alcohols decompose at temperatures as low as 50°C, whereas urethanes from primary and secondary alcohols may be stable up to 200°C.

Three general types of reactions have been reported.

1. Dissociation to Isocyanate and Alcohol

Dyer and Newborn (39) studied the degradation products from *n*-butyl, neopentyl, and benzyl biscarbamate of MDI, and the polycarbamate of hexamethylene glycol and 4,4′-diisocyanate diphenylmethane. These decompositions were carried out at 300°C under dry nitrogen at atmospheric pressure. This was the lowest temperature at which an appreciable rate of decomposition occurred. The decomposition products were carbon dioxide, a distillate fraction consisting of the parent alcohol, and a non-volatile residue.

The solids from *n*-butyl and neopentyl carbamates were intractable and infusible polymers insoluble in the common polymer solvents. Infrared spectra of these solids showed absorption at 4.7–4.8 μ, characteristic of the carbodiimide group. Dyer and Newborn suggested that these reactions proceed as

$$CH_2(C_6H_4NHCO_2R)_2 \rightarrow CH_2(C_6H_4NCO)_2 + 2ROH \tag{12}$$

$$nCH_2(C_6H_4NCO)_2 \rightarrow (\text{—}N{=}C{=}NC_6H_4CH_2C_6H_4\text{—})_n + nCO_2 \tag{13}$$

since the work of Stolle (40) had shown that phenyl isocyanate forms diphenylcarbodiimide when heated in a sealed tube at 300°C.

The residues from the benzyl and hexamethylene carbamates were found to be low melting basic solids in which the carbodiimides infrared spectra did not occur. All of the residues showed negligible absorption at 5.9 μ, indicating a loss of the carbamate carbonyl group and poorly defined absorption in the 2.9–3.6 μ and 6.1–6.9 μ regions.

Dyer and Read (41) studied the pyrolysis of two carbamates of naphthyl isocyanate. The products from pyrolysis at 255–275° were carbon dioxide, amine, alcohol, and the trimer of the isocyanate. Small amounts of disubstituted carbonate and disubstituted urea were formed at 137°C, the lowest temperature at which carbon dioxide was evolved.

The thermal dissociation of the urethanes of benzyl alcohol and cinnamyl alcohol of phenyl isocyanate in the presence of amines and fatty acids was studied by Mukaiyame and Hoshino (42). Mukaiyama et al. (43,44) had previously found that the rate of dissociation of urethanes and ureas increased as the acidity of the solvent increased. On this basis as well as the other investigation (42), it was concluded that the dissociation of the urethane proceeds through the double transfer of a proton under the influence of acidic or basic media.

The mechanisms of dissociation of urethane or urea were similar to those of carbonyl addition compounds, i.e., the carbonyl-forming elimination reaction. These amphoteric compounds dissociate in the presence of either basic or acidic catalysts. It is interesting to note that in these reverse reactions, the catalysts which are suitable for dissociation also accelerate addition by both basic catalysts such as amines (9) and acid catalysts such as aluminum chloride (45).

Bortnick et al. (46) obtained secondary and tertiary alkyl isocyanates in 50–80% yields from the base-catalyzed pyrolysis of *N*-alkylcarbamates. The dissociation of these compounds to isocyanate and alcohol was found to be difficult and required temperatures above 250°C.

2. Formation of Primary Amine and Olefin

$$\mathrm{RNHCOOCH_2CHR'} \rightarrow \mathrm{RNH_2} + \mathrm{CO_2} + \mathrm{R'CH{=}CH_2} \qquad (14)$$

Dyer and Wright (47) have proposed a reaction sequence including a cyclic structure.

3. Additional Reactions

Dyer, Newborn, and Wright (48) have reported the formation of primary and secondary amine, as well as olefin formation from urethane decomposition (Reaction 15).

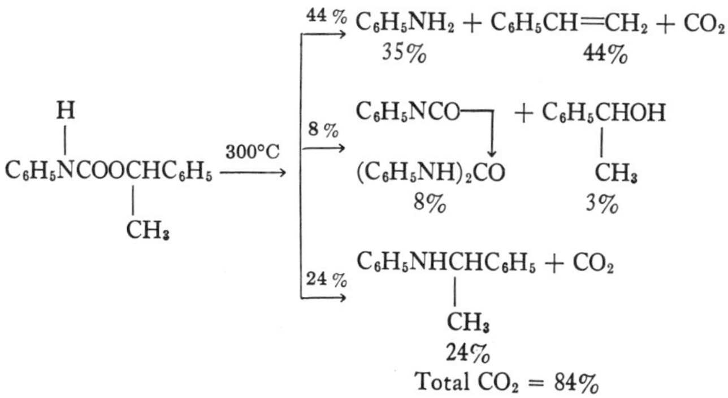

(15)

Bieber (49) found that the treatment of carbamic esters with 5% oleum led to the formation of ammonium bisulfate or amine acid sulfates. He concluded that these esters undergo alkyl–oxygen fission. When 30% and 60% acids were used, substantial amounts of sulfamic acids were formed which could be explained by *N*-sulfonation of the carbamic esters prior to their acidolysis.

E. Analytical Techniques and Characterization

1. Spectroscopic

The simple carbamates have been most often quantitatively characterized by elemental analysis and mixed melting point determinations. Decomposition studies often necessitated the infrared examination of residues and the quantitative determination of the carbon dioxide evolved by collection in Ascarite.

The carbonyl stretching region, occurring at 5.5–6.0 μ in the infrared, presents an excellent region from which identification of the particular species may be made. The amide I band of the urethanes (C=O) occurs at a different frequency from the C=O stretching vibration of other compounds such as the dimer and trimer. While it is not always possible to distinguish individual species in multicomponent mixtures because all of these bands tend to coalesce, some general information or clues as to identity can be obtained.

We have noted a slight difference in the urethane carbonyl absorption depending upon whether the compound being examined is a simple urethane (carbamate) or a polyurethane. Our own observations of simple carbamates indicate that this group absorbs between 1739 and 1700 cm^{-1} (5.75–5.88 μ). Occasional compounds have been found to fall slightly outside of this wavelength coverage, depending upon ring side groups which play a major role in influencing the molecular portion of the spectrum as well as the wavelength of specific carbonyl groupings.

Another factor which may influence the particular absorption frequency is the state of the sample. In general, mulls or melts appear to displace the wavelength to lower frequencies. However, the carbonyl band position, depending on structure in the ring and neighboring groups, is specific for constitution. Ring strain as in the case of dimers or trimers increases the frequency and hydrogen bonding decreases it.

For diurethanes of diisocyanates, band splitting occurs by coupling which also leads to a higher frequency which may go beyond the given limit for urethanes. However, in this case the higher frequency band is not the most intense. Table 4.1 shows the absorption frequencies for a number of carbamates.

Brumfield (51) studied the spectra of a number of urethanes, ureas, and the isocyanate reaction products of these materials. From basic correlations, common to the spectra studied, a method was devised to quantitatively measure the amounts of various group contributions in a compound, singly or in mixtures.

Brumfield's approach was to determine the contribution for each type of structure and its absorption wavelength. The absorptions in a spectrum

TABLE 4.1
Absorption Frequency for the Carbonyl Group in Simple Carbamates (50,51)

Compound	Frequency, cm^{-1}
Methyl phenyl carbamate	1706
Diphenyl carbamate	1718
2,4-Dimethyl tolyl carbamate	1720, 1697
2-4-Diphenyl tolyl carbamate	1758, 1708
1,5-Dimethyl naphthyl carbamate	1698
n-Butyl-*N*-phenyl carbamate	1717 (melt), 1700 (KBr pellet)
Ethyl *N*-phenyl carbamate	1700 (melt), 1708 (KBr pellet)
Ethyl carbamate	1710 (melt), 1693 (KBr pellet)
Toluene 2-isocyanato-4-methyl carbamate	1710

of a particular compound that contributed toward the total absorption of the molecule for each type of group or structure present were then measured at the point of maximum absorption for the particular frequency being considered. The amount of structure present, in %, was then equal to

$$\frac{\text{Total contributive absorption for structure} \times 100}{\text{Total contributive absorption for entire molecule}}$$

A Perkin-Elmer Model 221, equipped with a prism-grating interchange was utilized for all measurements. Liquids between salt plates or solids in KBr pellets or between salt plates were utilized and the sample thicknesses were adjusted to provide contributive absorption measurements between 0.15 and 0.8 absorption units. This was done to obtain optimum resolution and accuracy.

Table 4.2 was compiled by Brumfield from sources as indicated. It is included here because it served as a basis for Brumfield's initial spectral correlations and can also be useful to any investigator wishing to repeat or extend Brumfield's work.

Table 4.3 shows Brumfield's analysis and assignments of —NH bond positions. This table is included because of the specific usefulness and application to isocyanate derivatives.

The area of absorption in Table 4.3 show bonded NH (3400–3200 cm^{-1}) and free NH (3500–3400 cm^{-1}). The compounds studied had bonded —NH absorptions in the 3310–3380 cm^{-1} range. This area increased in absorption as the free —NH area (3500–3400 cm^{-1}) decreased during a urethane reaction. The carbamate structure was observed to have absorption only in the bonded —NH area (3320–3350 cm^{-1}) in the melt or liquid spectrum, but there was absorption in the free —NH or —OH range in the KBr spectrum (3490–3440 cm^{-1}).

TABLE 4.2

Infrared Band Assignments Useful for Evaluation of Urethanes by Infrared (51)

The following correlations were taken from Bellamy (84)

For NH—Stretching Modes	Frequency, cm^{-1}
Primary amides (free NH—)	near 3500 and 3400
Primary amides (bonded NH—)	near 3350 and 3180
Secondary amide (free NH—) (*trans-*)	near 3460 and 3400
Secondary amide (free NH—) (*cis*)	near 3440 and 3420
Secondary amide (bonded NH—) (*trans-*)	near 3320 and 3270
Secondary amide (bonded NH—) (*cis*)	near 3180 and 3140
Secondary amide (bonded NH—) (*cis* and *trans-*)	near 3100 and 3070
CO— Absorption (Amide I)	
Primary amides	solid near 1650, dilute solution near 1690
Secondary amides	solid 1680–1630, dilute solution 1700–1670
Tertiary amides	solid and dilute solution 1670–1630
Amide II	
Primary only	solid 1650–1620, solution 1620–1590
Secondary, noncyclic	solid 1570–1515, solution 1550–1510
Amide III	
Secondary only	near 1290

The following correlations were taken from A. Weissberger, Ed., *Technique of Organic Chemistry*, Vol. IX.

Type of absorption	Compound or group	Frequency, cm^{-1}
CH— stretching, asymmetrical	CH_2—	3040
CH— stretching, asymmetrical	CH_3—	2962
CH— stretching, asymmetrical	CH_2—	2926
CH— stretching, asymmetrical *tert.*	CH— group	2890
CH— stretching, symmetrical	CH_3—	2872
CH— bending	CH_2—	1467
CH— bending, asymmetrical	CH_3—	1460
CH— bending	CH_2—	1455
CH— bending	CH_2—	1452
CH— bending, symmetrical	CH_3—	1378
CH— wagging	CH_2—	1350 to 1150
CH—	C_6H_6	3089, 3070, 3063, 2905, 2886, 1617, 1585, 1482
CH—	$C_6H_5CH_3$: C_6H_5—	3090, 1605, 1497,
	CH_3	2964, 2874, 1460, 1379
CH—	$C_6H_5CH_2CH_3$: C_6H_5—	3067, 3034, 1606, 1497
	CH_3—	2968, 2874, 1454, 1377

(*continued*)

TABLE 4.2 (*Continued*)

Type of absorption	Compound or group	Frequency, cm^{-1}
	CH_2—	2933
	R—CO—NH_2[a]	1690
	R—CO—NH—R[a]	1680
	R—CO—NH—Ar[a]	1698
	R—CO—N—R_2[a]	1650
NH— stretching	Compounds of nitrogen	3500 to 3050
—OH . . . O	Compounds of nitrogen	3500 to 3200
—OH . . . OC	Compounds of nitrogen	3320 to 3240
—NH . . . N	Compounds of nitrogen	3300 to 3150

Non-associated NH— groups. NH— stretching is in the least perturbed form with dilute solution of amine in CCl_4. Primary aliphatic amine shows two bands near 3500 and 3400 cm^{-1} and secondary amines one band 3500 to 3400 cm^{-1}.

Amides. Primary amide gives two bands at 3530 and 3415 cm^{-1}.

[a] Where R is an alkyl group and Ar is an aryl group.

As an example of the calculations used by Brumfield to compute structure contribution for carbamates, *n*-butyl *N*-phenylcarbamate was evaluated as follows.

The structure was divided into carbamate, phenyl, and butyl groups.

Melt (Fig. 4.1)	KBr pellet (Fig. 4.2)
Carbamate (HNCOO)	
3331, 3200, 1700, and 1531 cm^{-1}	1699 cm^{-1}
"A" (carbamate) × 100	
Total "A" = 5.289	Total "A" = 0.950
$\frac{1.618 \times 100}{5.289} = 30.59\%$ (actual 30.52)	$\frac{0.286 \times 100}{0.950} = 30.11\%$ (actual 30.56)
Phenyl (C_6H_5)	
3064, 3023, 2903, 1611, 1593, 1499, and 1470 cm^{-1}	3065, 3045, 3904, 1599, 1498, 1497, and 1471 cm^{-1}
"A" (phenyl) × 100	
$\frac{2.114 \times 100}{5.289} = 39.94\%$ (actual 39.94)	$\frac{0.373 \times 100}{0.950} = 40.31\%$ (actual 39.89)
Butyl (C_4H_9)	
2973, 2938, and 2880 cm^{-1}	2988, 2965, 2939, 2878, and 2855 cm^{-1}
"A" (butyl) × 100	
$\frac{1.557 \times 100}{5.289} = 29.47\%$ (actual 29.54)	$\frac{0.281 \times 100}{0.950} = 20.58\%$ (actual 29.55)

TABLE 4.3
—NH Assignments (51)
(3400–3200, bonded NH; 3500–3400, free NH)

	Frequency, cm^{-1}		
Compound	Liquid	Melt	KBr pellet
n-Butyl 1,4-diphenylallophanate		(3320–3370)[a]	3490 (3300)
n-Butyl *N*-phenylcarbamate		3331	3450
Ethyl *N*-phenylcarbamate		3320	3440
Ethyl methylcarbamate	3352		
Ethyl carbamate		3353	3440
Ethyl ethylcarbamate	3330		
Urea			3350 (3440)
Cyanuric acid			none (3490)
Urethane reaction-X2-136B Resin (T-P Industries, Inc.)		3310	
1,3-Diphenyl 5-butylbiuret		3380 (3445)	 3442 (max)
Toluene diisocyanate	3390, 3500 (weak)		
1-Phenyl 3, 3-diethyl urea		3312	3442, 3425, 3325 (shoulder)
Diphenylurea			3449, 3329, 3305 (max)
1,2-Bis(2′-phenylureido) ethane			3447, 3329
Phenyl isocyanate dimer			3440 (max), 3300 (shoulder)

[a] The characteristic (—NH) absorption that was present in the carbamates shown below was not present in this melt spectrum. The spectrum had absorption peaks at 3220, 3255, and 3290 cm^{-1}.

Only 13 absorption frequencies were used for the KBr pellet instead of the 14 for the melt spectrum. This was done because the initial calculation was found to be too large for carbamate and too small for phenyl. Consequently, the 3450, 3200, and 1537 cm^{-1} were omitted and the 3045 cm^{-1} was added to give the correct total absorption for the molecule.

2. Gas Chromatography

Zielinski and Fishbein (52) recently demonstrated the separation and identification of simple carbamates by gas–liquid chromatography. Separations were carried out on two polar coatings (Carbowax 20M and Versamid 900), and one nonpolar coating (SE-30). Table 4.4 illustrates the contrast in carbamate separations that were obtained.

Elution was found to be in order of increasing chain lengths for the

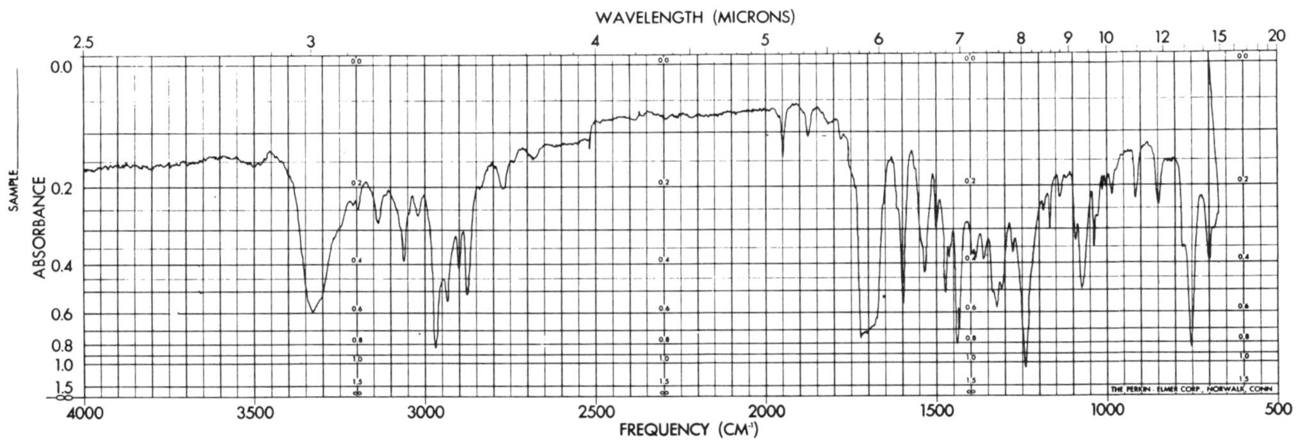

Fig. 4.1. Infrared spectrum of *n*-butyl-*N*-phenylcarbamate (melt) (51).

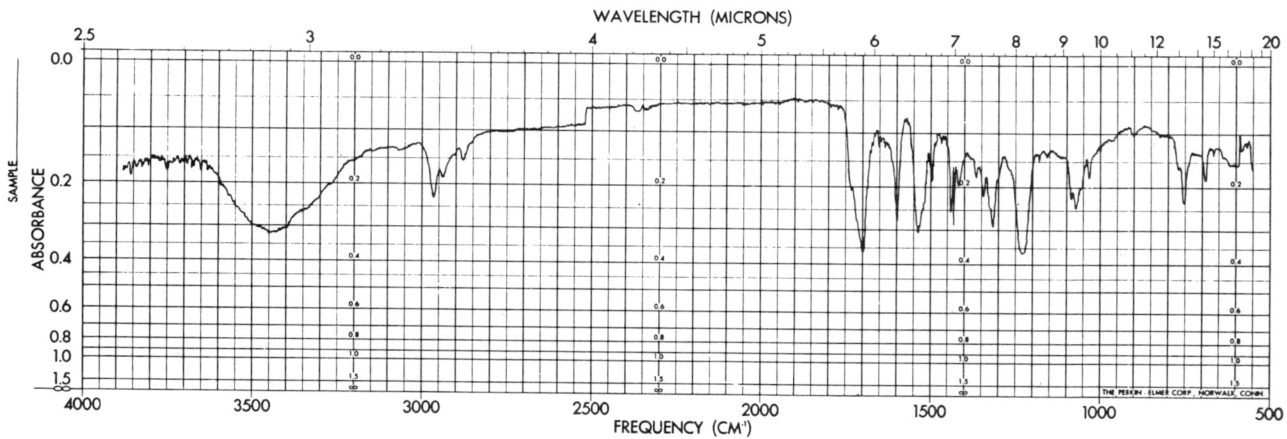

Fig. 4.2. Infrared spectrum of *n*-butyl-*N*-phenylcarbamate (0.108% in KBr Pellets) (51).

TABLE 4.4
Physical Property and GLC Retention Data by Simple Carbamates (52)

Compound	MW	MP, °C.	Relative elution[a]		
			Carbowax 20M[b]	Versamid 900[c]	SE-30[d]
Methyl carbamate	75	53–55	0.92	0.75	0.72
Ethyl carbamate (urethane)	89	51	1.0	1.0	1.0
Allyl carbamate	101	—[e]	1.9	1.7	1.8
Isopropyl carbamate	103	95–96	1.0	1.1	1.3
n-Propyl carbamate	103	52–53	1.5	1.6	1.9
Methallyl carbamate	115	51–52	2.6	2.5	3.2
sec-Butyl carbamate	117	93–94	1.5	1.7	2.7
Isobutyl carbamate	117	61–62	1.7	2.0	2.8
n-Butyl carbamate	117	53.5	2.1	2.5	3.4
β-Chlorethyl carbamate	124	71–73	0.45	0.55	0.32
Isopentyl carbamate	131	60–61	2.5	3.3	5.2
Amyl carbamate	131	56	3.0	3.9	7.0
Phenyl carbamate	138	149–150	3.7	6.2	2.3
Hexyl carbamate	145	67	4.2	6.0	12.4
Benzyl carbamate	152	86.5–87	3.0	3.4	3.6

[a] Relative elution to ethyl carbamate as 1.0. Elution of ethyl carbamate was 2.1 min on Carbowax 20M, 2.5 min on Versamid 900, and 1.2 min on SE-30.

[b] 10% w/w on 60–80 mesh Chromosorb W, 8 ft × 0.125 in. o.d. stainless steel. Operating conditions: column 185°C; injection port 265°C; detector block 250°C; filament current 150 mA; hot wire detector; helium carrier 22 ml/min.

[c] 8% w/w on 60–80 mesh Chromosorb W-HMDS treated, 8 ft × 0.125 in. o.d. stainless steel. Column conditions: column 185°C; injection port 225°C; detector 185°C; carrier flow 10 ml/min of nitrogen; 300 ml/min of air; 24 ml/min of hydrogen; flame ionization detector.

[d] 10% w/w on 60–80 mesh Chromosorb W, 8 ft × 0.125 in. o.d. stainless steel. Column conditions: column 100°C; injection port 265°C; detector block 250°C; helium carrier 24 ml/min of helium; filament current 150 mA; hot wire detector.

[e] Boiling point, 73–75°C at 2.0 mm.

n-alkyl derivatives. Unsaturation in the chain tended to lengthen the elution time on Carbowax while chain branching diminished the elution of derivatives containing the same number of carbon atoms in the alkyl chain for all three columns.

All elution pairs on Carbowax 20M were resolved on Versamid 900 and the converse for elution pairs on Versamid.

Zielinski and Fishbein found an order of 600–800 theoretical plates for the Carbowax and Versamid columns. Carbamate elution was found to be generally later on Versamid, which they suggested might be due to a slightly higher degree of solubility in an amide-type coating than in a glycol-type coating. By plotting the log of the relative retention times for the *n*-alkyl derivatives and examining the slope, it was found that a greater resolution was obtained on Versamid 900.

It appears that another factor which might explain the above factors is hydrogen bonding of the carbamate with the coating, since it has been demonstrated that the urethane group is capable of hydrogen bonding with amide-type linkages (29). This would have some confirmation in the fact that the nonpolar column had to be operated 85°C less than the two polar columns, since at identical operating temperatures the elution times were too fast. The lower column temperature has the advantage of greater thermal stability of the compound during analysis. Zielinski and Fishbein found some decomposition of the aliphatic carbamates and phenyl carbamate.

Perhaps a column could be designed that has the proper balance of polarity so that separation at lower temperatures could be achieved. It would appear that the block temperature utilized might be a little high for simple carbamates (see discussion on urethane degradation, this chapter). A chromatogram obtained by these workers on Carbowax 20M is shown in Figure 4.3.

A partial listing of carbamates and their physical properties is contained in Table 4.5.

III. ALLOPHANATES

A. Preparation

Allophanates are esters of allophanic acid

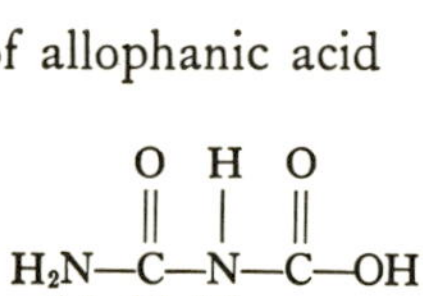

which is known only in the form of its derivatives.

The esters are frequently formed by the reaction of cyanic acid (HNCO, obtained by the depolymerization of cyanuric acid in a stream of CO_2 or

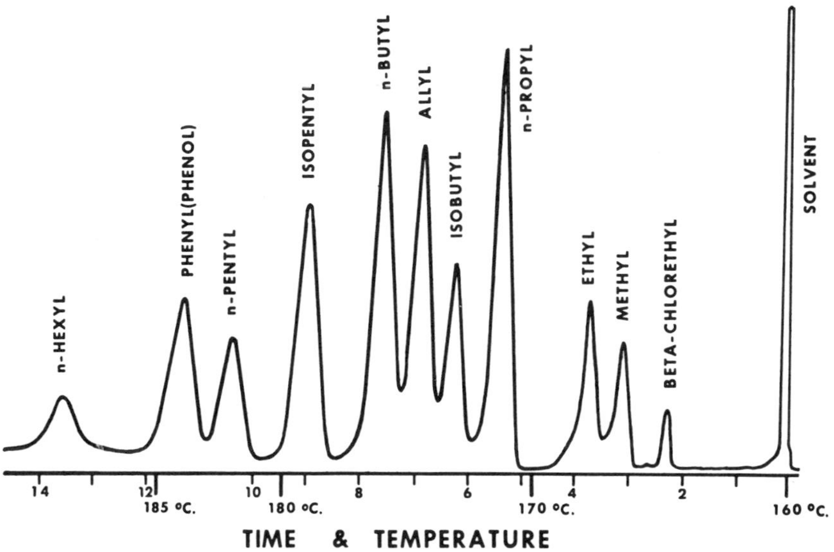

Fig. 4.3. GLC analysis of a carbamate mixture (52). Column: 10% Carbowax 20M on 60/80 mesh Chromosorb W, 8 ft × 0.125 in. o.d. stainless steel. Conditions: 30 psig helium, programmed 160–185°C at 2.1°C/min, injection port 265°C, detector block 250°C, filament current 150 mA, hot wire detector.

N_2) with alcohols and phenols. These reactions may be represented as

$$\underset{\text{Cyanuric acid}}{C_3H_3N_3O_3} \xrightarrow[CO_2 \text{ or } N_2]{360°C} \underset{\text{Cyanic acid}}{3HNCO} \tag{16}$$

$$2HNCO + ROH \rightarrow \underset{\text{Allophanate}}{H_2N{-}\overset{\overset{\displaystyle O}{\|}}{C}{-}\overset{\overset{\displaystyle H}{|}}{N}{-}\overset{\overset{\displaystyle O}{\|}}{C}{-}OH} \tag{17}$$

Numerous esters of primary, secondary, and tertiary alcohols and of phenols have been prepared and are used to characterize alcohols. Blohm and Becker have reviewed the literature through 1952 (53,54) and have described the properties of 297 derivatives. They reported on 10 additional compounds in 1956 (55).

Other unsubstituted allophanates (that is, simple esters of the parent allophanic acid, in which there are no substituents on the nitrogen atoms) have been prepared by Spielman, Barnes, and Close (56) and were tested

TABLE 4.5
Physical Properties of Urethanes (Carbamates)

I. *Urethanes Derived from Monoisocyanates*
  A. Alkyl
    1. Allyl *N-t*-octyl carbamate
    2. Ethyl *N*-2-octyl carbamate
    3. Ethyl *N-t*-octyl carbamate
    4. Methyl *N-t*-octyl carbamate
    5. Methyl *N*-octadecyl carbamate
  B. Aryl
    6. Benzyl *N*-phenyl carbamate
    7. Cinnamyl *N*-phenyl carbamate
    8. Phenyl carbanilate
  C. Mixed Alkyl and Aryl Carbamates
    9. Ethyl *N*-phenyl carbamate
    10. Ethyl-*N-p*-methoxyphenyl carbamate
    11. Ethyl-*N-p*-chlorophenyl carbamate
    12. Ethyl-*N-p*-nitrophenyl carbamate
    13. Ethyl-*N-o*-methoxyphenyl carbamate
    14. Ethyl-*N-o*-chlorophenyl carbamate
    15. Ethyl-*N-o*-nitrophenyl carbamate
    16. Ethyl-*N-o*-tolyl carbamate
    17. Ethyl-*N*-1-naphthyl carbamate
    18. Ethyl-*N*-diphenyl carbamate
    19. Ethyl-*N*-isopropyl-*N*-1-naphthyl carbamate
    20. Hexadecyl-*N*-1-naphthyl carbamate
    21. Hexadecyl-*N*-1-propyl-*N*-1-naphthyl carbamate
    22. Methylphenyl carbamate
    23. Methyl *p*-tolyl carbamate
  D. Heterocyclic
    24. Methyl-(5-methyl-2-furyl) carbamate
    25. Methyl-(2,4-dimethyl-3-furyl) carbamate
    26. Methyl-(2-methyl-2-furyl) carbamate
    27. Methyl-(2,5-dimethyl-3-furyl) carbamate
II. *Urethanes Derived from Diisocyanates*
  A. None
  B. Aryl
    28. Benzyl biscarbamate of MDI
    29. Hexamethylene polycarbamate of MDI
    30. *N*-Butyl biscarbamate of MDI
    31. Neopentyl biscarbamate of MDI
    32. Diurethane of the carbodiimide of 2,4-TDI
    33. 5,5′-Sulfonyl bis(2-*tert*-butyl) carbamic acid diethyl ester

(continued)

TABLE 4.5 (*Continued*)

No.	Name and structure	Analysis: Elem.	Analysis: Calc.	Analysis: Fd.	Infrared data: Solvent	Infrared data: Wavelength	Ref.
A-1.	Allyl-N-*t*-octyl carbamate	N	6.6	6.6			46, 49
	C_8H_{17}—NH—CO—O—CH_2—CH=CH_2						
	Formula $C_{12}H_{23}O_2N$ Mol. wt. 213.3	M.P.					
	Solvents for recrystallization	B.P. 122–124°C at 26 mm					
		n_D^{25} = 1.4417					
	Appearance	d_{25}^{25} = 0.9049					
A-2.	Ethyl *N*-2-octyl carbamate	N	7.0	6.9			9, 46
	C_6H_{13}—CH(CH_3)—NH—CO—O—C_2H_5						
	Formula $C_{11}H_{23}O_2N$ Mol. wt. 201.3	M.P.					
	Solvents for recrystallization	B.P. 88–89°C at 0.5 mm					
		n_D^{25} = 1.4371					
	Appearance	d_{25}^{25} = 0.9051					
A-3.	Ethyl *N*-*t*-octyl carbamate	N	7.0	6.9			8, 46
	C_8H_{17}—NH—CO—O—C_2H_5						

(*continued*)

TABLE 4.5 (*Continued*)

No.	Name and structure	Analysis			Infrared data		Ref.
		Elem.	Calc.	Fd.	Solvent	Wavelength	
	Formula $C_{11}H_{23}O_2N$ Mol. wt. 201.3	M.P. 20–23°C					
	Solvents for recrystallization	B.P. 123–125°C at 24 mm					
		$n_D^{25} = 1.4429$					
	Appearance	$d_{25}^{25} = 0.9227$					
A-4.	Methyl *N-t*-octyl carbamate	N	7.5	7.4			46, 70
	C_8H_{17}—NH—CO—O—CH_3						
	Formula $C_{10}H_{21}O_2N$ Mol. wt. 187.3	M.P.					
	Solvents for recrystallization	B.P. 125–127°C at 30 mm					
		$n_D^{25} = 1.4432$					
	Appearance	$d_{25}^{25} = 0.9368$					
A-5.	Methyl octadecyl carbamate						117
	$C_{18}H_{37}$—NH—CO—O—CH_3						
	Formula $C_{20}H_{41}O_2N$ Mol. wt. 324.5	M.P.					
	Solvents for recrystallization	B.P.					
	Appearance						

Compound						References
B-6. Benzyl *N*-phenyl carbamate 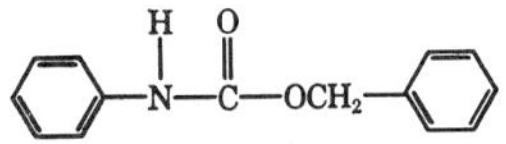						11, 42
Formula $C_{14}H_{13}O_2N$ Mol. wt. 227.2 Solvents for recrystallization Aqueous ethanol Appearance	M.P. 76°C B.P.					
B-7. Cinnamyl *N*-phenyl carbamate 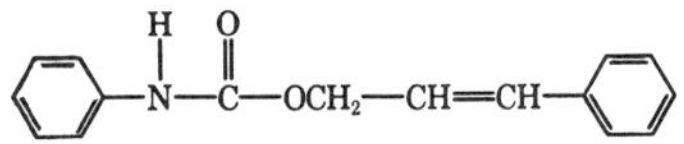						42, 46
Formula $C_{16}H_{15}O_2N$ Mol. wt. 253.3 Solvents for recrystallization Aqueous alcohol Appearance	M.P. 88–89°C B.P.					
B-8. Phenyl carbanilate 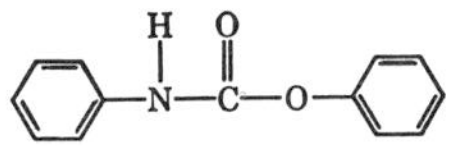						1, 70
Formula $C_{13}H_{11}O_2N$ Mol. wt. 213.2 Solvents for recrystallization Benzene Appearance	M.P. 127–128°C B.P.					

(continued)

TABLE 4.5 (*Continued*)

No.	Name and structure	Analysis: Elem.	Analysis: Calc.	Analysis: Fd.	Infrared data: Solvent	Infrared data: Wavelength	Ref.
C-9.	Ethyl *N*-phenyl carbamate				CCl_4	6756 cm^{-1}	7, 118
	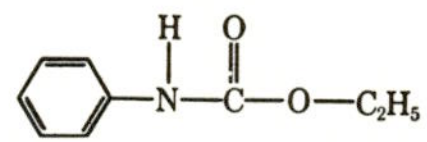				2.425	5045 cm^{-1}	
					moles/1	4970 cm^{-1}	
						4880 cm^{-1}	
	Formula $C_9H_{11}O_2N$ Mol. wt. 165.2	M.P. 48–49°C					
	Solvents for recrystallization	B.P.					
	Appearance						
C-10.	Ethyl-*N*-*p*-methoxyphenyl carbamate				CCl_4	6743 cm^{-1}	7, 118
	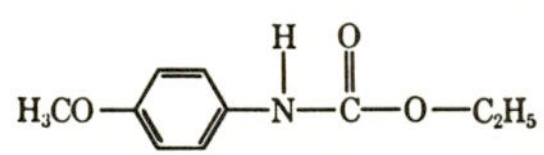				0.5020	5083 cm^{-1}	
					moles/1	4940 cm^{-1}	
						4850 cm^{-1}	
	Formula $C_{10}H_{13}O_3N$ Mol. wt. 195.2	M.P. 65°C					
	Solvents for recrystallization	B.P.					
	Appearance						
C-11.	Ethyl-*N*-*p*-chlorophenyl carbamate				CCl_4	6748 cm^{-1}	7, 118
	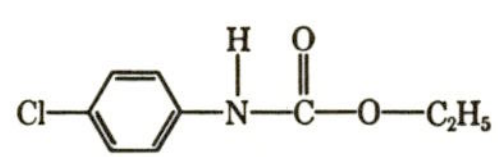				0.5140	5038 cm^{-1}	
					moles/1	4945 cm^{-1}	
						4855 cm^{-1}	

	Compound	Data	Ref.
	Formula $C_9H_{10}O_2NCl$ Mol. wt. 199.6	M.P. 66–67°C	
	Solvents for recrystallization	B.P.	
	Appearance		
C-12.	Ethyl-*N*-*p*-nitrophenyl carbamate 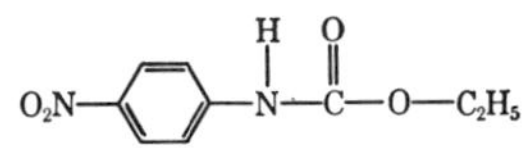		6
	Formula $C_9H_{10}O_4N_2$ Mol. wt. 210.2	M.P.	
	Solvents for recrystallization	B.P. 129–130°C	
	Appearance		
C-13.	Ethyl-*N*-*o*-methoxyphenyl carbamate 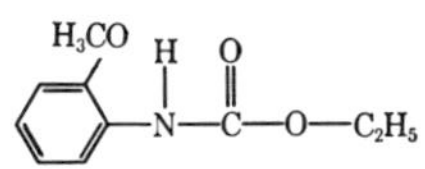		6
	Formula $C_{10}H_{13}O_3N$ Mol. wt. 195.2	M.P.	
	Solvents for recrystallization	B.P. 107–108°C at 1 mm	
	Appearance		
C-14.	Ethyl-*N*-*o*-chlorophenyl carbamate 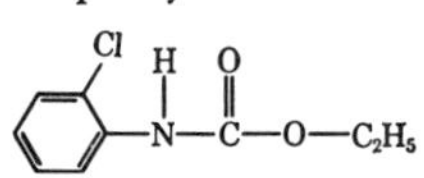		6

(*continued*)

TABLE 4.5 (*Continued*)

No.	Name and structure	Analysis			Infrared data		Ref.
		Elem.	Calc.	Fd.	Solvent	Wavelength	
	Formula $C_9H_{10}O_2NCl$ Mol. wt. 199.6	M.P.					
	Solvents for recrystallization	B.P. 119–120°C at 2 mm					
	Appearance						
C-15.	Ethyl-*N*-*o*-nitrophenyl carbamate 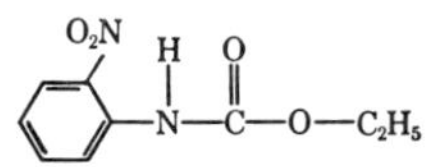	N	8.4	8.3			6
	Formula $C_9H_{10}O_4N_2$ Mol. wt. 210.2	M.P. 57–58°C					
	Solvents for recrystallization	B.P.					
	Appearance						
C-16.	Ethyl-*N*-*o*-tolyl carbamate 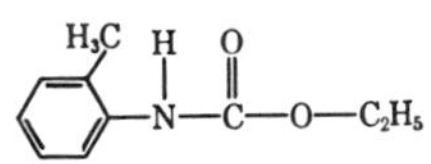						6
	Formula $C_{10}H_{13}O_2N$ Mol. wt. 179.2	M.P. 44–45°C					
	Solvents for recrystallization	B.P.					
	Appearance						

Compound						References
C-17. Ethyl-*N*-1-naphthyl carbamate 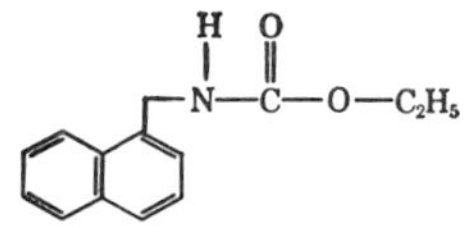						35, 37, 41
Formula $C_{13}H_{13}O_2N$ Mol. wt. 215.2	M.P.					
Solvents for recrystallization	B.P.					
Appearance						
C-18. Ethyl-*N*-diphenyl carbamate 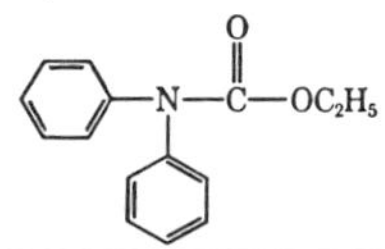						119
Formula $C_{15}H_{15}O_2N$ Mol. wt. 241.3	M.P. 72°C					
Solvents for recrystallization Water, petroleum ether, benzene, ether, alcohol	B.P. > 360°C at 760 mm					
Appearance						
C-19. Ethyl-*N*-1-propyl-*N*-1-naphthyl carbamate 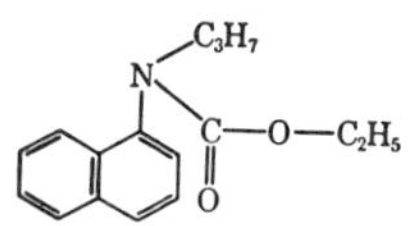	C	74.68	74.82			10, 41
	H	7.44	7.33			
	N	5.45	5.42			
Formula $C_{16}H_{19}O_2N$ Mol. wt. 257.320	M.P. 47.5–48°C					
Solvents for recrystallization *n*-Hexane	B.P. 143–144°C at 96 mm					
Appearance						

(continued)

TABLE 4.5 (*Continued*)

No.	Name and structure	Analysis			Infrared data		Ref.
		Elem.	Calc.	Fd.	Solvent	Wavelength	
C-20.	Hexadecyl-*N*-1-naphthyl carbamate 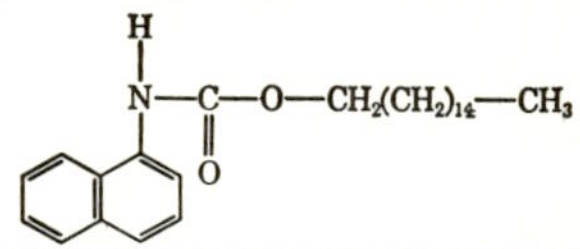						12, 41
	Formula $C_{27}H_{41}O_2N$ Mol. wt. 411.6	M.P.					
	Solvents for recrystallization	B.P.					
	Appearance						
C-21.	Hexadecyl-*N*-1-propyl-*N*-1-naphthyl carbamate 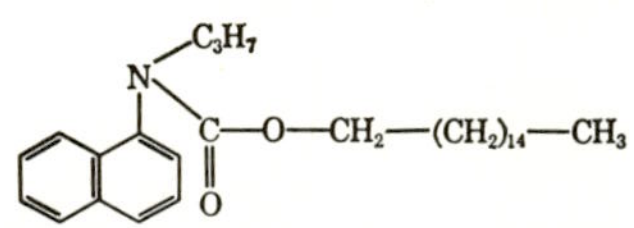	C	79.42	79.29			12, 41
		H	10.44	10.32			
		N	3.09	2.92			
	Formula $C_{30}H_{47}O_2N$ Mol. wt. 453.7	M.P.					
	Solvents for recrystallization	B.P. 240–242°C at 2–3 mm					
	Appearance						
C-22.	Methylphenyl carbamate 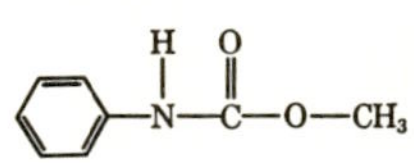						8, 39

Formula $C_8H_9O_2N$ Mol. wt. 151.2

Solvents for recrystallization Legroin

Appearance

M.P. 48°C

B.P.

C-23. Methyl *p*-tolyl carbamate

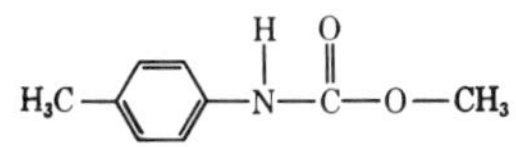

Ref. 6

Formula $C_9H_{11}O_2N$ Mol. wt. 165.2

Solvents for recrystallization

Appearance

M.P. 44–45°C

B.P.

D-24. Methyl 1-(5-methyl-2-furyl) carbamate

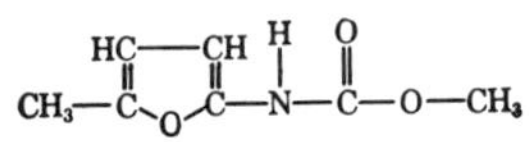

N	9.03	9.11			11, 41

Formula $C_7H_9O_3N$ Mol. wt. 155.2

Solvents for recrystallization

Appearance

M.P. 64–66°C

B.P.

D-25. Methyl (2,4-dimethyl-3-furyl) carbamate

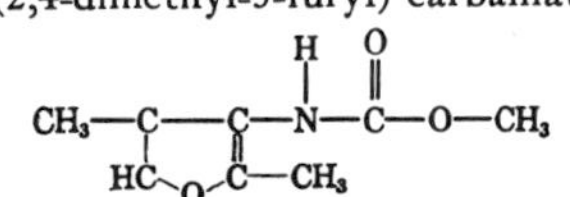

N	8.28	8.13			11, 41

(continued)

TABLE 4.5 (*Continued*)

No.	Name and structure	Analysis			Infrared data		Ref.
		Elem.	Calc.	Fd.	Solvent	Wavelength	
	Formula $C_8H_{11}O_3N$ Mol. wt. 169.2	M.P. 95–96°C					
	Solvents for recrystallization	B.P.					
	Appearance						
D-26.	Methyl (2-methyl-3 furyl) carbamate 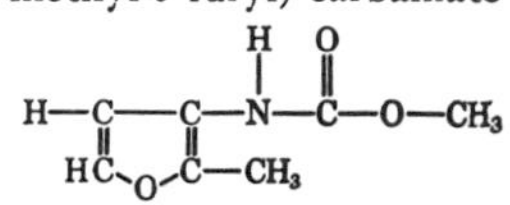	N	9.03	0.55			11, 41
	Formula $C_7H_9O_3N$ Mol. wt. 155.2	M.P.					
	Solvents for recrystallization	B.P. 128°C at 14 mm					
	Appearance						
D-27.	Methyl (2,5-dimethyl,3-furyl) carbamate 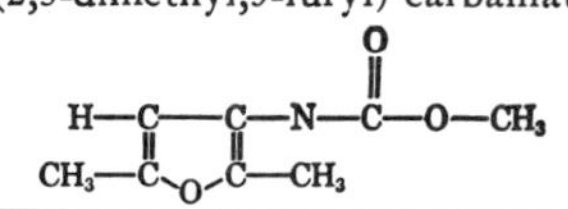	N	8.28	8.43			11, 41
	Formula $C_8H_{11}O_3N$ Mol. wt. 169.2	M.P.					
	Solvents for recrystallization	B.P. 100°C at 2.5 mm					
	Appearance						

D-28. Benzyl biscarbamate of MDI						
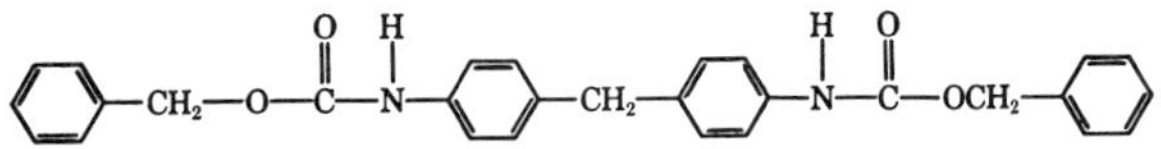	C	74.66	75.16			39
	H	5.62	5.65			
	N	6.01	5.83			
Formula $C_{29}H_{26}O_4N_2$ Mol. wt. 466.5	M.P. 198–199°C					
Solvents for recrystallization Acetone	B.P.					
Appearance						

D-29. Hexamethylene polycarbamate of MDI						
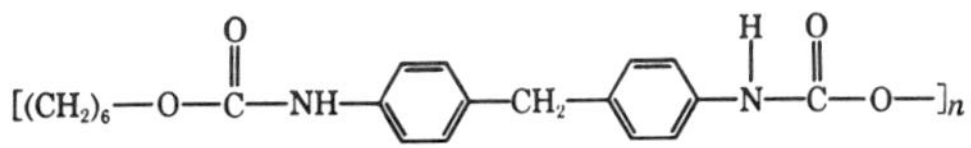	C	80.36	80.48			39
	H	4.96	5.37			
	N	13.05	13.01			
Formula Mol. wt.	M.P. 229–231°C; decomposes					
Solvents for recrystallization	B.P.					
Appearance						

D-30. *N*-Butyl biscarbamate of MDI						
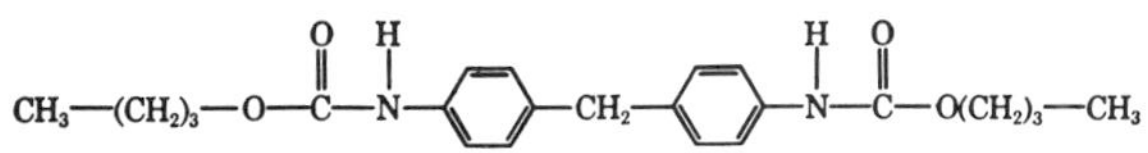	C	69.32	69.03			39
	H	7.59	7.46			
	N	7.03	7.03			
	O	16.06	—			
Formula $C_{23}H_{30}O_4N_2$ Mol. wt. 398.5	M.P. 119°C					
Solvents for recrystallization Ethanol	B.P.					
Appearance						

(*continued*)

TABLE 4.5 (*Continued*)

No.	Name and structure	Analysis			Infrared data		Ref.
		Elem.	Calc.	Fd.	Solvent	Wavelength	
D-31.	Neopentyl biscarbamate of MDI	C	70.39	70.84			39
	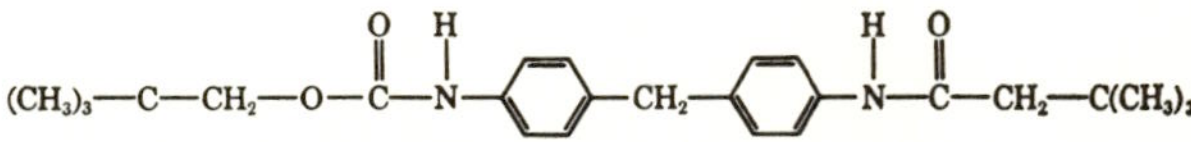	H	8.03	8.01			
		N	6.57	6.52			
		O	15.00	—			
	Formula $C_{25}H_{34}O_4N_2$ Mol. wt. 426.5	M.P. 128°C					
	Solvents for recrystallization	B.P.					
	Appearance						
D-32.	Diurethane of the carbodiimide of 2,4-TDI					—C=O of urethane,	120
	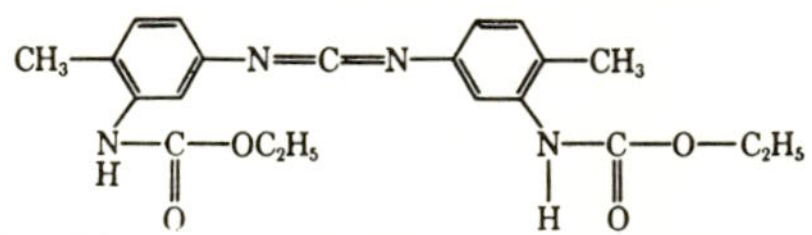					5.82 μ	
						—N=C=N—	
						4.70 μ	
	Formula $C_{21}H_{24}O_4N_4$ Mol. wt. 396.4	M.P.					
	Solvents for recrystallization	B.P.					
	Appearance						
D-33.	5,5′Sulfonyl bis(2-*tert*-butyl) carbamic acid diethyl ester	N	5.55	5.41			2, 121
	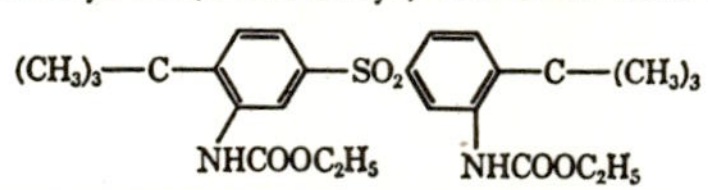	S	6.35	6.18			
	Formula $C_{26}H_{36}O_6N_2S$ Mol. wt. 504.6	M.P. 135°C					
	Solvents for recrystallization CCl_4	B.P.					
	Appearance						

for anticonvulsant activity for the treatment of epilepsy. Some 36 such compounds were described.

It is evident that substitution can also occur on the two nitrogen atoms labeled α and γ in the above structure. Such substituted allophanates frequently occur as reaction products of isocyanates. The initial product from the reaction between a urethane and an isocyanate is an allophanate.

$$\mathrm{RNCO + RNHCOOR' \rightarrow \underset{\underset{\text{Allophanate}}{\text{CONHR}}}{\underset{|}{\text{RNCOOR}'}}} \tag{18}$$

This reaction is not catalyzed strongly by tertiary amines, requiring stronger bases or requiring temperatures of approximately 120–140°C in uncatalyzed systems in order to provide a significant reaction rate.

Lakra and Dains (57) reacted phenyl isocyanate and ethylurethane for 6 hr at 140–160°C and isolated the four compounds (a, b, c, and d).

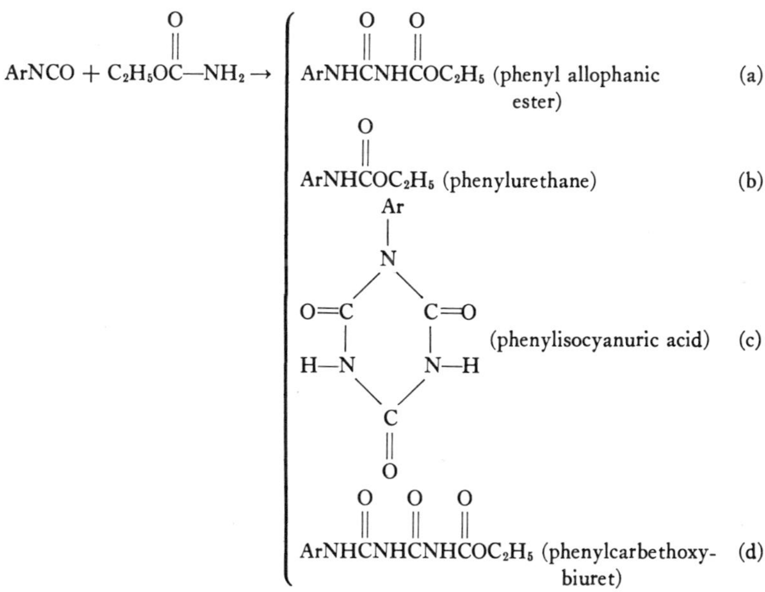

(19)

Saunders and Frisch (29) have pointed out that the allophanate is the normal addition product. The phenylurethane could result from

$$\mathrm{H_2N\overset{\overset{O}{\|}}{C}OC_2H_5 \underset{}{\overset{\text{decomposition}}{\rightleftharpoons}} HNCO + C_2H_5OH} \tag{20}$$

$$\mathrm{C_6H_5NCO + C_2H_5OH \rightarrow C_6H_5NHCOOC_2H_5} \tag{21}$$

The phenylisocyanuric acid could have been formed from

$C_6H_5NCO + 2HNCO \longrightarrow$ 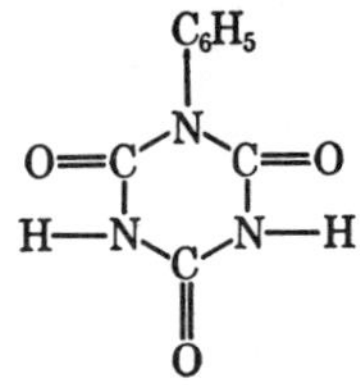 (22)

The phenylcarbethoxybiuret could have been formed by the stepwise reaction sequence

$$HNCO + H_2NCOOC_2H_5 \rightarrow H_2N\overset{O}{\overset{\|}{C}}NH\overset{O}{\overset{\|}{C}}C_2H_5 \qquad (23)$$

$$C_6H_5NCO + H_2N\overset{O}{\overset{\|}{C}}NH\overset{O}{\overset{\|}{C}}OC_2H_5 \rightarrow C_6H_5NH\overset{O}{\overset{\|}{C}}NH\overset{O}{\overset{\|}{C}}NH\overset{O}{\overset{\|}{C}}OC_2H_5 \qquad (24)$$

Kogon (6) studied reactions between phenyl isocyanate and ethyl alcohol and showed that α,γ-diphenylallophanate is formed when phenyl isocyanate and ethyl alcohol are heated at 125°C. The same allophanate is obtained from phenyl isocyanate and ethyl carbanilate.

The same reactants with *N*-methyl morpholine as a catalyst yield phenyl triisocyanurate. Evidence was presented to support the view that α,γ-diphenylallophanate and the dimer of phenyl isocyanate are intermediates in the formation of triphenylisocyanurate.

B. Catalysis and Reactions

The uncatalyzed isocyanate–urethane reaction is even slower than the isocyanate–urea reaction, so there may be even more possibilities for side reactions in the case of urethanes.

Bennet et al. (58) observed the reaction between monoisocyanates and model urethanes. The reaction was found to be much slower than the isocyanate–urea reaction, and the data did not give uniform rate constants with any kinetic equation. The calculated second-order rate constants became progressively smaller as the reaction proceeded.

The reaction between aryl isocyanates and ethyl-substituted carbanilates in carbon tetrachloride as the solvent was studied by Kogon (59). A large excess of isocyanate was used and the reaction was followed by measuring the rate of disappearance of the first overtone of the —NH stretching vibration which occurs near 6750 cm^{-1} in the near infrared. The data were found to fit a second-order rate plot and showed that an increase in reaction temperature from 106°C to 137°C resulted in a decrease of the value for $K_{eq.}$ (k_1/k_2). The rate constants for the reaction of phenyl isocyanate with ethyl carbanilate at 128°C were approximately 0.045×10^4 liter $mole^{-1}$ sec^{-1}. The presence of *N*-methylmorpholine was not found to have any appreciable catalytic effect on allophanate formation.

As the reaction approached equilibrium a large exotherm occurred followed by solidification of the solution to a mixture of triphenyl isocyanurate and ethyl carbanilate.

The rate measurements showed that substitution in the *para* position of the aromatic ring by a methyl group decreased the rate of reaction compared to the unsubstituted and *o*-methyl substituted carbanilate. Although this is contrary to what might be expected, it was explained by assuming ionization of the carbanilate occurs at elevated temperatures and the decreased rate of reaction of ethyl *p*-methylcarbanilate with phenyl isocyanate compared to ethyl carbanilate with phenyl isocyanate is probably due to retardation of the ionization of the carbanilate to the enol anion by the *p*-methyl group (60).

Although these data show that dissociation of the allophanate link occurs at 106°C and we would expect the allophanate crosslink to be relatively unstable, the stability of this crosslink might be reinforced in polymers from hydrogen bonding and molecular fit.

Cooper et al. (61) showed that urethanes prepared from *p*-phenylene diisocyanate and unsaturated glycols are much more reactive at 100°C than those from saturated glycols.

Kogon (62), in further studies on the isocyanate reactions, has shown that allophanates can be prepared at room temperature by use of a variety of metal carboxylates (naphthenates, hexanoates, and linoresinates) to catalyze the reaction between phenyl isocyanate and ethyl carbanilate. Lead and cobalt derivatives were found to be the most active. At room temperature the metal carboxylates, except zinc, were reported to catalyze the formation of triphenyl isocyanurate from phenyl isocyanate, although no data were given.

Thionyl chloride reacts with ethyl and *n*-butyl carbamate ($NH_2COOC_3H_9$) to give the corresponding allophanate ($H_2NCONHCOOC_3H_9$), according to Raiford and Freyermuth (63).

The rate of dissociation of allophanates was observed by Kogon (64) by following the appearance of the NH absorption band at 6750 cm^{-1}.

C. Analytical Techniques and Characterization

As in the case of urethanes, the simple allophanates have most often been identified by elemental analysis and mixed melting point determinations.

Kogon (6) has examined the infrared spectra of a number of α,γ-diarylallophanates. His studies have shown that these allophanates exhibit twin carbonyl absorption bands at 5.70 and 5.86 μ and similar bands at 5.74 and 5.90 μ for substituted allophanates. In isolated form these compounds are easily recognizable, but in mixtures of urethanes, isocyanurates, and allophanates, due to mutual interference and band overlapping they cannot be distinguished. However, in mixtures of urethanes and allophanates, a band present at the higher wavelength ($\approx$5.83 μ) indicates the presence of an allophanate.

Kogon (7) has pointed out that the hydroxyl group has an absorption band in the region of 7000 cm^{-1} due to the first overtone of the fundamental OH stretching vibration when no intramolecular hydrogen bonding is expected (65). Hilbert, Wulf, Hendricks, and Liddel (66,67) found that absorption in this region is absent for a large number of compounds which have configurations that favor the formation of the intramolecular hydrogen bond.

Kogon found that intramolecular hydrogen bonding could be detected by the near infrared absorption band of the first overtone of the fundamental stretching vibrations at 6700–6800 cm^{-1} in the spectra of α,γ-diarylallophanates and in *asym*-aryltrialkyl and *sym*-tetraalkyl biurets.

The aryl and alkyl ureas and aryl carbamates Kogon examined exhibited strong absorption bands in the region of 6750, 5000, 4900, and 4850 cm^{-1}. The absence of the absorption band at 6750 cm^{-1} in the spectra of allophanates and biurets suggests that these compounds are intramolecularly hydrogen bonded. This, of course, allowed Kogon (59) to study allophanate formation by observing the disappearance of the NH band at 6750 cm^{-1} as the carbanilate reacted with isocyanate to form the corresponding allophanate.

In the case of mixtures of allophanates and biurets, distinguishing is not possible in the carbonyl region due to band overlap as before. However, both compounds exhibit double bands of which the one with the higher frequency is more intense. Table 4.6 shows the carbonyl absorption frequencies for a number of allophanates.

Brumfield (51) also studied the spectra of several allophanates. The approach used in determining the contribution of the allophanates was identical to that previously described for carbamates. As an example of the calculations used by Brumfield to compute structure contribution for

TABLE 4.6
Absorption Frequency for the Carbonyl Group in Simple Allophanates (50,51)

Compound	Frequency, cm^{-1}
n-Butyl 1,4-diphenylallophanate	1729–1732 (melt) 1729–1732 (KBr pellet)
Methyl α,γ-dibutylallophanate	1725, 1712
Methyl α,γ-didecylallophanate	1723, 1693
Methyl α,γ-dihexamethylene-isocyanatoallophanate	1720, 1690
Methyl α,γ-diphenylallophanate	1725, 1691

allophanates, *n*-butyl 2,4-diphenylallophanate was evaluated as follows. The structure was divided into allophanate, phenyl, and butyl groups.

Melt Fig. (4.4)	KBr Pellet (Fig. 4.5)
Allophanate (HNCONCOO)	
3255, 1732, 1559 cm^{-1}	3220, 3200, 1547 cm^{-1}

$$\frac{\text{“A” (allophanate)} \times 100}{\text{Total “A”}}$$

Melt	KBr Pellet
Total “A” = 2.629	Total “A” = 1.185
$\frac{0.861 \times 100}{2.629} = 32.75\%$ (actual 32.40%)	$\frac{0.386 \times 100}{1.185} = 32.75\%$ (actual 32.40%)
Phenyl (C_6H_5)	
3218, 3150, 3097, 3067, 2904, 1601, 1593, 1499, 1483, and 1470 cm^{-1}	3064, 3023, 2903, 1611, 1593, 1499, and 1470 cm^{-1}

$$\frac{\text{“A” (phenyl)} \times 100}{\text{Total “A”}}$$

Melt	KBr Pellet
$\frac{1.292 \times 100}{2.629} = 49.20\%$ (actual 49.35%)	$\frac{0.582 \times 100}{1.185} = 49.20\%$ (actual 49.35%)
Butyl (C_4H_9)	
2974, 2939, 2929, 2887, and 2868 cm^{-1}	2988, 2964, 2937, 2878, and 2859 cm^{-1}

$$\frac{\text{“A” (butyl)} \times 100}{\text{Total “A”}}$$

Melt	KBr Pellet
$\frac{0.476 \times 100}{2.629} = 18.12\%$ (actual 18.25%)	$\frac{0.216 \times 100}{1.185} = 18.23\%$ (actual 18.25%)

While it is easily understandable and expected that exact frequencies will shift, depending on sample state (melt spectrum or KBr pellet) as well as on specific group influence, it is difficult to understand how such good results could be obtained in the case of the KBr pellet spectrum without utilizing the absorption of the Amide I band.

It has been pointed out previously that the Amide I band is specific for

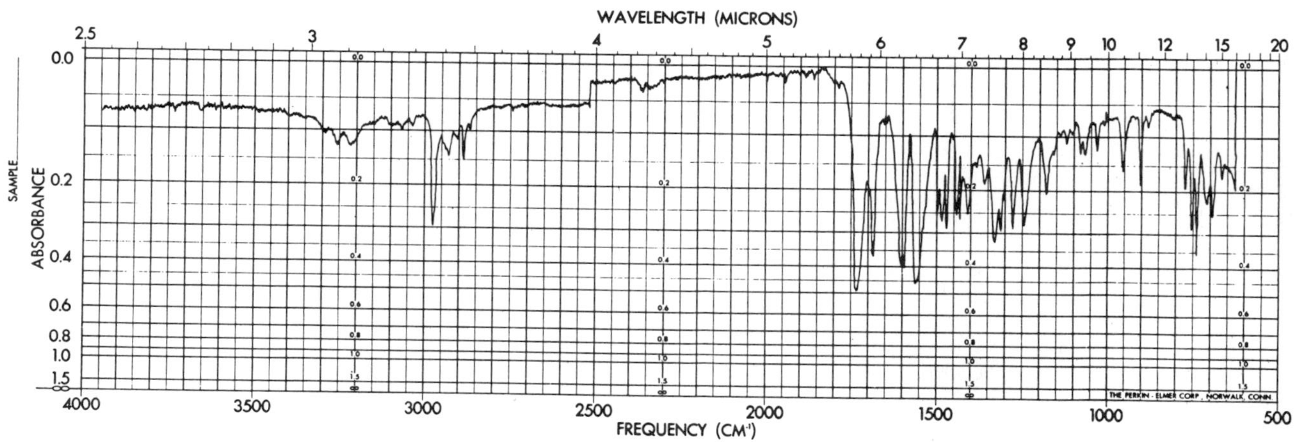

Fig. 4.4. Infrared spectrum of *n*-butyl 2,4-diphenylallophanate (melt) (51).

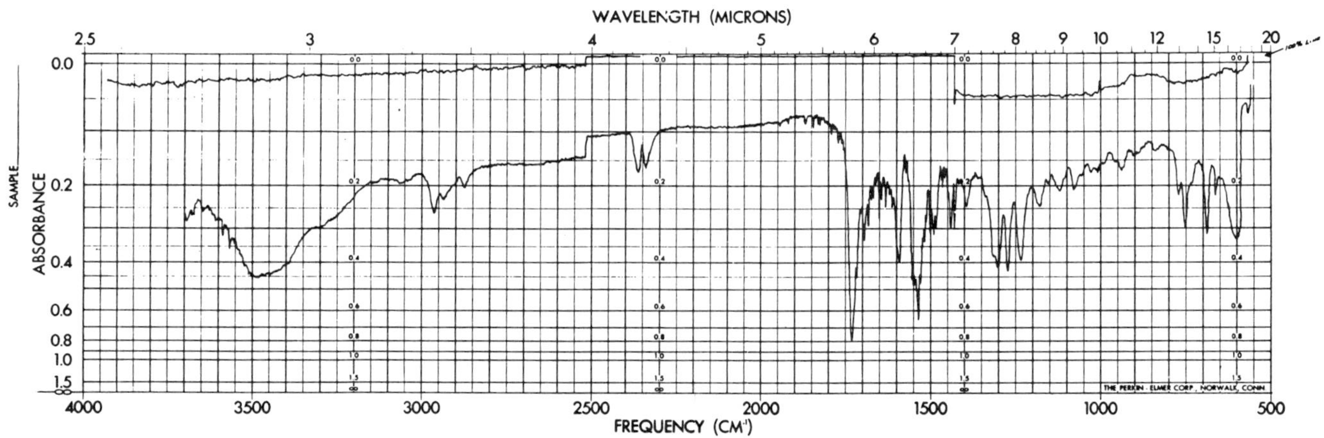

Fig. 4.5. Infrared spectrum of *n*-butyl 2,4-diphenylallophanate (0.11% n KBr Pellets) (51).

constitution, and since it contributes significantly to the allophanate group absorption it would be expected that it would have to be used. This band was utilized for evaluating the melt spectrum. The explanation appears to be that although this technique is useful and soundly based on previous literature correlations, it is highly empirical, and a good deal of trial and error must be used in determining which bands are specific for evaluation purposes. This would obviously limit this technique for use in polymer evaluations of structure contributions.

Brumfield has listed the following literature correlations with the bands used.

Literature correlations	
Free —NH	3530–3415 cm^{-1}
Amide I	1715–1675 cm^{-1}
Amide II	1560–1520 cm^{-1}
Phenyl	3089, 3070, 3063, 2905, 2886, 1617, 1605, 1585, and 1497 cm^{-1}
CH_3 stretching	2962–2972 cm^{-1}
CH_2 stretching	2916–2936 cm^{-1}
CH stretching	2880–2900 cm^{-1}
CH_3 stretching	2862–2882 cm^{-1}
CH_2 bending	1445–1485 cm^{-1}
CH_2 bending	1435–1475 cm^{-1}

The contribution of one part of a particular group to the entire group structure was also calculated by Brumfield. For example, the percent of

TABLE 4.7
Physical Properties of Various Allophanates

1. Ethyl α,γ-diphenyl allophanate
2. Ethyl α-*p*-methoxyphenyl α-phenyl allophanate
3. Ethyl α-*p*-chlorophenyl γ-phenyl allophanate
4. Methyl α-*p*-tolyl γ-phenyl allophanate
5. Ethyl α-phenyl γ-*o*-nitrophenyl allophanate
6. Ethyl α-phenyl γ-*p*-methoxyphenyl allophanate
7. Ethyl α,γ-di-*p*-methoxyphenyl allophanate
8. Ethyl α-*o*-methoxy, γ-phenyl allophanate
9. Ethyl α-*p*-chlorophenyl γ-*p*-tolyl allophanate
10. Ethyl α-*o*-chlorophenyl γ-*m*-methoxyphenyl allophanate
11. Ethyl α,γ-di-*p*-tolyl allophanate
12. Phenyl α,γ-diphenyl allophanate
13. Methyl allophanate
14. Methyl Cellosolve allophanate
15. Diethylene glycol diallophanate
16. Ethyl γ-phenyl allophanate
17. *n*-Butyl allophanate
18. Methyl α,γ-diphenyl allophanate
19. Ethyl α,γ-di-*p*-bromophenyl allophanate

TABLE 4.7 (*Continued*)

No.	Name and structure	Analysis			Infrared data		Ref.
		Elem.	Calc.	Fd.	Solvent	Wavelength	
1.	Ethyl α,γ-diphenyl allophanate 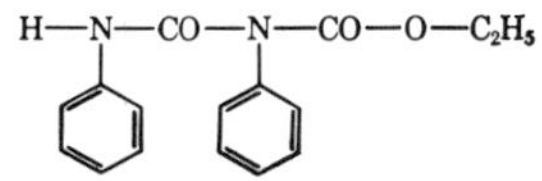				Benzene	4853 cm^{-1}	6, 7, 70
					CCl_4		
					Nujol	5.74 μ, 5.90 μ	
	Formula $C_{16}H_{16}O_3N_2$ Mol. wt. 284.30	M.P. 92–94°C					
	Solvents for recrystallization CCl_4 and petroleum ether	B.P.					
	Appearance						
2.	Ethyl α-*p*-methoxyphenyl γ-phenyl allophanate 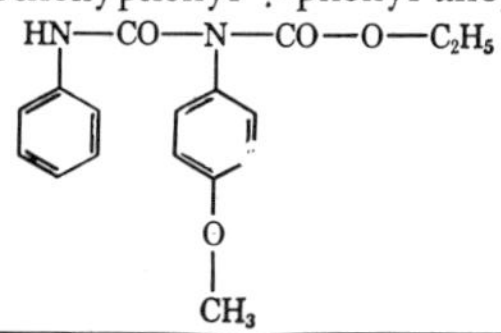	N	8.9	9.4	Benzene	4851 cm^{-1}	7
					Nujol mull	5.74 μ, 5.90 μ	
	Formula $C_{17}H_{18}O_4N_2$ Mol. wt. 314.33	M.P. 136–137°C					
	Solvents for recrystallization Ethyl alcohol	B.P.					
	Appearance						
3.	Ethyl α-*p*-chlorophenyl γ-phenyl allophanate 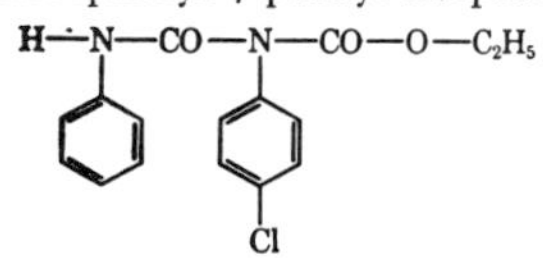	N	8.8	8.7	Benzene	4836 cm^{-1}	7
					Nujol mull	5.74 μ, 5.90 μ	

(*continued*)

TABLE 4.7 (*Continued*)

No.	Name and structure	Analysis			Infrared data		Ref.
		Elem.	Calc.	Fd.	Solvent	Wavelength	
	Formula $C_{16}H_{15}O_3N_2Cl$ Mol. wt. 318.75	M.P. 89–90°C					
	Solvents for recrystallization Ethyl alcohol	B.P.					
	Appearance						
4.	Methyl α-*p*-tolyl γ-phenyl allophanate 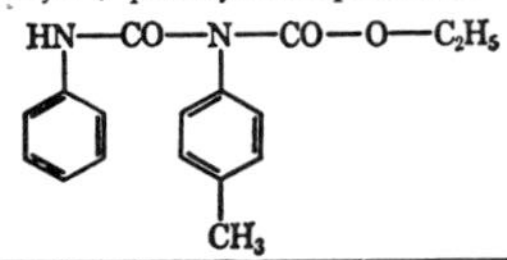	N	9.8	9.7	Nujol mull	5.74 μ, 5.90 μ	6
	Formula $C_{16}H_{16}O_3N_2$ Mol. wt. 284.30	M.P. 124–125°C					
	Solvents for recrystallization Hot water	B.P.					
	Appearance						
5.	Ethyl α-phenyl γ-*o*-nitrophenyl allophanate 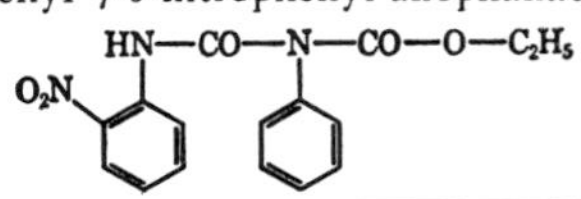	N	12.8	13.2	Nujol mull	5.74 μ, 5.90 μ	6
	Formula $C_{16}H_{15}O_5N_3$ Mol. wt. 329.30	M.P. 116–118°C					
	Solvents for recrystallization Ethyl alcohol	B.P.					
	Appearance						

Compound						
6. Ethyl α-phenyl γ-*p*-methoxyphenyl allophanate 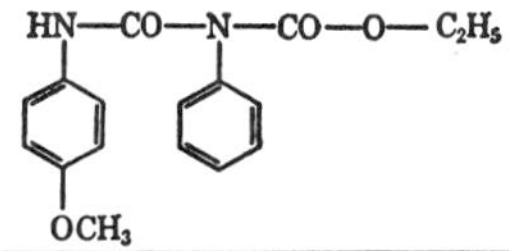	N	8.9	8.8			6
				Nujol mull	5.74 μ, 5.90 μ	
Formula $C_{17}H_{18}O_4N_2$ Mol. wt. 314.33	M.P. 116–118°C					
Solvents for recrystallization Ethyl alcohol	B.P.					
Appearance						
7. Ethyl α,γ-di-*p*-methoxyphenyl allophanate 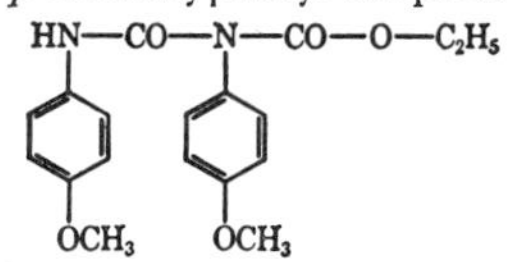	N	8.1	8.3	Nujol mull	5.74 μ, 5.90 μ	6
Formula $C_{18}H_{20}O_5N_2$ Mol. wt. 344.36	M.P. 130–131°C					
Solvents for recrystallization Ethyl alcohol	B.P.					
Appearance						
8. Ethyl α-*o*-methoxy, γ-phenyl allophanate 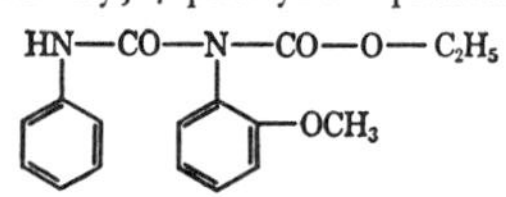	N	8.6	8.9	Nujol mull	5.92 μ, 5.94 μ	55, 122
Formula $C_{17}H_{18}O_4N_2$ Mol. wt. 314.33	M.P. 95°C					
Solvents for recrystallization Ethyl alcohol	B.P.					
Appearance						

(*continued*)

TABLE 4.7 (*Continued*)

No.	Name and structure	Analysis			Infrared data		Ref.
		Elem.	Calc.	Fd.	Solvent	Wavelength	
9.	Ethyl α-*p*-chlorophenyl γ-*p*-tolyl allophanate 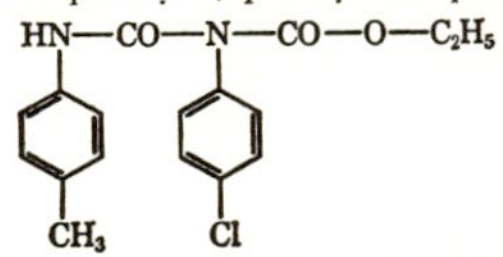	N	8.4	8.5	Nujol mull	5.82 μ, 5.95 μ	55, 122
	Formula $C_{17}H_{17}O_3N_2$ Mol. wt. 297.32	M.P. 96°C					
	Solvents for recrystallization Ethyl alcohol	B.P.					
	Appearance						
10.	Ethyl α-*o*-chlorophenyl γ-*m*-methoxyphenyl allophanate 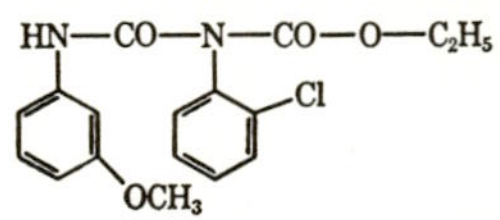	N	8.1	8.5	Nujol mull	5.83 μ, 5.96 μ	55, 122
	Formula $C_{17}H_{17}O_4N_2Cl$ Mol. wt. 348.78	M.P. 106°C					
	Solvents for recrystallization Ethyl alcohol	B.P.					
	Appearance						
11.	Ethyl α,γ-di-*p*-tolyl allophanate 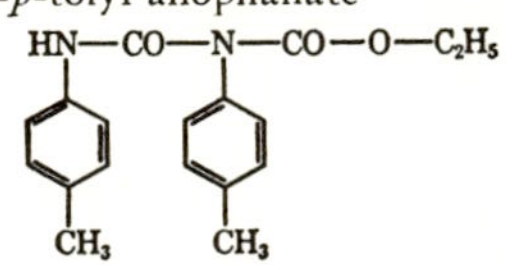	N	9.0	9.1	Nujol mull	5.73 μ, 5.88 μ	55, 122

Compound	Element	Calcd.	Found	Medium	Bands	Ref.
Formula $C_{18}H_{20}O_3N_2$ Mol. wt. 312.36	M.P. 104–106°C					
Solvents for recrystallization Ethyl alcohol	B.P.					
Appearance						
12. Phenyl α,γ-diphenyl allophanate 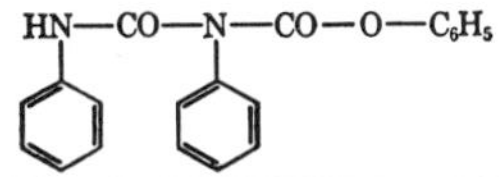	N	8.5	8.8	Nujol mull	5.75 μ, 5.85 μ	55, 122
Formula $C_{20}H_{16}O_3N_2$ Mol. wt. 332.34	M.P. 139–141°C					
Solvents for recrystallization Ethyl alcohol	B.P.					
Appearance						
13. Methyl allophanate $H_2N—CO—NH—CO—O—CH_3$				Nujol mull	5.70 μ, 5.85 μ	6
Formula $C_3H_6O_3N_2$ Mol. wt. 118.09	M.P. 218.5°C					
Solvents for recrystallization	B.P.					
Appearance						
14. Methyl Cellosolve allophanate $H_2N—CO—NH—CO—O—CH_2—CH_2—O—CH_3$				Nujol mull	5.70 μ, 5.85 μ	6

(continued)

TABLE 4.7 (*Continued*)

No.	Name and structure	Analysis			Infrared data		Ref.
		Elem.	Calc.	Fd.	Solvent	Wavelength	
	Formula $C_5H_{10}O_4N_2$ Mol. wt. 162.15	M.P. 161–162°C					
	Solvents for recrystallization Methyl alcohol	B.P.					
	Appearance						
15.	Diethylene glycol diallophanate	C	34.5	34.2	Nujol mull	5.70 μ, 5.85 μ	6
	CH_2—CH_2—O—CO—NH—CO—NH_2	H	5.06	5.10			
	O (bridging the two chains)	N	20.1	20.0			
	CH_2—CH_2—O—CO—NH—CO—NH_2						
	Formula $C_8H_{14}O_7N_4$ Mol. wt. 278.22	M.P. 237–238°C					
	Solvents for recrystallization Water, with difficulty	B.P.					
	Appearance						
16.	Ethyl γ-phenyl allophanate						10, 123
	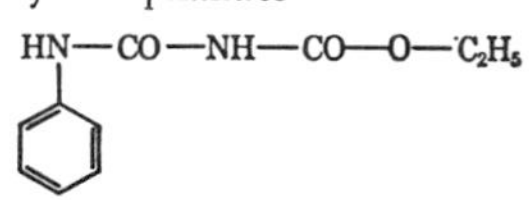						
	Formula $C_{10}H_{12}O_3N_2$ Mol. wt. 208.21	M.P. 106–107°C					
	Solvents for recrystallization	B.P.					
	Appearance						

Compound	Analysis	Calc.	Found			Ref.
17. n-Butyl allophanate $H_2N—CO—NH—CO—O—C_4H_9$	N	17.50	16.78			123, 124
Formula $C_6H_{12}O_3N_2$ Mol. wt. 160.17	M.P. 149–150°C					
Solvents for recrystallization Benzene	B.P.					
Appearance Colorless fluffy needles						
18. Methyl α,γ-diphenyl allophanate 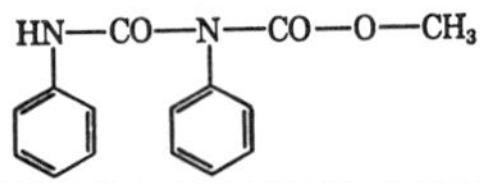	N	10.37	10.4			6, 124
	C	66.7	66.6			
	H	5.19	5.18			
Formula $C_{15}H_{14}O_3N_2$ Mol. wt. 270.28	M.P. 128–129°C					
Solvents for recrystallization Methanol	B.P.					
Appearance						
19. Ethyl α,γ-di-*p*-bromophenyl allophanate 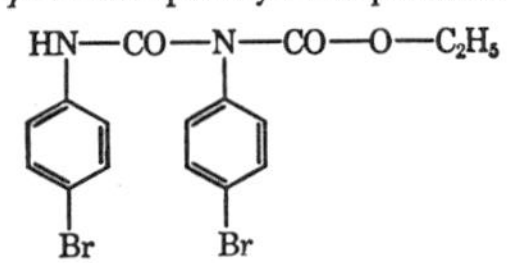	C	43.44	43.64			10
	H	3.17	3.40			
Formula $C_{16}H_{14}O_3N_2Br_2$ Mol. wt. 442.12	M.P. 153° C					
Solvents for recrystallization Ether	B.P.					
Appearance Fine white needles	Soluble in ether and alcohol					

the carbamate-type structure (—NCOO—) in the allophanate group (—NCONCOO—) was determined by means of the following:

$$\frac{\%\ \text{Carbamate (known)}}{\text{Abs. of carbamate at } 1600\ \text{cm}^{-1}} = \frac{\%\ \text{Carbamate in allophanate (unknown)}}{\text{Abs. of carbamate in allophanate at } 1699\ \text{cm}^{-1}}$$

A partial listing of allophanates and their physical properties is contained in Table 4.7.

IV. UREAS

A. Structural Relationships

The parent compound of this group of derivatives is urea, first prepared in 1828 by Wohler. It is manufactured in large quantities for use as a fertilizer component.

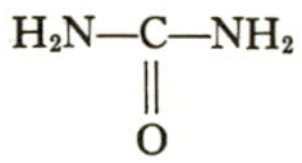

Structurally, urea may be considered as the diamide of carbonic acid as shown by the structures illustrated here.

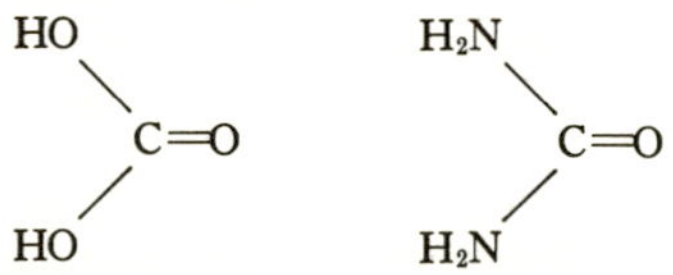

Carbonic acid Urea

Urea is sometimes called *carbamide*, which should not be confused with the name carbamate that is applied to urethane compounds.

Alkyl or aryl groups may replace the hydrogen atoms on the nitrogen to give a series of *N*-substituted ureas. These may be mono- or disubstituted on each or both of the nitrogen atoms, giving rise to symmetrical or asymmetrical compounds. If one of the hydrogen atoms is substituted with a phenyl group it is called phenyl urea, or phenyl carbamide.

$$C_6H_5—NH—CO—NH_2$$

If two phenyl groups replace two hydrogen atoms, it may be symmetrical or asymmetrical diphenyl urea; the symmetrical diphenyl urea is sometimes called *carbanilide*.

C_6H_5—NH—CO—NH—C_6H_5

(*sym*) Diphenyl urea
or carbanilide

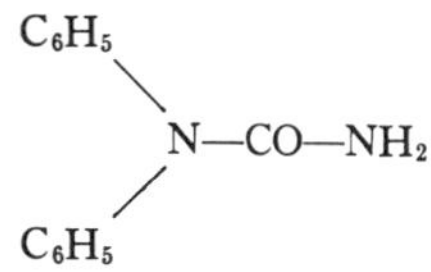

(*Asym*) Diphenyl urea

Further substitution products of the symmetrical diphenyl urea may also be called substituted *carbanilides*. To make the nomenclature even more confusing there are a series of compounds called *carbanilates* which are urethane compounds discussed in the urethane section.

Substitution may also occur on the oxygen atom to give *o*-substituted ureas. A large class of compounds called thioureas occur in which sulfur replaces the oxygen atom. Such compounds will not be considered in this book. Most of the compounds obtained from isocyanates are of the *N*-substituted type.

The substituted ureas are identified by the position of the substituent according to the following scheme:

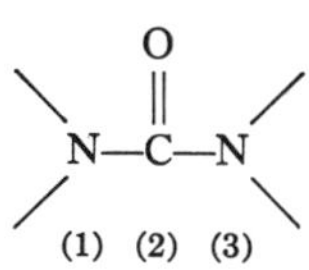

The substituents on the nitrogen atoms are numbered as 1,3-ureas. Sometimes these are identified as *N,N′* substitution.

A casual inspection of a recent *Chemical Abstracts Index* will disclose thousands of entries under the heading of Urea. For the most part the literature discussed herein has some reference to isocyanate chemistry.

B. Preparation of Substituted Ureas

1. By the Reaction of Isocyanates

Compounds containing the —NH group are potentially reactive with isocyanates, the most basic generally being the most reactive.

Disubstituted ureas are readily formed by the reaction of an isocyanate with an amine, represented by:

$$RNCO + R'NH_2 \rightarrow R\text{—}NH\text{—}CO\text{—}NH\text{—}R'$$

This reaction is frequently employed in the analysis of isocyanate products. An excess of dibutyl amine is added to a solution of the isocyanate in a suitable solvent, and allowed to react. The excess amine is determined by back-titration with standard hydrochloric acid. The dibutyl urea of

2,4-tolylene diisocyanate in which there are two urea groups (compound C-8 in Table 4.9) has been characterized by Saunders and Hardy (68). This is the only substituted urea found in the literature produced by this very common analytical procedure which has been identified.

Simons and Arnold (32) studied the relative reactivity of the two isocyanate groups in 2,4-tolylene diisocyanate by treating it with aniline to obtain a substituted urea formed from the isocyanate in the 4-position. The free isocyanate in the 2-position was hydrolyzed with HCl to obtain the 3-amino-4-methyl carbanilide (compound B-12). The infrared spectrum of this compound and a mixed melting point confirmed its identity with the compound prepared by treating phenyl isocyanate with 3-nitro-*para*-toluidine and subsequent conversion of the nitro group to amine. (See Compound B-13.) They prepared 2-methyl-5-amino-carbanilide, from the 5-nitro-*ortho*-toluidine in a similar manner, to show the lack of identity.

The readiness with which 2,4-TDI reacts with aniline to give a 68% yield of the *para*-substituted urea demonstrated the greater reactivity of the NCO group in the 4-position, whereas that in the 2-position did not react with the aniline and was subsequently converted to NH_2.

Substituted ureas are frequent contaminants of isocyanate products because of their rapid reaction with water. Moisture must be rigidly excluded to prevent their formation. This reaction may be represented as:

$$2RNCO + HOH \rightarrow R\text{—}NH\text{—}CO\text{—}NH\text{—}R + CO_2 \left\{ \begin{array}{l} \underset{\text{H}\ \ \text{O}}{\text{RNC—OH}} \rightarrow RNH_2 + CO_2 \\[2em] RNH_2 + RNCO \rightarrow \overset{\text{H O H}}{\text{RNCNR}} \end{array} \right. \qquad (25)$$

In the case of diisocyanates, water reacts with both isocyanate groups, tending to give noncrystalline or crystalline polymeric products, depending upon the isocyanate, none of which can be easily identified.

A process for preparing monomeric ureas from certain diisocyanates has been patented by Pelley (69). According to this patent, the diisocyanates having substituents (such as methyl, methoxy, or chlorine) in the 1-position shields the NCO group in the 2-position permitting the formation of a urea linkage by the NCO group in the 4-position.

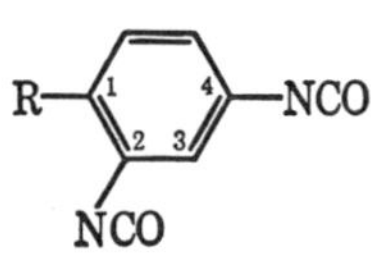

By this process the urea of 2,4-TDI can be obtained.

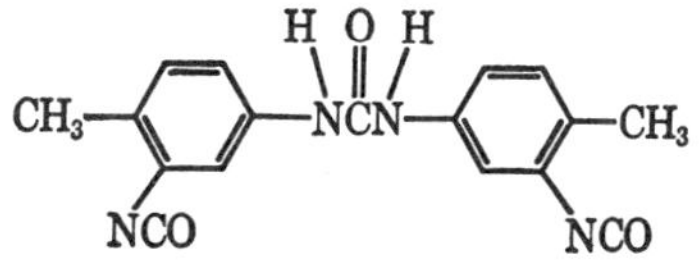

Diisocyanates such as *meta*- or *para*-phenylene do not possess this shielding effect and form polymeric materials.

The di(isocyanatophenyl)ureas were prepared by reacting the isocyanate with water in the molar ratio of approximately 2:1. If too great an amount of water is used, the isocyanate group in the 2-position reacts and polymeric ureas form. The temperature is maintained between 0 and 30°C, but generally at 20–25°C.

Balon et al. (70) describe essentially the same procedure for the preparation of ureas from diisocyanates including that from 2,6-TDI and from MDI. These latter compounds are numbered B-20 and B-23 in Table 4.9. Their melting points are not reported.

The present authors found that octadecyl isocyanate reacts slowly with water to form the urea without the aid of a catalyst. Triethylamine permitted the reaction to proceed smoothly in benzene solvent. The product (compound A-10) was identical to the compound formed from the isocyanate and amine in infrared pattern and in melting point.

The ureas containing the furan ring (compounds D-4 to D-9 in Table 4.9) were prepared by reacting the corresponding isocyanates (see Chapter VII) with water.

In general, the isocyanate dimers (described under Dimers) can be dissociated to the monomer providing a sufficiently high boiling solvent is used and sufficient time allowed for the dissociation to take place. Saunders and Hardy (68) studied the rate of reaction of the dimer of 2,4-TDI in *o*-dichlorobenzene by reacting it in the presence of dibutyl amine and determining the excess amine present at various periods of time. The dibutyl urea formed was identified and appears as No. C-8.

Balon et al. (70) likewise prepared diphenyl urea (compound B-3) from phenyl isocyanate dimer and aniline. Apparently, this dimer does not dissociate appreciably in benzene or toluene at their boiling points (80° and 110°).

Ureas are also formed by the reaction of Grignard reagents with an isocyanate dimer, as well as by hydrolysis of the dimer in the presence of alcoholic KOH, according to Raiford and Freyermuth (63). The reactions between phenyl isocyanate and nitroparaffins (except nitromethane) in the presence of catalytic amounts of trialkyl amines proceed rapidly at room temperature to give diphenyl urea, carbon dioxide, and furoxanes, according to Mukaiyama and Hoshino (71). A typical reaction is:

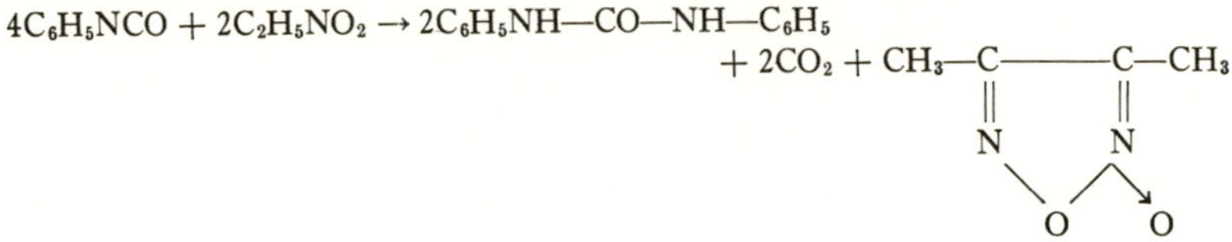

(26)

The reactions between an isocyanate, primary nitroparaffin, and unsaturated compounds in the presence of tertiary alkyl amines were also studied (including styrene acrylonitrile, vinyl acetate, and others). In this case diphenyl urea, carbon dioxide, and isoxaline or oxadiazolines are formed, for example:

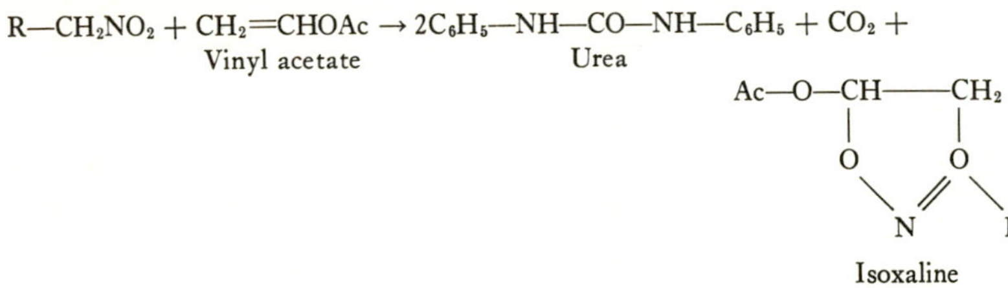

(27)

Several isoxalines and oxadiazoxolines are given in the original paper. Diethyl urea is listed as compound A-3.

2. Preparation from Urea or Substituted Ureas

Davis and Underwood (72) studied "the urea dearrangement" and obtained a number of mono- and disubstituted ureas. They pointed out the nature of the equilibrium of isocyanic acid and ammonia with urea in aqueous solution. Davis and Blanchard (73) prepared numerous substituted ureas by direct reaction of urea with amines in aqueous solution.

The papers by Hale (74) entitled "Four-Membered Cyclic Ureas. I. History and Nomenclature," and by Hale and Lange (75), "Four-Membered Cyclic Ureas. II. The Condensation of Isocyanic Acid with a Schiff Base," describe early work on the ureas giving rise to isocyanate dimers. These should be consulted for further details on the reactions of ureas.

Replacement of one of the ethyl groups in diethyl urea with a phenyl group by reaction with aniline was described by Davis and Blanchard (73).

Further substitution in a substituted urea by an alkyl group in which a propyl group is added to 1,3-di(1-naphthyl) urea to form 1,3-di(1-naphthyl) 1-propyl urea is described by Dyer and Read (41).

Furfural reacts with urea to form a difurfuryl triureide (compound D-8) according to Dunlop and Peters (12).

Acyl ureas of the type R—CH_2—CO—NH—CO—NH_2 have been prepared by Spielman, Geiszler, and Close (76) by the reaction of various acyl chlorides with urea. Some 32 compounds are described in the original paper. These compounds were prepared and tested for anti-convulsant activity for the treatment of epilepsy.

3. Miscellaneous Methods of Preparation

Mukaiyama and Nohira (77), in their study of the dehydrations of hydroxamic acids and of aldoximes by isocyanate, obtained *sym*-disubstituted ureas. These ureas were obtained in the course of their study of a new method for the preparation of isocyanates.

Mukaiyama and Hoshino (42) briefly describe 1-phenyl-3,3-dibenzyl urea as a product of the reaction of benzyl *N*-phenyl carbamate and dibenzyl amine in tetralin. This compound was prepared as part of their study and the effects of solvents (fatty acids and amines) on the thermal dissociation of urethanes.

Bortnick (46) prepared 1,3-tertiary octyl urea by the reaction of potassium cyanate and *t*-octyl amine. Furfuryl amine and phosgene condense under the influence of KOH with the formation of *sym*-difuryl urea according to Dunlop and Peters (12).

C. Reactions of Substituted Ureas

Although our interest in ureas, so far as this work is concerned, is associated primarily with derivatives of isocyanates, a number of studies have been made of the reverse reaction by which an isocyanate is the product of the thermal dissociation of a substituted urea. That is, the reaction

$$\text{R—NH—CO—NH—R}' \rightarrow \text{RNCO} + \text{R}'\text{NH}_2 \qquad (28)$$

affords another route for the preparation of isocyanates, although the recombination of the isocyanate with the amine has limited its usefulness.

Bennet, Saunders, and Hardy (78) found that some substituted ureas (diphenyl urea and *N*-phenyl urea) are dissociated (99% and 97%, respectively) in the vapor phase at 370°C. By introducing HCl to the vapor of disubstituted urea to combine with the amine and cooling, they were able to recover 57–71% of the isocyanate. In a similar manner, ethyl isocyanate and cyclohexyl isocyanate were obtained. They also studied the vapor-phase reaction of amines and amine salts with urea and cyanuric acid and found that the yield of isocyanates was lower.

Bortnick et al. (46) studied the preparation of *t*-alkyl isocyanates and isothiocyanates by the pyrolysis of the corresponding ureas. The cleavage of most 1,3-dialkyl ureas requires extreme conditions and the addition of strong acids either to the vapor space or to the reaction vessel. Distillation of di-2-octyl urea at 275° yielded no liquid products. Di-*t*-octyl urea (see compound A-7) distilled in the presence of mineral acids or phosphorous pentoxide, resulted in charred products. By contrast this urea yields *t*-octyl isocyanate by distillation at 220°C and collection of the vapors in cold dilute aqueous acid to prevent the recombination of the *t*-carbinamines with the desired product. Similar treatment of di-*t*-butyl urea resulted in sublimation (at its melting point, about 250°C) of the starting material, with only slight dissociation even in the presence of alkali. The authors conclude, "Although the method lacks generality, the cleavage of substituted ureas is attractive because it is based on cheap reagents. It is a convenient route to *t*-octyl isocyanate and *t*-butyl isothiocyanate." The di-1,3-*t*-octyl urea was prepared from *t*-octyl amine and urea.

Hoshino, Mukaiyama, and Hoshino (79) studied the thermal degradation of diphenyl urea in the presence of various fatty acid solvents. The following reactions were proposed as representing the dissociation steps in the presence of acetic acid:

$$C_6H_5\text{—NH—CO—NH—}C_6H_5 \underset{k_2}{\overset{k_1}{\rightleftharpoons}} C_6H_5NCO + C_6H_5NH_2 \tag{29}$$

$$C_6H_5NCO + CH_3COOH \xrightarrow{k_2} C_6H_6NH\text{—CO—O—CO—}CH_3 \tag{30}$$

(The product of equation 30 is a mixed anhydride of phenyl-carbamic and acetic acids.)

$$C_6H_5\text{—NH—CO—O—CO—}CH_3 \xrightarrow{k_3} \underset{\text{Acetanilide}}{C_6H_5\text{—NH—CO—}CH_3} + CO_2 \tag{31}$$

$$C_6H_5NH_2 + CH_3COOH \rightleftharpoons C_6H_5\text{—NH—CO—}CH_3 + H_2O \tag{32}$$

The rate of dissociation can be established if reaction 29 is rate determining by measuring the CO_2 evolved in reaction 31.

They found that the rate of dissociation of diphenyl urea in anhydrous acetic acid is four times faster than when 5% water is present. They considered the dielectric constant of four fatty acids and found that there was some correlation—that is, the higher the dielectric constant the greater the rate of dissociation. The approximate rate constants (interpolated from their data) are compared with the dielectric constant in the following tabulation.

	k, min^{-1}	Ratio	Dielectric constant
Butyric acid	0.0021	1	2.8
Caproic acid	0.0031	1.5	3.2
Propionic acid	0.0090	4	3.2
Acetic acid	0.0310	15	6.4

In their study of the thermal degradation of carbamates (urethanes), Dyer and Read (41) subjected 1-naphthylisocyanate to degradation conditions and obtained the carbodiimide and carbon dioxide. The rate of CO_2 evolution was greatly increased when the isocyanate was pyrolyzed in the presence of added carbodiimide. They concluded that the decomposition was autocatalytic due to the presence of the carbodiimide. They likewise studied the effect of 1,3-dinaphthyl urea on the pyrolysis of the isocyanate and state (although no data is presented) that "1,3-dinaphthyl urea also acted as a catalyst for the decomposition of the isocyanate."

These studies on the thermal dissociation of substituted ureas have an important bearing on the industrial preparation of isocyanates, which they frequently contaminate, and the process of distillation subjects them to thermal degradation conditions.

D. Catalysis and Kinetics

1. Amines

The work of Davis and Ebersole (80) has shown that the more basic amines react fastest when differences in steric hinderance are not great. Naegeli et al. (81) carried out studies on the effect of nuclear substitution in aromatic amines and isocyanates on reaction rates. The results showed that electron withdrawal from the isocyanate group promoted the reaction, and electron withdrawal from the amine group (reduction of basicity) reduced the reaction rate.

Craven (82) studied the reaction between phenyl, *o*- and *p*-tolyl isocyanates and several amines including aniline, *o*- and *p*-toluidine, and *o*-chloroaniline. The isocyanate concentration was followed as a function of time by reaction of a portion of the removed sample with *n*-butylamine in THF and back-titration with HCl. At concentrations below $0.02N$, the aromatic amines did not interfere with the titration of the *n*-butylamine using bromphenol blue as indicator. The p*K* values of the aromatic and aliphatic amines are sufficiently different so as to permit this method of determination. However, in some solvent systems one could experience difficulty with certain mixed aryl and alkyl amines using this technique.

Autocatalysis by the product urea, is evidenced by an inflection point in the rate curves in most reactions. This was further demonstrated by adding varying amounts of *N*-phenyl-*N′*-*o*-tolylurea to reaction mixtures.

Craven found that substituents in the *ortho* position of the amine or isocyanate decreased reactivities from steric effects, while a *para* methyl group on the amine increased the rate and an *ortho* chlorine atom significantly decreased the rate.

The order of reactivity increased as the base strength or nucleophility of the amine increased and decreased as the electrophilicity of the isocyanate decreased. This is evident from the reaction of phenyl isocyanate with *p*-toluidine, which showed the fastest reaction time.

Craven has pointed out that any proposed mechanism must explain why (*1*) the initial reaction rates show a greater dependence on amine than on isocyanate concentration and are proportional to the first power of the isocyanate concentration, but to a power, between first and second, of the amine concentration, and (2) the reactions are catalyzed by the product ureas.

The following sequence was proposed

I.

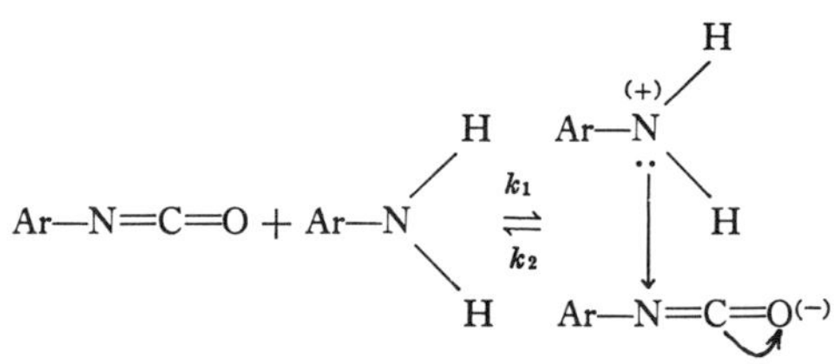

(33)

II.

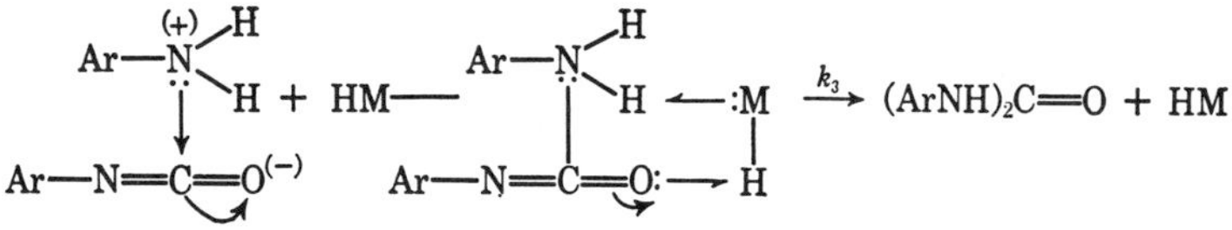

I. followed by

III.

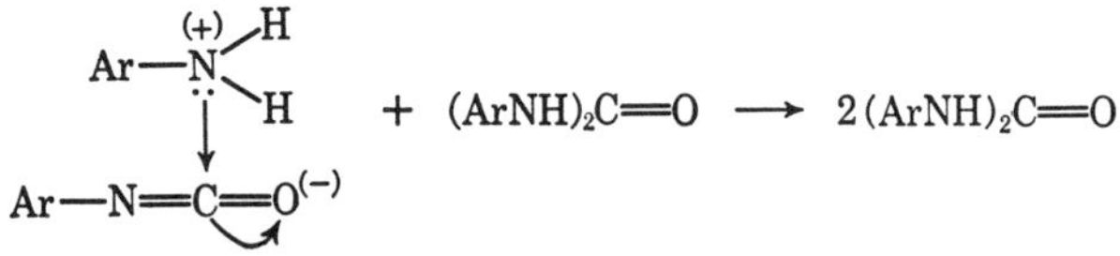

(35)

where HM represents some proton acceptor–donor such as amine, urea, or carboxylic acid, which have been reported as catalysts for this reaction (81–83).

The rate expression used to obtain linear plots was

$$\frac{ia}{-di/dt} = \frac{k_2}{k_1k_3'}\frac{1}{u + (k_3/k_3')a} + \frac{1}{k_1}$$

where a and i are amine and isocyanate concentrations respectively, and u the urea concentration at corresponding times.

The expression relating initial concentrations was found to be $R_0 = ki_0a_0n$, where $1 < n < 2$. The rate constant here is not a true rate constant since this expression applies only to initial concentrations.

2. Water Reaction

The mechanisms proposed by Naegeli (81) for the isocyanate water reaction are shown here as reaction 36. Saunders and Frisch (29) have

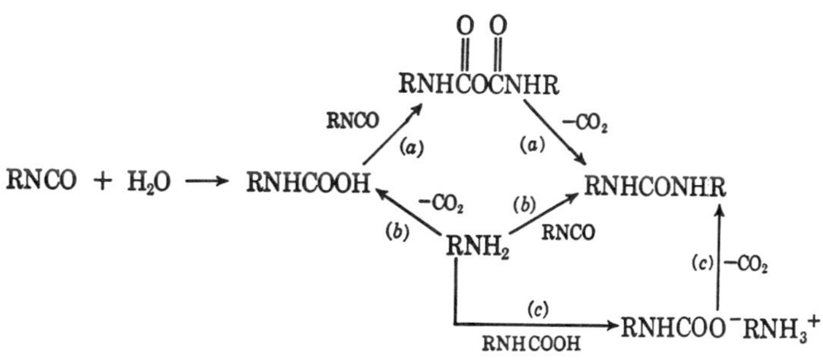

(36)

pointed out that (*a*) would be favored when the carbamic acid is relatively stable but is reactive toward the isocyanate. Sequence (*b*) would be favored when the carbamic acid decomposes rapidly and the liberated amine reacts quickly with the isocyanate. Sequence (*c*) might increase when the carbamic acid is not stable but the amine and isocyanate react only at a slow rate.

The effect of substituents on the isocyanate–water reaction is similar to that observed for the isocyanate reactions with other active hydrogen compounds. Negative substituents promote the addition of water to isocyanate, but the same groups on the resulting amine retard the addition to the isocyanate.

Tertiary amines, certain organometallic compounds, and tin catalysts have been found to catalyze this reaction to varying degrees.

E. Analytical Techniques and Characterization

As in the previous cases, ureas have most often been identified by elemental analysis and mixed melting point determinations.

Bellamy (84) has pointed out that the spectra of substituted derivatives have been examined by Thompson et al. (85), Randall et al. (86), Boivin and Boivin (87,88), and Becher (89,90). Although the spectra are complex and it is not possible to interpret the results fully, Bellamy (84) has concluded that the carbonyl absorption appears near 1660 cm^{-1} in fully substituted ureas, and absorption in this same region is also shown by mono- and disubstituted products.

This band is distinct from the other isocyanate derivatives containing the carbonyl group whose frequencies have been listed previously. The lower frequency of the urea carbonyl is probably due to increased hydrogen bonding, which would lower the frequency as opposed to higher frequencies exhibited by carbonyl attached to a ring since ring strain increases the frequency. This band can be readily recognized in mixtures since it occurs at 1670–1630 cm^{-1}. Table 4.8 shows the carbonyl absorption frequencies for a number of substituted ureas.

The calculations used by Brumfield (51) to compute the structure contribution for the substituted urea, 1-phenyl-3,3-diethyl urea, were as follows.

TABLE 4.8
Absorption Frequency for the Carbonyl Group in Substituted Ureas (51,91)

Compound	Frequency, cm^{-1}
1,3-Phenyl butyl urea	1664
1,4-Diethyl urea	1634
1,3-Diphenyl urea[a]	1658
1,3-Diphenyl urea[a]	1643 (melt)
1-Phenyl-3,3-diethyl urea	1632 (melt)
	1635 (KBr pellet)

[a] The difference in absorption frequency between the two 1,3-diphenyl ureas listed is undoubtedly due to sample state since in one case the frequency was measured on the melt and in the other case on a powdered sample between two salt plates.

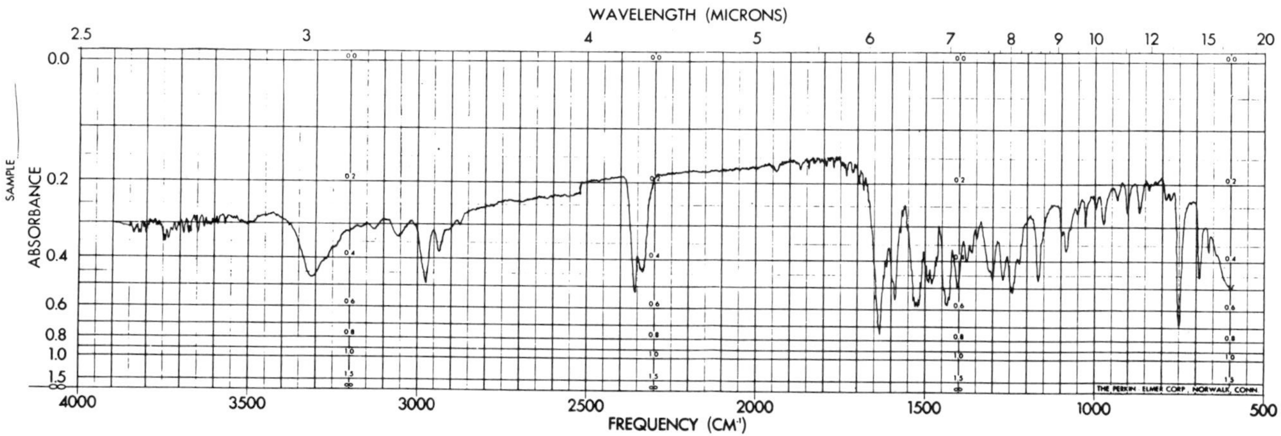

Fig. 4.6. Infrared spectrum of 1-phenyl 3,3-diethyl urea (melt) (51).

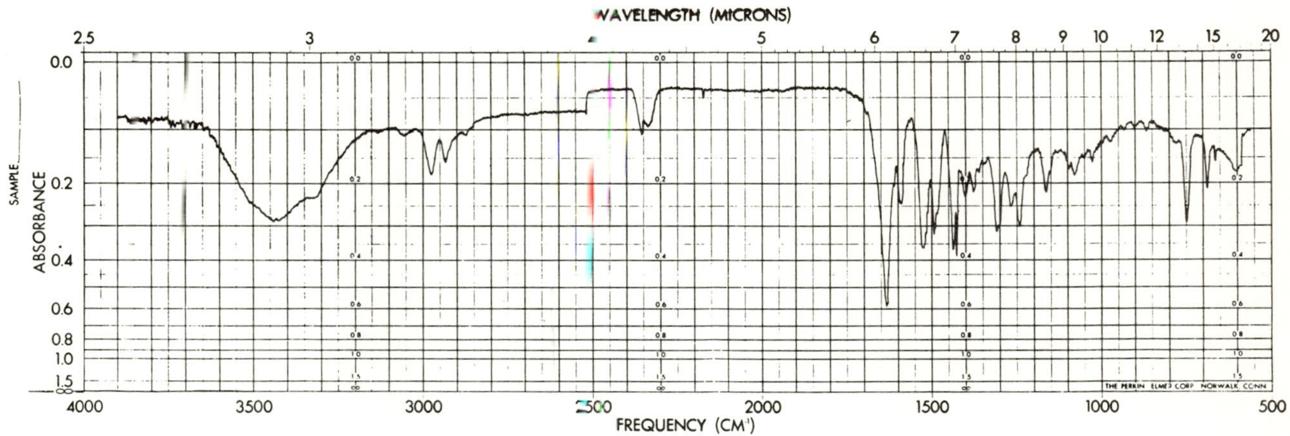

Fig. 4.7. Infrared spectrum of 1-phenyl 3,3-diethyl urea (0.11% in KBr Pellets) (51).

The structure was divided into urea, phenyl, and ethyl groups.

Melt (Fig. 4.6)	KBr pellet (Fig. 4.7)

Urea (—NCON)

Melt (Fig. 4.6)	KBr pellet (Fig. 4.7)
3312, 3200, and 1632 cm^{-1}	1635 cm^{-1}
Total "A" = 2.784	Total "A" = 1.791

$$\frac{\text{"A" (urea)} \times 100}{\text{Total "A"}}$$

$\frac{0.804 \times 100}{2.784} = 28.90\%$ (actual 29.65%) $\qquad$ $\frac{0.528 \times 100}{1.791} = 29.49\%$ (actual 29.65%)

Phenyl (C_6H_5—)

Melt (Fig. 4.6)	KBr pellet (Fig. 4.7)
3170, 3132, 3125, 3059, 2904, 1588, 1500, 1490, and 1470 cm^{-1}	3059, 3045, 2905, 1590, 1493, 1483, and 1470 cm^{-1}

$$\frac{\text{"A" (phenyl)} \times 100}{\text{Total "A"}}$$

$\frac{1.126 \times 100}{2.784} = 41.10\%$ (actual 40.10%) $\qquad$ $\frac{0.729 \times 100}{1.791} = 40.70\%$ (actual 40.10%)

Ethyl (C_2H_5)

Melt (Fig. 4.6)	KBr pellet (Fig. 4.7)
2975, 2957, 2936, 2875, 1467, and 1435 cm^{-1}	2976, 2957, 2934, 2874, 1467, and 1437 cm^{-1}

$$\frac{\text{"A" (ethyl)} \times 100}{\text{Total "A"}}$$

$\frac{0.834 \times 100}{2.784} = 29.92\%$ (actual 30.25%) $\qquad$ $\frac{0.534 \times 100}{1.791} = 29.80\%$ (actual 30.25%)

The physical properties of various substituted ureas are shown in Table 4.9.

TABLE 4.9
Physical Properties of Various Substituted Ureas

A. Alkyl
   1. 1,3-Diamyl urea
   2. 1,3-Di-*n*-butyl urea
   3. 1,3-Diethyl urea
   4. 1,1,3-Triethyl urea
   5. 1,3-Dimethyl urea
   6. 1-*tert*-Octyl urea
   7. 1,3-*tert*-Octyl urea
   8. 1,1-Dimethyl-3-*tert*-octyl urea
   9. 1-Isopropyl-3-*tert*-octyl urea
   10. 1,3-Dioctadecyl urea

B. Aryl
   1. 1-Phenyl urea
   2. 1-Benzyl urea

TABLE 4.9 (*Continued*)

3. 1,3-Phenyl urea
4. 1,3-Dibenzyl urea
5. 1,3-*o*-Tolyl urea
6. 1,3-*p*-Tolyl urea
7. 1,3-(1-Naphthyl) urea
8. 1,3-(2-Naphthyl) urea
9. 1-Phenyl-3-benzyl urea
10. 1-Phenyl-3,3′-benzyl urea
11. 1-Phenyl-3-*p*-methoxyphenyl urea
12. 1-Phenyl-3-(3-amino-4-methylphenyl) urea
13. 1-Phenyl-3-(3-nitro-4-methylphenyl) urea
14. 1-Phenyl-3-(5-amino-2-methylphenyl) urea
15. 1-Phenyl-3-(5-nitro-2-methylphenyl) urea
16. 1,3-Di(4-ethoxyphenyl) urea
17. 1,3-Di(3-isocyanato-4-methylphenyl) urea
18. 1,3-Di(3-isocyanato-4-methoxyphenyl) urea
19. 1,3-Di(3-isocyanato-4-chlorophenyl) urea
20. 1,3-Di(3-isocyanato-2-methylphenyl) urea
21. 1,1-Diphenyl urea
22. 1-(1-Naphthyl)-3-phenyl urea
23. 1,3-Di(4,4′-diphenyl methane diisocyanate) urea

C. Mixed alkyl and aryl
1. 1-Phenyl-3-butyl urea
2. 1-Phenyl-3-ethyl urea
3. 1-Phenyl-3-*t*-octyl urea
4. 1,3-Diphenyl-1-methyl urea
5. 1-Phenyl-1-ethyl urea
6. 1-Phenyl-1-methyl urea
7. 1,3-Di(1-naphthyl)-1-propyl urea
8. 1-Tolyl-2,4-di(3,3-dibutyl) urea

D. Heterocyclic ureas
1. 1-Cyclohexyl-3-phenyl urea
2. 1-Cyclohexyl-3-*t*-octyl urea
3. 1,3-Dicyclohexyl urea
4. 1,3-Di(5-methyl-2-furyl) urea
5. 1,3-Di(2,4-dimethyl-3-furyl) urea
6. 1,3-Di(2-methyl-3-furyl) urea
7. 1,3-Di(2,5-dimethyl-3-furyl) urea
8. Difurfurylidene triureide urea
9. 1,3-Difurfuryl urea

TABLE 4.9 (*Continued*)

No.	Name and structure	Analysis			Infrared data		Ref.
		Elem.	Calc.	Fd.	Solvent	Wavelength	
A-1.	1,3-Diamyl urea						73, 76
	C_5H_{11}—NH—CO—NH—C_5H_{11}						
	Formula $C_{11}H_{24}N_2O$ Mol. wt. 200.32	M.P. 92.8°C					
	Solvents for recrystallization Hot water	B.P.					
	Appearance						
A-2.	1,3-Di-*n*-butyl urea						73
	C_4H_9—NH—CO—NH—C_4H_9						
	Formula $C_9H_{20}N_2O$ Mol. wt. 172.27	M.P. 70.5–71.0°C					
	Solvents for recrystallization Benzene, sparingly soluble in hot water	B.P.					
	Appearance White flakes						
A-3.	1,3-Diethyl urea						73, 76
							32, 69
	C_2H_5—NH—CO—NH—C_2H_5						

(*continued*)

TABLE 4.9 (*Continued*)

No.	Name and structure	Analysis			Infrared data		Ref.
		Elem.	Calc.	Fd.	Solvent	Wavelength	
	Formula $C_5H_{12}N_2O$ Mol. wt. 116.16	M.P. 112°C					73, 76
		108–110°C					32, 69
	Solvents for recrystallization	B.P.					
	Appearance						
A-4.	1,1,3-Triethyl urea $C_2H_5—N(C_2H_5)—CO—NH—C_2H_5$				CCl_4	6780 cm^{-1}	7, 42
					0.75 m/1	4974 cm^{-1}	
						4915 cm^{-1}	
						4865 cm^{-1}	
	Formula $C_7H_{16}N_2O$ Mol. wt. 144.21	M.P. 62°C					
	Solvents for recrystallization	B.P.					
	Appearance						
A-5.	1,3 Dimethyl urea $CH_3—NH—CO—NHCH_3$						73, 76
	Formula $C_3H_8N_2O$ Mol. wt. 88.11	M.P. 99.5–100°C					
	Solvents for recrystallization	B.P.					
	Appearance Colorless flakes						

A-6. 1-*tert*-Octyl urea	N	16.3	16.1			46
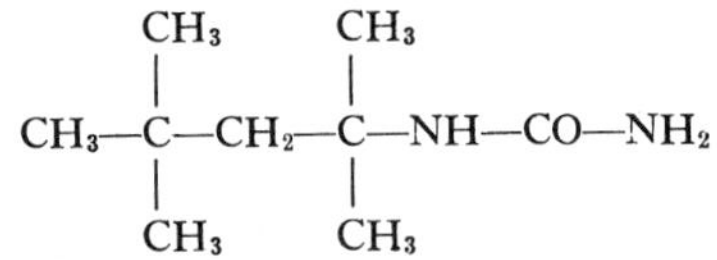						
Formula $C_9H_{20}N_2O$ Mol. wt. 172.27	M.P. 72–73°C					
Solvents for recrystallization Petroleum ether	B.P.					
Appearance						
A-7. 1,3 *Di-tert*-Octyl urea	N	9.9	9.7			46
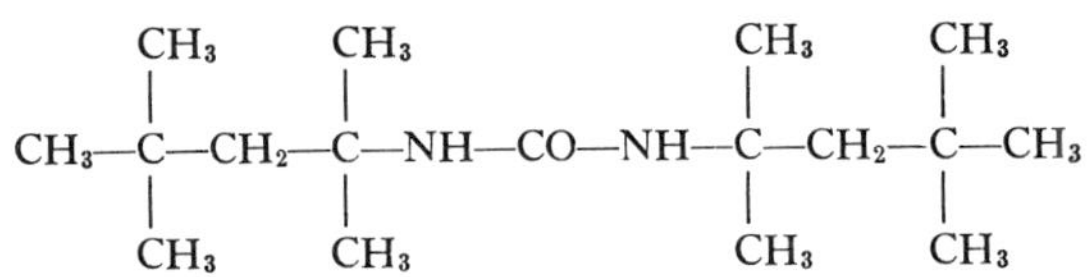						
Formula $C_{17}H_{36}N_2O$ Mol. wt. 284.47	M.P. 152–153°C					
Solvents for recrystallization Petroleum ether, 80% ethanol	B.P.					
Appearance						
A-8. 1,1-Dimethyl-3-*tert*-octyl urea	N	14.0	13.7			46, 70
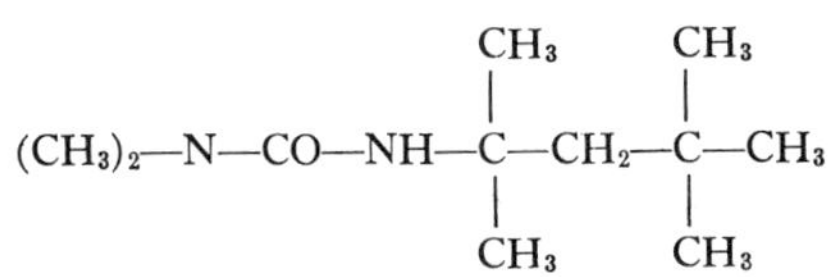						

(continued)

TABLE 4.9 (*Continued*)

No.	Name and structure	Analysis: Elem.	Analysis: Calc.	Analysis: Fd.	Infrared data: Solvent	Infrared data: Wavelength	Ref.
	Formula $C_{11}H_{24}N_2O$ Mol. wt. 200.32	M.P. 84–85°C					
	Solvents for recrystallization 50% Ethanol	B.P.					
	Appearance						
A-9.	1-Isopropyl-3-*tert*-octyl urea 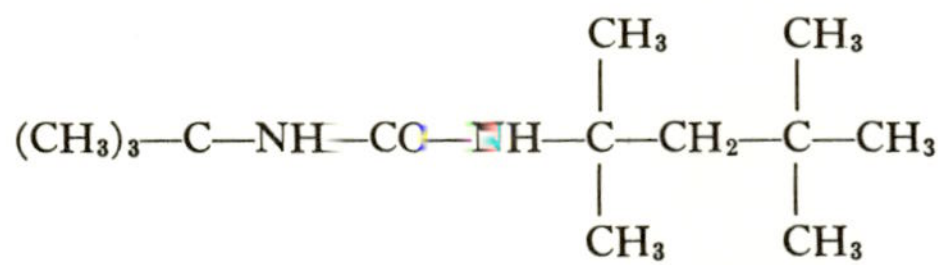	N	12.3	12.2			46, 70
	Formula $C_{13}H_{28}N_2O$ Mol. wt. 228.37	M.P. 153–155°C					
	Solvents for recrystallization 50% Ethanol	B.P.					
	Appearance						
A-10.	1,3-Dioctadecyl urea $C_{18}H_{37}$—NH—CO—NH—$C_{18}H_{37}$						46, 119

	Compound						Ref.
	Formula $C_{37}H_{76}N_2O$ Mol. wt. 565.00	M.P. 114–115°C					
	Solvents for recrystallization Benzene, carbon tetrachloride	B.P.					
	Appearance						
B-1.	1-Phenyl urea 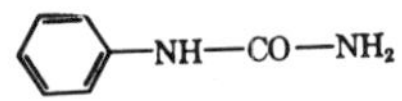						72, 73
	Formula $C_7H_8N_2O$ Mol. wt. 136.15	M.P. 147°C					
		Sp. gr. 1.302					
	Solvents for recrystallization Hot water, ether	B.P.					
	Appearance Needles						
B-2.	1-Benzyl urea 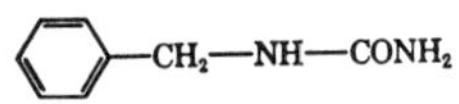						73
	Formula $C_8H_{10}N_2O$ Mol. wt. 150.18	M.P. 146.6°C					
	Solvents for recrystallization	B.P.					
	Appearance						
B-3.	1,3-Phenyl urea 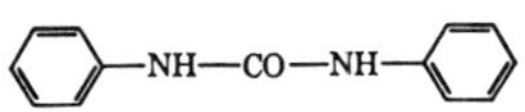	N	13.20	13.28			42, 70, 63, 32, 73, 77, 119

(continued)

TABLE 4.9 (*Continued*)

No.	Name and structure	Analysis			Infrared data		Ref.
		Elem.	Calc.	Fd.	Solvent	Wavelength	
	Formula $C_{13}H_{12}N_2O$ Mol. wt. 212.24	M.P. 238°C; 237–238°C					63, 32, 73,
		240°C; 241–242°C					77, 119
	Solvents for recrystallization Ethanol						
	Appearance						
B-4.	1,3-Dibenzyl urea 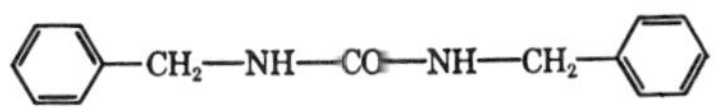						73
	Formula $C_{15}H_{16}N_2O$ Mol. wt. 240.29	M.P. 167°C					
	Solvents for recrystallization Alcohol	B.P.					
	Appearance						
B-5.	1,3-*o*-Tolyl urea 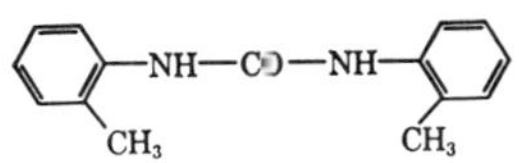						72
	Formula $C_{15}H_{17}N_2O$ Mol. wt. 241.30	M.P. 248°C					
	Solvents for recrystallization	B.P.					
	Appearance						

B-6. 1,3-*p*-Tolyl urea

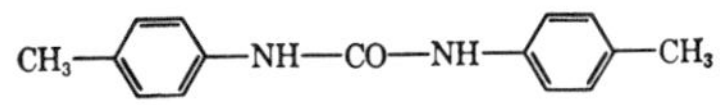

						Ref.
Formula $C_{15}H_{17}N_2O$ Mol. wt. 241.30						72
Solvents for recrystallization Alcohol	M.P. 264°C					
Appearance White flakes	B.P.					

B-7. 1,3-(1-Naphthyl) urea

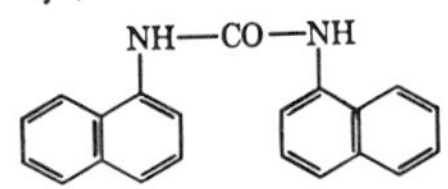

						Ref.
	N	8.87	9.12			72
Formula $C_{21}H_{16}N_2O$ Mol. wt. 312.35	M.P. 286°C					
Solvents for recrystallization Ethyl acetate	B.P.					
Appearance						

B-8. 1,3-(2-Naphthyl) urea

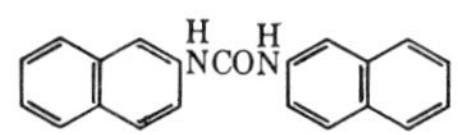

						Ref.
						72, 73
Formula $C_{21}H_{16}N_2O$ Mol. wt. 312.35	M.P. 296°C, decomposes					
Solvents for recrystallization Acetone, amyl acetate, ethyl acetate	B.P.					
Appearance White needles						

(continued)

TABLE 4.9 (*Continued*)

No.	Name and structure	Analysis			Infrared data		Ref.
		Elem.	Calc.	Fd.	Solvent	Wavelength	
B-9.	1-Phenyl-3-benzyl urea 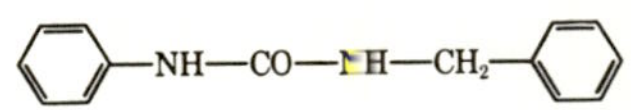						72
	Formula $C_{14}H_{14}N_2O$ Mol. wt. 226.27	M.P.					
	Solvents for recrystallization	B.P.					
	Appearance						
B-10.	1-Phenyl-3,3′-di-benzyl urea 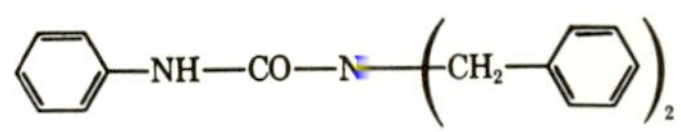						42
	Formula $C_{21}H_{20}N_2O$ Mol. wt. 316.39	M.P. 125°C					
	Solvents for recrystallization	B.P.					
	Appearance						
B-11.	1-Phenyl-3-*p*-methoxyphenyl urea 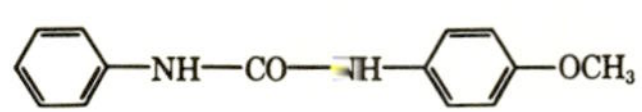						77

	Formula $C_{14}H_{14}N_2O$ Mol. wt. 226.27	M.P. 189–190°C					
	Solvents for recrystallization 95% Ethanol	B.P.					
	Appearance						
B-12.	1-Phenyl-3-(3-amino-4-methylphenyl) urea 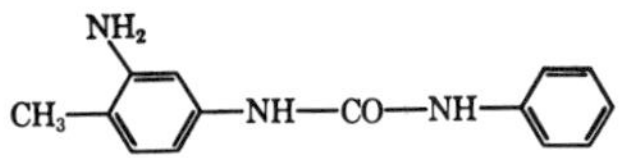	N	17.4	17.5 17.3			32, 78, 125
	Formula $C_{14}H_{15}N_3O$ Mol. wt. 241.28	M.P. 198–201°C					
	Solvents for recrystallization	B.P.					
	Appearance						
B-13.	1-Phenyl-3-(3-nitro-4-methylphenyl) urea 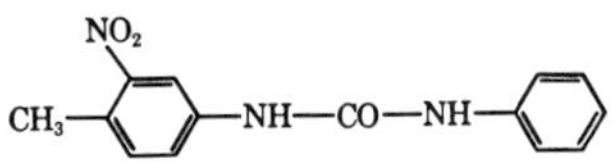	N	15.5	15.4			125, 32
	Formula $C_{14}H_{13}N_3O_3$ Mol. wt. 271.27	M.P. 218–219°C					
	Solvents for recrystallization	B.P.					
	Appearance						
B-14.	1-Phenyl-3-(5-amino-2-methylphenyl) urea 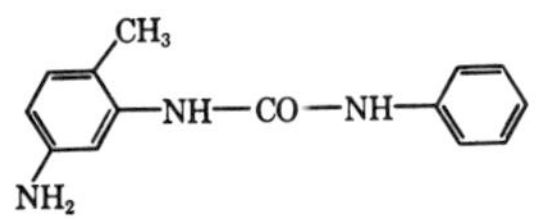	N	17.4	17.0			32, 125

(continued)

TABLE 4.9 (*Continued*)

No.	Name and structure	Analysis			Infrared data		Ref.
		Elem.	Calc.	Fd.	Solvent	Wavelength	
	Formula $C_{14}H_{15}N_3O$ Mol. wt. 241.28	M.P. > 200°C					
	Solvents for recrystallization	B.P.					
	Appearance						
B-15.	1-Phenyl-3-(5-nitro-2-methylphenyl) urea 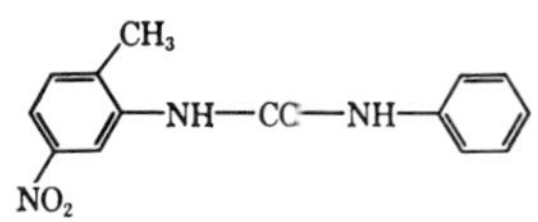	N	15.5	15.7			32, 125
	Formula $C_{14}H_{13}N_3O_3$ Mol. wt. 271.27	M.P.					
	Solvents for recrystallization	B.P.					
	Appearance						
B-16.	1,3-Di(4-ethoxyphenyl) urea 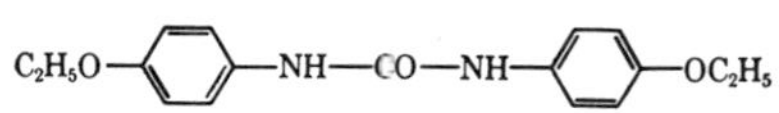						32, 68
	Formula $C_{17}H_{20}N_2O_3$ Mol. wt. 300.46	M.P. 225–226°C					
	Solvents for recrystallization Acetic acid	B.P.					
	Appearance Colorless needles						

Compound						Ref.
B-17. 1,3-Di(3-isocyanato-4-methylphenyl) urea 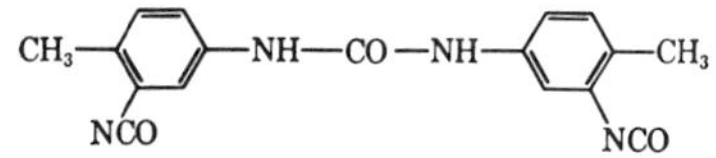	NCO	26.1	25.8			69, 75
						79, 70
Formula $C_{17}H_{14}N_4O_3$ Mol. wt. 322.31	M.P. 175–179°C					
Solvents for recrystallization Benzene	B.P.					
Appearance						
B-18. 1,3-Di(3-isocyanato-4-methoxyphenyl) urea 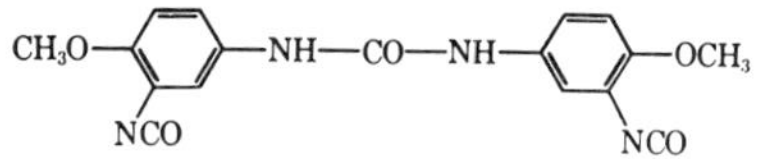	NCO	23.7	23.7			69, 75
	N	15.8	15.1			79, 70
Formula $C_{17}H_{14}N_4O_5$ Mol. wt. 354.31	M.P. 241–245°C					
Solvents for recrystallization	B.P.					
Appearance						
B-19. 1,3-Di(3-isocyanato-4-chlorophenyl) urea 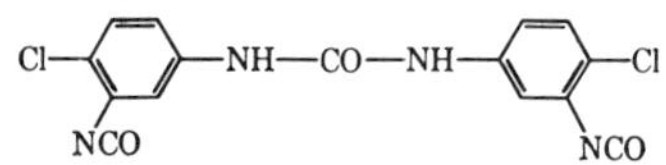	NCO	23.2	23.7			69, 75
Formula $C_{15}H_8N_4O_3Cl_2$ Mol. wt. 363.16	M.P.					
Solvents for recrystallization	B.P.					
Appearance						

(*continued*)

233

TABLE 4.9 (*Continued*)

No.	Name and structure	Analysis			Infrared data		Ref.
		Elem.	Calc.	Fd.	Solvent	Wavelength	
B-20.	1,3-Di(3-isocyanato-2-methylphenyl) urea 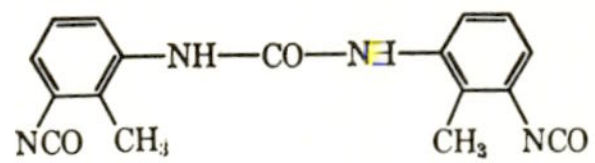	NCO	26.1	25.4			72, 70
	Formula $C_{17}H_{14}N_4O_3$ Mol. wt. 322.31	M.P.					
	Solvents for recrystallization	B.P.					
	Appearance						
B-21.	1,1-Diphenyl urea 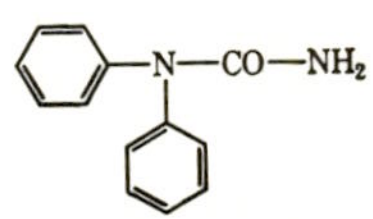						119
	Formula $C_{13}H_{12}N_2O$ Mol. wt. 212.24	M.P. 189°C					
	Solvents for recrystallization	B.P.					
	Appearance						
B-22.	1-(1-Naphthyl)-3-phenyl urea 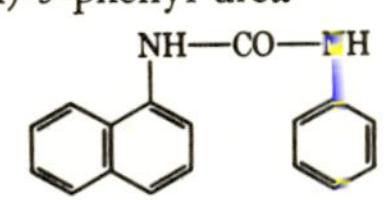						72, 78

Compound						Ref.
Formula $C_{17}H_{14}N_2O$ Mol. wt. 262.30	M.P. 222–224°C					
Solvents for recrystallization	B.P.					
Appearance						
B-23. 1,3-Di(4,4′-diphenyl methane diisocyanate) urea	NCO	17.5	17.4			7, 70
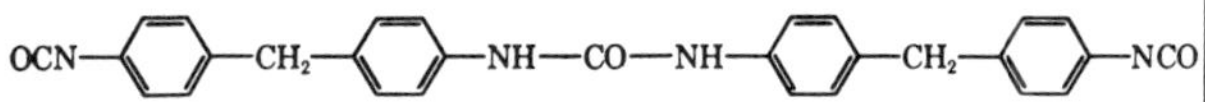						
Formula $C_{29}H_{22}N_4O_3$ Mol. wt. 474.50	M.P.					
Solvents for recrystallization	B.P.					
Appearance						
C-1. 1-Phenyl-3-butyl urea						79, 63
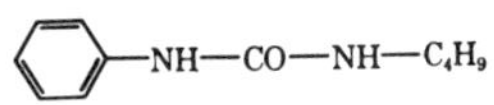						
Formula $C_{11}H_{16}N_2O$ Mol. wt. 192.25	M.P.					
Solvents for recrystallization	B.P.					
Appearance						
C-2. 1-Phenyl-3-ethyl urea				CCl_4	6725 cm^{-1}	7
				0.05 m/1	4915 cm^{-1}	
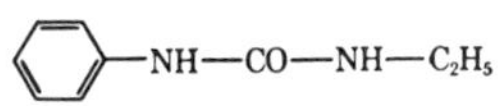					4900 cm^{-1}	

(continued)

TABLE 4.9 (*Continued*)

No.	Name and structure	Analysis			Infrared data		Ref.
		Elem.	Calc.	Fd.	Solvent	Wavelength	
	Formula $C_9H_{12}N_2O$ Mol. wt. 164.20	M.P. 96–97°C					73
			98–99.5°C				7
	Solvents for recrystallization	B.P.					
	Appearance						
C-3.	1-Phenyl-3-*t*-octyl urea 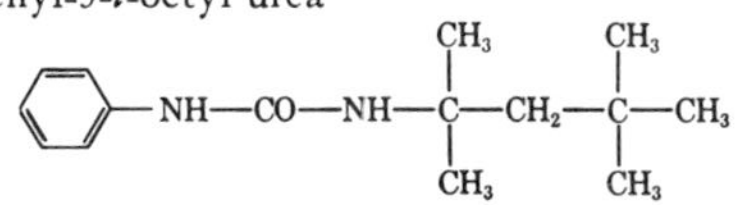	N	11.3	11.2			70, 46
	Formula $C_{15}H_{24}N_2O$ Mol. wt. 248.36	M.P. 136–138°C					
	Solvents for recrystallization 50% Ethanol	B.P.					
	Appearance						
C-4.	1,3-Diphenyl-1-methyl urea 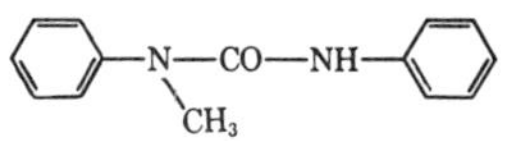				CCl_4	6710 cm^{-1}	7
					0.15 m/1	5022 cm^{-1}	
						4946 cm^{-1}	
						4867 cm^{-1}	
	Formula $C_{14}H_{14}N_2O$ Mol. wt. 226.27	M.P. 104°C					
	Solvents for recrystallization	B.P.					
	Appearance						

Compound						
C-5. 1,Phenyl-1-ethyl urea 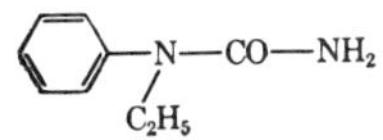						73
Formula $C_9H_{12}N_2O$ Mol. wt. 164.20	M.P. 62°C					
Solvents for recrystallization Ligroin	B.P.					
Appearance White lustrous plates						
C-6. 1-Phenyl-1-methyl urea 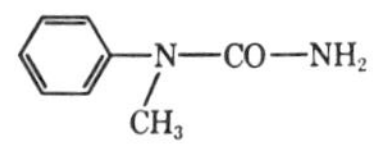						73
Formula $C_8H_{10}N_2O$ Mol. wt. 150.18	M.P. 150°C					
Solvents for recrystallization Ligroin	B.P.					
Appearance Thin plates						
C-7. 1,3-Di(1-naphthyl)-1-propyl urea 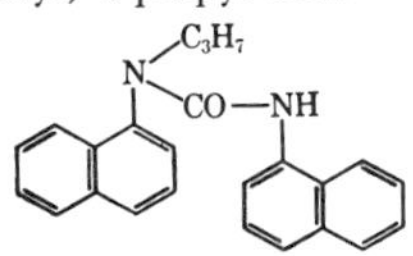	C	81.32	81.14			41
	H	6.26	6.17			
	N	7.91	8.11			
Formula $C_{24}H_{22}N_2O$ Mol. wt. 354.43	M.P. 142–143°C					
Solvents for recrystallization Ether	B.P.					
Appearance						

(continued)

TABLE 4.9 (*Continued*)

No.	Name and structure	Analysis			Infrared data		Ref.
		Elem.	Calc.	Fd.	Solvent	Wavelength	
C-8.	1-Tolyl-2,4-di(3,3-dibutyl) urea 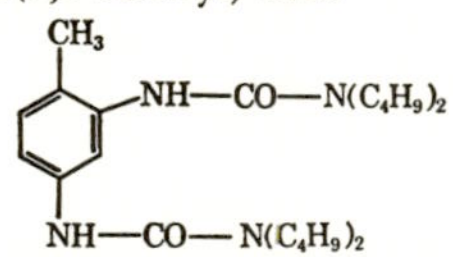						68, 71 11
	Formula $C_{25}H_{44}N_4O_2$ Mol. wt. 432.63	M.P. 111.1–111.7°C					
	Solvents for recrystallization Hexane	B.P.					
	Appearance						
D-1.	1-Cyclohexyl-3-phenyl urea 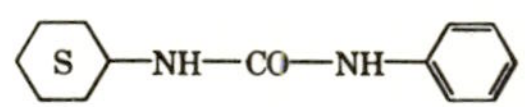						41, 78
	Formula $C_{13}H_{18}N_2O$ Mol. wt. 218.29	M.P. 149–150°C					
	Solvents for recrystallization	B.P.					
	Appearance						
D-2.	1-Cyclohexyl-3-*t*-octyl urea 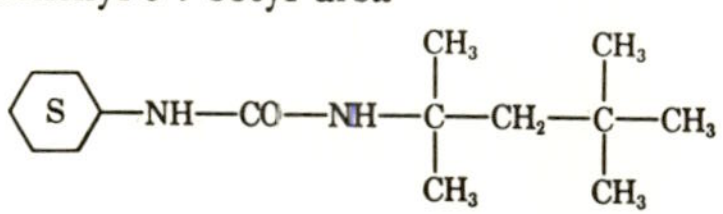	N	11.0	10.9			70, 46

	Formula $C_{15}H_{30}N_2O$ Mol. wt. 254.41	M.P. 148–148.5°C					
	Solvents for recrystallization 50% Ethanol	B.P.					
	Appearance						
D-3.	1,3-Dicyclohexyl urea 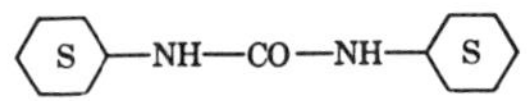						73, 78
	Formula $C_{13}H_{24}N_2O$ Mol. wt. 224.34	M.P. 227–227.5°C					
	Solvents for recrystallization	B.P.					
	Appearance						
D-4.	1,3-Di(5-methyl-2-furyl) urea 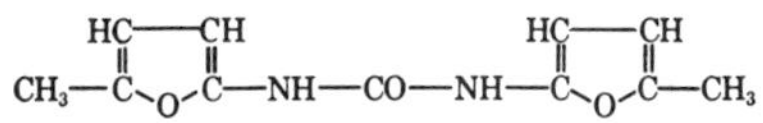	N	12.72	12.65			126, 11
	Formula $C_{11}H_{12}N_2O_3$ Mol. wt. 220.22	M.P. 182°C					
	Solvents for recrystallization 50% Ethanol	B.P.					
	Appearance						
D-5.	1,3-Di(2,4-dimethyl-3-furyl) urea 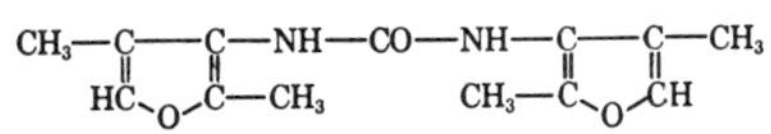	N	11.29	11.20			126, 11

(continued)

TABLE 4.9 (*Continued*)

No.	Name and structure	Analysis			Infrared data		Ref.
		Elem.	Calc.	Fd.	Solvent	Wavelength	
	Formula $C_{13}H_{16}N_2O_3$ Mol. wt. 248.2	M.P. 285°C					
	Solvents for recrystallization 95% Ethanol	B.P.					
	Appearance						
D-6.	1,3-Di(2-methyl-3-furyl) urea 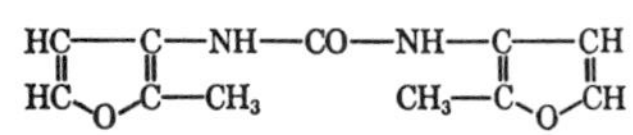	N	12.72	12.90			126, 11
	Formula $C_{11}H_{12}N_2O_3$ Mol. wt. 220.23	M.P. 227°C					
	Solvents for recrystallization 95% Ethanol	B.P.					
	Appearance						
D-7.	1,3-Di(2,5 dimethyl-3-furyl) urea 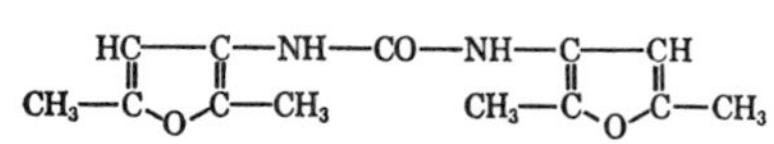						126, 11
	Formula $C_{13}H_{16}N_2O_3$ Mol. wt. 248.27	M.P. 247°C					
	Solvents for recrystallization 95% Ethanol	B.P.					
	Appearance						

D-8. Difurfurylidene triureide urea | 12

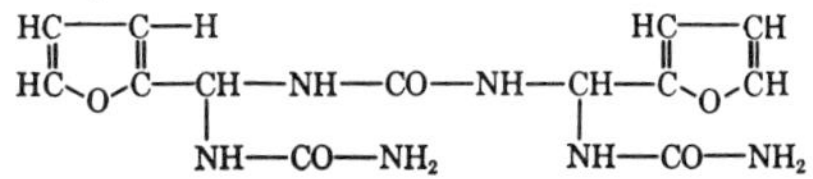

Formula $C_{13}H_{16}N_6O_5$ Mol. wt. 336.31 | M.P. 168–169°C

Solvents for recrystallization | B.P.

Appearance

D-9. 1,3 Difurfuryl urea | 12

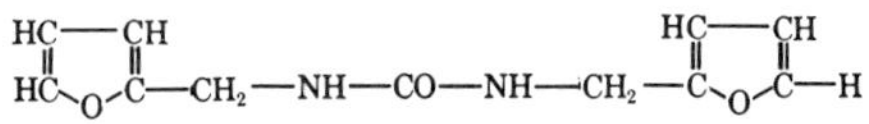

Formula $C_{11}H_{12}N_2O_3$ Mol. wt. 220.22 | M.P. 128°C

Solvents for recrystallization | B.P.

Appearance

V. BIURETS

A. Structural Relationships

Biuret (**1**) is related to several simpler compounds. It is formally derived from carbonic acid by the replacement of the two hydroxyl groups by one amine and one ureido group. It may also be regarded as the ureide of the hypothetical carbamic acid or as the amide of allophanic acid (**2**).

$$\overset{1}{N}H_2\overset{2}{C}O\overset{3}{N}H\overset{4}{C}O\overset{5}{N}H_2 \qquad \overset{4}{N}H_2\overset{3}{C}O\overset{2}{N}H\overset{1}{C}OOH$$

(**1**) Biuret (**2**) Allophanic acid

The name biuret, given to the parent compound of Wiedemann, who discovered it in 1847 (92), expressed its formation from two molecules of urea. The numbering system as shown in (**1**) has been adopted by *Chemical Abstracts* and provides an unambiguous name for any biuret or thiobiuret derivative. Triurets are similarly defined (**3**).

$$—\overset{1}{N}H_2\overset{2}{C}O\overset{3}{N}H\overset{4}{C}O\overset{5}{N}H\overset{6}{C}O\overset{7}{N}H_2$$

(**3**) Triuret

B. Preparation

Trisubstituted biurets were prepared as early as 1858 by the hydrolysis of *N,N,N*-trialkylisocyanuric acids (93,94). A general synthesis of biuret by the ammonolysis of allophanic esters was discovered by Hofmann in 1871 (95,96) and was extended to substituted homologs the same year.

Small quantities of biuret are formed when cyanic acid gas is passed into melted urea

$$NH_2CONH_2 + HNCO \rightarrow NH_2CONHCONH_2 \tag{37}$$

or by the treatment of ethyl allophanate with aqueous ammonia at 100°C in closed vessels (97).

$$NH_2CONHCOOC_2H_5 + NH_3 \rightarrow NH_2CONHCONH_2 + C_2H_5OH \tag{38}$$

or

$$RNHCONHCOOC_2H_5 + NH_3 \rightarrow RNHCONHCONH_2 + C_2H_5OH \tag{39}$$

Biuret can also be obtained from cyanourea by the action of dilute sulfuric or nitric acid (98).

$$NH_2CONHCN + H_2O \xrightarrow{H_2SO_4} NH_2CONHCONH_2 \tag{40}$$

or

$$RNCONHCN + H_2O \rightarrow RNHCONHCONH_2 \tag{41}$$

for substituted biurets or the treatment of nitrobiuret with concentrated ammonia in closed vessels at 100°C (99).

$$NH_2CONHCONHNO_2 \rightleftharpoons NH_2CONCO + NH_2NO_2 \tag{42}$$

$$NH_2CONCO + NH_3 \rightarrow NH_2CONHCONH_2 \tag{43}$$

This reaction, which occurs equally readily with primary and secondary amines, is a convenient route to substituted biurets. An excellent review of the chemistry of biuret and related compounds has been prepared by Kurzer (100).

The biuret formed by the reaction of an isocyanate with urea is of interest because of its relation to crosslink formation in many urethane polymers where water or amine is employed.

$$RNCO + R'NHCONHR'' \rightarrow R'N(CONHR)CONHR'' \qquad (44)$$

1,5-Diarylbiurets have been prepared by the condensation of aromatic ureas and isocyanates at 120–150°C. At lower temperatures some monosubstituted biuret is also formed (57). The condensation of substituted ureas and isocyanates has also been used to prepare 1,3,5-trisubstituted biurets (57).

Lakra and Dains (57) heated phenyl isocyanate and urea at 100°C for 7 hr and obtained the 1-phenyl biuret as the main product (Eq. 45).

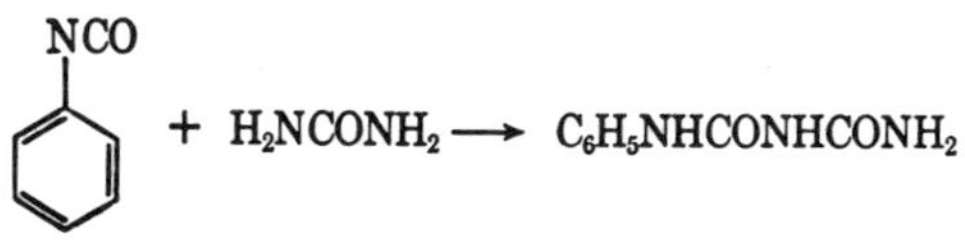

(45)

When the isocyanate and urea were reacted at 120°C for 4 hr, ammonia, urea, biuret, diphenylurea, monophenylurea, phenylisocyanuric acid, and isocyanuric acid were obtained. These products were explained by reactions 46–50.

$$NH_2CONH_2 \rightleftharpoons NH_3 + HNCO \qquad (46)$$

$$2C_6H_5NCO + 2NH_3 \rightarrow 2C_6H_5NHCONH_2 \rightarrow (C_6H_5NH)_2CO + NH_2CONH_2 \qquad (47)$$

$$NH_2CONH_2 + HNCO \rightarrow NH(CONH_2)_2 \qquad (48)$$

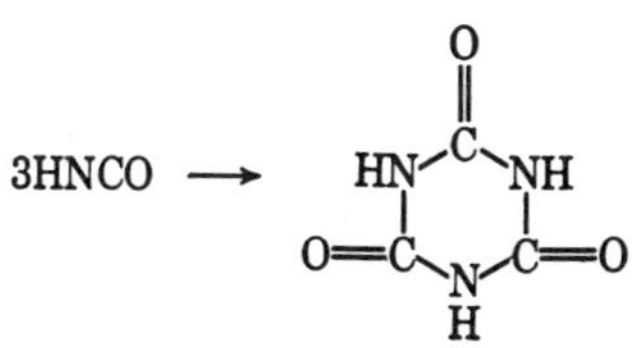

(49)

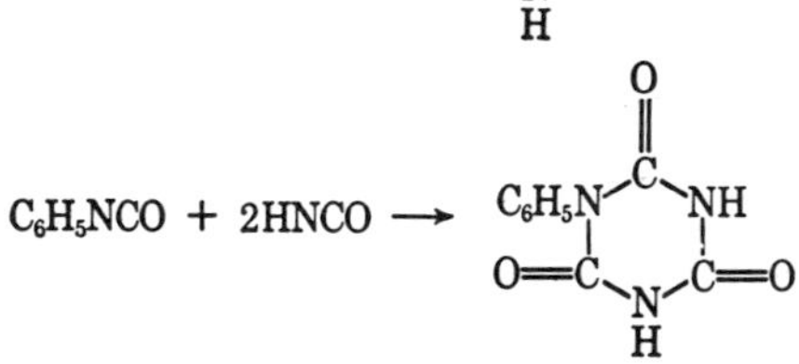

(50)

Lakra and Dains also reacted phenyl isocyanate and monophenyl urea at 100°C for 4 hr and obtained diphenyl urea, 1,5-diphenyl biuret, and monophenyl biuret, indicating that the isocyanate can add to the phenyl urea at position 3, yielding the 1,5-diphenyl biuret, and position 1, to give the isomeric biuret $C_6H_5NHCON(C_6H_5)CONH_2$. This dissociates into diphenyl urea and isocyanic acid, which then adds to the unchanged phenyl urea. Carbanilide and phenyl isocyanate were found to yield the 1,3,5-triphenyl biuret.

Trisubstituted biurets may also be prepared by hydrolyzing *N*,*N*,*N*-trialkylisocyanuric acids with boiling barium hydroxide or 1*N* potassium hydroxide at 30°C (Eq. 51). (101)

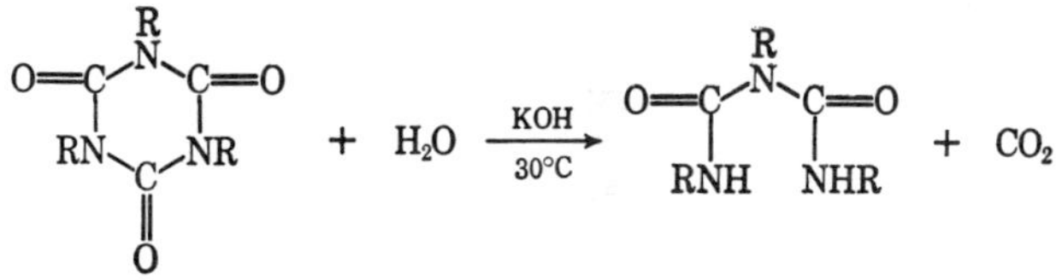

(51)

Gatewood (87) prepared 1-phenyl-5-methyl biuret by treating γ-phenyl-α-ethyl allophanate with methylamine as in reaction 52.

$$C_6H_5NHCONHCOOC_2H_5 + CH_3NH_2 \rightarrow C_6H_5NHCONHCONHCH_3 + C_2H_5OH \quad (52)$$

Uretediones can be converted into biurets by reaction with ammonia or substituted amines (reaction 53) (95).

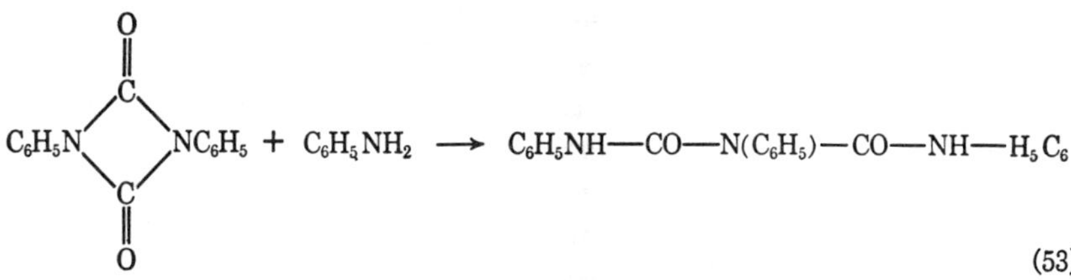

(53)

C. Reactions and Catalysis

The relatively high temperatures required for the isocyanate-urea reaction, as well as the many possible side reactions that occur, make the kinetics of this reaction difficult to follow.

The reaction of phenyl and *o*-tolyl isocyanates with phenyl and *sym*-diphenyl ureas was followed by dibutylamine titration of unreacted isocyanate and found to fit second-order kinetic calculations over the approximate range of 10–60% reaction (58,102). The rate data obtained from 60–140°C did not agree with the Arrhenius rate equation, which indicates a change in the course of the reaction at higher temperatures.

Some authors have reported a moderate catalytic effect from tertiary amines on this reaction (103,104), while other workers have not found this to be true (58,102). Tin catalysts have also been reported to catalyze this reaction (104).

Kogon (64) studied the dissociation of biurets by following the appearance of the NH band at 6750 cm^{-1}. This has been explained in Section III-C. The results of this study show that alkyl-substituted biurets have greater stability than aryl-substituted and that biuret crosslinks in a polymer would not have good thermal stability since decomposition of the aryl biurets dissociated at 130–145°.

D. Analytical Techniques and Characterization

Analytically, the biurets have been characterized by mixed melting points and elemental analysis.

1-Substituted biurets act as weak acids and can be generally dissolved by dilute alkalies and reprecipitated by mineral acids. The lower homologs are readily soluble in boiling water, acetic acid, and most organic solvents (100).

Substituted biurets decompose at or above their melting points, and special precautions are necessary to ensure the reproducibility of the results of melting point determinations since the rate of heating affects the results.

Biuret itself gives the characteristic red color upon addition of copper sulfate and alkali that has been referred to as the biuret reaction (105).

Schiff (106) came to the conclusion that only those substituted biurets gave the red color that had both of the original amine groups intact. This would mean that unsymmetrical biurets would not give the reaction. For example, *asym*-phenyl biuret does not give the biuret reaction.

We would expect the carbonyl absorption band to occur in the same approximate frequency range as that of ureas since these two classes of compounds resemble each other closely. This indeed proves to be the case. Biurets show characteristic double bands, as allophanates do, at 1720–1690 cm^{-1}, the higher frequency band being more intense and sometimes hiding the lower frequency band. This latter occurrence is largely affected by sample state, solvent effects, etc., which may or may not restrict the group vibration.

These compounds, as before, are easily recognizable but are difficult, if not impossible, to distinguish in mixtures of isocyanurates, allophanates, and ureas because of band overlap. It is especially true in the case of allophanates which exhibit bands at approximately 1690 and 1725 cm^{-1}. (See Table 4.6.)

This problem is compounded by the many possible side reactions during

TABLE 4.10
Absorption Frequency for the Carbonyl Group in Simple Biurets (50,51)

Compound	Frequency, cm^{-1}
1,3,5-Triphenylbiuret	1710, 1678
1,3-Diphenyl 5-butylbiuret	1708 (melt)
	1708 (KBr pellet)

preparation, as previously pointed out, or the many possible products that may be present in mixtures. These difficulties may partly account for the paucity of property and infrared data concerning biurets. Table 4.10 shows the carbonyl absorption frequencies of two biurets.

Brumfield (51) employed the following wavelengths and calculations in order to compute the structure contribution for 1,3-diphenyl 5-butyl-biuret.

The structure was divided into biuret, phenyl, and butyl groups.

Melt (Fig. 4.8)	KBr pellet (Fig. 4.9)

Biuret (*HNCONCONH—*)

Melt	KBr pellet
3382, 1716, 1553, 1548, and 1535 cm^{-1}	1553 and 1542 cm^{-1}
Total "A" = 10.382	Total "A" = 1.492

$$\frac{\text{"A" (biuret)} \times 100}{\text{Total "A"}}$$

$\frac{3.254 \times 100}{10.382}$ 31.35% (actual 32.15%) $\qquad \frac{0.478 \times 100}{1.492} = 32.05\%$ (actual 32.15%)

Phenyl (C_6H_5)

Melt	KBr pellet
3170, 3130, 3089, 3069, 2905, 1589, 1500, 1491, and 1484 cm^{-1}	3062, 3035, 2903, 1594, and 1502 cm^{-1}

$$\frac{\text{"A" (phenyl)} \times 100}{\text{Total "A"}}$$

$\frac{5.120 \times 100}{10.382} = 49.4\%$ (actual 49.60%) $\qquad \frac{0.737 \times 100}{1.492} = 49.40\%$ (actual 49.50%)

Butyl (C_4H_9)

Melt	KBr pellet
2974, 2938, 2878, and 1430 cm^{-1}	2960, 2933, and 2874 cm^{-1}

$$\frac{\text{"A" (butyl)} \times 100}{\text{Total "A"}}$$

$\frac{2.008 \times 100}{10.382} = 19.28\%$ (actual 18.37%) $\qquad \frac{0.277 \times 100}{1.492} = 18.55\%$ (actual 18.35%)

As in the case of the allophanate evaluation, it is difficult to understand why the Amide I band was used in evaluation of the melt spectrum (1716 cm^{-1} band) but was not used for evaluation of the KBr pellet spectrum. This is especially noteworthy since this is a strong band whose wavelength is specific for constitution.

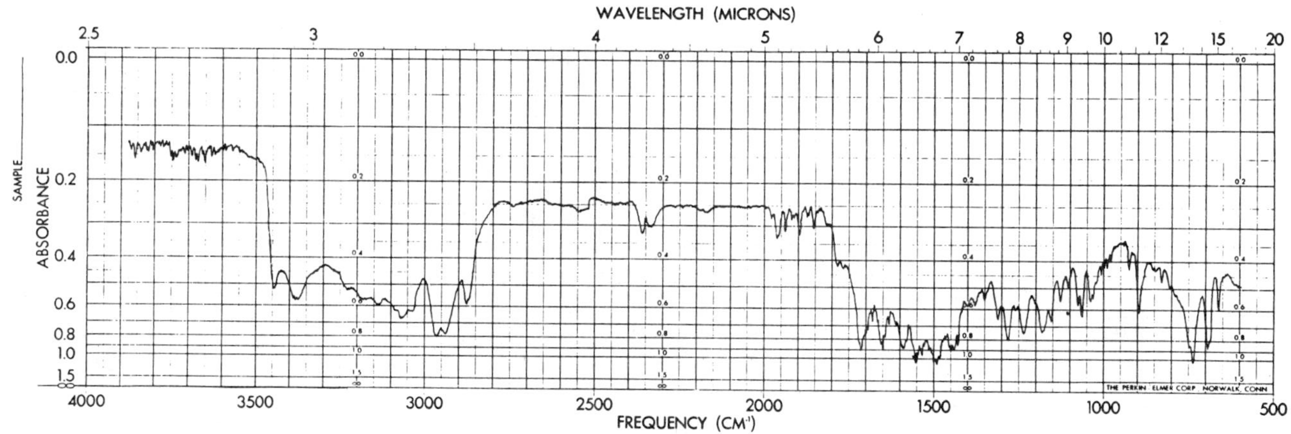

Fig. 4.8. Infrared spectrum of 1,3-diphenyl 5-butyl biuret (melt) (51).

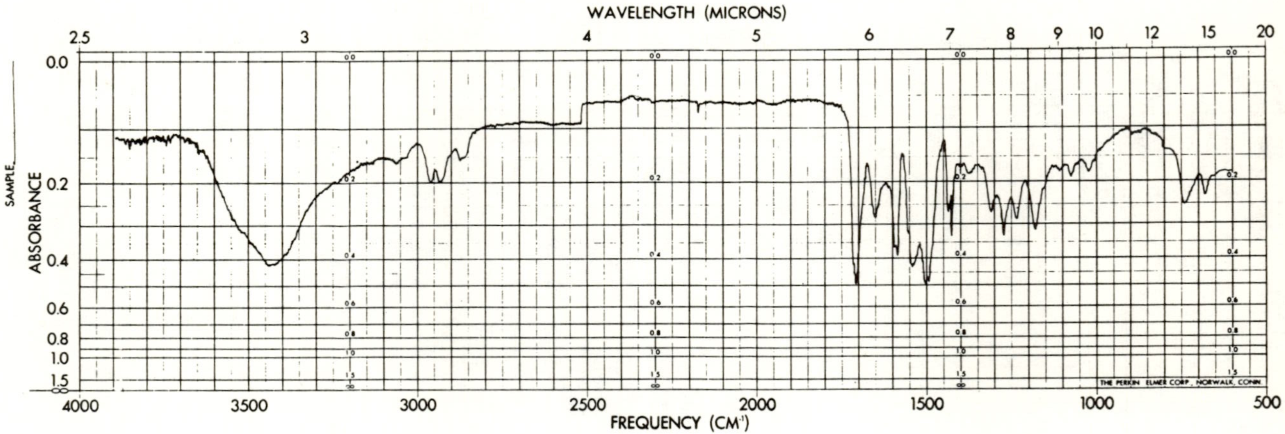

Fig. 4.9. Infrared spectrum of 1,3-diphenyl 5-butyl biuret (0.11% in KBr Pellets) (51).

TABLE 4.11
Physical Properties of Various Substituted Biurets

1. 1,1,3-Triethyl-5-phenyl biuret
2. 1,1,3,5-Tetraethyl biuret
3. 1,3,5-Triphenyl biuret
4. 1,3,5-Triethyl biuret
5. 1-Phenyl biuret
6. 1,5-Diphenyl biuret
7. 1,5-Di-*t*-ioctyl biuret
8. 1-Methyl-5-phenyl biuret
9. 1-Phenyl-3-methyl biuret
10. 1,3-Diphenyl-5-methyl biuret
11. 1,3-Diphenyl-5-ethyl biuret
12. 1,3-Diphenyl-5-*n*-butyl biuret
13. 1,3-Diphenyl-5-*n*-propyl biuret
14. 1,3-Diphenyl-5-isoamyl biuret
15. 1,3-Diphenyl-5-allyl biuret
16. 1,3-Diphenyl-5-piperidyl biuret
17. 1,3-Di(3-methylphenyl) 5-*n*-butyl biuret
18. 1,3-Di(4-methylphenyl) 5-*n*-butyl biuret
19. 1,3-Di(4-methylphenyl)-5-(4-tolyl) biuret
20. 1,3-Di(4-ethoxyphenyl) 5-methyl biuret
21. 1,3-Di(3-chlorophenyl)-5-*n*-butyl biuret
22. 1,3-Di(4-chlorophenyl)-5-*n*-butyl biuret
23. 1,3-Di(4-bromophenyl)-5-*n*-butyl biuret

The physical properties of various substituted biurets are shown in Table 4.11.

E. Summary

It can be readily seen from the previous sections that a large number of reaction products in mixtures of isocyanates and active hydrogen compounds are possible because of the reactivity of the isocyanate group. This fact and thermal stability problems that may be associated with an analytical procedure, as well as relatively high molecular weights in many cases, limit the analytical chemist in the analytical approach he may wish to take.

Primary emphasis has been placed on spectroscopic methods because of the relative experimental ease in handling and the ability to provide useful information for characterization.

It should be noted that, perhaps for the above reasons, spectroscopic tools have been most often employed and more information is available for

TABLE 4.11 (*Continued*)

No.	Name and structure	Analysis			Infrared data		Ref.
		Elem.	Calc.	Fd.	Solvent	Wavelength	
1.	1,1,3-Triethyl-5-phenyl biuret				Benzene	4877 cm^{-1}	7
	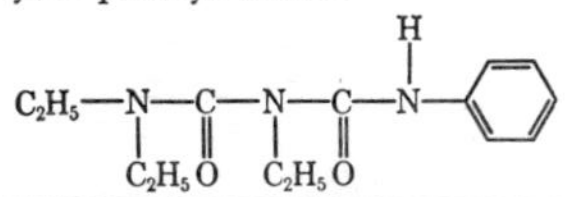				0.38 M/1		
	Formula $C_{14}H_{21}N_3O_2$ Mol. wt. 263.34	M.P. 113–113.5°C					
	Solvents for recrystallization	B.P.					
	Appearance						
2.	1,1,3,5-Tetraethyl biuret	N	19.4	19.6	Benzene	4877 cm^{-1}	7
	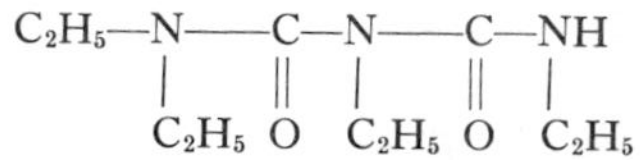				0.08 M/1		
	Formula $C_{10}H_{21}N_3O_2$ Mol. wt. 215.30	M.P.					
	Solvents for recrystallization	B.P.					
	Appearance						
3.	1,3,5-Triphenyl biuret	N	12.68	16.34	Carbon	6711 cm^{-1}	70
	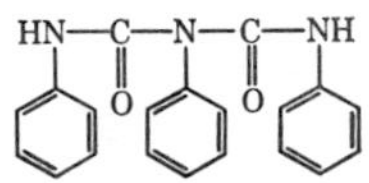				Tet.	5276	7
					0.16 M/1	4943	
						4870	

Compound						Ref.
Formula $C_{20}H_{17}N_3O_2$ Mol. wt. 331.37	M.P. 150–151°C					57
	147°C					7
Solvents for recrystallization	B.P.					57
Appearance						
4. 1,3,5-Triethyl biuret 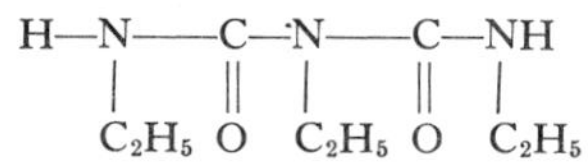				Benzene	6725 cm^{-1}	7
				0.16 M/1	5000	
					4920	
					4825	
Formula $C_8H_{17}N_3O_2$ Mol. wt. 187.24	M.P.					
Solvents for recrystallization	B.P.					
Appearance						
5. 1-Phenyl biuret 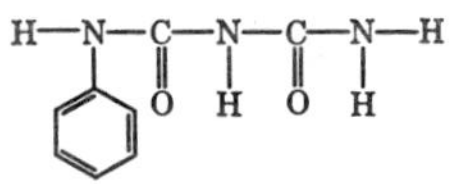	N	23.48	23.60			127, 128
						129, 130
Formula $C_8H_9N_3O_2$ Mol. wt. 179.18	M.P. 156–157°C					57
	165°C					123
Solvents for recrystallization Hot water and dilute alcohol						
	B.P.					
Appearance						
6. 1,5-Diphenyl biuret 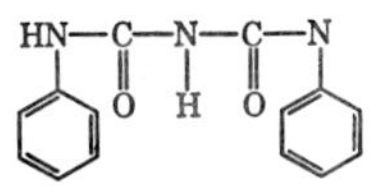	N	16.47	16.34			132

(continued)

TABLE 4.11 (*Continued*)

No.	Name and structure	Analysis			Infrared data		Ref.
		Elem.	Calc.	Fd.	Solvent	Wavelength	
	Formula $C_{14}H_{13}N_3O_2$ Mol. wt. 255.27	M.P. 210°C					57
			196–197.5°C				131
	Solvents for recrystallization	B.P.					
	Appearance						
7.	1,5 Di-*t*-octyl biuret 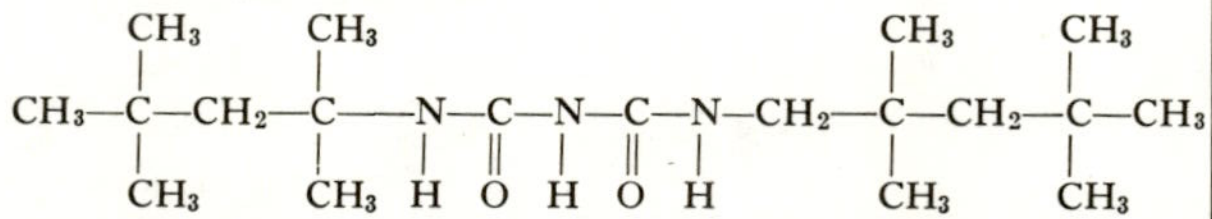						46
	Formula $C_{18}H_{37}N_3O_2$ Mol. wt. 327.51	M.P. 120–122°C					
			124–125°C				
	Solvents for recrystallization Ethylene dichloride isooctane						
		B.P.					
	Appearance						
8.	1-Methyl-5-phenyl biuret 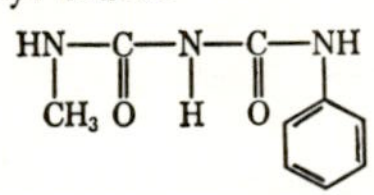	C	55.94	55.99			123
		H	5.74	5.70			
		N	21.76	22.01			
	Formula $C_9H_{11}N_3O_2$ Mol. wt. 193.20	M.P. 133°C					
	Solvents for recrystallization Methanol	B.P.					
	Appearance						

9. 1-Phenyl-3-methyl biuret

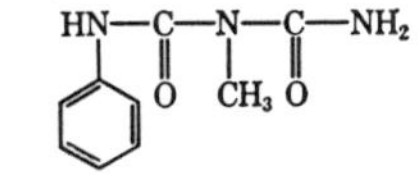

Formula $C_9H_{11}N_3O_2$ Mol. wt. 193.20

Solvents for recrystallization

Appearance

	Calcd.	Found			Ref.
C	55.94	55.73			123
H	5.74	5.50			
N	21.76	21.47			

M.P. 183°C

B.P.

10. 1,3-Diphenyl-5-methyl biuret

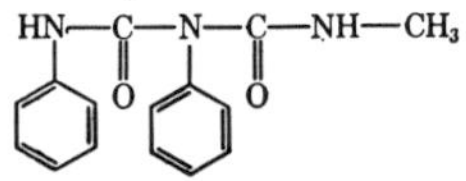

Formula $C_{15}H_{15}N_3O_2$ Mol. wt. 269.30

Solvents for recrystallization Ethanol

Appearance Colorless needles

	Calcd.	Found			Ref.
N	15.61	15.57			63
Mol. wt.	268	271			

M.P. 144–145°C

B.P.

11. 1,3-Diphenyl-5-ethyl biuret

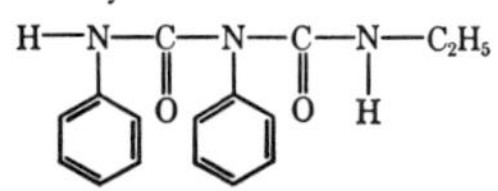

Formula $C_{16}H_{17}N_3O_2$ Mol. wt. 283.33

Solvents for recrystallization Ethanol

Appearance Colorless needles

	Calcd.	Found			Ref.
N	14.84	14.90			63
Mol. wt.	283	284			

M.P. 88–89°C

B.P.

(continued)

TABLE 4.11 (*Continued*)

No.	Name and structure	Analysis			Infrared data		Ref.
		Elem.	Calc.	Fd.	Solvent	Wavelength	
12.	1,3-Diphenyl-5-*n*-butyl biuret 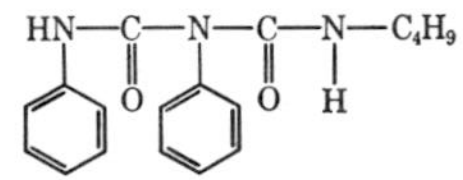	N	13.50	13.50			63
		Mol. wt.	311	308			
	Formula $C_{18}H_{21}N_3O_2$ Mol. wt. 311.38	M.P. 79–80°C					
	Solvents for recrystallization Ethanol	B.P.					
	Appearance Colorless needles						
13.	1,3-Diphenyl-5-*n*-propyl biuret 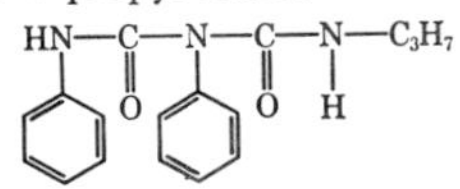	N	14.14	14.13			63
		Mol. wt.	297	284			
	Formula $C_{17}H_{19}N_3O_2$ Mol. wt. 297.35	M.P. 115–116°C					
	Solvents for recrystallization Ethanol	B.P.					
	Appearance Colorless needles						
14.	1,3-Diphenyl-5-isoamyl biuret 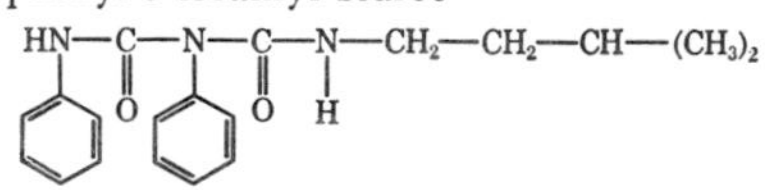	N	12.92	13.07			63
		Mol. wt.	325	332			

Compound						Ref.
Formula $C_{19}H_{23}N_3O_2$ Mol. wt. 325.41	M.P. 62–64°C					
Solvents for recrystallization	B.P.					
Appearance Colorless prisms						
15. 1,3-Diphenyl-5-allyl biuret 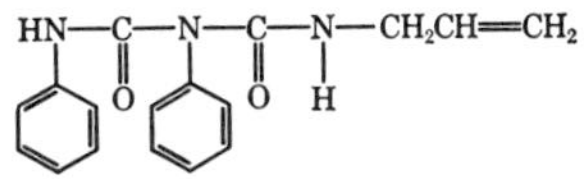	N	14.23	14.05			63
	Mol. wt.	295	285			
Formula $C_{17}H_{17}N_3O_2$ Mol. wt. 295.34	M.P. 94–95°C					
Solvents for recrystallization	B.P.					
Appearance Large colorless prisms						
16. 1,3-Diphenyl-5-piperidyl biuret 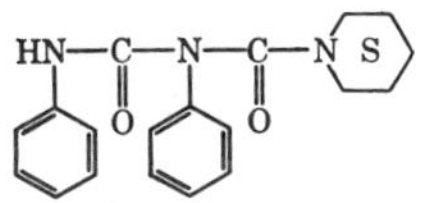	N	13.00	12.76			63
	Mol. wt.	323	316			
Formula $C_{19}H_{21}N_3O_2$ Mol. wt. 323.39	M.P. 135–136°C					
Solvents for recrystallization	B.P.					
Appearance Colorless needles						
17. 1,3-Di(3-methylphenyl)-5-*n*-butyl biuret 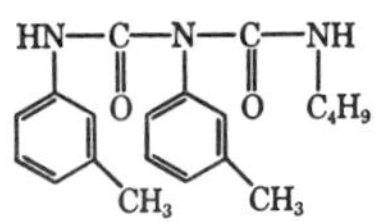	N	12.38	12.46			63
	Mol. wt.	339	332			

(continued)

TABLE 4.11 (*Continued*)

No.	Name and structure	Analysis			Infrared data		Ref.
		Elem.	Calc.	Fd.	Solvent	Wavelength	
	Formula $C_{20}H_{25}N_3O_2$ Mol. wt. 339.43	M.P. 102–103°C					
	Solvents for recrystallization Ethanol	B.P.					
	Appearance Colorless needles						
18.	1,3-Di(4-methylphenyl)-5-*n*-butyl biuret 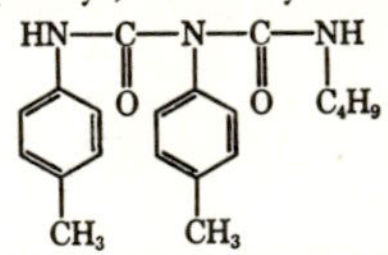	N	12.38	12.46			63
		Mol. wt.	339	339			
	Formula $C_{20}H_{25}N_3O_2$ Mol. wt. 339.44	M.P. 131–132°C					
	Solvents for recrystallization Ethanol	B.P.					
	Appearance Colorless needles						
19.	1,3-Di(4-methylphenyl)-5-(4-tolyl) biuret (1,3,5-*p*-tolyl biuret) 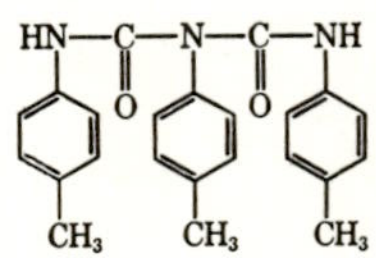	N	11.26	11.40			63

No.	Compound		Calcd.	Found			Ref.
	Formula $C_{23}H_{23}N_3O_2$ Mol. wt. 373.46	M.P. 265°C					
	Solvents for recrystallization Ethanol	B.P.					
	Appearance Colorless needles						
20.	1,3-Di(4-ethoxyphenyl)-5-methyl biuret 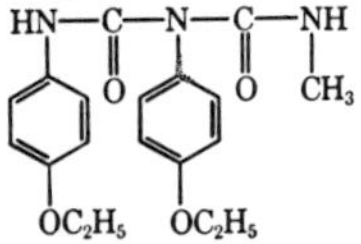	N	11.76	11.54			63
		Mol. wt.	357	346			
	Formula $C_{19}H_{23}N_3O_4$ Mol. wt. 357.41	M.P. 110–111°C					
	Solvents for recrystallization Ethanol	B.P.					
	Appearance Nearly colorless needles						
21.	1,3-Di(3-chlorophenyl)-5-*n*-butyl biuret 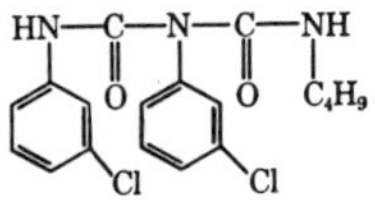	N	11.05	11.05			63
		Cl	18.67	18.61			
	Formula $C_{18}H_{19}Cl_2N_3O_2$ Mol. wt. 380.27	M.P. 119–120°C					
	Solvents for recrystallization Ethanol	B.P.					
	Appearance Colorless needles						

(continued)

TABLE 4.11 (*Continued*)

No.	Name and structure	Analysis			Infrared data		Ref.
		Elem.	Calc.	Fd.	Solvent	Wavelength	
22.	1,3-Di(4-chlorophenyl)-5-*n*-butyl biuret	N	11.05	11.04			63
	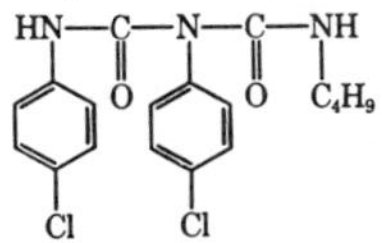	Cl	18.67	18.67			
		Mol. wt.	380	373			
	Formula $C_{18}H_{19}Cl_2N_3O_2$ Mol. wt. 380.27	M.P. 104–105°C					
	Solvents for recrystallization Ethanol	B.P.					
	Appearance Colorless square columns						
23.	1,3-Di(4-bromophenyl)-5-*n*-butyl biuret	N	8.95	8.74			63
	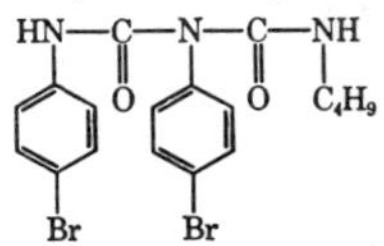	Br.	34.11	34.03			
		Mol. wt.	469	415			
	Formula $C_{18}H_{19}Br_2N_3O_2$ Mol. wt. 469.19	M.P. 118–120°C					
	Solvents for recrystallization Ethanol	B.P.					
	Appearance Small colorless prisms						

TABLE 4.12
Wavelengths of Most Intense Bands of Isocyanate Derivatives

Group	Maxima μ	Maxima cm^{-1}
—NCO in isocyanates	4.40–4.46	2270–2240
—N=C=N—in carbodiimides	≈4.72	≈2120
—C=O in uretidione ring (dimer)	5.61–5.65	1783–1770
—C=O in isocyanurate ring (trimer)	5.85–5.92	1709–1689
—C=O in carbamates (urethanes)	5.75–5.88	1739–1700
—C=O in allophanates	5.71–5.81[a]	1751–1721[a]
	5.81–5.90	1721–1695
—C=O in ureas	5.99–6.13	1670–1630
—C=O in biurets	5.81–5.92[a]	1720–1690[a]
	5.85–6.05	1708–1653
—NH deformation (Amide II)	6.41–6.58	1560–1520
—NH stretch (bonded)	2.94–3.13	3400–3200
—NH stretch (free)	2.86–2.94	3500–3400
—C=N in carbodiimide dimer	≈5.95	≈1681
—C=N in carbodiimide trimer	≈5.99	≈1669

[a] Higher frequency band is more intense.

this technique. Table 4.12 is presented to summarize the most useful wavelengths for characterization of these isocyanate compounds.

VI. MISCELLANEOUS REACTIONS

A. Amides

Amides react at moderate rates at about 100°C forming the acylurea because of the electronegative nature of the substituent on the NH group (2,107,108),

$$RNCO + R'CONH_2 \rightarrow RNHCONHCOR' \quad (54)$$

under normal conditions (108).

B. Phenols

Phenols react more slowly with isocyanates than aliphatic alcohols because they are more acidic and a catalyst such as a tertiary amine or aluminum chloride is generally used to promote this reaction (107,109).

$$RNCO + ArOH \rightarrow RNHCOOAr \quad (55)$$

C. Carboxylic Acids

Carboxylic acids show variation in reaction rate depending upon their acidic strength (110). The reaction has been found to be catalyzed by bases such as tertiary amines and metal compounds.

Aliphatic isocyanates and aliphatic acids usually give mixed anhydrides that decompose to substituted amides

$$\mathrm{RNCO + R'COOH \rightarrow RNHCOOCOR \rightarrow RNHCOR' + CO_2} \quad (56)$$

However, aromatic isocyanates react with aliphatic acids or weak aromatic acids, at moderate temperatures, to give other anhydrides which go to acid anhydride, urea, and carbon dioxide.

$$\mathrm{ArNCO + R'COOR \rightarrow ArNHCOOCOR} \quad (57)$$

$$\mathrm{ArNHCOOCOR \rightarrow ArNHCOOCONHAr + R'COOCOR'} \quad (58)$$

$$\downarrow$$

$$\mathrm{ArNHONHAr + CO_2}$$

D. Miscellaneous Hydroxy Compounds

Hydroxylamines react preferentially at the amino group if the reaction is carefully cooled and controlled, but react at both the amino and hydroxyl groups if sufficient isocyanate is present (111).

$$\mathrm{RNCO + C_6H_5CONHOH \rightarrow C_6H_5CONHOOCNHR} \quad (59)$$

Oximes also react to give urethanes, but generally decompose at their melting points (107).

$$\mathrm{RNCO + HON{=}CR'_2 \rightarrow RNHCOON{=}CR'_2} \quad (60)$$

$$\mathrm{RNHCOON{=}CR'_2 \rightarrow 2RCN + RNHCONHR + CO_2} \quad (61)$$

Sulfuric acid reacts with aliphatic isocyanates to give the sulfamic acid or in the presence of water gives the amine salt. Phenyl isocyanate gives sulfanilic acid (49).

$$\mathrm{RNCO + H_2SO_4 \rightarrow RNHCOOSO_3H \rightarrow RNHSO_3H + CO_2} \quad (62)$$

$$\mathrm{RNHSO_3H + H_2O \rightarrow RNH_2{\cdot}H_2SO_4} \quad (63)$$

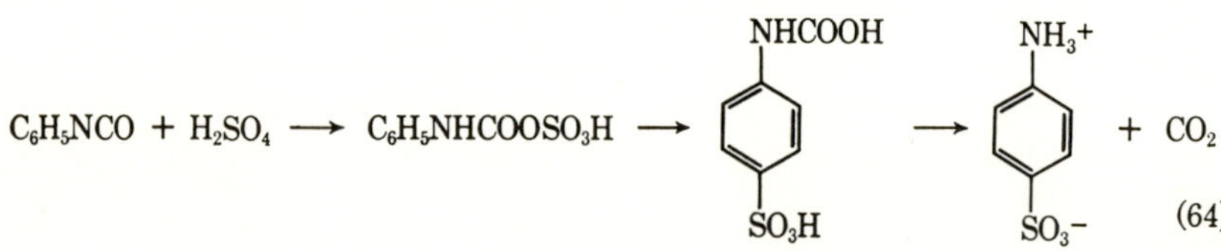

(64)

E. Reactions with Compounds Containing the C—H Group

Compounds containing C—H bonds in which the hydrogen may be replaced readily by sodium may react with isocyanates to give substituted amides.

Isocyanates may be used as acylating agents in the usual Friedel-Crafts technique, giving aryl amides (112,113) at moderate temperatures (reaction 65).

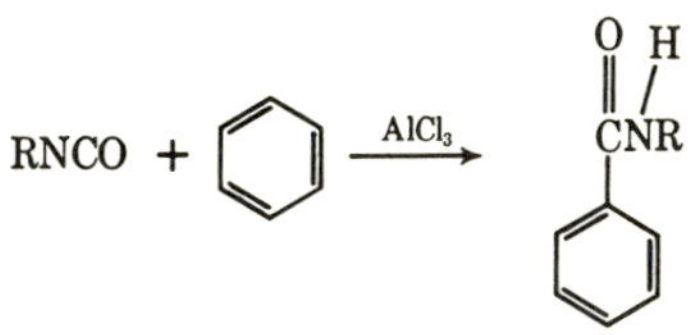

(65)

Hydrogen cyanide and hydrogen halides add to isocyanates (reactions 66 and 67) (113–115).

$$\text{RNCO} + \text{HCN} \underset{120\text{–}130^\circ\text{C}}{\rightleftharpoons} \text{RNHCOCN} \tag{66}$$

$$\text{RNCO} + \text{HX} \underset{90\text{–}110^\circ\text{C}}{\rightleftharpoons} \text{RNHCOX} \tag{67}$$

F. Reactions of Carbodiimides

Carbodiimides react readily with water to give ureas (29).

$$\text{RN}{=}\text{C}{=}\text{NR} + \text{H}_2\text{O} \rightarrow \left[\text{RNH}\text{—}\overset{\overset{\text{OH}}{|}}{\text{C}}{=}\text{NH}\right] \rightarrow \text{RNHCONHR} \tag{68}$$

Alcohols react with carbodiimides at elevated temperatures

$$\text{RN}{=}\text{C}{=}\text{NR} + \text{ROH} \rightarrow \text{RNH}\overset{\overset{\text{OR}}{|}}{\text{C}}{=}\text{NR} \tag{69}$$

whereas carboxylic acids react under mild conditions and rearrange to give acylureas in excellent yields.

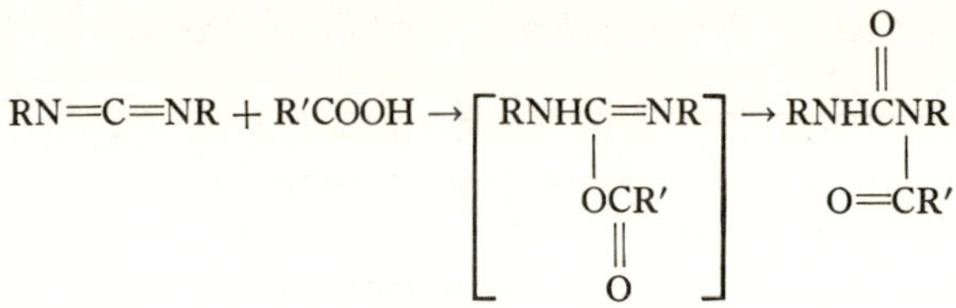

(70)

Carbodiimides form polymers under varying conditions depending upon the temperature and the substituents on the ring. Both dimers and trimers have been isolated (see reactions 71 and 72).

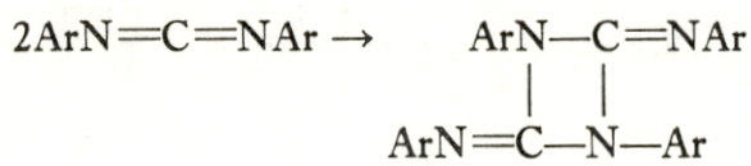

(71)

The carbodiimide dimer absorbs in the infrared near 5.95 μ (116). The

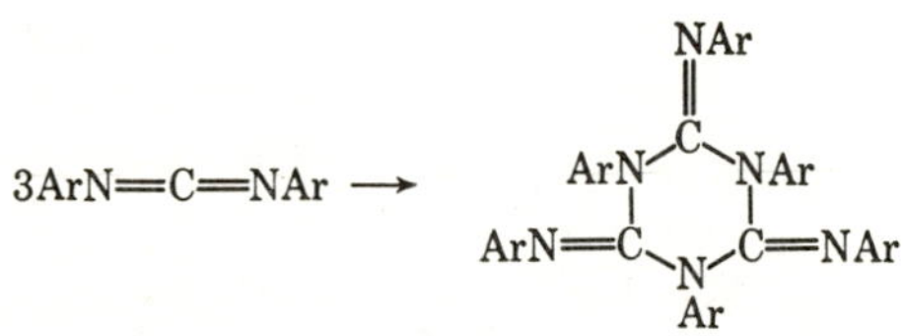

(72)

carbodiimide trimer absorbs in the infrared near 5.99 μ (116). These absorptions are due to the —C═N group.

G. Summary

The above sections are by no means intended to represent all of the possible isocyanate reactions since many additional reactions are possible and have been reported. These reactions and their chemistry have been treated extensively by Saunders and Frisch (29).

It was intended to show only the more prevalent reactions which may provide a background to the analytical chemist. Of interest is the fact that many of the products of the additional reactions may be characterized by the techniques described previously, that is by titration, infrared spectra, etc., since these products contain functional groups whose characteristic absorption frequencies are specific for constitution. Obviously, substitution differences or unusual products will have to be characterized by the usual generally available information or individual studies.

References

1. Bickel, V. T., and French, H. E., *J. Am. Chem. Soc.*, **48**, 747 (1926).
2. French, H. E., and Wirtel, A. F., *J. Am. Chem. Soc.*, **48**, 1736 (1926).
3. McKinley, J. B., Nickels, J. E., and Sidhu, S. S., *Ind. Eng. Chem., Anal. Ed.*, **16**, 304 (1944).
4. Tarbell, D. S., Mallatt, R. C., and Wilson, J. W., *J. Am. Chem. Soc.* **64**, 2229 (1942).
5. Lyons, J. M., and Thomson, R. H., *J. Chem. Soc.*, **1950**, 1971.
6. Kogon, I. C., *J. Am. Chem. Soc.*, **78**, 4911 (1953).
7. Kogon, I. C., *J. Am. Chem. Soc.*, **79**, 2253 (1957).
8. Baker, J. W., and Holdsworth, J. B., *J. Chem. Soc.*, **1947**, 713.
9. Baker, J. W., and Gaunt, J., *J. Chem. Soc.*, **1949**, 9.
10. Dennstedt, M., *Ber.*, **13**, 228 (1880).
11. Bloomquist, A. T., and Stevenson, H. B., *J. Am. Chem. Soc.*, **56**, 146 (1934).
12. Dunlop, A. P., and Peters, F. N., *The Furans* Reinhold, New York, 1953, p. 374.
13. Stork, G., *J. Am. Chem. Soc.*, **67**, 884 (1945).
14. Hofmann, K., and Bridgwater, A., *J. Am. Chem. Soc.*, **67**, 738 (1945).
15. Hofmann, K., and Bridgwater, A., *J. Am. Chem. Soc.*, **67**, 1165 (1945).
16. Hofmann, K., *J. Am. Chem. Soc.*, **67**, 1459 (1945).
17. Hofmann, K., Chen, C., Bridgwater, A., and Axelrod, A. E., *J. Am. Chem. Soc.*, **69**, 191 (1947).
18. Hofmann, K., Bridgwater, A., and Axelrod, A. E., *J. Am. Chem. Soc.*, **69**, 1550 (1947).
19. Hofmann, K., Bridgwater, A., and Axelrod, A. E., *J. Am. Chem. Soc.*, **71**, 1253 (1949).
20. Hofmann, K., U.S. Pat. 2,506,594 (to Ciba Pharmaceutical Products, Inc.) (May 9, 1950).
21. Hofmann, K., U.S. Pat. 2,520,404 (to Ciba Pharmaceutical Products, Inc.) (Aug. 29, 1950).
22. Hofmann, K., U.S. Pat. 2,453,627 (to Ciba Pharmaceutical Products, Inc.) (Nov. 9, 1948).
23. Hofmann, K., Brit. Pat. 615,908 (Jan. 13, 1949), *Chem. Abstr.* 43:5805.
24. Karmas, G., U.S. Pat. 2,574,484 (to Ortho Pharmaceutical Corp.) (Nov. 13, 1951).
25. Saunders, J. H., and Slocombe, R. J., *Chem. Rev.*, **43**, 203 (1948).
26. Rekker, R. F., Faber, A. C., Tom, D. H. E., Verleur, H., and Nauta, W. Th., *Rec. Trav. Chim.*, **70**, 113 (1951); *Chem. Abstr.* **45**:9528g.
27. Iwakura, Y., and Taneda, Y., *J. Org. Chem.*, **24**, 1992 (1959).

28. Iwakura, Y., Sato, M., and Ikegami, A., *Nippon Kagaku Zasshi* **80,** 635 (1959); *C.A.* **55,** 5386a (1961).
29. Saunders, J. H., and Frisch, K. C., *Polyurethanes, Chemistry and Technology*, Part I, *Chemistry*, Interscience, New York, 1962.
30. Stagg, H. E., *Analyst* **71,** 557 (1946).
31. Bailey, M. E., Kirss, V., and Spaunburgh, R. G., *Ind. Eng. Chem.*, **48,** 794 (1956).
32. Simons, D. M., and Arnold, R. G., *J. Am. Chem. Soc.*, **78,** 1658 (1956).
33. Morton, M., and Deisz, M. A., Paper No. 34, presented at the American Chemical Society Meeting, Atlantic City, September, 1956, Preprint, **16,** No. 3, p. 15.
34. Deisz, M. A., and Ohta, M., "Degradation Studies on Condensation Polymers," U.S. Dept. of Commerce Rept. PB-131795, Mar. 31, 1957; *Chem. Abstr.* **54:**10385e.
35. Burkus, J., *J. Org. Chem.*, **27,** 474 (1962).
36. Frisch, K. C., Reegen, S. L., and Thir, B., Paper sponsored by IUPAC, presented at the International Symposium on Macromolecular Chemistry, Prague, Aug.–Sept., 1965; *J. Poly. Sci.*, Pt. C, Polymer Symposia, **16,** Pt. 4, Interscience.
37. Baker, J. W., Davies, M. M., and Gaunt, J., *J. Chem. Soc.*, **1949,** 24.
38. Robertson, W. G. P., and Stutchbury, J. E., *J. Chem. Soc.*, **1964,** 4000.
39. Dyer, E., and Newborn, G. E., Jr., *J. Am. Chem. Soc.*, **80,** 5495 (1958).
40. Stolle, R., *Ber.*, **41,** 1125 (1908).
41. Dyer, E., and Read, R. E., *J. Org. Chem.*, **26,** 4388 (1961).
42. Mukaiyama, T., and Hoshino, Y., *J. Am. Chem., Soc.*, **78,** 1946 (1956).
43. Mukaiyama, T., *Bull. Chem. Soc. Japan* **28,** 253 (1955).
44. Mukaiyama, T., and Fujita, Y., *Bull. Chem. Soc. Japan* **29,** 54 (1956); *Chem. Abstr.* **50:**13775g.
45. Leuckart, R., *J. Prakt. Chem.*, **41,** 318 (1890).
46. Bortnick, N., Luskin, L. S., Hurwitz, M. D., and Rytina, A. W., *J. Am. Chem. Soc.*, **78,** 4358 (1956).
47. Dyer, E., and Wright, G. C., *J. Am. Chem. Soc.*, **81,** 2138 (1959).
48. Dyer, E., Newborn, G. E., Jr., and Wright, G. C., "Thermal Degradation of Carbamates," Delaware Chemical Symposium, Feb. 1958.
49. Bieber, T. I., *J. Am. Chem. Soc.*, **75,** 1409 (1953).
50. Pestemer, M., Farbenfabriken Bayer, personal communication, 1957.
51. Brumfield, H. L., "Infrared spectrophotometric analysis of urethane compounds," Sandia Corporation, Livermore Report SC-R-322; U. S. Atomic Energy Commission TID-4500; *Chem. Abstr.* **60:**15151b.
52. Zielinski, W., Jr., and Fishbein, L., *J. Gas Chromatog.* **3,** 142 (1965).
53. Blohm, H. W., and Becker, E. I., *Chem. Rev.*, **51,** 471 (1952).
54. Blohm, H. W., and Becker, E. I., American Document Institute, Document No. 2858 (1950); *Chem. Abstr.* **44:**8284i.
55. Blohm, H. W., and Becker, E. I., *J. Am. Chem. Soc.*, **72,** 5342 (1950).
56. Spielman, M. A., Barnes, J. D., and Close, W. J., *J. Am. Chem. Soc.*, **72,** 2520 (1950).
57. Lakra, H., and Dains, F. B., *J. Am. Chem. Soc.*, **51,** 2220 (1929).

58. Bennet, W. B., Saunders, J. H., and Hardy, E. E., "Kinetics of the reaction of mono-isocyanates with substituted ureas and urethanes," paper presented at the Annual Meeting of the Alabama Academy of Sciences, Tuscaloosa, April 2, 1954.
59. Kogon, I. C., *J. Org. Chem.*, **24,** 83 (1959).
60. Mukaiyama, T., and Iwanami, M., *J. Am. Chem. Soc.*, **79,** 73 (1957).
61. Cooper, W., Pearson, R. W., and Darke, S., *Ind. Chemist* **36,** 121 (1960); *Chem. Abstr.* **54:**11546d.
62. Kogon, I. C., *J. Org. Chem.*, **26,** 3004 (1961).
63. Raiford, C. L., and Freyermuth, H. B., *J. Org. Chem.*, **8,** 230 (1943).
64. Kogon, I. C., *J. Org. Chem.*, **23,** 1594 (1958).
65. Mecke, R., *Discussions Faraday Soc.*, 1950, No. 9, 161, *Chem. Abstr.* **46:**3408t.
66. Hilbert, G. E., Wulf, O. R., Hendricks, S. B., and Liddel, U., *J. Am. Chem. Soc.*, **58,** 548 (1936).
67. Hendricks, S. B., Wulf, O. R., Hilbert, G. E., and Liddel, U., *J. Am. Chem. Soc.*, **58,** 1991 (1936).
68. Saunders, J. H., and Hardy, E. E., *J. Am. Chem. Soc.*, **75,** 5439 (1953).
69. Pelley, R. L., U.S. Pat. 2,757,184, (to E. I. du Pont de Nemours and Co.) (July 31, 1956).
70. Balon, W., Langerak, E. O., Simons, D. M., and Stallman, O., Paper No. 40, Pt. I, presented at the American Chemical Society Meeting, Atlantic City, September, 1956, Preprint, **16,** No. 3, p. 67.
71. Mukaiyama, T., and Hoshino, T., *J. Am. Chem. Soc.*, **82,** 5339 (1960).
72. Davis, T. L., and Underwood, H. W., Jr., *J. Am. Chem. Soc.*, **44:** 2595 (1922).
73. Davis, T. L., and Blanchard, K. C., *J. Am. Chem. Soc.*, **45,** 1816 (1923).
74. Hale, W. J., *J. Am. Chem. Soc.*, **41,** 370 (1919).
75. Hale, W. J., and Lange, N. A., *J. Am. Chem. Soc.*, **41,** 379 (1919).
76. Spielman, M. A., Geiszler, A. O., and Close, W. J., *J. Am. Chem. Soc.*, **70,** 4189 (1948).
77. Mukaiyama, T., and Nohira, H., *J. Org. Chem.*, **26,** 782 (1961).
78. Bennet, W. B., Saunders, J. H., and Hardy, E. E., *J. Am. Chem. Soc.*, **75,** 2101 (1953).
79. Hoshino, T., Mukaiyama, T., and Hoshino, H., *J. Am. Chem. Soc.*, **74,** 3097 (1952).
80. Davis, T. L., and Ebersole, F., *J. Am. Chem. Soc.*, **56,** 885 (1934).
81. Naegeli, C., Tyabji, A., and Conrad, L., *Helv. Chim. Acta* **21,** 1127 (1938); *Chem. Abstr.* **33:**5404.
82. Craven, R. L., Paper No. 33, presented at the American Chemical Society Meeting, Atlantic City, September, 1956, Preprint, **16,** No. 3, p. 2.
83. Baker, J. W., and Bailey, D. N., *J. Chem. Soc.*, **1957,** 4649, 4652, 4663.
84. Bellamy, L. J., *The Infra-red Spectra of Complex Molecules*, Methuen, London, 1958.
85. Thompson, H. W., Nicholson, D. L., and Short, L. N., *Discussion Faraday Soc.*, 1950, No. 9, 222, *Chem. Abstr.* **46:**3404c.
86. Randall, H. M., Fowler, R. G., Dangle, J. R., and Fuson, N., *Infra-red Determination of Organic Structures*, Van Nostrand, Princeton, N.J., 1949.
87. Boivin, J. L., Boivin, P. A., *Can. J. Chem.*, **32,** 561 (1954).
88. Boivin, P. A., Bridgeo, W., and Boivin, J. L., *Can. J. Chem.*, **32,** 242 (1954).

89. Becher, H. J., and Griffel, F., *Naturwissenschaften*, **43**, 467 (1956).
90. Becher, H. J., Chem. *Ber.*, **89**, 1593 (1956).
91. Bonar, F. M., and David, D. J., personal communication, 1965.
92. Wiedemann, G., *Ann.*, **68**, 324 (1848); *Poggendorff's Ann.*, **74**, 67 (1849).
93. Habich, A., and Limpricht, H., *Ann.*, **109**, 101, 104 (1859).
94. Limprickt, H., *Ann.*, **105**, 395 (1858).
95. Hofmann, A. W., *Ber.*, **4**, 246 (1871).
96. Hofmann, A. W., *Ber.*, **4**, 262 (1871).
97. Finckh, C., *Ann.*, **124**, 331 (1862).
98. Baumann, E., *Ber.*, **8**, 708 (1875).
99. Davis, T. L., and Blanchard, K. C., *J. Am. Chem. Soc.*, **51**, 1801 (1939).
100. Kurzer, F., *Chem. Rev.*, **56**, 95 (1956).
101. Fischer, E., *Ber.*, **31**, 3266 (1898).
102. Saunders, J. H., *Rubber Chem. Technol.*, **32**, 337 (1959).
103. Cox, E. F., and Hostettler, F., Paper No. 186, presented at the American Chemical Society Meeting, Boston, April, 1959.
104. Hostettler, F., and Cox, E. F., *Ind. Eng. Chem.*, **52**, 609 (1960).
105. Gatewood, E. S., *J. Am. Chem. Soc.*, **45**, 146 (1923).
106. Schiff, H., *Ann.*, **352**, 72 (1907); *Chem. Abstr.* **1**:1700.
107. Petersen, S., *Ann.*, **562**, 205 (1949); *Chem. Abstr.* **44**:115b.
108. Wiley, P. F., *J. Am. Chem. Soc.*, **71**, 1310 (1949).
109. Leuckart, R., and Schmidt, M., *Ber.*, **18**, 2330 (1885).
110. Dieckmann, W., and Breest, F., *Ber.*, **39**, 3052 (1906).
111. Marguis, R., *Compt. Rend.*, **143**, 1163 (1906); *Chem. Abstr.*, **1**:730.
112. Butler, J. M., *J. Am. Chem. Soc.*, **71**, 2578 (1949).
113. Leuckart, R., *Ber.*, **18**, 873 (1885).
114. Dieckmann, W., and Kämmerer, K., *Ber.*, **40**, 3737 (1907).
115. Buckley, G. D., Piggott, H. A., and Welch, A. J. E., *J. Chem. Soc.*, **1945**, 864.
116. David, D. J., Bonar, F. M., private communication, 1963.
117. Bischoff, C., *Ber.*, **7**, 1078 (1874).
118. Brewster, J. H., *J. Am. Chem. Soc.*, **73**, 366 (1951).
119. Lange, N., *Handbook of Chemistry*, 8th Ed., Handbook Publishers, Inc., Sandusky, Ohio, 1952.
120. Smeltz, K. C., U.S. Pat. 2,840,589 (to E. I. du Pont de Nemours and Co.) (June 24, 1958).
121. Hart, W. F., McGreal, M. E., and Thurston, P. E., *J. Org. Chem.*, **27**, 338 (1962).
122. Kogon, I. C., *J. Org. Chem.*, **26**, 3004 (1961).
123. Gatewood, E. S., *J. Am. Chem. Soc.*, **47**, 407 (1925).
124. Raiford, L. C., and Freyermuth, H. B., *J. Org. Chem.*, **8**, 174 (1943).
125. Barthel, E., Kehr, C. L., Langerak, E. O., Pelley, R. L., and Smeltz, K. C., Paper No. 40, Pt. II, presented at the American Chemical Society Meeting, Atlantic City, September, 1956, Preprint, **16**, No. 3, p. 78.
126. Mukaiyama, T., and Matsunaga, T., *J. Am. Chem. Soc.*, **75**, 6209 (1953).

127. Pickard, R. H., and Carter, W., *J. Chem. Soc.*, **79,** 841 (1901).
128. Pickard, R. H., Allen, C., Bowdler, W. A., and Carter, W., *J. Chem. Soc.*, **81,** 1563 (1902).
129. McKee, R. H., *Am. Chem. J.*, **26,** 254 (1901); see Abstr. in Review Section *J. Am. Chem. Soc.*, **23,** 169 (1901).
130. Dains, F. B., and Wertheim, E., *J. Am. Chem. Soc.*, **42,** 2307 (1920).
131. Moore, F. J., and Gatewood, E. S., *J. Am. Chem. Soc.*, **45,** 135 (1923).
132. Kuhn, B., and Henschel, E., *Ber.*, **21,** 504 (1888).

CHAPTER V

POLYOLS

I. INTRODUCTION

The two polyols most useful in urethane chemistry are polyethers and polyesters. The extensive literature concerning the preparation and some of the pertinent reactions will be reviewed in this chapter so that the analytical problems can be better appreciated.

These compounds have generally been characterized by classical analytical methods well-known in the literature. The most useful of these will be presented in detail, together with newer methods based on instrumental techniques.

II. POLYETHERS

A. General Reactions

Polyethers are prepared from alkylene oxides by two different series of reactions. The first results from the reaction of the epoxide group in the alkylene oxide with other compounds having a labile hydrogen, such as water, alcohols, phenols, ammonia, amines, amides and acids, and yields oxyalkylated derivatives containing from two to approximately 100 alkylene oxide units. The second reaction may be described as a catalytic polymerization of the epoxide group, and occurs in the presence of catalysts, such as Lewis acids and bases, and salts. These yield polyalkylene oxides having molecular weights up to several million. The literature concerning these reactions has been well-covered by St. Pierre (1).

1. Acid-Catalyzed Hydration

A reaction between hydrochloric acid and ethylene oxide was reported by Hantsch and Hibbert (3). It was subsequently found that hydrochloric acid also catalyzed the addition of water to form ethylene glycol (4). In the absence of a large excess of water, polymerization of the oxide occurs rather than hydration.

$$H_2O + C_2H_4O \rightarrow HO(C_2H_4O)H \qquad (1)$$

$$HO(C_2H_4O)H + {}_nC_2H_4O \rightarrow HO(C_2H_4O)_{n+1}H \qquad (2)$$

Bronsted et al. (5) studied acid addition and acid-catalyzed hydrolysis and showed that there was indeed a hydrogen ion catalysis of water addition with the rate dependent upon the hydrogen ion concentration.

The mechanism proposed by Bronsted et al. can be represented as involving as the rate-determining step the simultaneous attack of a hydrogen ion and its conjugate base on the epoxide (reaction 3).

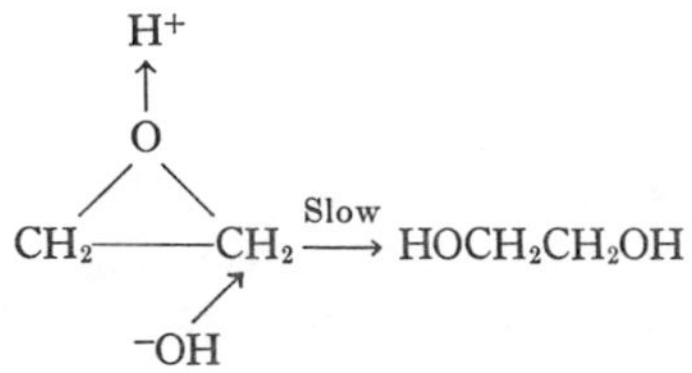

(3)

Ingold (6) has suggested an alternative mechanism which involves the rearrangement of the conjugate acid as the rate-determining step (reaction 4).

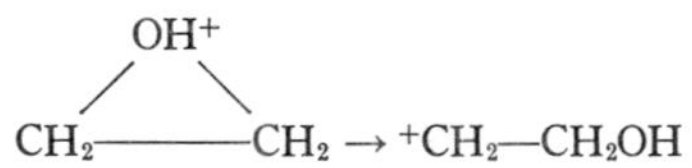

(4)

Long and Pritchard (7) found that with propylene oxide the addition of $H_2^{18}O$ in acid solution gave 70% the secondary $-^{18}OH$ product. From further studies of these and similar reactions (8), they proposed the following mechanism:

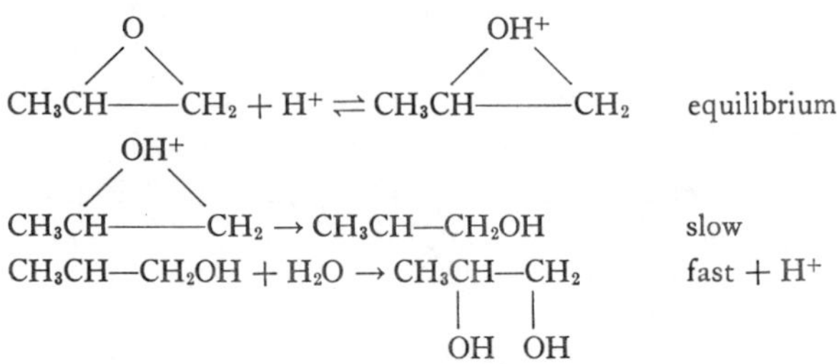

(5)

(6)

(7)

Addition occurs at either the primary or secondary carbon. Increasing the concentration of epoxide in either acid or base solvents results in the formation of dimers and trimers and finally results in polymer formation (9,10).

2. Base-Catalyzed Hydration

Basic catalysis of the hydration of ethylene oxide was first reported by Porret (11) and subsequently investigated by Lichtenstein and Twigg (12). The reaction fits the S_N2 type, with the predominant attack occurring at the sterically favored primary carbon and with the rate-determining step being the hydroxyl ion combination with the epoxide.

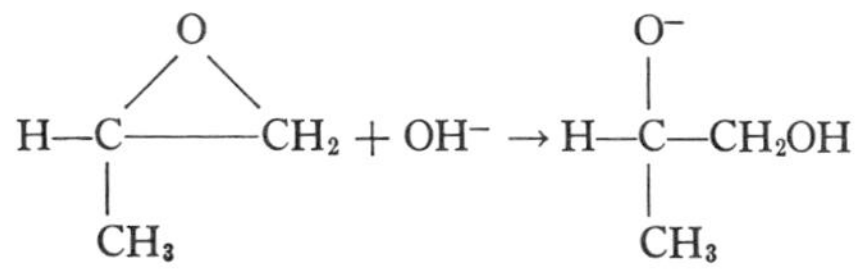

slow (8)

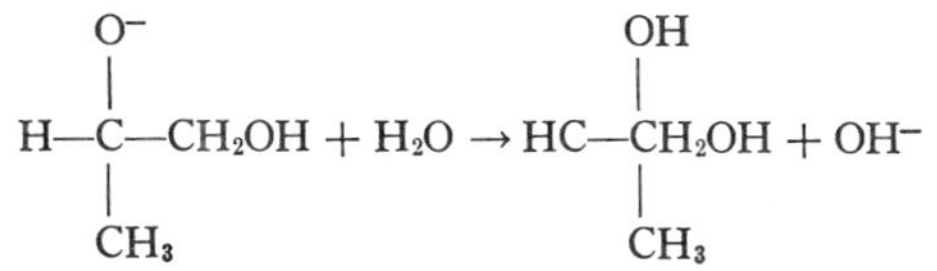

fast (9)

Whether the reaction occurs at the primary or at the secondary carbon is of considerable importance because the relative concentrations of primary and secondary hydroxyl groups influence reaction rates with isocyanates and properties of the end polyurethane and, in some cases, affect the reliability of the analytical results for the hydroxyl number determination.

Long and Pritchard (7), using water that had been oxygen tagged and propylene oxide, determined the relative quantities of the primary and secondary oxygen tagged OH. They found about 80% of the tagged hydroxyl groups attached to the primary carbon and 20% to the secondary carbon, using sodium hydroxide in concentration of 0.2–1.1*M*. Neutral solutions gave 61–67% of the primary product and 33–39% of the secondary. Similar results were obtained when the reactions were carried out at 45 and 100°C, showing that temperature had a negligible effect on the point of attack of the hydroxide ion.

B. Polymerization of Ethylene and Propylene Oxides

The first polymer from ethylene oxide was reported by Wurtz (13), who prepared it by reacting the oxide with ethylene glycol. Lourenco at the same time reported (14) a polyethylene glycol synthesis.

This work lay dormant for a long period until Staundinger (15,16) began his classic studies. He investigated various catalysts such as amines, alkali metals, and salts and oxides of zinc and the alkaline earths. Molecular weights from 60,000 to 120,000 were prepared by using the oxides of zinc, strontium, and calcium as catalysts.

A comprehensive discussion of polyoxyethylene glycols, their preparation, and their properties is given by Curme and Johnston (17).

Present-day commercial polymerization techniques are carried out using sodium hydroxide as catalyst at 170°C and a pressure of 35–40 psig to yield polymers in the 2000–20,000 molecular weight range (1). Current techniques have been described by Malkemus (18).

Propylene oxide was first prepared by Oser in 1861 (19) and the first polymerizations of propylene oxide were reported by Levene and Walti in 1927 (20).

Schlosser and Gray (21) reported the preparation of poly(oxypropylene) glycol by the polymerization of propylene oxide in the presence of water, with sodium hydroxide as catalyst. The polymerization was carried out by heating the monomer in a closed system at approximately 100°C until the pressure decreased to atmospheric.

1. Mechanisms

Polymerization of ethylene oxide and propylene oxide may be considered to occur by an ionic mechanism under the influence of strong acids or bases. In the case of strong bases, the anion formed by the action of the basic catalyst on an initiator is the chain propagator (1,22,23).

$$\mathrm{HO\overset{\overset{CH_3}{|}}{C}HCH_2OH + B^{\ominus} \rightleftharpoons HO\underset{\underset{CH_3}{|}}{C}HCH_2O^{\ominus} + BH} \qquad (10)$$

$$\mathrm{HOC\overset{\overset{CH_3}{|}}{C}HCH_2O^{\ominus} + \underbrace{CH_2\text{—}\overset{\overset{CH_3}{|}}{C}H}_{O} \rightarrow HO\overset{\overset{CH_3}{|}}{C}HCH_2OCH_2\overset{\overset{CH_3}{|}}{C}HO^{\ominus}} \qquad (11)$$

$$\mathrm{HO\overset{\overset{CH_3}{|}}{C}HCH_2OCH_2\overset{\overset{CH_3}{|}}{C}HO^{\ominus} + \underbrace{CH_2\text{—}\overset{\overset{CH_3}{|}}{C}H}_{O} \rightarrow HO\overset{\overset{CH_3}{|}}{C}HCH_2OCH_2\overset{\overset{CH_3}{|}}{C}HOCH_2\overset{\overset{CH_3}{|}}{C}HO^{\ominus}} \qquad (12)$$

This base-catalyzed reaction involves a nucleophilic attack of an alkoxyl ion at the primary carbon atom (Walden inversion), which has lower electron density, which allows the primary carbon–oxygen bond to break and leaves the secondary carbon–oxygen bond intact (24,25). The chain is terminated by the deactivation of the anion by a proton.

$$\mathrm{HO\overset{\overset{CH_3}{|}}{C}HCH_2OCH_2\overset{\overset{CH_3}{|}}{C}HOCH_2\overset{\overset{CH_3}{|}}{C}HO^{\ominus} + BH \rightarrow HO\overset{\overset{CH_3}{|}}{C}HCH_2OCH_2\overset{\overset{CH_3}{|}}{C}HOCH_2\overset{\overset{CH_3}{|}}{C}HOH + B^{\ominus}} \qquad (13)$$

In the acid-catalyzed reaction it has been proposed that a proton adds to the oxygen bridge, weakening the carbon–oxygen bonds as has been mentioned previously.

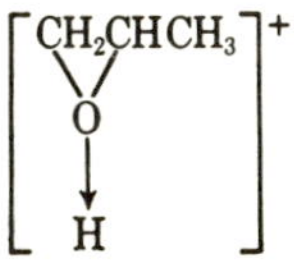

(14)

Attack by an alcohol group then readily occurs at both the primary and secondary carbon atom (24,25).

Mechanisms for the polymerization of propylene oxide have been proposed by Colclaugh et al. (26) and Price and Osgan (27), with ferric chloride as the Lewis acid catalyst. Price and Osgan proposed that polymerization proceeded by a mechanism involving polarization of the propylene oxide C—O bond but not through an actual carbonium ion mechanism. They observed a rate dependence upon the initial catalyst concentration but no dependence upon the monomer concentration.

Price and Osgan (27) polymerized 1-propylene oxide with finely powdered solid potassium hydroxide and obtained a low molecular weight, optically active polymer in contrast to the liquid polymer from *dl*-propylene oxide under the same experimental conditions. The following mechanism was postulated:

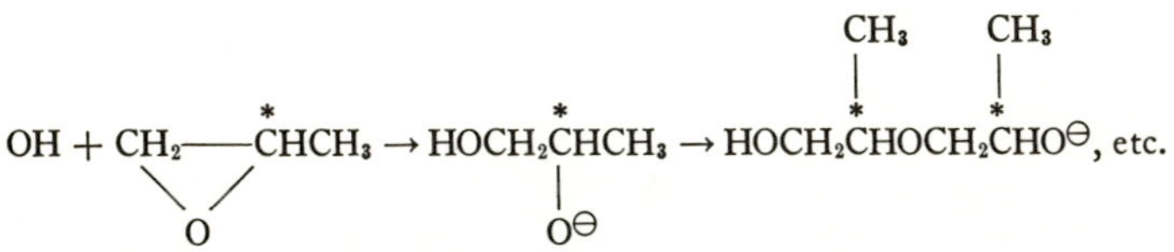

(15)

When 1-propylene oxide and *dl*-propylene oxide were polymerized using the same catalyst and experimental conditions, no detectable differences between polymers were observed except for optical activity. The crude polymer from 1-propylene oxide was separated into amorphous and crystalline portions. The crystalline fraction had the same rotation as that of the polymer from 1-propylene oxide using KOH as catalyst. This indicated that the polymerization with KOH proceeded with a much higher retention of asymmetry. In the case of ferric chloride catalysis, considerable racemization of asymmetric centers is involved.

Unlike polyethylene oxide, polypropylene oxide can possess several types of monomer arrangements which all have an effect on the properties. Some of the possibilities are (1):

A. Head-to-tail

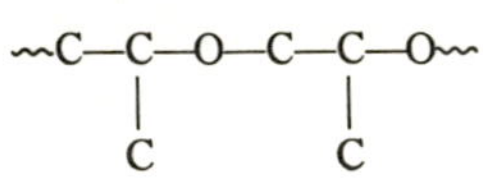

B. Head-to-head and tail-to-tail

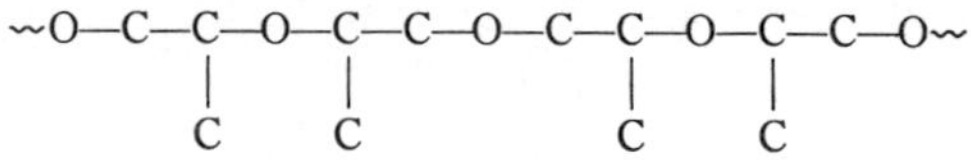

C. Random structure

$$\sim C—C—O—C—C—O—C—C—O\sim$$

D. Isotactic structure

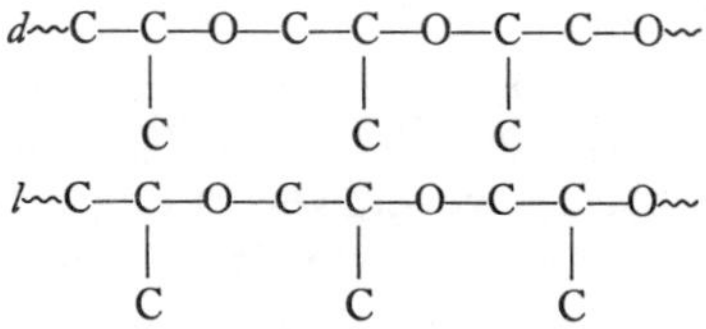

E. Isotactic structure—optically active

Exclusively *d* or *l*

Neither homogeneous acid- nor base-catalyzed polymerization occurs with 100% stereospecificity and it is more likely that commercial polymers have a random structure.

An important aspect to be noted which may affect the polyurethane performance and properties is the possible occurrence of a side reaction during polymerization of propylene oxide that leads to the formation of terminal unsaturated polyethers.

Simons and Verbanc (28) reported that at a constant polymerization temperature the concentration of double bonds increases rapidly with the molecular weight of the glycol produced. Also, with increasing temperature there is a rapid increase in the amount of double bond formed. These investigators suggested reactions 16 and 17. Thus, monofunctional

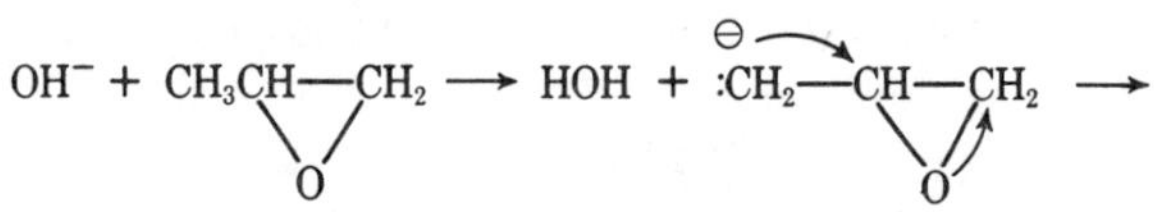

(16)

$$CH_2{=}CH—CH_2O^- \xrightarrow{PO} \text{monofunctional chain} \qquad (17)$$

polymer chains are formed that are terminated at one end with an allyl or propenyl group and hydroxyl terminated at the other end.

C. Other Polymers

Some of the first polyethers used in the manufacture of polyurethanes were block copolymers of ethylene oxide and propylene oxide (29). Lundsted (30) and Vaughn and Jackson (31) have described the preparation of this type of polyol which can be represented by:

$$HO(CH_2CH_2O)_a(\underset{\displaystyle CH_3}{\underset{|}{CH_2CHO}})_b(CH_2CH_2O)_cH$$

These polymers are prepared by the reaction of propylene oxide with propylene glycol or water in the presence of a basic catalyst to form a poly(oxypropylene) homopolymer. This polymer is then reacted with ethylene oxide to form a block copolymer. This type of polyether has a higher incidence of primary hydroxyl groups than does a poly(oxypropylene) ether of comparable molecular weight.

Random copolymers of ethylene and propylene oxide have also been prepared by reacting mixtures of ethylene oxide and propylene oxide with aliphatic diols such as propylene glycol (32).

Poly(oxytetramethylene) glycols, having terminal hydroxyl groups, are also used in polyurethane preparations. These glycols are prepared by the polymerization of tetrahydrofuran. Various catalysts and methods of preparation have been reported in the literature (33–35).

Poly(oxybutylene) glycols are prepared by the base-catalyzed reaction of either 1,2- or 2,3-butylene oxide or mixtures of the isomeric oxides with an initiator molecule (2).

Poly(oxypropylene) triols are prepared from propylene oxide with low molecular weight triols such as trimethylolpropane, glycerol, and 1,2,6-hexanetriol (36–38). Triol-based polyethers are used extensively in the manufacture of polyurethanes.

Higher functionality polyethers are used for urethane applications and are used mainly in rigid foam applications. A number of preparations have been described (2,39–41).

D. Reactions

Polyether polyols undergo the characteristic reactions of hydroxyl compounds. The polyethers, in general, are relatively inert and exhibit good stability. However, they are susceptible to oxidation, particularly at elevated temperatures (42,43) since they contain tertiary hydrogen atoms adjacent to ether oxygen and are subject to peroxide formation. The oxidation proceeds through a hydroperoxide intermediate, as reported from a study of diethylene glycol (44). The peroxide decomposes to form aldehydes, organic acids, and esters. These are important aspects to be noted as representing possible interferences to analytical techniques and contaminants, in addition to unsaturation previously mentioned, that may necessitate additional analyses.

The oxidation and weight loss at elevated temperatures can be arrested and/or reduced by the addition of a number of phenolic or amine-type antioxidants (45).

Decomposition products, identified in studies of the thermal degradation of propylene oxide polymers, were low molecular weight hydrocarbons ketones, aldehydes, and ethers (46,47).

III. POLYESTERS

Before the commercialization of the polyurethanes, foams, and elastomers, the polyesters utilized in plastic applications were generally unsaturated, designed for air or peroxide cure, and often copolymerized with other materials. However, for urethane formation the desired polyesters should be saturated and contain hydroxyl groups available for the isocyanate reaction, thereby providing chain extension and molecular weight increase (see Chapter 6 on polyurethanes).

The chemistry, preparation, manufacture, and applications of the unsaturated type have been well covered (48).

A. Preparation

Linear polyesters of approximately 2000 MW are preferred for elastomer preparation while slightly branched and more highly branched polyesters are required for flexible and rigid foams, respectively.

For these purposes polyesters are prepared from adipic acid, phthalic anhydride, and dimerized linoleic acid for the acid portion while the glycol portion frequently consists of ethylene, propylene, and 1,3- and 1,4-butylene glycols. The triols are frequently glycerin, 1,2,6-hexanetriol, trimethylolpropane, and trimethylolethane. More recently, polyesters derived from ϵ-caprolactone have appeared on the market.

Water generated by the reaction

$$(n+1)\mathrm{R(OH)_2} + n\mathrm{R'(COOH)_2} \rightarrow \mathrm{HO}\left[\mathrm{RO\overset{\overset{\displaystyle O}{\|}}{C}{-}R'{-}\overset{\overset{\displaystyle O}{\|}}{C}{-}O}\right]_n\mathrm{H} + n\mathrm{H_2O} \tag{18}$$

must be removed to drive the equilibrium to the right, which results in the desired molecular weight buildup in a reasonable time. The water content of the finished polyester should be low (preferably less than 0.1); otherwise, undesirable properties of the polyurethane may result. This is especially true in the case of elastomers where bubbles may form (see Chapter 6) which may ruin the appearance of the finished elastomer and contribute to poor processability and properties.

Along with this, due consideration must be given to control of the process to limit the free acid and glycol present which also will adversely affect the properties of the urethane. Hence, low monomer content of both acid and glycol are desired.

The preparation may be carried out without solvents in basically two steps. A single glycol or mixed glycols may be charged, the acid component added and heated with stirring under vacuum to remove the water. The reaction is forced to completion by increasing the heat and reducing the pressure further until the acidity and water content are low and the hydroxyl number is in the desired range.

Catalysts such as acids (e.g., *p*-toluenesulfonic acid) or metal oxides (e.g., lead oxide) may be used to shorten the esterification time (2). However, the utilization of catalysts may affect the reaction time with the isocyanate (residual catalytic effects), as well as influence the thermal stability and permanence properties of the polyurethane.

A number of interesting polyester preparations that are suitable for urethanes have been recently described (49–54). In addition, processes for the continuous preparation of polyesters have been reported (55,56).

B. Reactions

One reaction of polyesters that is of particular interest for those materials to be used for polyurethane preparation is decarboxylation.

$$\text{HO—}\left[\text{RO}\overset{\overset{\text{O}}{\|}}{\text{C}}\text{R}'\overset{\overset{\text{O}}{\|}}{\text{C}}\text{O}\right]_n\text{RO}\overset{\overset{\text{O}}{\|}}{\text{C}}\text{R}'\text{COOH} \rightarrow \text{HO}\left[\text{RO}\overset{\overset{\text{O}}{\|}}{\text{C}}\text{R}'\overset{\overset{\text{O}}{\|}}{\text{C}}\text{O}\right]_n\text{RO}\overset{\overset{\text{O}}{\|}}{\text{C}}\text{R}'\text{H} + \text{CO}_2 \qquad (19)$$

Since this reaction produces monofunctional constituents, chain growth will be prevented during urethane formation reaction and low molecular weight species will result which adversely affect the physical properties of the finished material. These reactions have been reported to occur at approximately 290°C for a variety of acids (57).

One inherent drawback of urethanes prepared from polyesters, as compared to polyethers, is their tendency to undergo hydrolysis, especially under more strenuous conditions of humidity and temperature (58).

$$\text{—O}\overset{\overset{\text{O}}{\|}}{\text{C}}\text{R}'\overset{\overset{\text{O}}{\|}}{\text{C}}\text{OR—O—} + \text{H}_2\text{O} \rightarrow \text{—O—}\overset{\overset{\text{O}}{\|}}{\text{C}}\text{R}'\text{COOH} + \text{HORO—} \qquad (20)$$

Residual catalysts remaining in the polyesters may promote this reaction. Thermal treatment of polyesters at approximately 300°C has been shown to result in formation of cyclic esters and a variety of unidentified products (59). Other polyesters and reactions have been comprehensively reviewed by Saunders and Frisch (2).

IV. ANALYTICAL TECHNIQUES AND CHARACTERIZATION

A. General (Polyethers and Polyesters)

For both polyethers and polyesters that are to be utilized for polyurethane formation the acid and hydroxyl content are of prime concern.

In spite of the numerous techniques that have been reported in the literature for the determination of hydroxyl content the older techniques such as phthalation and acetylation appear to be the most universally employed methods.

Of course, in formulating for polyurethane applications, all active hydrogens that take part in the urethane reaction need to be known as accurately as possible. The active hydrogen contributors of major interest are water, acid, and hydroxyl content. Unless these variables can be closely controlled, accurate NCO/OH ratios will not be possible and it will prove difficult to manufacture polyurethanes of consistent quality with the attendant desired good properties.

B. Wet Methods

1. Hydroxyl Number

The term universally used to describe the hydroxyl content of polyols is "hydroxyl number." Hydroxyl number is defined as the number of milligrams of potassium hydroxide equivalent to the hydroxyl content of one gram of the sample.

Verley and Bölsing (60) developed a method for determining the hydroxyl number of fats and fatty material. This procedure was modified by West, Hoagland, and Curtis (61), who used a 1:7 (v/v) mixture of acetic anhydride in pyridine, followed by hydrolysis with hot water and titration of the acid formed with alcoholic alkali solution.

Different ratios of acetic anhydride in pyridine have been recommended and reported. For example, Peterson and West (62) used 1:2 volumes of acetic anhydride to pyridine; Marks and Morrell (63) used 1:3 volumes of acetic anhydride to pyridine.

These ratios are important since they affect reaction time and therefore the quantitative results. The total number of hydroxyl groups that can be quantitatively acetylated with a given volume of reagent and the ratios also have some effect on quantitative accuracy.

The type and amounts of hydroxyl groups, such as primary and secondary hydroxyl groups, also influence accuracy since they do not acetylate with the same ease.

Malm, Genung, and Williams (64) studied the effects of time and temperature and concentration of acetic anhydride in pyridine on acetylation of cellulose derivatives. They found relatively long times required for complete acetylation (24 hr) when relatively dilute solutions of acetic anhydride in pyridine were used (1:19).

Ogg, Porter, and Willits (65) reported both macro and semimicro methods for the determination of the hydroxyl content of hydroxylated fatty acids and alcohols. One part of acetic anhydride in three parts of pyridine was used for acetylation, followed by hot water hydrolysis and homogenization with *n*-butanol and titration with alcoholic 0.5*N* NaOH.

The feature of the macro procedure that is noteworthy is that a modified iodine flask was used which possessed two side arms. All three openings of the flask were fitted with ground glass stopcocks so that the entire sample introduction, reagent addition, digestion, and subsequent potentiometric titration could be carried out without having to make any transferals and with a minimum of moist air exposure. The semimicro procedure utilizes the same reagents but a smaller pear-shaped flask and decreased volumes.

The effect of strength and age of the acetylating reagent was studied. In general, closer agreement of experimental values with theoretical and better stability of fresh, as opposed to four-day-old, reagent were found for the 1:3 (acetic anhydride to pyridine) than for the 1:7 mixture.

In many cases where the color of the material prevents sharp visual color changes at the pH endpoint, potentiometric titrations must be used. We have found a different pH endpoint using phenolphthalein indicator, as opposed to a potentiometric endpoint, preventing absolute comparisons between techniques. However, the visual endpoint using a suitable indicator such as phenolphthalein provides good accuracy and reproducible results on samples analyzed by the same technique providing the color of the sample does not interfere with endpoint detection.

Fritz and Schenk (66) reported a procedure employing acid-catalyzed acetylation using perchloric acid. The acetylation was performed on various alcohols and using ethyl acetate as the solvent required only 5 min at room temperature.

Primary and secondary alcohols were determined by acetylation in either ethyl acetate or pyridine using perchloric acid to catalyze the reaction. The pyridine–acetic anhydride–perchloric acid reagent had the disadvantages of requiring a longer reaction time and must be prepared fresh daily, whereas the ethyl acetate system was stable for two weeks. An acetylating solution that was 0.15*M* in perchloric acid was found to be optimum.

The procedure involves acetylating for 5 min at room temperature 3–4 mmoles of hydroxyl in a 125 ml glass-stoppered flask with 5 ml of 2*M* acetic anhydride in ethyl acetate or pyridine (3:1 ratio of acetic anhydride to solvent) that contains 0.15*M* perchloric acid. Water (1–2 ml) is added, the mixture shaken, and then 10 ml of 3:1 pyridine-water is added. After 5 min, the solution is titrated with 0.55*M* sodium hydroxide using a mixed indicator of 1 part of 0.1% neutralized aqueous cresol red with 3 parts of 0.1% neutralized thymol blue. Presumably, a potentiometric titration would serve equally as well or better. A blank is run omitting the sample and the hydroxyl number is calculated from the titration difference. Ethyl alcohol–water mixtures were found to acetylate but it was found necessary to have a definite excess of anhydride over water present.

The proposed mechanism for the catalyzed acetylation is (66–68): In ethyl acetate

$$\mathrm{Ac_2O + H^+ \rightarrow Ac_2OH^+} \tag{21}$$

$$\mathrm{Ac_2OH^+ \rightleftharpoons Ac^+ + HOAc} \tag{22}$$

$$\mathrm{Ac^+ + ROH \rightarrow ROAc + H^+} \tag{23}$$

In pyridine

$$\mathrm{C_6H_5\overset{+}{N}H + Ac_2O \rightleftharpoons (C_5H_5NAc)^+ + AcOH} \tag{24}$$

The pyridine-acetylium ion then reacts with the alcohol to form the ester and regenerate the pyridinium ion.

$$\mathrm{(C_5H_5NAc)^+ + ROH \rightarrow ROAc + C_5H_5\overset{+}{N}H} \tag{25}$$

A number of compounds were found to offer mild interferences which may be circumvented by means of temperature control of the acetylation and/or solvent system. However, it should be noted that the applicability of this procedure to polyesters has not been demonstrated and possible interferences from trace contaminants in these compounds may preclude its use.

Siggia, Hanna, and Culmo (69) have described a procedure for the determination of hydroxy and amino compounds using pyromellitic dianhydride. In this procedure, 50 ml of 0.5*M* pyromellitic dianhydride in tetrahydrofuran along with 10 ml of pyridine are added to a 250 ml flask containing 0.010–0.015 equivalents of alcohol or amine. The flask is placed on a steam bath for 2 min and an electric hot plate for 5 min. An additional 20 ml of pyridine is added and the heating continued for 3 min. The mixture is cooled and titrated with 1*N* sodium hydroxide using phenolphthalein. Aldehydes and phenols were not found to interfere, but the effect of water which may be present was not reported.

The method appears to be slightly longer and the heating step is a disadvantage over the method of Fritz and Schenk (66). The nonvolatility of the pyromellitic dianhydride is a definite advantage, however. Again, the application of this method to high molecular weight polyols was not demonstrated.

An acid-catalyzed acetylation method directly applicable to the determination of hydroxyl numbers of polyethers was reported by Stetzler and Smullen (70). This method employs 2*M* acetic anhydride in ethyl acetate and 0.15–0.20*M* *p*-toluenesulfonic acid as the catalyst. This catalyst system must be used in place of perchloric acid because perchloric acid attacks the ether groups in polyethers (70,71) and therefore cannot be used.

Five to six millimoles of hydroxyl is acetylated with 5 ml of the 2*M* acetic anhydride in ethyl acetate. The mixture is heated for 15 min at 50°C with swirling after 5 min. Then 2 ml of distilled water is added and the mixture swirled. Ten ml of 3:1 pyridine–water solution is added. After cooling for 5 min, the mixture is titrated with 0.56*N* potassium hydroxide using the mixed indicator of Fritz and Schenk (66). A blank is run omitting the sample, and the final calculation is corrected for the presence of any basic or acidic components determined on a separate sample.

The effect of water in the sample of the determination was not reported. The absolute accuracy of the method appears to be on the order of $\pm 1\%$ absolute in the range from 76 to 431 hydroxyl numbers.

An excellent study of base-catalyzed and salt-catalyzed acetylation of hydroxy groups was reported by Schenk, Wines, and Majzis (72). These investigators pointed out that pyridine is so often used as the solvent and/or base catalyst for many systems because it is a good enough nucleophile to catalyze the ionization of acetic anhydride, and a poor enough base to avoid interference at the endpoint with the basic titrants used to determine the extent of the reaction.

These investigators (72) studied three methods in detail: (*1*) A macro method employing 2.5*M* acetic anhydride–0.19*M* triethylenediamine in ethyl acetate, (*2*) a semimicro method employing 0.6*M* acetic anhydride–0.28*M* triethylenediamine in ethyl acetate, and (*3*) a macro method employing 2.5*M* acetic anhydride–0.5*M* tetraethylammonium bromide in acetic acid.

The sample is acetylated at reflux (80°C) for 15 or 20 min. Primary and secondary alcohols as well as phenols can be quantitatively determined. Moisture should be excluded and a blank must be run as before. Aniline (25 ml) is added after acetylation and the mixture back-titrated potentiometrically with 0.2*M* perchloric acid. Methyl violet is used as a visual indicator.

The method was shown to be applicable to polymers such as Span 65 and Tween 20 but apparently there is no great improvement of the accuracy over previous methods, the reproducibility being about 1%. NMR and IR established a definite triethylenediamine–acetic anhydride interaction, which appeared to be an intermediate complex involving a tertiary nitrogen and a carbonyl group consistent with the following:

$$R_3N\!: + Ac_2O \rightleftharpoons R_3NAc^+ + OAc^- \qquad (26)$$

$$R_3NAc^+ + :OR \rightarrow R_3NH^+ + ROAc \qquad (27)$$

A rapid hydroxyl method applicable to the determination of hydroxyl content of primary alcohols was reported by Floria, Dobratz, and McClure (73). The method utilizes *N,N*-dimethylformamide as a solvent, 0.2*M* 3-nitrophthalic anhydride in *N,N*-dimethylformamide as the esterification reagent, triethylamine as the catalyst, 0.1*N* tetrabutylammonium hydroxide as the titrant, and 1% thymolphthalein as the indicator.

The method requires only 10 min of esterification time at room temperature but has the disadvantage that it can be used only for primary alcohols. However, application to compounds in the molecular weight range from 32 to 20,000 was demonstrated with precision comparable to the standard acetic anhydride method.

An interesting and noteworthy observation was made during the studies inherent in the development of this method. These investigators observed an indication of a sample size–hydroxyl value dependence. As the equivalent ratio of reagent to sample was varied from 5:1 to 20:1, the hydroxyl values varied from 0.095 to 0.160 meq/g.

Further studies showed that in order to obtain accurate hydroxyl analyses, a 5:1 equivalent reagent/sample ratio was required. Our own observations agree with this (74) and it is felt that this point is often overlooked and may account, in part, for inter- and intralaboratory inaccuracies and nonreproducibilities.

Bush, Kunzelsauer, and Merrill (75) developed a method utilizing phosgene for the determination of organic hydroxyl groups. The hydroxyl-containing compound reacts with phosgene to form chloroformate.

$$ROH + COCl_2 \rightarrow RO\overset{\overset{\displaystyle O}{\|}}{C}Cl + HCl \qquad (28)$$

The excess reagent is removed by volatilization, the chloroformate is hydrolized with aqueous alkali, and the chloride content of the hydrolysate, which is equivalent to the original hydroxyl content, is titrated potentiometrically with silver nitrate.

The method has been shown to be particularly applicable to the determination of hydroxyl end groups in polymers. Water does not interfere and the relative error of the method was found to be $\pm 2\%$.

The disadvantages of the method are the time, equipment, and safety requirements. These investigators allowed the excess phosgene to evaporate overnight, and then removed the last traces under a vacuum for 1 hr. Then, of course, the sample must be solubilized and the titration carried out. Amines and unsaturated compounds were found to offer varying degrees of interference.

Reed, Critchfield, and Elder (76) developed a method for the determination of hydroxyl content of materials and in particular the polyoxyalkylene compounds. Since a method was desired that would more nearly define the amount of an isocyanate required in a urethane foam formulation, phenyl isocyanate is used as a reactant and catalyzed with stannous octoate. The excess isocyanate is then reacted with dibutylamine (see Chapter 3) and back-titrated with perchloric acid.

The procedure is as follows: Twenty-five milliliters of dimethylformamide is pipetted into each of four 500 ml glass-stoppered Erlenmeyer flasks. Into two of the flasks samples are introduced containing not more than 11 meq of hydroxyl compound. The flasks are swirled, 20.0 ml of 1*N* phenyl isocyanate in toluene is pipetted, and 1.0 ml of stannous octoate reagent (21.0 g stannous octoate diluted to 100 ml with toluene) is added. The samples are allowed to react 30–40 min at room temperature and then 20.0 ml of dibutylamine is pipetted into each flask and the mixture is allowed to stand 15 min. Then 100 ml of methyl Cellosolve is added along with 0.5 ml of 0.1% bromocresol green in methanol indicator. The samples are then titrated with 0.5*N* perchloric acid in methyl Cellosolve to the first definite yellow endpoint.

Stannous octoate was chosen as a catalyst because of its specificity in catalyzing the urethane reaction. The optimum quantity, under the reaction conditions employed, was found to be 0.2–0.5 g depending upon the exact nature of the hydroxyl compound.

The solvents used must be dried since water reacts quantitatively to consume 2 moles of isocyanate. The advantage of the method itself is that any water in the polyether will also consume isocyanate during foaming and thus this method more accurately represents the hydroxyl content which will consume isocyanate during urethane preparation.

Aromatic hydroxyl compounds were found not to react. The precision of the method was found to be ± 0.9 hydroxyl numbers, which appears to be in the same range as that reported for many of the previous methods.

It was observed that the phenyl isocyanate method agreed more closely with the results obtained by the acetylation method rather than phthalation after correction for water. For example, phenyl isocyanate 111.3, acetylation 111.5, and phthlation 113.6 for the propylene oxide adduct of sorbitol.

Even though phenyl isocyanate is monofunctional, which should minimize side reactions, aromatic isocyanates are quite reactive due to the ring resonance effects, and because of the excess used in the procedure the isocyanate may undergo some side reaction. Also, assuming the proper ratio of acetic anhydride to sample is used, the importance of which was previously pointed out, it has been our experience that acetylation is not so good as phthlation for materials containing secondary hydroxyl groups, generally requiring slightly longer reaction times. The lower results and agreement of the phenyl isocyanate and acetylation procedures as opposed to phthlation may indicate that 100% of secondary hydroxyl groups, when present in high molecular weight polyols do not undergo the straightforward urethane reaction under the test conditions of the phenyl isocyanate method.

Otey et al. (77) described a method that utilizes toluene diisocyanate as the reactant for the determination of hydroxyl numbers of polyoxyalkylene ethers. The remainder of the method is similar to the previous (76) except for solvent systems, catalyst (triethylenediamine in this case), and the use of 1*N* hydrochloric acid for back-titration.

The hydroxyl numbers obtained by this procedure were reported at being comparable to those obtained by the acetic anhydride–pyridine method although precision data were not given.

Due to the difunctionality of the isocyanate employed in this method, and therefore the many types of side reactions possible such as allophanate formation, dimer formation, and trimer formation, which are described in detail in Chapter 4, some error and nonreproducibility might be expected in an analytical procedure employing this isocyanate.

Theoretically, an ideal isocyanate for such a procedure, which has a good practical basis as has been mentioned previously and pointed out by some investigators (76,77), would be a monofunctional aliphatic isocyanate possessing a high equivalent weight because this type of compound has a limited tendency to undergo side reactions and would provide the necessary accuracy. Unfortunately, most of these compounds are difficult to obtain pure and/or possess a high volatility and are excellent lachrymators which preclude their use in routine control procedures.

A thorough review of the older methods for hydroxyl determination has been given by Mehlenbacher (78).

2. Determination of Primary and Secondary Hydroxyl Groups

As was pointed out previously, the amount of primary and secondary hydroxyl groups present in polyoxyalkylene ethers may be of interest to one concerned with polyurethanes, since they may influence reaction rates and physical properties of the finished polymer.

Not all the methods reported in the literature included here have been applied directly to the analysis of high molecular weight polyols. However, the methods or the basis of the method may be applicable and for this reason a brief description of the methods are included here.

Siggia and Hanna (79) demonstrated the use of differential reaction rates to analyze mixtures of alcohols for the primary and secondary hydroxyl contents. One of the methods reported was based on the differences in reaction rate of different alcohols with acetic anhydride. A standard linear plot for second-order reactions makes possible the analysis of mixtures containing the same functional group. The integrated form of the equation describing second-order reactions is

$$k = \frac{2.303}{t(b - a)} \log \frac{a(b - x)}{b(a - x)}$$

Plots of log $[(b - x)/(a - x)]$ vs t in the conventional manner yield straight lines for second-order reactions. When two second-order reactions are proceeding in the same mixture, a curve with two straight-line portions is obtained if the reaction rates are sufficiently different.

A sample containing approximately 0.5 mole of hydroxyl was acetylated with acetic anhydride in pyridine, and aliquots were removed at known time intervals, hydrolized, and back-titrated with base. A blank is treated similarly and log $[(b - x)/(a - x)]$ was plotted vs t, where x is the decrease in concentration of reactant in time t, and a and b are the initial concentrations of alcohol and anhydride, respectively.

A plot of the second-order reaction curve of 1-butanol and 2-butanol is shown in Figure 5.1. The line representing the slope of the less reactive alcohol (slope *2*) was extrapolated to point A at zero time. A line AB was drawn parallel to the time axis. The time, t, at point B of intersection between this line and slope *1* is read. The concentration of the more reactive alcohol is then the concentration of alcohol reacted at this time.

From the curve in Figure 5.1, t was found to be 59.5 min. From a plot of x vs. t the concentration of alcohol reacted in this time was 0.0398 mole/liter. The total moles per liter of butanol present was 0.2000.

$$\frac{0.0398}{0.2000} \times 100 = 19.9\% \text{ 1-butanol found vs. } 19.9\% \text{ present}$$

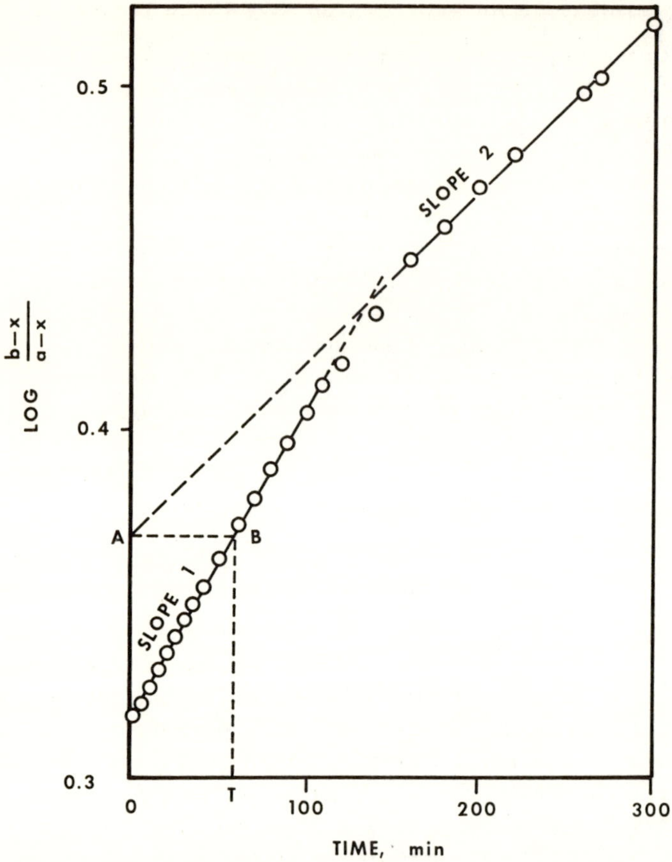

Fig. 5.1. Second-order reaction curve for mixtures of butanols (79).

If mixtures containing more than 70% of a primary alcohol (more reactive) were analyzed, difficulty in plotting was experienced. This difficulty was overcome by two techniques: (*1*) adding a known amount of the less reactive alcohol and subsequently correcting the final result, or (*2*) employing a larger sample and having the alcohol in excess of the anhydride.

This same technique was successfully applied to the analysis of polypropylene glycols for primary and secondary hydroxyl group contents (80).

A simplified technique, based on the analysis of binary mixtures of Siggia and Hanna (79) by second-order differential reaction rates using single- and double-point methods, was reported by Papa, Mark, and Reilley (81). In the method of Siggia and Hanna (79) the initial reagent concentration b (acetic anhydride, for alcohols) cannot equal a (the total sample concentration) if the plots are to be applicable. Reilley and Papa (82) proposed three methods, each of which is applicable to irreversible second-order kinetics in the case where $a = b$. This equality necessitated a prior determination of total sample content. The methods were demonstrated by the determination of 1-butanol and 2-butanol in a mixture of the two.

Crummet (83) developed a spectrophotometric method for the determination of polyglycols which was based on the work of Shmulyakovskii (84,85). The basis for the method is the ultraviolet absorption of the nitrites of the alcohols at 355 and 385 μ. Crummett found that the primary nitrite of a polyglycol is unstable in the presence of air. Therefore, he designed a reaction system to exclude air and allow introduction of the nitrited alcohols into the cell in the absence of air. Good agreement of experimental results with theoretical was found and the precision was reported to be $\pm 2\%$.

An infrared method for the analysis of primary alcohols in polyglycols based on the reaction of the polyol with triphenylchloromethane was reported by Hendrickson (86). The reagent consists of trityl chloride, with tris(dimethyl amino)-phosphine oxide in quinoline as a solvent. This reagent is added to the sample, scanned from 3700 to 3100 cm^{-1} to establish the total hydroxyl absorbance, and the reaction is then carried out in a glass-stoppered flask in a thermostatted bath at 75°C. At appropriate times, aliquots are removed and scanned, the absorbance being measured at 3280 cm^{-1}.

A rate plot is made and extrapolated back to zero time. The rate of disappearance of the secondary hydroxyl band is measured, and the quantity of primary alcohol is determined by difference. The method is based on the fact that the reaction rate of triphenylchloromethane with primary alcohols is 25–100 times faster than the reaction rate with secondary alcohols.

Willeboordse and Critchfield (87) developed a method for determining primary, secondary, and tertiary alcohols in mixtures. It is based on the difference in reaction rates between primary, secondary, and tertiary hydroxyl groups with phenyl isocyanate. The reaction is followed by the disappearance of the NCO band at 4.42 μ in the infrared. The differential reaction rate method is generally limited to binary mixtures. Good agreement between the results by this method and the method of Hanna and Siggia (80) were obtained on polypropylene glycols.

TABLE 5.1
Specifications and Properties of Polyoxypropylene Glycol (MW 2000)
(ASTM D-1786-60T)

Specifications	
Hydroxyl No.	54.5–57.5
Acid No., max	0.1
Modified APHA color, max	50
Unsaturation, meq/g, max	0.04
Water, %, max	0.1
Sodium and potassium, ppm, max	10
Properties:	
Sp. gr., 25/25°C.	1.003–1.006
Index of refraction, n_D^{23}	1.448–1.449
Copper, %	<0.002
Manganese, %	<0.001
Flash point, Cleveland open cup, °C	232

3. General Analytical Methods (Wet)

The following methods are offered as general methods for the analysis of commercially available poly(oxypropylene) glycols. Typical specifications and properties of glycols in the 2000 molecular weight range are shown in Table 5.1.

Some methods, such as refractive index, distillation range, and flash point, have not been included since methods of this type are well known and normally in routine use. In addition, methods such as these must be set up on a highly individual basis depending upon the specific material.

Rather, it is the intention here, as throughout the book, to present methods of general interest and applicability.

Polyethers—Measurement of pH (Apparent) (88)

Scope

This method can be used as a rapid check to determine gross levels of acidic or basic contamination of the poly(oxypropylene)glycol.

Principle

The pH is determined on a solution of the sample in a solvent containing 10 parts 99% isopropanol to 6 parts of boiled distilled water.

Equipment

(*a*) Beckman pH meter Model G or equivalent.

(*b*) Glass and calomel electrodes.

(*c*) Buffer solutions as supplied and prepared according to the manufacturer's direction.

(*d*) Beaker, 150 ml.

(*e*) Solvent solution: 10 parts by volume of 99% isopropanol to 6 parts by volume of boiled distilled water.

(*f*) Hydrochloric acid, 0.01*N*.

(*g*) Sodium hydroxide, 0.01*N*.

(*h*) Magnetic stirrer.

PROCEDURE

(*1*) Set up the pH meter as recommended by the manufacturer and standardize carefully with the prepared buffer solutions.

(*2*) Weigh, by difference, on an analytical balance 10 ± 0.05 of sample into a 150 ml beaker.

(*3*) Neutralize 60 ml of the solvent (10:6 isopropanol–water) to a pH of 7.0 on the pH meter using 0.01*N* hydrochloric acid or 0.01*N* sodium hydroxide as required.

(*4*) Add the 60 ml of solvent along with a Teflon stirring bar to the beaker containing the sample. Stir until completely in solution.

(*5*) Transfer the solution to the pH meter and determine the pH at 25°C. The solution should not be stirred with the magnetic stirrer while taking the reading. Instead, agitate gently by hand.

(*6*) The results should be reported to the nearest 0.1 pH unit.

Determination of Hydroxyl Number of Polyethers (89)

SCOPE

This method describes a procedure for the determination of hydroxyl groups in polymeric aliphatic glycols. It is applicable to polyether glycols containing primary or mixtures of primary and secondary hydroxyl groups. The precision at the 95% confidence level is:

Repeatability. Duplicate results by the same analyst should not vary more than ±1% relative.

Reproducibility. The average result reported by one laboratory should not vary from that reported by another laboratory by more than ±1.5% relative.

PRINCIPLE

Hydroxyl number is defined as the number of milligrams of potassium hydroxide equivalent to the hydroxyl content of one gram of the sample. The hydroxyl group is esterified with a solution of phthalic anhydride in pyridine. The excess reagent is back-titrated with standard sodium hydroxide solution and a blank is run on the reagents to determine the amount of anhydride consumed (see reactions 29 and 30).

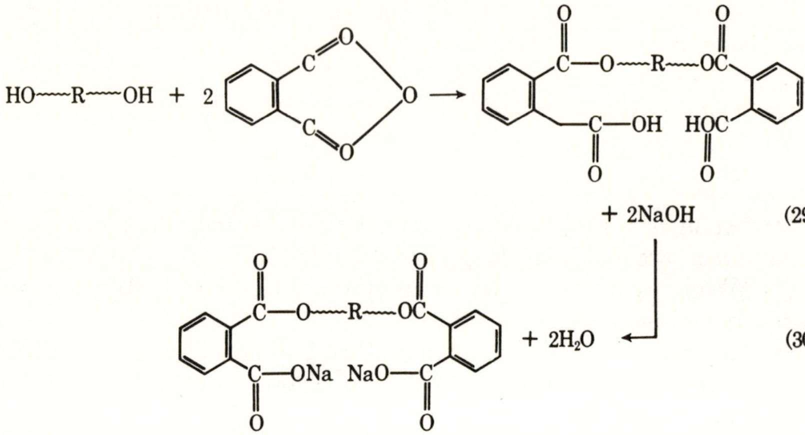

(29)

(30)

Excessive amounts of water in the sample will interfere by destruction of the esterification reagent and contribution to a high unreal hydroxyl number value. If the water content of the sample exceeds 0.2%, it is recommended that the sample be dried. Primary and secondary amines and higher fatty acids also interfere since they react with the reagent to form stable compounds and would be included in the analysis.

Neither the phthalic anhydride or pyridine and/or the phthalic anhydride–pyridine reagent should be allowed to stand open to the air or used when there is evidence of discoloration.

EQUIPMENT

(*a*) Flasks, 250 ml capacity, soil digestion, with 24/40 joints and air condensers.

(*b*) Normax bulb buret, 100 ml.

(*c*) A 25 ml precision pipet.

(*d*) Phthalating reagent: (42.0 g of phthalic anhydride dissolved in 300 ml freshly distilled pyridine). Allow to stand overnight before use.

(*e*) Phenolphthalein (1% solution in pyridine).

(*f*) Standardized 0.5*N* NaOH.

(*g*) Pyridine A.R. If exposed to air for long periods of time, purify by refluxing with barium oxide and distilling. This reagent should contain less than 0.1% water.

(*h*) Oil bath maintained at 115 ± 2°C.

PROCEDURE

Caution: Pyridine and phthalating reagent should always be handled in the hood. If these reagents come in contact with the skin, wash immediately with soap and water.

(*1*) All equipment must be thoroughly washed and dried before use. Remove flasks and condenser tubes from the oven and allow to cool.

(*2*) Weigh on analytical balance an appropriate size sample. The following is offered as a guideline only.

Approx. OH No.	Sample size g
390	1.1
210	2.1
165	2.7
45	7.0

No material must be allowed to touch the neck of the flask.

(*3*) Pipet 25 ml of phthalating reagent into each flask and put the air condensers in place.

(*4*) Place flasks into an oil bath (115 ± 2°C) for 1 hr. Keep sufficient oil in the bath to cover approximately one-half of the flask.

(*5*) After heating, remove from the bath and cool to room temperature. If the contents of the flask have turned dark, discontinue the analysis and rerun using freshly distilled pyridine.

(*6*) Add 15 ml A.R. pyridine to each flask, rinsing down the condenser.

(*7*) Titrate with 0.5N NaOH to a faint pink endpoint that lasts for 15 sec using five drops of phenolphthalein in pyridine as an indicator or titrate potentiometrically.

(*8*) Blanks are run in the same manner omitting the sample. Blanks must check within ± 0.10 ml. It is essential that the sample titration require at least ¾ of the titration of the blank.

CALCULATIONS

$$\text{OH No.} = \frac{56.1 \times N.\ \text{NaOH} \times (\text{ml. blank} - \text{ml. sample})}{\text{Sample weight}}$$

OH No. (corrected for Acid No.) = OH No. + Acid No.

Report results to the nearest 0.1 unit.

Determination of Unsaturation (90)

SCOPE

This method describes a procedure for the determination of unsaturated compounds in polyether-type urethane raw materials.

The precision (95% confidence level) is:

(*a*) *Repeatability*. Duplicate results by the same analyst should not vary by more than 0.002 units.

(*b*) *Reproducibility*. The average result obtained by one laboratory should not vary from that obtained by another laboratory by more than 0.004 units.

PRINCIPLE

Carbon-to-carbon unsaturated compounds in the sample are reacted with mercuric acetate and methanol in a methanolic solution to produce acetoxy-mercuricmethoxy compounds and acetic acid. The amount of acetic acid released in this equimolar reaction, which is determined by titration with standard alcoholic potassium hydroxide, is a measure of the unsaturation originally present. Because the acid cannot be titrated in the presence of excess mercuric acetate due to the formation of insoluble mercuric oxide, sodium bromide is added to convert the mercuric acetate to the bromide, which does not interfere.

Inasmuch as the method is based on an acidimetric titration, a suitable correction must be applied if the sample is not neutral to phenolphthalein indicator. Care must be taken to exclude carbon dioxide which of course titrates as an acid and gives erroneous results.

EQUIPMENT

(*a*) Pipet, 50 ml capacity.

(*b*) Erlenmeyer flasks, 250 ml glass-stoppered.

(*c*) Balance, triple beam.

(*d*) Buret, 50 ml capacity.

(*e*) Mercuric acetate, 0.12*M* in methanol. Dissolve 40 g of CP mercuric acetate in sufficient methanol to make 1 liter of solution and add sufficient glacial acetic acid to require a blank titration of 1–10 ml of standard 0.1*N* alcoholic potassium hydroxide per 50 ml of reagent. Usually, three to four drops of acid are sufficient. Prepare the reagent fresh weekly and filter before using.

(*f*) Alcoholic potassium hydroxide, 0.1*N*.

(*g*) Phenolphthalein indicator 1% in methanol.

(*h*) Sodium bromide CP crystals.

(*i*) Methanol.

PROCEDURE

(*1*) Add 50 ml of methanol to a sufficient number of 250 ml Erlenmeyer flasks to determine the acidity of each sample in duplicate. Neutralize to a faint pink endpoint, using a few drops of phenolphthalein indicator and 0.1*N* alcoholic KOH.

(*2*) Add 30 g of the sample weighed to the nearest 0.1 g to each flask and swirl to effect complete solution.

(*3*) Titrate with standard alcoholic KOH to a pink endpoint permanent for at least 15 sec and record the volume of titrant as "A."

(*4*) Pipet 50 ml of the mercuric acetate reagent into each of a sufficient number of 250 ml Erlenmeyer flasks to make all blank and sample determinations in duplicate. Reserve two of the flasks for the blank determinations.

(*5*) Into each of the other flasks introduce 30 g of the sample weighed to the nearest 0.1 g and swirl to effect complete solution.

(*6*) Allow the samples to stand together with the blanks at room temperature for 30 min. Swirl the flasks occasionally.

(*7*) Add 8–10 g sodium bromide crystals to each flask and swirl to mix thoroughly.

(*8*) Add approximately 1 ml of phenolphthalein indicator and titrate immediately with standard $0.1N$ alcoholic KOH to a pink endpoint permanent for at least 15 sec. Record the volume of titrant used for the sample as D and that used for the blank as E. The sample titration should not exceed 50 ml of standard $0.1N$ alcoholic KOH.

CALCULATIONS

Acidity correction:

$$\frac{A \times B}{W} = C \text{ (acidity, meq/g)}$$

A = ml. of $0.1N$ alc. KOH used for neutralization
B = normality of alc. KOH
E = grams of sample

Terminal unsaturation:

$$\frac{(D - E) \times B}{W} - C = \text{Unsaturation, meq/g}$$

D = ml. of $0.1N$ alc. KOH required for sample.
E = average ml of $0.1N$ alc. KOH required for titration of blanks.
B = normality of alc. KOH.
C = meq of acidity per gram of sample.
W = grams of sample used.

INTERFERENCES

The method does not apply to compounds in which the unsaturation is conjugated with carbonyl, carboxyl, or nitrate groups. Because water presumably hydrolyzes the reaction product to form basic mercuric salts, quantitative results are obtained only when the system is essentially anhydrous. Acetone in low concentrations does not interfere significantly, although its presence may be detrimental to the endpoint. Inorganic salts, especially halides, must be absent from the sample because even small amounts of salts may nullify the reaction of the mercuric acetate with the unsaturated compound.

Determination of Peroxides Content (91)

SCOPE

This method permits the determination of the peroxide content of polyethers which may form under certain conditions, such as oxidative, during manufacture.

PRINCIPLE

When KI solution reacts with peroxides, free iodine is liberated. The amount liberated is directly proportional to the amount of peroxides present. The free iodine can be titrated with sodium thiosulfate solution. The peroxide content is then calculated as ppm H_2O_2.

EQUIPMENT

(*a*) Sodium thiosulfate solution, 0.01*N*.
(*b*) KI solution, 10%.
(*c*) Glacial acetic acid, A.R.
(*d*) Iodine flasks, 250 ml.
(*e*) Graduated pipette, 10 ml.

PROCEDURE

(*1*) Transfer 100 ml of glacial acetic acid to each of two 250 ml iodine flasks.

(*2*) Reserve one of the flasks as a blank. Into the second flask introduce 25 gms of sample weighed to the nearest 0.1 gm. Mix thoroughly.

(*3*) To each flask add 2 ml of 10% KI solution by means of a transfer pipette.

(*4*) Mix thoroughly and allow to stand in the dark for 15 minutes.

(*5*) Titrate with standard 0.01*N* sodium thiosulfate to a colorless endpoint. Do not use an indicator. Titrate the blanks first and then titrate the sample to exactly the same endpoint.

CALCULATIONS

$$\frac{(\text{Sample titre} - \text{blank titre}) \times N \times 1.7 \times 10{,}000}{\text{Sample weight}} = H_2O_2 \text{ ppm}$$

Report the results to the nearest ppm.

Determination of Acid Number (92)

Scope

This method is suitable for the determination of acidic constituents in polyethers, resins, and other materials soluble in mixtures of benzene, isopropanol, and methanol. It is useful as an indication of the extent of a reaction involving acids, and the results are utilized in calculating the true hydroxyl number.

The precision (95% confidence level) is:

Repeatability. Duplicate results by the same analyst should not vary more than 1% relative.

Reproducibility. The average result reported by one laboratory should not vary from that reported by another laboratory by more than 4% relative.

Principle

Acid number is defined as the quantity of base, expressed in milligrams of KOH, that is required to titrate acidic constituents present in 1 g of sample.

The sample is dissolved in a 1:1:1 mixture of benzene, isopropyl alcohol, and methanol and the resulting single-phase solution is titrated at room temperature with standard alcoholic KOH, to the endpoint indicated by the color change of the added phenolphthalein.

Equipment

(*a*) Dispensing buret, 50 ml.

(*b*) Erlenmeyer flasks, 250 ml.

(*c*) Acid number solvent (333 ml of reagent grade benzene mixed with 335 ml of reagent grade isopropanol, and diluted to 1 liter with reagent grade methanol).

(*d*) Standard 0.1*N* alcoholic KOH.

(*e*) Magnetic stirrer and bar.

(*f*) Koch microburet, 10 ml.

Procedure

(*1*) Weigh a 40 g sample into a 250 ml Erlenmeyer flask.

(*2*) Add, by dispensing buret, 50 ml of acid number solvent.

(*3*) Stir, using a magnetic stirrer, until the sample is completely in solution.

(*4*) Add 5 drops of phenolphthalein indicator and titrate with 0.1*N* alcoholic KOH to the first faint pink endpoint.

(*5*) Determine a blank the same way, omitting the sample.

CALCULATIONS

Make the necessary temperature corrections.

$$\text{Acid No.} = \frac{\text{ml KOH for sample} - \text{ml KOH for blank}}{\text{Sample weight}} \times N\text{KOH} \times 56.1$$

Determination of Water in Polyethers by a Modified Karl Fischer Technique (93)

SCOPE

Varying amounts of water may be present which may adversely affect other analytical techniques as well as influence the properties of the polyurethane prepared from the polyether since these systems are sensitive to small amounts of water. In addition, an inaccurate knowledge of water content of materials of this type will affect the accuracy of the formulation since the NCO/OH ratio will be uncertain.

If larger than trace amounts of water are present, or analytical time requirements must be kept to an absolute minimum, then any one of a number of standard Karl Fischer methods, such as those given in Chapter 2 for DNT and those described by Mitchell and Smith (94), can be used.

PRINCIPLE

The sample is titrated in a modified Sattler-Geyer moisturimeter, as shown in Figure 5.2, which is a closed system designed to protect the sample from atmospheric moisture. The conventional titration of water with K.F. reagent uses a two electrode system whereby a potential is applied across the electrodes. At the endpoint the potential is about 0.1 V. In the polarization technique, a three-electrode system is used. Alternating current supplied through the transformer from the 60 cycle, 12 V supply creates a polarization voltage at the indicator electrode E_C (see Fig. 5.3). This electrode is connected to the input of a high impedance vacuum tube voltmeter. Near the equivalence point the iodide ion of the K.F. reagent increases sharply, causing a large change in the polarization resistance and a polarization voltage decay. At the equivalence point the voltage is in the order of 0.1 mv and remains at a minimum almost indefinitely. This increased sensitivity provides a highly reproducible endpoint in solutions of low water content.

EQUIPMENT

(*a*) Sattler-Geyer Moisturimeter (see Fig. 5.2).

(*b*) Power unit, No. 2606K, 115 V, 60 cycles ac (Welch Scientific Company).

(*c*) Vacuum tube voltmeter, Model 400 D (Hewlett Packard Co.).

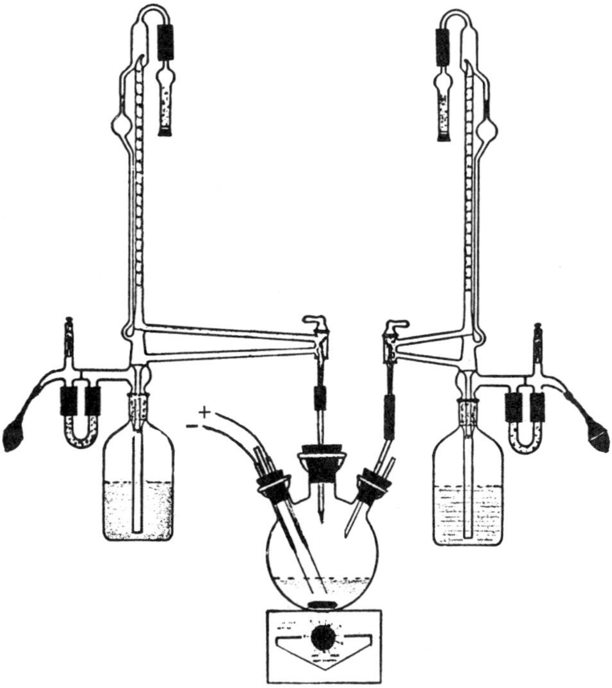

Fig. 5.2. Polarization titration apparatus.

(*d*) Karl Fischer reagent (4 parts K.F. reagent mixed with 6 parts of absolute methanol).

(*e*) Methyl alcohol, reagent grade.

(*f*) Dioxane

(*g*) Syringe with 3 in. needle, 1 cc.

PART A. STANDARDIZATION

(*1*) Place two clean 1 ml syringes and needles and a 25 ml volumetric flask in a 105–110°C oven for 1 hr.

(*2*) Remove the above materials from the oven and allow to cool to room temperature in a desiccator.

(*3*) Weigh the 25 ml volumetric flask on an analytical balance.

(*4*) Fill the 1 ml syringe and needle with distilled water and add 0.25 ml of water to the bottom of the 25 ml volumetric flask. Do not get water on the neck of the flask.

(*5*) Reweigh the volumetric flask to obtain the weight of H_2O added.

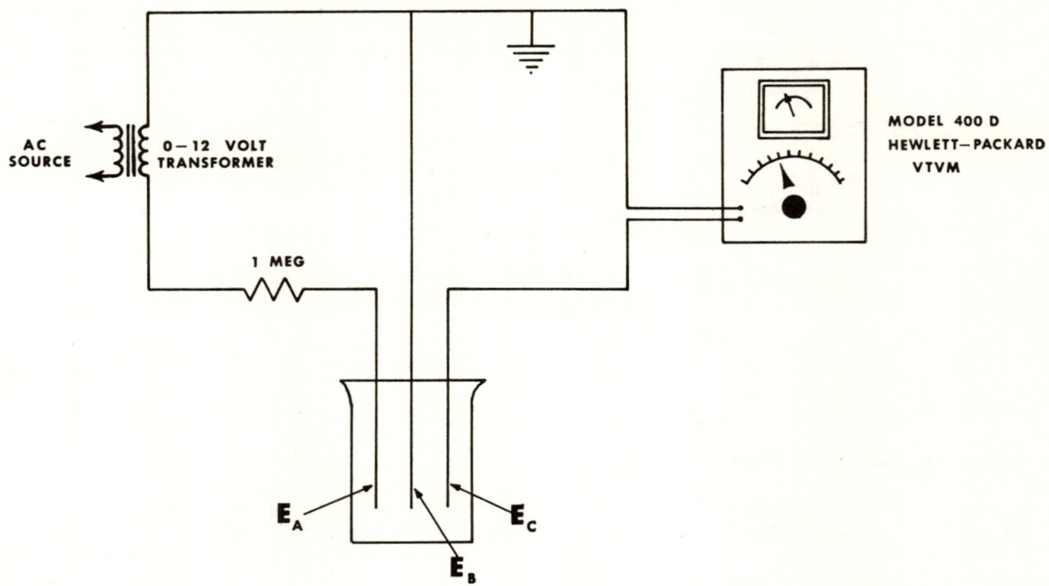

Fig. 5.3. Circuit for polarization titrations showing electrode assembly.

(*6*) Dilute to volume with dioxane from the dispensing buret after flushing the buret several times. Stopper the flask tightly and shake well.

(*7*) Follow steps *1–5* and *9–12* of Part B using a clean, dry, 1 ml syringe and needle to transfer 1 ml of the water–dioxane standard to the titration flask in step *9* of Part B.

(*8*) Follow steps *9–12* of Part B, this time adding 10 ml of dioxane from the dispensing buret, after flushing the buret several times, to the titration flask in step *9* of Part B.

(*9*) Run the standard and dioxane a second time (steps *7* and *8* above).

(*10*) If no sample is to be run immediately, follow steps *13* and *14* of Part B.

CALCULATION OF FACTOR

$$\text{Factor} = \text{mg of } H_2O \text{ per ml of K.F. reagent}$$

$$\text{Factor} = \frac{\text{g } H_2O \times 40}{\text{ml K.F.} - \text{ml K.F. for 1 ml dioxane}}$$

PART B. MOISTURE DETERMINATION

(*1*) Activate the voltmeter and turn the range selector to the 0.001 V position.

(*2*) The titration flask should contain approximately 20 ml of methanol which contains an excess of K.F. reagent (dark red); this is to keep the system moisture-free while not in use.

(*3*) Drain the dark red solution from the titration flask and flush twice with methanol using 10-ml portions.

(*4*) The electrodes should be evenly spaced and not touching one another. Add 20 ml of methanol to the flask. The voltmeter needle will deflect to the extreme right and remain there.

(*5*) Fill the K.F. buret, flushing several times before zeroing. Blank out the methanol by titrating with K.F. until the needle remains steady (does not drift to the right) for 45 sec.

(*6*) Transfer 10 ml of molten sample using a 10 ml syringe and Teflon needle into a clean, dry, tared 25 ml volumetric flask. Draw the sample into the syringe from the center portion of the sample container. If the sample is too viscous, pour the sample into the syringe inside an oven or drybox to prevent moisture pickup.

(*7*) Stopper flask, cool and reweigh and then add 14 ml of dioxane from the dispensing buret.

(*8*) Stopper flask and heat gently to aid solution. Cool to room temperature and dilute to volume. Record the total volume of dioxane, stopper, and mix well.

(9) Remove the drying tube and add by means of a dry pipet an appropriate size sample. Replace the drying tube.

(*10*) Fill the K.F. buret and titrate the sample until the needle remains steady (does not drift to the right) for 45 sec.

(*11*) Record the ml K.F. reagent used.

(*12*) Drain the flask and flush twice with 10-ml portions of methanol. If the sample remains in the flask, flush with dioxane or pyridine until clean.

(*13*) Add 20 ml methanol to the flask and add K.F. reagent until the solution is dark red.

(*14*) Drain the reagents back into their reservoir.

CALCULATIONS

Undiluted samples:

$$\%\ H_2O = \frac{\text{ml K.F.} \times \text{factor}}{\text{Sample weight} \times 10}$$

Samples diluted to 25 ml:

$$\%\ H_2O = \frac{\text{Factor}\ [A - (B \times C \times D)] \times 100}{E \times C \times 1000}$$

where

$$\text{Factor} = \text{mg } H_2O/\text{ml K.F.}$$

A = ml of K.F.
B = ml of solvent used for dilution
C = correction factor for volume and weight of sample used
$= \dfrac{\text{ml solution titrated}}{25 \text{ ml}}$
D = ml K.F. required for the residual water content of 1 ml of the dioxane used to dilute the sample.
E = Sample weight in g.

The sensitivity of the method, or the minimum amount of water that can be determined, is 0.002%. The precision of the method is ±0.001% at the 95% confidence level.

Other methods that may be of interest such as color, specific gravity, viscosity, etc., can be found in any good textbook such as Scott's *Standard Methods of Chemical Analysis* (95) or the Kolthoff-Elving series (96).

Essentially the same methods described for polyethers can be applied directly to polyesters except for the methods for peroxides and terminal unsaturation.

In addition to these methods, the determination of hydroxyl content can be performed using acetylation since secondary hydroxyl groups are not experienced here as in the case of polyethers. The acetylation procedure described below has the advantage that it requires a minimum of time to perform and provides sufficient accuracy and reproducibility.

Rapid Determination of Hydroxyl Number (97)

SCOPE

This method describes a rapid procedure for determining the total hydroxyl content of a polyester using acetic anhydride as an acetylating agent and pyridine as a solvent. The method is reproducible to approximately $\pm 1\%$ of the value obtained.

PRINCIPLE

Hydroxyl groups on primary and secondary carbon atoms and phenolic hydroxyl groups can be determined by esterification with an organic acid. Since direct use of an organic acid would liberate water and result in an equilibrium which would prevent the reaction from going to completion, the acid anhydride is used.

$$(CH_3CO)_2 + ROH \xrightarrow{\text{pyridine}} CH_3COOR + CH_3COOH \qquad (31)$$

The anhydride consumed can be calculated from a titration of the excess with standard base. The acid number value must be added to the hydroxyl value in order to obtain a correct hydroxyl number.

It has been found that excess moisture seriously interferes with the reaction and results in incomplete acetylation and gives low results. Neither the acetic anhydride or pyridine and/or the acetic anhydride–pyridine reagent should be allowed to stand open to the air or used when there is evidence of discoloration of the reagent or sample after acetylation.

EQUIPMENT

(*a*) Flasks, 250 ml capacity, soil digestion with 24/40 joints and 650 mm tubes.

(*b*) Normax bulb buret, 100 ml.

(*c*) Precision pipet, 10 ml.

(*d*) Acetylating reagent (70 ml pyridine to 30 ml acetic anhydride).

(*e*) Phenolphthalein (1% in pyridine).

(*f*) Standard 1.0*N* NaOH.

(*g*) Carborundum boiling chips.

(*h*) Pyridine, analytical reagent containing less than 0.1% water.

(*i*) Hot plate.

(*j*) Boiled distilled water.

PROCEDURE

(*1*) Clean and dry all equipment to be used.

(*2*) Weigh on an analytical balance a proper size sample into the iodine flask. Material should not be allowed to touch the neck of the flask. The sample size guide given in the polyether method should be consulted or as a general guide the following formula can be used:

$$\text{Sample weight, g} = \frac{56.1 \times 8}{\text{Approx. OH No.}}$$

(*3*) By means of a pipet, add 10 ml of acetylating reagent and then add several nonporous (Carborundum) boiling chips.

(*4*) Fit the greased, ground-glass-jointed condenser tube into place.

(*5*) Heat the flask and contents to a slow boil on a hot plate, swirling the flask until all the sample is in solution, and maintain at a slow boil for 10 min. The vapor should be allowed to condense before reaching the air condenser.

(*6*) Prepare blanks in the same manner, omitting the sample.

Note: The boiling point of the acetylating reagent is about 120°C.

(*7*) Place the flask complete with condenser into an ice-water bath.

(*8*) From a graduate, rinse down the condenser and flask with 25 ml of pyridine and add 50 ml of cold (0–5°C) boiled, distilled water.

(*9*) Remove the condenser and rinse the ground glass joint with boiled, distilled water.

(*10*) Titrate with 1.0N NaOH to a faint pink endpoint using phenolphthalein in pyridine as an indicator.

(*11*) Blanks must check within ±0.10 ml. It is essential that the sample titrations require more than ¾ of the titration of the blank.

Calculations

$$\text{OH No.} = \frac{56.1 \times N\ \text{NaOH} \times (\text{ml blank} - \text{ml sample})}{\text{Sample weight}}$$

$$\text{OH No. (corrected for Acid No.)} = \text{OH No.} + \text{Acid No.}$$

The results should be reported to the nearest 0.1 unit.

The analysis of polyesters and polymers for moisture and volatiles has also been reported (98). This technique utilizes a vapor pressure measurement which is proportional to the amount of moisture present.

The organics and moisture present are expelled under heat and vacuum. When the organics are sufficiently volatile so as to provide interference, both the moisture and organics are pumped through calcium hydride, condensed, and remeasured. The difference in the pressure readings represents the true amount of moisture which can be obtained from the calibration curve.

In the case of polyesters, small amounts of free glycol did not interfere since its vapor pressure was sufficiently low at room temperature. For stable polymers, a precision (95% confidence level) of ±0.001% was reported.

V. SPECTROSCOPIC TECHNIQUES

A. General

The major portion of spectroscopy studies has been devoted to the measurement of the hydroxyl content. This is by far the most useful application of spectroscopy to polyethers and polyester other than qualitative identification.

Qualitative identification of polyester or polyether can be accomplished simply by comparing infrared spectra of the two materials and observing the characteristic differences in the carbonyl, ester, and ether regions as shown in Figures 5.4 and 5.5.

Lucifer and Beatty (99) have studied the structure of saturated aliphatic esters in the infrared region 65.0–260 cm^{-1}. Although it was found that acyl components having more than seven carbon atoms were difficult to characterize in this region, it may be that the region has some usefulness to the common types of polyesters used for urethanes.

Primarily, the useful spectrum, for these types of materials, is limited to the carbonyl stretching frequency (1736–1750 cm^{-1}), the alpha methylene C—H scissor mode (1400–1430 cm^{-1}), and the C—O stretching frequency (1150–1270 cm^{-1}). Information obtained from these regions sometimes aids in establishing gross features of molecular structure.

Procedures for analyzing polyesters for their constituents are given in Chapter 6. Additional procedures for identifying dicarboxylic acids in polymeric esters have been reported (100,101).

B. Hydroxyl Number Determination

Burns and Muraca (102) have reported an infrared method for the determination of the hydroxyl content in polypropylene glycols. In this method the baseline technique was used to evaluate the absorbance of the hydroxyl stretching vibration which occurred at 2.84 μ. The total absorption is a function of the total amount of hydroxyl and water present

$$A = A_{\mathrm{OH}} + A_{\mathrm{H_2O}} = a_{\mathrm{OH}}C_{\mathrm{OH}} + a_{\mathrm{H_2O}}C_{\mathrm{H_2O}} \qquad (32)$$

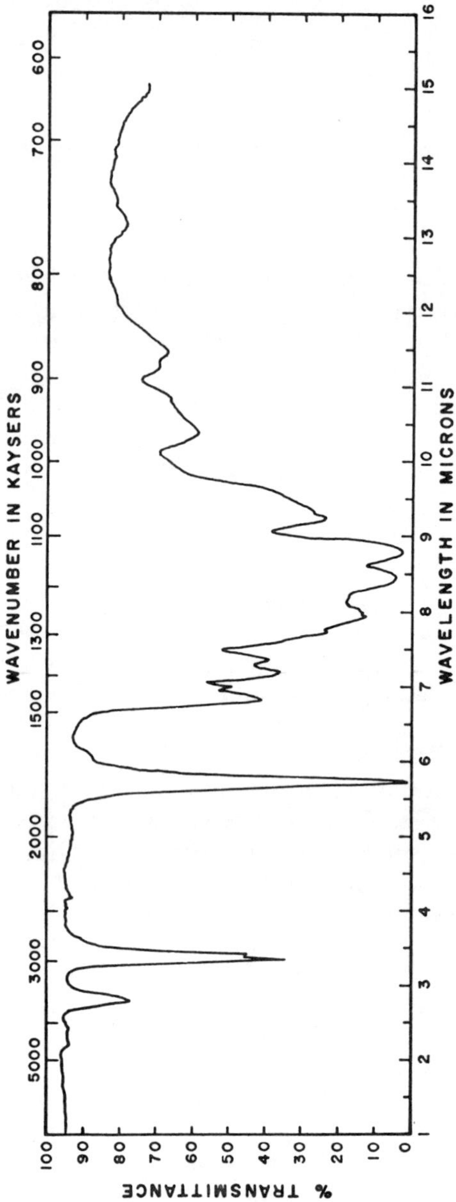

Fig. 5.4. Infrared spectrum of polyester polyol.

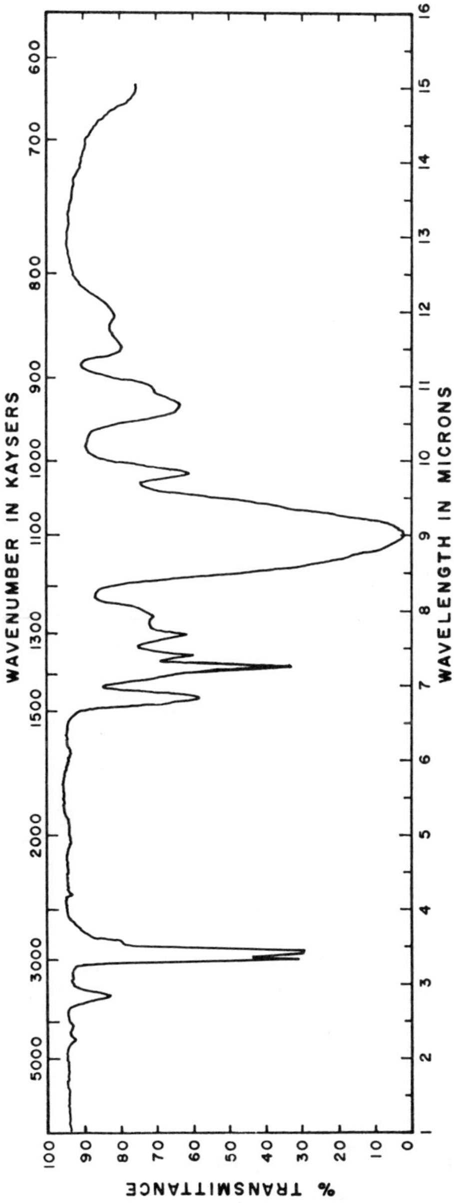

Fig. 5.5. Infrared spectrum of polyether polyol.

where a_{OH} and a_{H_2O} represent proportionality constants related to the molar absorptivities of the species denoted by the respective subscripts, and C_{OH} and C_{H_2O} represent concentrations.

Different known amounts of water were introduced into a sample and absorbance plotted as a function of water content. The intercept was taken as a function of hydroxyl content alone and the following equation was obtained:

$$a_{OH} = \frac{A_{OH}}{C_{OH}} = \frac{A - (0.0755)(\%\ H_2O)}{C_{OH}} \tag{33}$$

where A_{OH} = g/mmole of OH and C_{OH} = mmoles of OH/g.

The values of A_{OH}/C_{OH} were found to be essentially constant and if the samples contained less than 0.25% water, the water correction could be omitted.

For highest accuracy, the water content of the sample was determined by the Karl Fischer technique, and this value used for determining the actual molar absorptivity of the hydroxyl group according to:

$$C_{OH} = \frac{A - 1.86b(\%\ H_2O)}{6.74b} \tag{34}$$

where A = observed absorbance, b = path length in mm, and C_{OH} = hydroxyl concentration in millimoles/g. The method was reported to have a coefficient of variation of 2.2%.

Hilton (103) has described a similar method, using the hydroxyl absorption frequency, but utilizing a Beckman Model DK-2 spectrophotometer.

Free carboxylic acid groups were found to interfere and a separate titration had to be carried out to determine the amount of free acid necessary.

The total contributive absorption, which is proportional to the amount of each functional group present, leads to the following:

$$a_s\text{polyester} = a_s\text{OH}\ (X) + a_s \left(-\text{C}\begin{matrix}\nearrow \text{O} \\ \searrow \text{OH}\end{matrix} \right) (Y) + a_s\text{remainder}\ (1\text{-}X\text{-}Y) \tag{35}$$

$$a_s\text{polyether} = a_s\text{OH}\ (X) + a_s\text{remainder}\ (1\text{-}X) \tag{36}$$

By utilizing experimental data for percent hydroxyl and carboxyl (where applicable) in the above equations, the absorptivities of the groups were found. The final equations obtained were:

(Polyester) Hydroxyl No. = 1337 a_s 2.83 − 0.396 Acid No. − 38.94 (37)

(Polyether) Hydroxyl No. = 1294 a_s 2.87 − 19.1 (38)

It was found that a rise in temperature of 1°C lowered the absorbance 0.010 and, therefore, the cell compartment had to be thermostatted to obtain reproducible results. The accuracy of the procedure was found to be ±0.5% relative.

One- to two-gram samples were dissolved in 10.0 ml of chloroform and diluted to 100 ml with carbon tetrachloride and 1 g of anhydrous sodium sulfate was added to remove any moisture present.

A near-infrared method utilizing the first overtone of the primary hydroxyl stretching vibration (2.8 μ) in the infrared was applied to the determination of the hydroxyl value of aliphatic primary alcohols by Crisler and Burrill (104). These investigators used carbon tetrachloride or tetrachloroethylene as solvents and measured absorbance using the baseline technique on a Cary Model 14 spectrophotometer.

It was observed that the OH overtone band at 1.4 μ and the combination bands at 2.0 μ were present in the solution spectrum but were weak or missing in spectra of pure materials. Unlike the bands in the fundamental region, the bonded OH absorption is much weaker than the free OH absorption. Because of hydrogen bonding effects, Beer's law was followed at low concentrations only. Above 2 mg of OH per ml of solution nonlinearity occurred.

The effects of hydrogen bonding may contribute significantly to error, in a method such as this, unless suitable precautions are taken as these workers did. The overtone band has been previously used in hydrogen bonding studies (105–107).

Although the applicability of this method to the determination of hydroxyl content of polyethers and polyesters was not demonstrated, it may provide a greater accuracy and reproducibility than the present infrared methods. For these materials the same considerations should apply since hydrogen bonding of the polyethers and polyesters could also contribute to method error. Some slight effects due to hydrogen bonding, depending upon solvent, concentration, molecular configuration, etc., may also contribute to error in the infrared region and may be partly responsible for the change in molar absorptivity from compound to compound which necessitates a different calibration curve for each new polyester or polyether (103).

In the method of Crisler and Burrill (104), as before, any water present was removed with anhydrous sodium sulfate. The precision was quite good, having been found to be ±0.27 in the range 85–115 mg OH/g sample.

Kyriacou (108) has reported a method which utilizes a direct spectrophotometric titration of the hydroxyl compound with a standard solution of acetyl chloride.

Six or seven samples of poly(propylene) glycol weighing about 10 g are reacted with 0.6*M* acetyl chloride in toluene, added from a buret. Various amounts of acetylating reagent are chosen to obtain absorbances between 0.2 and 0.8. Approximately 10 g zinc was used as a catalyst and the final volumes must be constant.

After 40–45 min in a bath maintained at 37°C, they are cooled and the absorbance measured at 2.875 μ. The absorbance is plotted vs. milliliters of acetyl chloride solution. The break in the curve shows the number of milliliters required and the hydroxyl number can be calculated.

$$\text{Hydroxyl No.} = \frac{\text{ml} \times M \times 56.1}{W}$$

where M = molarity of acetyl chloride and W = weight of sample.

The reaction is:

$$CH_3\overset{\overset{\displaystyle O}{\|}}{C}\text{—}Cl + ROH \xrightarrow[\text{zinc}]{37°C} CH_3\overset{\overset{\displaystyle O}{\|}}{C}\text{—}OR + HCl \qquad (39)$$

The reproducibility of this method appears to be slightly poorer than present spectrophotometric procedures. However, it has the advantage that a calibration curve need not be established.

C. Miscellaneous Techniques

Page and Breslet (109) have developed a technique for determining the number average molecular weight by end-group analysis of polyalkylene glycols and glycol polyesters using NMR.

These investigators found that methylene and methine groups attached to a hydroxyl group could be differentiated from those attached to an ether oxygen. This made possible the calculation of number-average molecular weight without using the —OH resonance area and thus circumvented water interference.

The technique involved using the complex formed between pyridine and —OH groups to shift the —CH_2OH resonance from the —CH_2—O— resonance. A trace of HCl gas was also used to increase the exchange rate between —OH protons on adjacent molecules and effectively decouple the spin of the hydroxyl proton from those of the adjacent methylene protons which gave a sharper, better resolved spectrum.

This technique was applied to polyethylene glycols, polypropylene glycols, and polydiethylene glycol succinate. In the case of polypropylene glycols the ratio of primary to secondary —OH end groups, as well as number average molecular weight could be calculated. The results reported showed good agreement with conventional techniques.

Mlejnik (110) identified polyhydric alcohols in polyesters using gas chromatography. The alcohols were converted to acetates by reaction with excess acetic acid, with 1% concentrated sulfuric acid as a catalyst and then identified by gas chromatography.

References

1. St. Pierre, L. E., in *Polyethers*, Part I, *Polyalkylene Oxides and Other Polyethers*, N. G. Gaylord Ed., Interscience, New York, 1963, Chap. III.
2. Saunders, J. H., and Frisch, K. C., *Polyurethanes Chemistry and Technology*, Part I, *Chemistry*, Interscience, New York, 1962.
3. Hantzsch, A., and Hibbert, H., *Ber.* **40,** 1514 (1907).
4. Henry, L., *Compt. Rend.*, **144,** 1404 (1907); *Chem. Abstr.* **1**:2555.
5. Bronsted, J. N., Kilpatrick, M., and Kilpatrick, M., *J. Am. Chem. Soc.*, **51,** 428 (1929).
6. Ingold, C. K., *Structure and Mechanism in Organic Chemistry*, Cornell Univ. Press, Ithaca, N.Y., 1953.
7. Long, F. A., and Pritchard, J. G., *J. Am. Chem. Soc.*, **78,** 2663 (1956).
8. Pritchard, J. G., and Long, F. A., *J. Am. Chem. Soc.*, **78,** 2667 (1956).
9. Robeson, M. O., and Webb, T. P., U.S. Pat. 2,623,909 (to Celanese Corp. of America) (Dec. 30, 1952).
10. Levene, P. A., and Walti, A., *J. Biol. Chem.*, **73,** 263 (1927).
11. Porret, D., *Helv. Chim. Acta*, **27,** 1321 (1944).
12. Lichtenstein, H. J., and Twigg, G. H., *Trans. Faraday Soc.*, **44,** 905 (1948); *Chem. Abstr.* **43**:4549a.
13. Wurtz, A., *Compt. Rend.*, **49,** 813 (1859).
14. Lourenco, A., *Compt. Rend.*, **49,** 619 (1859).
15. Staudinger, H., *Die Hochmolekularen Organischen Verbindungen*, Springer, Berlin, 1932.
16. Staudinger, H., and Lohmann, H., *Ann.*, **505,** 41 (1933); *Chem. Abstr.* **27**:4790.
17. Curme, G. L., Jr., and Johnston, F., *Glycols*, Reinhold, New York, 1952.
18. Malkemus, J. D., *J. Am. Oil Chem. Soc.*, **33,** 571 (1956).
19. Oser, V., *Ann.* (*Suppl.*) **1,** 253 (1861).
20. Levene, P. A., and Walti, A., *J. Biol. Chem.*, **73,** 263 (1927).
21. Schlosser, P. H., and Gray, K. R., U.S. Pat. 2,362,217 (to Rayonier, Inc.) (Nov. 7, 1944).
22. Schildknecht, C. E., *Polymer Processes*, Interscience, New York, N.Y. (1956).
23. Flory, P. J., *Principles of Polymer Chemistry*, Cornell University Press, Ithaca, N.Y., (1953).
24. Chitwood, H. C., and Freure, B. T., *J. Am. Chem. Soc.*, **68,** 681 (1946).
25. Reeve, W., and Sadle, A., *J. Am. Chem. Soc.*, **72,** 1251 (1950).
26. Colclaugh, R. O., Gee, G., Higginson, W. C. E., Jackson, J. B., and Litt, M., *Plastics*, **23,** 381 (1958); *J. Poly. Sci.* **34,** 171 (1959); *Chem. Abstr.* **53**:7744c.
27. Price, C. C., and Osgan, M., *J. Am. Chem. Soc.*, **78,** 4787 (1956).
28. Simons, D. M., and Verbanc, J. J., *J. Polymer Sci.*, **44,** 303 (1960).
29. Staff report, *Chem. Eng. News*, **35,** (No. 3), 78 (1957).
30. Lundsted, L. G., U.S. Pat. 2,674,619 (to Wyandotte Chemicals Corp.) (Apr. 6, 1954).
31. Vaughn, T. H., Jackson, D. R., and Lundsted, L. G., *J. Am. Oil Chemists' Soc.*, **29,** 240 (1952).

32. Toussaint, W. J., and Fife, H. R., U.S. Pat. 2,425,845 (to Union Carbide) (Aug. 19, 1947).
33. Meerwein, H. D., Delfs, D., and Morschel, H., *Angew. Chem.*, **72**, 927 (1960).
34. Brit. Pat. 854,958 (to Farbenfabriken Bayer) (Nov. 23, 1960).
35. Murbach, W. J., and Adicoff, A., *Ind. Eng. Chem.*, **52**, 772 (1960).
36. Lundsted, L. G., U.S. Pat. 2,674,619 (to Wyandotte Chemicals Corp.) (Apr. 6, 1954).
37. Pruitt, M. E., and Rogers, W. A., Jr., U.S. Pat. 2,948,757 (to Dow Chemical Co.) (Aug. 9, 1960).
38. DeGroote, M., U.S. Pat. 2,626,911 (to Petrolite Corp.) (Jan. 27, 1953).
39. Griffin, W. C., U.S. Pat. 2,673,882 (to Atlas Powder Co.) (Mar. 30, 1954).
40. DeGroote, M., U.S. Pat. 2,554,667 (to Petrolite Corp.) (May 29, 1951).
41. Monson, L. T., and Dickson, W. J., U.S. Pat. 2,766,292 (to Petrolite Corp.) (Oct. 9, 1956).
42. Patton, J. T., U.S. Pat. 2,786,080 (to Wyandotte Chem. Corp.) (Mar. 19, 1957).
43. Leis, D. G., and Stout, E. C., U.S. Pat. 2,942,033 (to Union Carbide Corp.) (June 21, 1960).
44. Lloyd, W. G., *J. Am. Chem. Soc.*, **78**, 72 (1956).
45. "Polypropylene Glycols," Form No. 125-129-1256, Dow Chemical Company bulletin, Midland, Michigan, 1957.
46. St. Pierre, L. E., and Price, C. C., *J. Am. Chem. Soc.*, **78**, 3432 (1956).
47. Madorsky, S. L., and Straus, S., *J. Polymer Sci.*, **36**, 183 (1959).
48. Bjorksten, J., Tovey, H., Harker, B., Henning, J., *Polyesters and Their Applications*, Reinhold, New York, 1956.
49. Kerschner, P. M., and Greenwald, B. W., U.S. Pat. 3,086,044 (to Cities Service Research and Development Co.) (Apr. 16, 1963).
50. Goodyear Tire and Rubber Co., Brit. Pat. 971,628 (Sept. 30, 1964).
51. Sorokin, M. F. and Khinchina, E. L., *Lakokrasochnye Materialy i ikh Primenenie*, **1964**(**5**), 15 (in Russian); *Chem. Abstr.* **62**:6563a.
52. Rehfuss, C., U.S. Pat. 3,172,904 (to Allied Chem. Corp.) (Mar. 9, 1965).
53. Nurushina, K., Takehisa, Y., and Ichikawa, T., Japan Pat. 20,533 (to Toyo Rayon Co. Ltd., (Sept. 19, 1961).
54. Allied Chemical Corp., Neth. Pat. Appl. 6,410,818 (Mar. 19, 1965); *Chem. Abstr.* **63**: 10092f.
55. Seiner, J. A., Fr. Pat. 1,367,995 (to Pittsburgh Plate Glass Co.) (July 24, 1964); *Chem. Abstr.* **62**:9260f.
56. Chemische Werke Huels A.-G., Fr. Pat. 1,367,278 (Cl. CO7c) (July 17, 1964).
57. Korshak, V. V., and Rogozhin, S. V., *Khim. i Fiz-Khim. Vysokomolekul. Soedinenii, Doklady 7-oi Konf. po Vysokomolekul. Soedineniyam*, **1952**, 11; *Chem. Abstr.*, **48**: 3912c (1954).
58. Saunders, J. H., Steingiser, S., Gemeinhardt, P. G., Morecroft, A. S., and Hardy, E. E., *Chem. Eng. Data Ser.*, **3**, 153 (1958).
59. Hill, J. W., and Carothers, W. H., *J. Am. Chem. Soc.*, **55**, 5031 (1933).
60. Verley, A., and Bölsing, F., *Ber.*, **34**, 3354 (1901).
61. West, E. S., Hoagland, C. L., and Curtis, G. H., *J. Biol. Chem.*, **104**, 627 (1934).
62. Peterson, V. L., and West, E. S., *J. Biol. Chem.*, **74**, 379 (1927).
63. Marks, S., and Morrell, R. S., *Analyst* **56**, 428 (1931).
64. Malm, C. J., Genung, L. B., and Williams, R. F., Jr., *Ind. Eng. Chem., Anal. Ed.*, **14**, 935 (1942).
65. Ogg, C. L., Porter, W. L., and Willits, C. O., *Ind. Eng. Chem., Anal. Ed.*, **17**, 394 (1945).
66. Fritz, J. S., and Schenk, George H., *Anal. Chem.*, **31**, 1808 (1959).
67. Burton, H., and Praill, P. F. G., *J. Chem. Soc.*, **1950**, 1203.

68. Burton, H., and Praill, P. F. G., *J. Chem. Soc.*, **1951**, 522.
69. Siggia, S., Hanna, J. G., and Culmo, R., *Anal. Chem.*, **33**, 900 (1961).
70. Stetzler, R. S., and Smullin, C. F., *Anal. Chem.*, **34**, 194 (1962).
71. Critchfield, F. E., "Indirect Acid Base Methods," Div. Anal. Chem., 139th Meeting, ACS, St. Louis, Mo., March 1961.
72. Schenk, G. H., Wines, P., and Mojzis, C., *Anal. Chem.*, **36**, 914 (1964).
73. Floria, J. A., Dobratz, I. W., and McClure, J. H., *Anal. Chem.*, **36**, 2053 (1964).
74. DeGrazzio, A. R., David, D. J., and Baumann, G. F., Mobay Chemical Company, Res. Rept. No. 62-4, December 1961, unpublished.
75. Bush, D. G., Kunzelsauer, L. J., and Merrill, S. H., *Anal. Chem.*, **35**, 1250 (1963).
76. Reed, D. H., Critchfield, F. E., and Elder, D. K., *Anal. Chem.*, **35**, 571 (1963).
77. Otey, F. H., Zagoren, B. L., and Mehltretter, C. L., *J. Appl. Polymer Sci.*, **8**, 1985 (1964).
78. Mehlenbacher, V. C., *Organic Analysis*, Vol. I, Interscience, New York, 1953, pp. 1–38.
79. Siggia, S., and Hanna, J. G., *Anal. Chem.*, **33**, 896 (1961).
80. Hanna, J. G., and Siggia, S., *J. Polymer Sci.*, **56**, 297 (1962).
81. Papa, L. J., Mark, H. B., Jr., and Reilley, C. N., *Anal. Chem.*, **34**, 1513 (1962).
82. Reilley, C. N., and Papa, L. J., *Anal. Chem.*, **34**, 801 (1962).
83. Crummett, W. B., *Anal. Chem.*, **34**, 1147 (1962).
84. Shmulyakovskii, Y. E., *Khim. i. Tekhnol. Topliv i Masel*, **4**, No. 12, 46 (1959); *Chem. Abstr.* **54**:7437g.
85. Shmulyakovskii, Y. E., *Zhur. Prikl. Khim.*, **32**, 2513 (1959); *Chem. Abstr.* **54**:7437h.
86. Hendrickson, J. G., *Anal. Chem.*, **36**, 126 (1964).
87. Willeboordse, F., and Critchfield, F. E., *Anal. Chem.*, **36**, 2270 (1964).
88. Malec, E. J., and Christman, R. A., "Mobay Method for the Measurement of pH of Polyethers," 1958, Mobay Chemical Co., unpublished.
89. Sherwood, N., and Malec, E. J., "Mobay Method for the Determination of Hydroxyl Number of Polyethers," 1958, Mobay Chemical Co., unpublished.
90. Malec, E. J., and Christman, R. A., "Mobay Method for the Determination of Terminal Unsaturation of Polyethers," 1958, Mobay Chemical Co., unpublished.
91. Malec, E. J., and Christman, R. A., "Mobay Method for the Determination of Peroxides Content of Polyethers," 1958, Mobay Chemical Co., unpublished.
92. Sherwood, N., Malec, E. J., and Christman, R. A., "Mobay Method for the Determination of Acid Number of Polyethers," 1958, Mobay Chemical Co., unpublished.
93. Coen, C. S., and David, D. J., "Application of a Polarization Titration Technique to the Quantitative Determination of Water in Organics Using Karl Fischer Reagent," Mobay Chemical Co., Res. Rept. No. 62-18, 1961, unpublished.
94. Mitchell, J., Jr., and Smith, D. M., *Aquametry*, Interscience, New York, 1948.
95. Scott, W. W., *Standard Methods of Chemical Analysis*, Vols. I and II, 5th ed., Van Nostrand, New York, 1939.
96. Kolthoff, I. M., and Elving, P. J., *Treatise on Analytical Chemistry*, Part II, Vols. I–IX, Interscience, New York, 1961–1967.
97. Malec, E. J., and Christman, R. A., "Mobay Method for the Rapid Determination of Hydroxyl Numbers in Polyesters," 1961, Mobay Chemical Co., unpublished.
98. David, D. J., Baumann, G. F., and Steingiser, S., *SPE Trans.*, **2**, 1 (1962).
99. Lucifer, J. J., and Bentley, F. F., *Spectrochim. Acta*, **20**, 1 (1964).
100. Stafford, R. W., Francel, R. J., and Shay, J. F., *Anal. Chem.*, **21**, 1454 (1949).
101. Stafford, R. W., Shay, J. F., and Francel, R. J., *Anal. Chem.*, **26**, 656 (1954).
102. Burns, E. A., and Muraca, R. F., *Anal. Chem.*, **31**, 397 (1959).
103. Hilton, C. L., *Anal. Chem.*, **31**, 1610 (1959).

104. Crisler, R. O., and Burrill, A. M., *Anal. Chem.*, **31,** 2055 (1959).
105. Bellamy, L. J., *Infrared Spectra of Complex Molecules*, Wiley, New York, 1954, p. 83.
106. Goldman, I. M., and Crisler, R. O., *J. Org. Chem.*, **23,** 751 (1958).
107. Pauling, L., *The Nature of The Chemical Bond*, Cornell Univ. Press, Ithaca, N.Y., 1940, p. 316.
108. Kyriacou, D., *Anal. Chem.*, **33,** 153 (1961).
109. Page, T. F., Jr., and Bresler, W. E., *Anal. Chem.*, **36,** 1981 (1964).
110. Mlejnek, O., *Chem. Prumysl*, **13,** 105 (1963); *Chem. Abstr.* **59**:1069b.

CHAPTER VI

POLYURETHANES

I. INTRODUCTION

Polyurethane polymers present difficult analytical problems because of the variety of reactions which can take place between the ingredients of a formulation. Methods for the identification of several of the possible structures and analytical methods which apply to such polymers will be discussed in this chapter. Convenient methods for calculating the formula for a particular polymer, based on the analysis of the components are discussed, since they emphasize the practical importance of these analytical values.

II. CELLULAR POLYURETHANES

A. Preparation

Rigid urethane foams were reported first by O. Bayer (2) in 1947 and a flexible urethane foam system was reported by Höchtlen (3) in 1952. There are three basic methods used in the preparation of flexible foam: (*1*) prepolymer, (*2*) semi- or quasi-prepolymer, and (*3*) one-shot system.

The prepolymer is prepared by reacting a polyether or polyester with an excess of isocyanate to form an —NCO terminated prepolymer. Foaming is accomplished in the second step by reacting with water to form carbon dioxide for cell formation and urea linkages. Appropriate catalysts and surfactants are utilized in the second step in order to control the reaction rate and cell size.

Foam prepolymers and quasi-prepolymers will not be considered here since this method of preparation and application has been given over, largely, to the one-shot system. Also, the analytical techniques necessary for these prepolymers, such as percent —NCO, are essentially the same as for elastomeric prepolymers.

Since elastomeric prepolymers are still used extensively, the analytical techniques for prepolymers will be discussed in the elastomer section of this chapter.

Urethane foams have been prepared from a variety of polyesters (4–11) and polyethers (12–16). Rigid foams (17–20) and flame resistant foams (21,22) have been reported based on a variety of systems.

Whether or not a prepolymer or one-shot flexible foam system is used, the reactions illustrated in Scheme I occur and are the basic reactions to foam formation.

Because of the relatively faster rate of reaction of isocyanate with urea groups, a polymer containing equal quantities of urea and urethane groups may be branched principally by biuret formation.

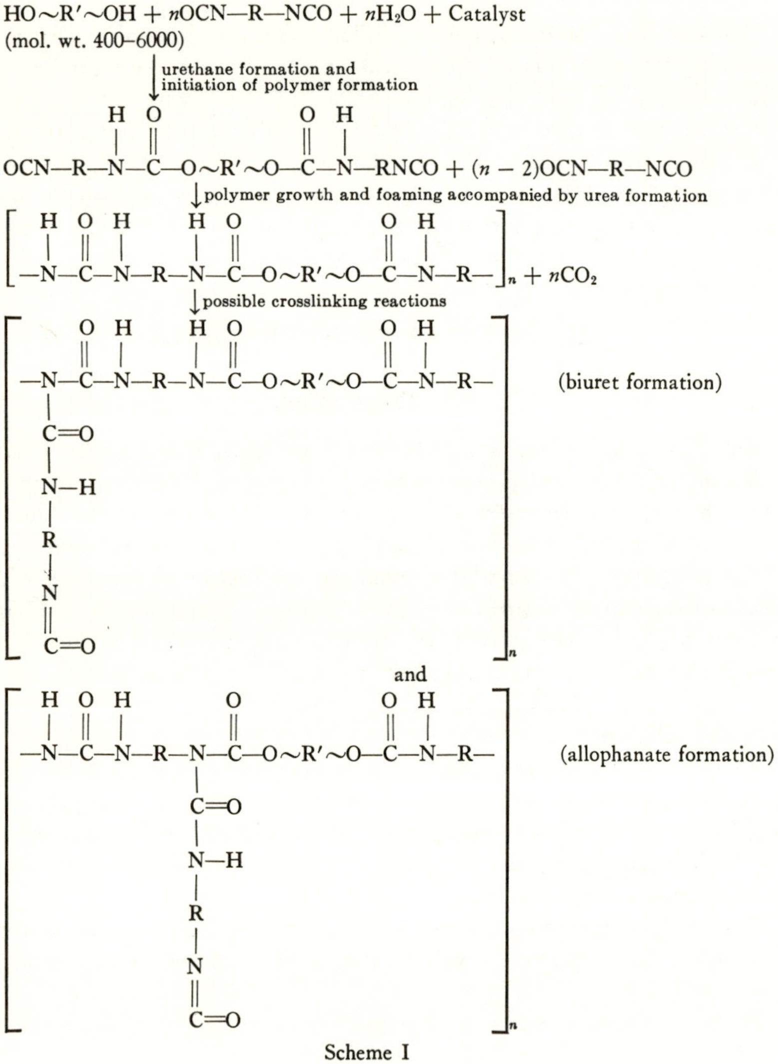

Scheme I

These reactions are controlled by the use of suitable and/or selective catalysts, the use of inert gases or volatile materials to supplement foam formation or replace the carbon dioxide obtained from the isocyanate–water reaction, the index of the foam formulation (NCO/OH), and the use of diols, triols, or tetrols whose hydroxyl number and therefore equivalent weight may be selected.

It is immediately apparent that in order to control these reactions the isocyanate content and contributions of the total active hydrogen content must be accurately known. This, of course, is also true for prepolymer and elastomer preparations.

The possible impurities in the isocyanates and polyols and the analytical methods that can be employed for their determination have been detailed previously. The influence of these impurities as well as the effects and significance of the accuracy of the active hydrogen contributors are as follows:

1. Effect and Use of Polyol Analytical Data

Effect of Hydroxyl Number and Acid Number. This affects the formulation since the index (NCO/OH) will influence viscosity, molecular weight, cell size, and crosslinking reactions.

Commercial polyols in the molecular weight range 400–6000 are generally employed or triols of molecular weights of about 3000. Polyols with increasing functionality give foams with higher compression modulus but with a tendency toward decreasing tensile strength, and decreased tear strength, and elongation.

Many widely used triols of 3000 MW have predominantly secondary hydroxyls, and increasing the primary hydroxyl content increases the rate of cure.

Also of prime importance is the presence of trace metals, acid or alkaline materials that may remain in the polyols from the manufacturing process. These substances may provide an unwanted or uncontrolled secondary catalytic effect. It has been pointed out previously (23) that it is important to know the sum total basic, acidic, and metallic impurities in the system, and not only those impurities that may be present in the isocyanate from the manufacturing step.

2. Effect and Use of Isocyanate Analytical Data

a. Isocyanate Isomer Ratio

The tolylene diisocyanates normally employed are the 80/20 and 65/35 mixture of the 2,4 and 2,6 isomers. The analytical methods for isocyanates have been given in Chapter 3. In identical flexible foam formulations the lower 2,4 isomer mixture will result in a foam with slightly higher density and compression modulus.

Gmitter and Gruber (24) investigated the effect of TDI isomer ratio on properties of flexible one-shot polyester foams. These investigators found the highest compression set values for the 76/24 isomer range and the lowest for the 80–85% range. Variations were noted in tear values, tensile strength, and stiffness. Slight variations in many properties were also reported for 79% and 81% 2,4 isomer.

These data illustrate the importance and significance of isomer control of the TDI.

b. Effects of Isocyanate and Isocyanate Impurities

The effect of acidic materials, such as hydrolyzable chloride content, on prepolymer formation was also investigated by Kane et al. (25). These investigators found an effect on the preparation of a prepolymer from propylene glycol and TDI, depending upon the hydrolyzable chloride content.

At an NCO/OH ratio of 1.5, a normal reaction occurred when the hydrolyzable chloride content was 0.01%. At a content of 0.005% or below, gelation took place and if adipoyl chloride was added to raise the hydrolyzable chloride content from 0.001% to 0.01%, a normal reaction again occurred. These results indicated the presence of traces of alkaline catalysts in the polyether.

Scholten et al. (26) found that the net CPR (net controlled polymerization rate), determined by titration of basic and/or acidic materials in the polyol or isocyanate, and defined as the difference between the reserve basicity of the polyols and the active hydrolyzable chlorides of the diisocyanates, affected both viscosity and reaction time.

The presence of other possible impurities in the isocyanate is unwanted since these materials may have an undesired, unknown, or uncontrollable effect upon the foaming reaction. This, of course, is also true in the case of polyols.

It is of paramount importance to know the percent NCO present in the isocyanate, since formulations of accurate NCO/OH are important in order to achieve the desired properties in the foam.

Polyether foams generally contain an excess of 3–7% excess TDI. Greater amounts than this produce a greater proportion of large and closed cells with a corresponding effect on tensile and elongation. At less than stoichiometric amounts of TDI, the foam exhibits poor strength and appearance and is difficult to reproduce.

In addition to the two primary urethane ingredients, a number of additional materials are employed. The quantity of water used largely determines the foam density. In addition to carbon dioxide generated from the water isocyanate reaction, auxiliary blowing agents such as trichlorofluoromethane may be used to further reduce the foam density.

One or more catalysts are generally employed to promote the reactions between prepolymer and water, or between isocyanate, resin, and water at such rates that the foam rises and cures sufficiently fast to prevent collapse of the foam. In other words, all reactions must be balanced by accounting for differences of ring substituents activity, and resin and water reactivity, to provide the proper polymer growth and gas evolution.

Surfactants are employed to promote intimate mixing of the components, thus regulating and controlling cell size, and to impart stability to the foam system. Many different classes of surfactants are used, such as organic nonionic and anionic surfactants and silicones.

The use of these components and specific applications as well as the effects of agitation and other variables has been treated extensively by Saunders and Frisch (27).

B. Catalyses

Metalloorganic compounds such as organotin compounds in combination with tertiary amines are generally employed. The one-shot polyether foams generally require more catalyst than the prepolymer or semiprepolymer or one-shot polyester foam systems.

Tertiary amines and stannous salts, such as stannous octoate, are often used in combination for the preparation of one-shot polyether foams. An antioxidant is used to control the effects of high exotherm ($>120°C$) in thick foam blocks.

The tin catalyst promotes the isocyanate–hydroxyl reaction primarily, while the tertiary amine catalysts favor the isocyanate–water reaction.

The structure of the tertiary amine has a considerable influence on its catalytic effects and usefulness for foam production (24,28). The catalytic strength generally increases as the basicity of the amine increases and steric shielding of the amine nitrogen decreases (29). The tertiary amine catalysts are satisfactory for one-shot polyester or polyether prepolymer systems because of high initial viscosity and the fact that the polyesters contain primary hydroxyl groups. The tertiary amines are not adequate for one-shot polyether foam systems largely because of the low viscosity of the polyethers used.

This drawback of the tertiary amine catalysts led to the development of the tin catalysts. Britain and Gemeinhardt (30) and Hostettler and

Cox (31) independently investigated the catalytic effect of many metal compounds.

Dibutyltin dioctoate, dibutyltin dilaurate, stannous oleate, and stannous octoate have a much greater catalytic effect upon the isocyanate–hydroxyl reaction than tertiary amines but do not appear to be strong catalysts for the isocyanate–water reaction.

Thus, by proper choice of catalyst system, the properties of the foam can be controlled. Sandridge et al. (32) have shown that the choice of catalyst concentrations has a marked effect on the open or closed cell, and therefore on the foam properties.

C. Formulation and Useful Terms

Quite often the numbers that the analytical chemist "generates" are applied in a manner that is not easily understood by someone who is not familiar with their application through constant use. In addition, there seems to be some disagreement over term usage, definition, and origin of the calculations utilized in urethane formulations.

Although the usage of simplified calculations and variations for specific formulations meet the needs of the application chemist, there is some confusion and misuse because the original fundamental concepts have not been passed on or presented. Therefore, this section is presented to acquaint the analytical chemist with the practical application of analytical data and to act as a reference for anyone who wishes to formulate.

It seems appropriate to include the definition of the more important, frequently used terms. These and additional definitions and examples have been described in an excellent *Technical Bulletin* (33). For the determination of molecular weight and a more complete discussion of the significance of associated terms, a number of excellent works are available (1,34–36).

1. Number Average Molecular Weight

Number average molecular weight equals the total polymer weight divided by the total number of moles of polymer molecules. The usual definition is: Number average molecular weight equals the sum of the number of moles of polymer molecules of each species multiplied by the molecular weight of each corresponding species, divided by the total number of moles of polymer molecules. This average, where each molecule contributes equally independent of its weight, is most useful in polyurethane formulation calculations since it is inversely proportional to the number of active groups, that is, the Hydroxyl Number.

This can be mathematically described as follows:

(*a*) General Form

(*1*) $\bar{M}_n = \sum_{i=1}^{\infty} W_i \Big/ \sum_{i=1}^{\infty} n_i$

or

$$\bar{M}_n = \sum_{i=1}^{\infty} n_i \cdot \bar{M}_i \Big/ \sum_{i=1}^{\infty} n_i$$

$$n_i\bar{M}_i = W_i$$

(*2*) For a two-component mixture

$$\bar{M}_{nl} = \frac{n_j \cdot \bar{M}_j + n_k \cdot \bar{M}_k}{n_j + n_k},$$

where l = mixture and j,k = individual components.

(*b*) Molecular Weight for a Two-Component Mixture Given: Number average molecular weights ($\bar{M}_n$)

(*1*) Molecular weight of mixture

$$\bar{M}_{nl} = \frac{\bar{M}_{nj} \cdot \bar{M}_{nk}}{(\bar{M}_{nk})(\alpha) + (\bar{M}_{nj})(1 - \alpha)}$$

where α = fractional weight of first component, and $(1 - \alpha)$ = fractional weight of second component.

(*2*) Component weights for Desired $\bar{M}_n$

$$\alpha = \left[\frac{\bar{M}_{nj}}{\bar{M}_{nl}}\right] \left[\frac{\bar{M}_{nk}}{\bar{M}_{nk}}\right] \left[\frac{\bar{M}_{nl}}{\bar{M}_{nj}}\right]$$

(*c*) Molecular weight from Hydroxyl Number

(*1*) Molecular weight of mixture

$$\bar{M}_{nl} = \frac{56100 f_j f_k}{(\mathrm{OH}_j)(f_k)(\alpha) + (\mathrm{OH}_k)(f_j)(1 - \alpha)}$$

where OH = hydroxyl number and f = functionality

(*2*) Component Weights for Desired $\bar{M}_n$

$$\alpha = \frac{(f_i)}{(\bar{M}_n)} \left[\frac{(56100)(f_k) - (\bar{M}_{nl})(\mathrm{OH}_k)}{(\mathrm{OH}_j)(f_k) - (\mathrm{OH}_k)(f_j)}\right]$$

(*3*) Hydroxyl Number of Mixture

$$\mathrm{OH}_l = (\mathrm{OH}_j)(\alpha) + (\mathrm{OH}_k)(1 - \alpha)$$

(*4*) Functionality of Mixture

$$f_l = \frac{(\bar{M}_{nl})(\text{OH}_l)}{56100}$$

(*5*) Functionality of Mixture from Hydroxyl Number

$$f_l = \frac{(\text{OH}_l)(f_j)(f_k)}{(\text{OH}_j)(f_k)(\alpha) + (\text{OH}_k)(f_j)(1 - \alpha)}$$

2. Weight Average Molecular Weight

Weight average molecular weight equals the sum of the weights of each species multiplied by the molecular weight of each corresponding species, divided by the total polymer weight. The contribution of each molecular weight class to the weight average molecular weight is proportional to the weight of the molecule in that class. This average is seldom used in polyurethane calculations, and then only when the molecular weight is calculated from viscosity, light scattering, and ultracentrifuge data. Weight average and number average molecular weights cannot be used interchangeably; they are equivalent only in the case of pure compounds.

The weight average molecular weight can be described mathematically as follows:

(*a*) General Form

$$\bar{M}_w = \sum_{i=1}^{\infty} W_i\bar{M}_i \Big/ \sum_{i=1}^{\infty} W_i$$

or

$$\bar{M}_w = \sum_{i=1}^{\infty} n_i(\bar{M}_i)^2 \Big/ \sum_{i=1}^{\infty} n_i\bar{M}_i$$

$$n_i\bar{M}_i = W_i$$

(*b*) Component System

(*1*) Molecular Weight of Mixture

$$\bar{M}_{wl} = (\bar{M}_{wj})(\alpha) + (\bar{M}_{wk})(1 - \alpha)$$

(*2*) Component Weights for a Desired $\bar{M}_{wl}$

$$\alpha = \frac{\bar{M}_{wl} - \bar{M}_{wk}}{\bar{M}_{wj} - \bar{M}_{wk}}$$

(*3*) *Functionality.* Functionality is the number of chemically active atoms or groups per molecule for the considered reactions. This is an average value for polymers.

(4) *Equivalent Weight.* Equivalent weight is the weight of compound or polymer per active group for a specific reaction. This is an average value for polymers. This can be described mathematically, as follows:

$$\frac{\text{Total sample wt.}}{\text{Equivalent wt.}} = \text{Total equivalents}$$

$$E = \frac{\bar{M}_n}{f}$$

where E = equivalent weight and f = functionality.
For polyols or OH-terminated prepolymers:

$$E = \frac{56.1 \times 1000}{\text{OH No.}}$$

$$\bar{M}_n = \frac{56.1 \times 1000 \times f}{\text{OH No.}}$$

For isocyanates or NCO-terminated prepolymers:

$$E = \frac{4200}{\%\text{NCO}}$$

(5) *Hydroxyl Number.* Hydroxyl Number is defined as the number of milliweights (or milligrams) of potassium hydroxide equivalent to the active functions (hydroxyl contant) of 1 g of the compound or polymer.

(6) *Amine Equivalent.* Amine equivalent is defined as the weight of sample which will combine with 1 g-equiv weight of dibutylamine.

$$\text{Theoretical AE} = \frac{\bar{M}_n}{\text{No. of reactive groups } (f)}$$

(7) *Percent NCO.* Percent NCO is defined as the percent by weight of NCO groups present. Free or uncombined NCO is usually expressed on the same basis.

$$\text{AE} = \frac{4200}{\%\text{NCO}}$$

$$\text{where } \%\text{NCO} = \frac{42 \times (f) \text{ No. of Reactive Groups} \times 100}{\bar{M}_n}$$

$$E = \frac{4200}{\%\text{NCO}}$$

(8) *Isocyanate Equivalent.* The isocyanate equivalent is defined as the weight of sample which will combine with 1 g-equiv weight of phenyl isocyanate. The sample is generally a polyol, amine, or other compound that possesses groups capable of reacting with an isocyanate.

(9) *Isocyanate Index.* The isocyanate index is the NCO/OH multiplied by 100. This term is universally used for foams in the United States.

While there are other terms that are used and required, the above are believed to be the most important and frequently used ones.

The following example calculations are included to show a convenient, easily understandable method for formulating urethanes.

Example Calculation for a Typical Medium Density Slab Foam Formulation (27)

	Parts by wt.
Polyether triol, 3000 MW	100
Silicone L-520 or SF 1034	1.0
N-Ethylmorpholine	0.4
Dabco	0.05
Stannous octoate	0.45
Water	3.3
TDI, 80/20	X
	105.20

Given: The above formulation

Find: The amount of TDI 80/20 required to provide a TDI index of 100, 105, and 110.

It is advisable to first calculate the equivalents of OH and NCO that are being utilized.

In this case we have decided to use the above weights of polyol, stabilizer, and catalysts. Making use of the previously defined relationships.

For Polyol

$$E = \frac{\bar{M}_n}{f} = \frac{3000}{3} = 1000; \quad \frac{100}{1000} = 0.1 \text{ total equivalent}$$

For Water

$$E = \frac{\bar{M}_n}{f} = \frac{18}{2} = 9; \frac{3.3}{9} = 0.36 \text{ total equivalent}$$

If amines or diols had been used, the equivalents would have had to be calculated as above. If OH No. had been known, the relationship

$E = 56100/(\text{OH No.})$ would have been used. Therefore,

> Total parts by weight for OH-containing compounds is 105.20 and the total equivalents are 0.46.

The equivalent weight of the mixture for reaction with an isocyanate is 105.20/0.46 equiv = 228 g. That is, 228 g will react with 87 g of TDI on an equivalent basis. But in this case we wish to keep the total weight to that originally specified. Therefore, for the above weight 0.46×87 or 40 parts by weight of TDI should be used for a 100 index foam and 1.05 and 1.10 times this amount for the 105 and 110 TDI index foams.

Let us assume a prepolymer is going to be prepared and that 8% free NCO is required. For example purposes, we will use a mixed polyol. We proceed as before except in this case we must add sufficient equivalents of TDI to balance the OH equivalents before calculating the excess necessary.

	Parts by wt.	Equiv.	Group
Triol, 3000 MW	100	0.1	OH
Diol, 2000 MW	100	0.1	OH
TDI, 80/20	17.4	0.2	NCO

The next step is to determine the weight of isocyanate that will provide 8% free NCO in the final mixture.

Let us call the weight of isocyanate required X, then $X/87$ is the number of equivalents added and $(X/87)\ 42$ is the weight of isocyanate that will give 8% excess NCO.

Therefore,

$$\frac{(X/87)\ 42}{217.4 + X} = 0.08$$

To the diol and triol is added a total of 60.6 g of TDI.

This same technique of calculations can be used for hydroxyl-terminated prepolymers or for mixtures of many active hydrogen components, not necessarily OH-containing, as long as the total active hydrogen equivalents and NCO equivalents are calculated and kept separate.

III. ANALYTICAL TECHNIQUES AND CHARACTERIZATION

A. General

Under certain circumstances, it may be desirable to analyze a polyurethane for a variety of elements. A number of methods are available for both qualitative and quantitative analysis. It is not the intention of this chapter to detail these methods since this has been done, but to summarize the techniques that are available and applicable to polyurethanes.

It is, however, the intent to provide information which will allow the chemist to identify and characterize, with guidance, by the most rapid and simplest means available.

Analytical methods, applicable to the qualitative analysis of polymeric materials, have been reported for carbon (37,38), hydrogen (39), oxygen (40), halogens, sulfur, and nitrogen (38). These methods have been summarized and presented by Brauer and Horowitz (41).

In cases where quantitative analysis are desired for nitrogen, chlorine, etc., the standard methods appear to give satisfactory results. Most of the usual methods are applicable to polymers with little change.

Methods used on polymers have been reported by Brauer and Horowitz (41). Specific methods for certain polymers have been reported for carbon and hydrogen (42,43), sulfur (44), nitrogen by Kjeldahl (45–49) and Dumas (41, 50–52), chlorine by Carius wet combustion, dry combustion, or the Paar bomb technique.

Total chlorides have been determined by the Schöniger method which is rapid and utilizes small samples (53), and by the Wickbold method which can be utilized for polymers as solids (54,55) or liquids which may be aspirated into the combustion chamber as solutions (56) in a suitable solvent. The Paar bomb or Wickbold combustion techniques are particularly amenable to the analysis of urethanes because of their low density (foams) and generally low chlorine levels.

Newer methods of analysis are already beginning to replace the older, more established methods. Outstanding examples of this are the instrumental carbon-hydrogen-nitrogen analyzers, which provide precision, convenience, and speed. Analyzers of this type have been described by Hinsvark and Waltz (57).

Polymers are often analyzed for metals which can occur as a result of residual catalysts from manufacture of the intermediates or finished polymer or as a result of unknown contamination during processing.

A number of approaches have been employed for the analysis of polymers, for trace metals. These methods usually employ wet ashing or dry ashing techniques (45). The ashed residue is then solubilized by the desired solvent followed by colorimetric (58–60), polarographic (61), or emission spectroscopic techniques.

Atomic absorption is replacing some of these analyses and will undoubtedly continue to do so. Many polymers can be analyzed directly after a proper solution is prepared, without prior ashing. For polymers that are difficult to solubilize or that provide a solution too viscous to aspirate into the flame, partial degradation by hydrolysis, thermal treatment, or solvent extraction may be used to effectively alleviate these difficulties. For atomic absorption of polymer solutions directly, the premix type of burner (laminar flow) does not appear to offer the advantages of the total consumption burner due to solvent fractionation and subsequent polymer residue formation within the premix chamber.

B. Identification

Polymeric materials such as elastomers, foams, and coatings may be characterized as urethanes by means of the infrared bands previously given which are characteristic of this generic class of materials.

Swann and Esposito (62) have reported a method for detecting isocyanate and urethane resins based on a color reaction. The particular application was to coating resins although it can be applied to other urethane polymers with certain limitations.

p-Dimethylaminobenzaldehyde was used to determine the composition of a number of different resins. The color of the product after reagent addition determined the type. Urea, melamine, isocyanate, and urethane resins were differentiated in this manner. The reagent develops a yellow color with NCO groups. When urea resins are hydrolyzed with mineral acids, a positive test is obtained but with unhydrolyzed urea resins the test is negative.

With isocyanate and polyurethane resins, which contain —NCO groups, the test is positive without prior treatment. A positive test is obtained even after these resins are extended with polyols. In glacial acetic acid the yellow color that forms immediately with the isocyanate resin provides a specific test for identification.

It was reported that even blocked isocyanates resins give a positive test at room temperature. Since it requires a considerable period of time and/or elevated temperatures to react every single —NCO group in a primary urethane reaction or allophanate-urea-biuret crosslinking–branching reactions, it is interesting to speculate if a positive test would be obtained on polyurethanes that were quite old or held for extended periods above room temperatures. This may be a limitation of this method since a basis for the reaction appears to require that perhaps a few free —NCO groups be present.

Spot Test for Isocyanate or Urethane Resins

Procedure

The test is conducted by dissolving the liquid or solid in 5–10 ml of glacial acetic acid with or without heating. A suitable solvent may be utilized first for dissolving those solid urethanes not soluble in glacial acetic. *m*-Cresol, dimethylsulfoxide, or dimethylacetamide are solvents which will often effect solution. Approximately 0.1 g of *p*-dimethylaminobenzaldehyde reagent is added and a bright yellow color develops in a few minutes at room temperature if either isocyanate or urethane resins are present.

A spot test developed by Baumann and Steingiser (163) the following year permits polyurethanes prepared from polyethers and those prepared from polyesters to be differentiated. The test is based on the following reactions where R and R′ are di- or polyfunctional:

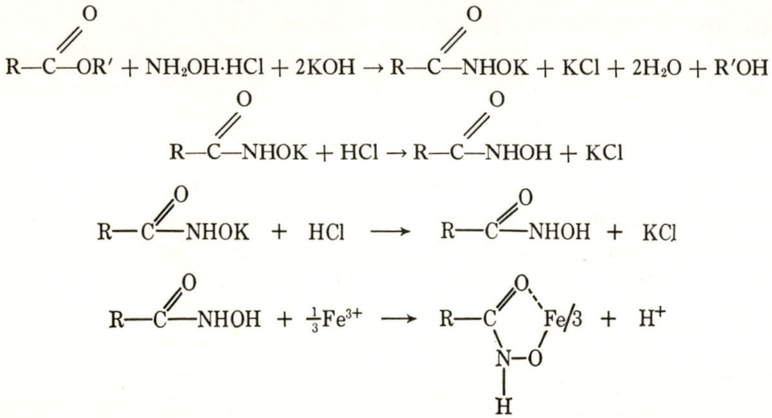

The hydroxamic acid from the polyester reacts with the ferric ion to form an inner complex salt which is water-soluble and violet or purple in color for most hydroxamic acids.

Spot Test for Polyester- or Polyether-Based Polyurethanes

REAGENTS

(*a*) A 2*N* solution of KOH in methanol plus phenolphthalein to a dark pink color.

(*b*) A saturated solution of $NH_2OH \cdot HCl$ in methanol.

(*c*) A 1*N* hydrochloric acid solution.

(*d*) A 1–3% solution of $FeCl_3$ in water.

PROCEDURE

Approximately 50 mg of small pieces of the polymer are treated with a few drops of reagent (*a*) and then a few drops of reagent (*b*) are added. The resulting mixture must be alkaline as indicated by the phenolphthalein. For highly crosslinked polymers it is desirable to heat the mixture for 20–40 sec in order to increase the reactivity. After 30–60 sec reaction time, the mixture is acidified with reagent (*c*) after which one drop of reagent (*d*) is added. If esters are present, the characteristic violet color will appear immediately.

Esters present as solvents, plasticizers, etc., interfere. The functional groups possible of forming during urethane formation, which have been pointed out previously, do not interfere with the color test under the test conditions. Ethers do not give the characteristic color reaction.

Polyesters prepared from castor oil or dimerized fatty acids will give a brownish or brownish-violet color which is nevertheless distinct from the yellow-brownish color resulting from the ferric chloride alone. On flexible foam, the test may be carried out directly on the foam by adding the reagents in the proper order.

Polymers can be differentiated readily, in most cases, as urethanes and further as polyester or polyether urethanes by infrared. The presence of the urethane structure can be ascertained by examination of the spectrum for bands at approximately 6.5 μ [Amide II (—NH deformation)], 5.8 μ [Amide I, (—C═O stretch)], and 3.0 μ (—NH stretch).

The presence of polyester or polyether linkages in the polymer chain can be established by examining the relative intensity of the rather large diffuse bands, one of which occurs near 8.0 μ due to the $\overset{\overset{\displaystyle \mathrm{O}}{/\!\!/}}{\text{—C}}\text{—O—C—}$ group, and the other of which occurs near 9.0 μ due to the —C—O—C— group. In the case of ester polyurethanes, the 8.0 μ band is much more intense and broad, and for ether polyurethanes the band intensity is reversed. Figures 6.1 and 6.2, which are spectra of ether- and ester-based polyurethanes, clearly show this.

These spectra were obtained by utilizing ATR (Attenuated Total Reflectance). Better spectra can usually be obtained using FMIR (Frustrated Multiple Internal Reflectance), although either technique has been adequate for identification purposes. Excellent transmission spectra of elastomeric polyurethanes can be obtained by cutting thin sections with a scalpel and utilizing a microbeam condenser. This technique is not very practical for foamed polyurethanes due to their flexibility, which makes it difficult to obtain thin sections, and the effects of light scattering due to the cell structure. On the other hand, many elastomeric polyurethanes are difficult to microtome because of their high elasticity and tensile strength but relatively small thin sections can be cut easily with a scalpel. These samples are amenable to analysis using a microbeam condenser.

Plasticizers and fillers, if the foam is heavily loaded, may interfere. Fillers, such as titanium dioxide, absorb in the 810 cm^{-1} to 650 cm^{-1}

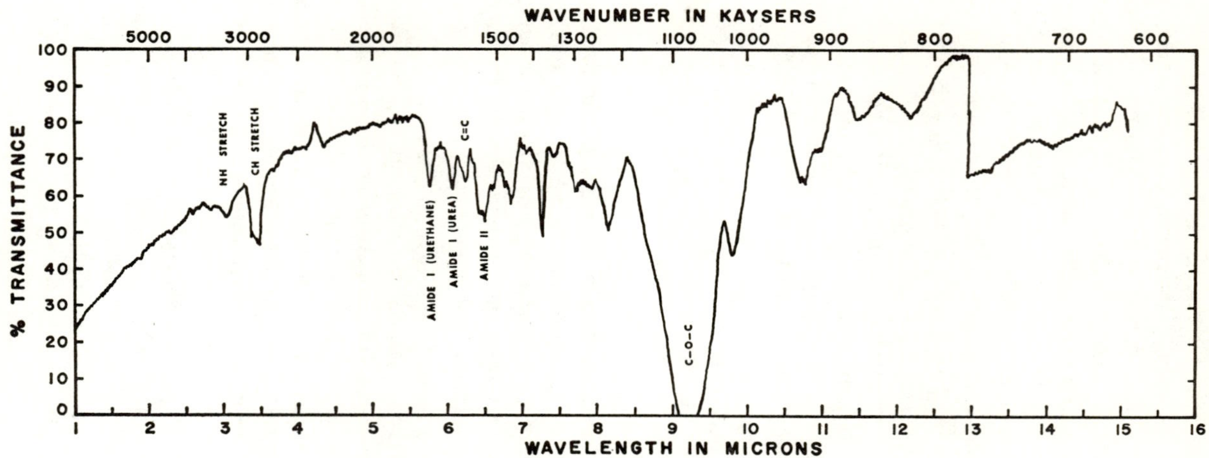

Fig. 6.1. ATR spectrum of polyether flexible foam.

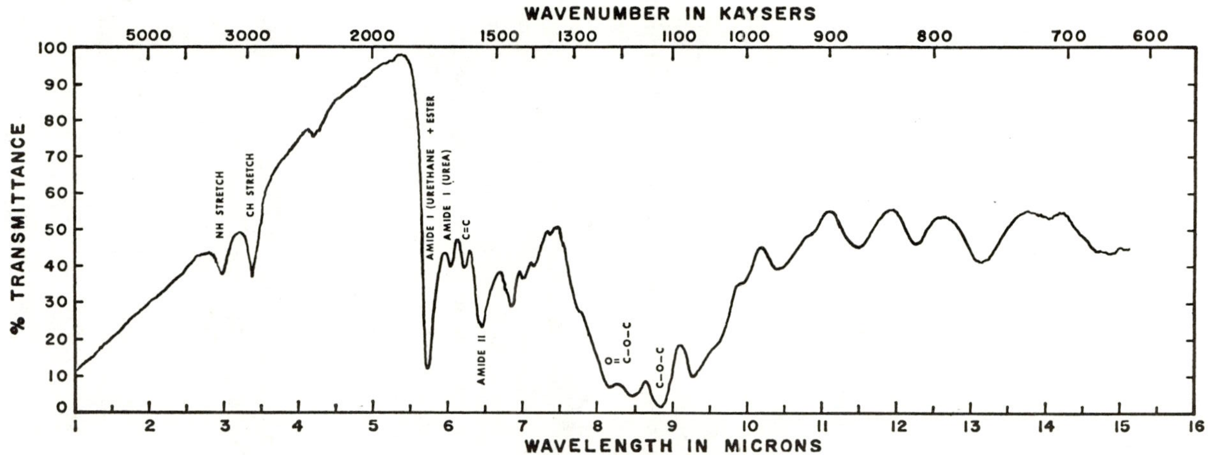

Fig. 6.2. ATR spectrum of polyester flexible foam.

region, silica in the 1100 cm^{-1} to 800 cm^{-1} region, clays in the 1040 cm^{-1} region, and zinc stearate shows a strong absorption at 1040 cm^{-1} (64).

The presence and identity of a plasticizer may be determined by extracting with a Soxhlet extraction apparatus approximately 1–2 g of foam with acetone. The acetone can then be evaporated and a spectrum of the residue can be obtained simply by preparing a salt plate smear of the residue. The infrared spectrum can then be compared with common plasticizers such as 2-diethylhexyl phthalate.

The fillers can be isolated by ashing a quantity of foam that will yield the desired quantity of ash for spectroscopic analysis. Quite often the filler can be identified by infrared without any elemental analyses. Usable spectra of inorganic compounds may be obtained by grinding with a mortar and pestle to an impalpable powder, spreading uniformly between two salt plates, and gently moving one plate over the other until good contact between plates and sample is achieved. A demountable cell is normally required to center the plates in the spectrophotometer beam.

Corish (65) and Corish and Davison (66,67) have examined a number of polyurethanes by infrared spectroscopy. Microtomed sections, 15 $\times$ 5 mm in area and 10–20 μ thick, from cast or molded samples were utilized for the studies of Corish (65). These sections were stretched and fastened onto a sample cell with tape and masked with aluminum foil. It is not always possible to prepare samples in this manner, as in the case of roams and certain types of elastomers.

Corish (65) reported that it is possible to identify the acids used in the ester portion of the polyurethane by stretching the polyurethane in order to induce crystallinity. Other work has shown that the acid portion of the polyester will produce crystallinity changes in the 700–750 cm^{-1} region.

The most frequently encountered acid is adipic but others are also employed in polyurethanes. Polyurethanes prepared from poly(ethylene glycol) adipate and 1,5-naphthalene diisocyanate produced two strong absorptions at 750 cm^{-1} and 735 cm^{-1} while the urethane from poly(ethylene glycol) sebacate and 1,5-naphthalene diisocyanate showed crystallinity bands at 722, 729, 754, 860, 881, and 895 cm^{-1}.

Some information can be obtained as to the identity of the isocyanate used by careful examination of the spectrum but the glycol portion is difficult if not impossible to identify directly from the spectrum. For positive identification of both isocyanate and glycol, the sample should be hydrolyzed into its component parts so that each can be examined separately by infrared.

The procedure we utilize is a modification of the Kappelmeir method (68) which has been described and discussed by Parker (69). This procedure can be used for polyesters and polyethers, but in the case of polyethers cleavage occurs only at the urethane linkage. This has the drawback of not obtaining the parent material which may be examined directly. By comparing the spectra of the parent glycol or polyether with reference spectra, positive identification can often be made.

In the case of polyester foams, hydrolysis with caustic occurs according to Scheme I.

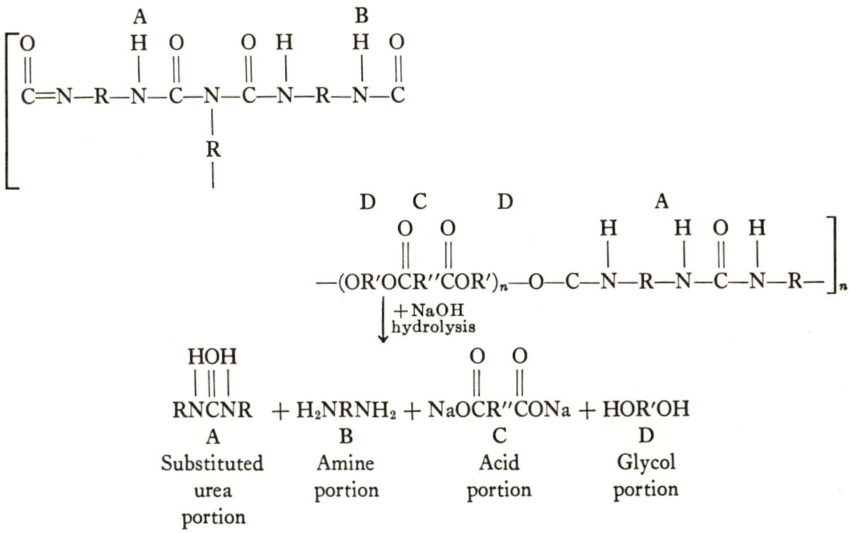

Scheme I

For polyethers, cleavage occurs only at the urethane bond according to Scheme II, where R′ and R″ may be the same.

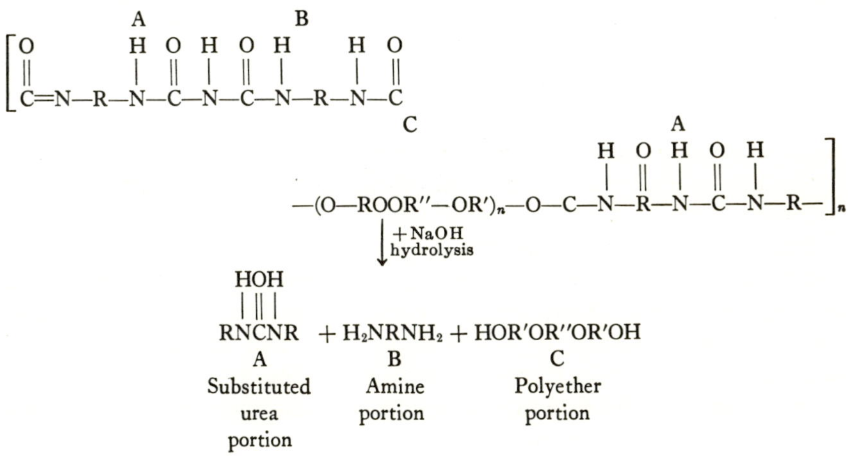

Scheme II

Procedure for Identification of Polyurethane Components (70)

Approximately 1 g of polyester or polyether foam is shredded or diced into small pieces and placed in a 125 ml erlenmeyer flask. Sufficient 40% aqueous NaOH is added to cover the foam. The flask is heated gently and then refluxed for 30 min. The solution is cooled to room temperature and filtered if a precipitate is observed. The precipitate will generally be substituted ureas.

POLYESTERS

If the diamine portion of the polyurethane is desired, the solution is extracted with several 25 ml portions of diethyl ether. These are combined, the ether removed, and the spectrum of the diamines can then be identified from reference infrared spectra and correlation tables (71). The solution is then neutralized and acidified slightly with nitric or hydrochloric acid and the glycols remaining in solution from saponification of a polyester foam must be steam distilled to effect recovery.

The acid portion will remain as a crystalline precipitate which can be water washed, dried, and identified by infrared as in the other two cases. Other procedures for the identification of dicarboxylic acids in polymeric esters have been reported (72,73).

POLYETHERS

The solution from the hydrolysis is neutralized and acidified slightly with hydrochloric acid and the polyether is extracted with several 25 ml portions of diethyl ether. These are combined, the ether removed, and the spectrum of the polyether can then be identified from reference spectra. The presence of traces of diamine will not interfere, normally, with the polyether identification.

The aqueous solution is then made basic again with sodium hydroxide and extracted with 25 ml portions of diethyl ether. These are combined, the ether removed, and the spectrum of the diamines can then be identified from reference infrared spectra and correlation tables (71).

An alternate procedure for identification of the diamine portion is as follows:

Extract with diethyl ether, the aqueous hydrolysis product of the polyurethane. This extract will contain both diamine and polyether. After evaporation of the ether, dissolve the extracted material in a suitable solvent (approximately 10% solution). Spot and analyze the sample solution by thin layer chromatography simultaneously with solutions of known amines. The R_f values and/or elution distances can then be compared to obtain the identification.

CONDITIONS

Coating: 250 Silica Gel G

Solvent system: 50 ml monochlorobenzene, 50 ml heptane, 21 ml methanol

Development: Diazotization in nitrous acid vapor, and coupling with *N*-1-naphthylethylene diamine·2HCl (1 g in 50:50 mixture of 20% HCl and dimethylformamide)

It should be noted that the above procedure is not quantitative, and is only intended to represent a relatively simple, fast qualitative procedure. For example, a dibasic acid has very limited solubility in diethyl ether and/or water, so that small amounts will be lost or contaminate the ether portion. This interference is normally not enough to cause difficulties during spectroscopic identification. Also, low yields of glycols will be experienced.

However, if precise quantitative results are required in order to determine the ratio of components, the procedure of Kappelmeir (68) should be followed.

While some information can be gained by infrared examination of the polymer spectrum in the ring substitution region, polyurethanes exhibit a complex spectrum that generally precludes a positive identification as to the isocyanate used. This is not true in the case of prepolymers in which case identification of the isocyanate from substitution bands can often be made.

Figure 6.3 is a spectrum of the diethyl ether extract (after hydrolysis) of the polyether foam shown in Figure 6.1. This foam was treated according to the above procedure except that the diamine portion was not separated from the polyether portion by acidification and extraction as evidenced by the band at approximately 6.2 μ. However, the omission of this step does not prevent identification. The spectrum of the polyether matches that of a 3000 mw poly(oxypropylene)triol from which the foam was prepared.

An alternate procedure for determining the parent isocyanate is as follows.

Colorimetric Isocyanate Test (70)

COLORIMETRIC REAGENT

1% Solution of *p*-nitrobenzenediazonium fluoroborate in distilled water.

PROCEDURE

Wet a strip of filter paper with the colorimetric reagent. Heat a glass stirring rod to "red hot" and touch it to the polymer. Hold wet filter paper directly over the polymer and glass rod to allow the fumes to contact the paper. The following characteristic colors will develop: TDI, reddish brown; MDI, yellow; and NDI, violet.

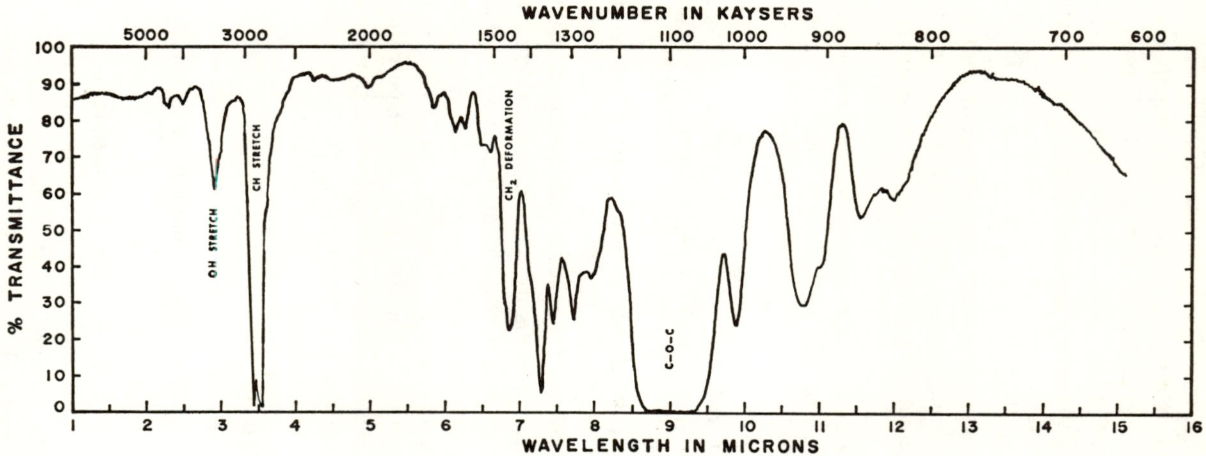

Fig. 6.3. Infrared spectrum of ether extract of polyether foam after hydrolysis.

A clear, relatively simple method that utilizes gas chromatography was recently presented by Wittendorfer (74) for the analysis of polyesters and polyurethane foams for trimethylolpropane. Although the method was mainly applied to the determination of small amounts of trimethylolpropane, which is a frequent ingredient in polyurethane formulations, it might also be applied to the determination of diethylene glycol and/or other glycols or polyols whose acetate can be determined by GLC as suggested by Wittendorfer.

The following procedure is similar to that reported by Kappelmeir et al. (75), Esposito and Swann (76), and Percival (77).

PROCEDURE

Approximately 5 g of sample are weighed accurately and refluxed with 10 ml of phenethylamine (bp 198–200°C) for 3 hr. The shredded foam must be added to the refluxing amine. If this is not done, the foam absorbs the amine and upon degradation the amine chars on the bare flask surface. The reaction mixture is cooled and 25 ml of acetic anhydride and 2 drops of water are added. The mixture is refluxed for another 90 min, cooled, and 35 ml of water is added; it is refluxed for an additional 5 min and cooled again. The mixture is then extracted with several portions of chloroform. The chloroform extracts are combined and extracted 5 times with 5.0 ml portions of water and then the chloroform is diluted to 50.00 ml in a volumetric flask. The chloroform is dried with anhydrous sodium sulfate. A suitable size sample, depending upon instrument sensitivity and concentration of the component being determined, is injected into the chromatograph and the area of the component compared against a standard prepared similarly.

Wittendorfer (74) used a 14 ft copper tubing column ($\frac{3}{16}$ in. o.d.), packed with 20% Ucon 50HB2000 on 30–60 mesh Celite 545. The instrument was operated isothermally with column, injector, and detector temperatures of 220°C. The helium gas flow and detector current can be adjusted to fit individual needs for resolution and sensitivity.

Schroeder (78) has employed acid hydrolysis of polyurethanes prepared from hexamethylene dissocyanate and 1,4-butane diol in order to quantitatively determine the hexamethylene diamine. The glycol portion was determined by hydrolyzing the polymer with water in an autoclave.

C. Analysis and Characterization

1. INFRARED

The basic polymer structure which results from the formation of urethane groups and may be linear or crosslinked largely determines the polymer's properties. However, other than these primary reactions, secondary reactions such as biuret, allophanate, dimer, and trimer formation will also have an effect on the properties of the polymer.

Many of these reactions can be successfully followed using the data given in Chapter 4 and Table 4.12. Brumfield (79) has laid valuable groundwork for a natural extension of his work from model compounds to polymeric materials (see Chapter 4).

Merten, Braun, and Lauerer (80) studied the structure of urethane foams by infrared spectroscopy. Rigid foams were examined using KBr pellet techniques while flexible foams were analyzed by ATR techniques. This rather comprehensive IR investigation of urethane foams showed that many assumptions which have been made concerning foam structure can be questioned based on the included data.

Fluorocarbon blown rigid foams based on Desmodur 44V (undistilled *p,p'*-diphenylmethane diisocyanate) showed basically the expected urethane and ester structural units. These materials contained rather high free —NCO contents, especially toward the top of the foam blocks. It was speculated that this increase was caused by diffusion. Temperature may be another factor that influences —NCO concentration.

An increased index in the foams led to an increase of free NCO at the center of the blocks but not at the tops and bottoms. Increased index was not found to result in an increased trimer content.

Water-blown rigid foams showed structure similar to the fluorocarbon-blown foams. The expected urea group absorptions were not evidenced in these foams. This may be accounted for because of sensitivity of the carbonyl–urea band or the urea may have acted as an intermediate to form a biuret structure.

Carboxylic acid groups did not show amide absorptions when reacted with the NCO. The amide apparently also further reacted to acylurea.

Reactions such as biuret were speculated to lead to more crosslinking in foams, but on the other hand the isocyanate used in biuret formation might not be available for chain propagation through the polyol and water reactions and result in early chain termination.

Foams based on Desmodur T-65/65/35 isomer ratio, 2,4/2,6 were generally similar except they were found to contain less free NCO and smaller amounts of trimer and carbodiimide. This is shown in Figures 6.4 and 6.5, which show polyester rigid and flexible foams, respectively. In general, TDI-based foams contained less free —NCO than MDI (*p,p'*-diphenylmethane diisocyanate) based foams.

Flexible foam formulations resulted in a considerably increased urea absorption. This was explained by pointing out that at 100 index there

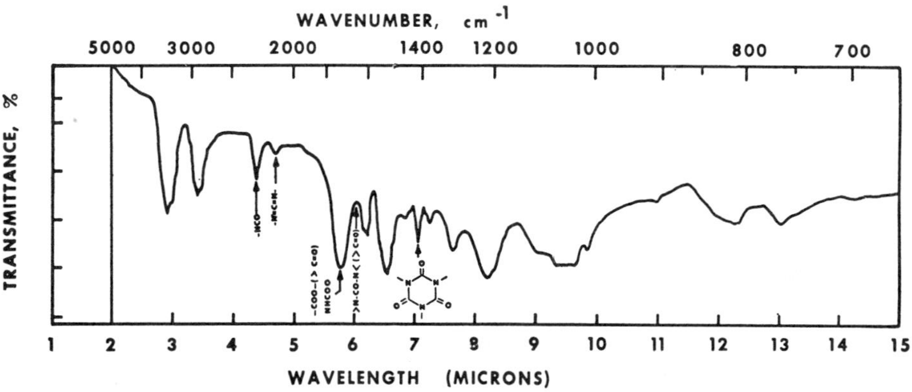

Fig. 6.4. Infrared spectrum of polyester rigid foam based on MDI (80).

is not enough NCO to convert the urea to biuret. An increase of index leads to further decrease of urea absorption so that at 115 index only a small amount can be found and at 135 index no urea absorption exists. Decreasing the index tended to increase the urea absorption to a maximum at 90 index after which it decreases as the index decreases.

The outer portion of flexible polyester foam blocks (index 50–150) showed only small amounts of urea absorption and the trimer content was found to be independent of index. These foams showed only small amounts of free NCO. These results were verified with flexible polyether foams. The absence of ester C=O allowed a clearer determination of urea groups. Urea was found to be clearly defined.

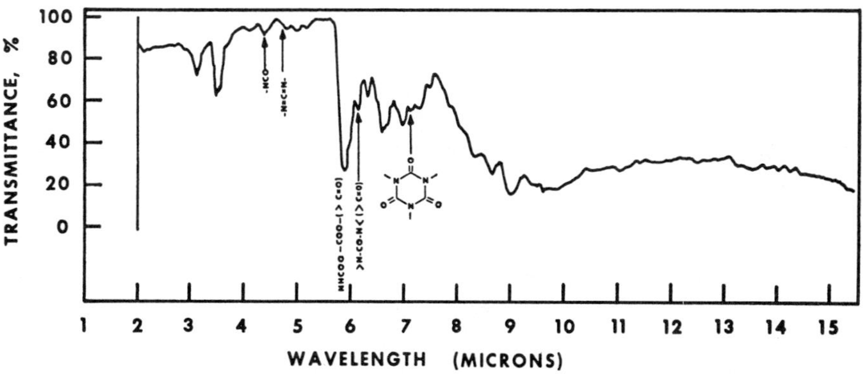

Fig. 6.5. Infrared spectrum of polyester flexible foam based on TDI (80).

In semirigid ester foams, the addition of high hydroxyl number polyols resulted in a decrease of the urea absorption.

These considerations were confirmed by low molecular weight model compounds and a comparison of the physical properties of flexible foams. Absorption bands and wavelengths previously discussed in Chapter 4 were utilized for this study.

In a more recent study, Niederdellmann, Lauerer, and Merten (81) quantitatively determined the molecular ratios in which urethane and urea groups can be determined in the presence of each other. The techniques of ATR and KBr pellets were both employed, using the following amounts of low molecular weight model compounds.

(*1*) Pure phenylurethane
(*2*) 20 moles of phenylurethane/1 mole diphenyl urea
(*3*) 10 moles of phenylurethane/1 mole diphenyl urea
(*4*) 5 moles of phenylurethane/1 mole diphenyl urea
(*5*) 1 mole of phenylurethane/1 mole diphenyl urea
(*6*) 1 mole of phenylurethane/5 moles diphenyl urea
(*7*) 1 mole of phenylurethane/10 moles diphenyl urea
(*8*) Pure diphenyl urea

The results of the infrared analysis using KBr pellets are shown in Figure 6.6. The urea band at 1640 cm^{-1} is well defined and can be measured quantitatively in concentrations as low as 5 mole %. However, the spectra show that small concentrations of urethane groups cannot be easily determined in the presence of large amounts of urea. These results were confirmed by quantitatively building urea and urethane groups into macromolecules.

Murphy and O'Neil (82) studied the structure of polyurethane foam by infrared and dielectric methods of analysis. Infrared analysis of rigid polyether foams was carried out utilizing KBr pellets. These foams were prepared from a polyether triol and 2,4-tolylene diisocyanate.

From a study of the spectrum of this polyurethane foam, Murphy and O'Neil made the assignments shown in Table 6.1. These assignments are consistent with those reported previously by other workers (83) although not as detailed. They are included here since they represent, in general fashion, the wavelengths of the more important bands, and implement qualitative polyurethane characterization.

An important aspect of the work of these investigators was the development of a method for the determination of unreacted —NCO groups in foam. This technique would be especially useful in quality control applications, or in the study of curing as a function of foam block thickness, time, or temperature.

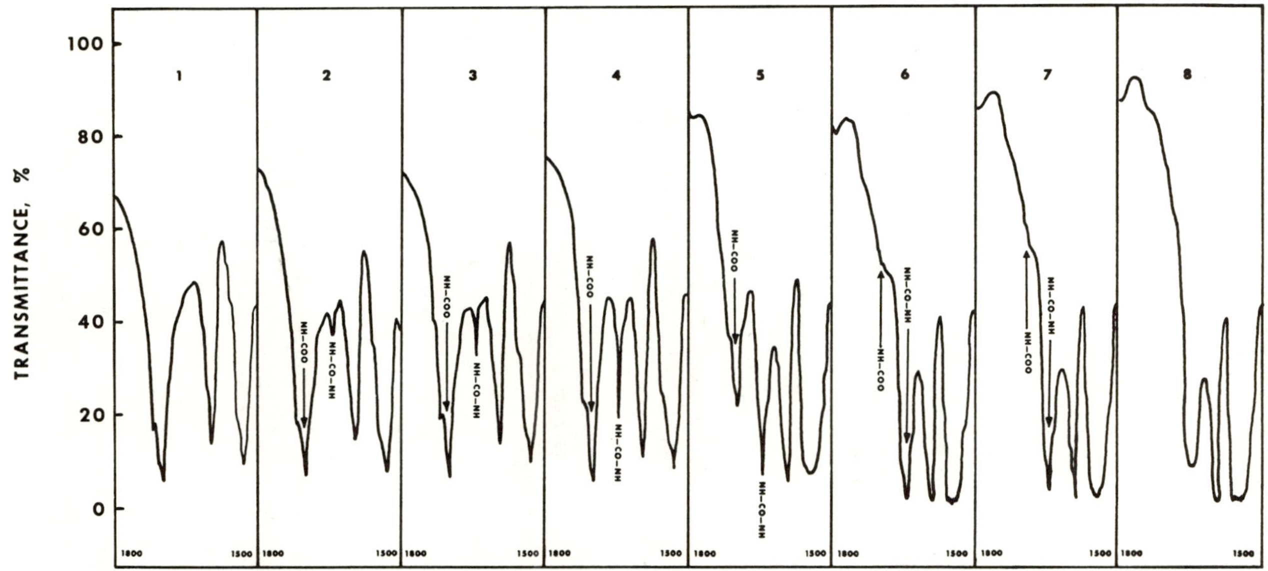

Fig. 6.6. Infrared spectra of the carbonyl region of various urethane–urea mole ratios (81). (*1*) Pure phenylurethane. Phenylurethane/diphenyl urea mole ratios for curves *2–7* are: (*2*) 20/1; (*3*) 10/1; (*4*) 5/1; (*5*) 1/1; (*6*) 1/5; (*7*) 1/10. (*8*) Pure diphenyl urea.

TABLE 6.1
Polyether—Polyurethane Infrared Assignments (82)

Microns	cm^{-1}	Assignment
2.90–2.95	3448–3990	—OH stretch; hydroxyl groups probably associated with intermolecular hydrogen bonds; —NH stretch of *cis* form of 2° amides
3.40	2941	CH_2, CH_3 stretch
4.40	2270	—NCO absorption
5.80	1724	Carbonyl C═O
		Amide I (C═O) band of polyurethanes
6.10	1639	Amide I (C═O) absorption
6.20	1613	C═C skeletal stretch of benzene ring
6.50	1538	Amide II (NH deformation)
6.65	1504	C═C skeletal stretch of benzene ring
6.85	1460	CH_2 deformation; C—CH_3 (asymmetrical) deformation: C═C skeletal stretch of benzene ring
7.25	1379	C—CH_3 (symmetrical) deformation
8.15	1227	C—O stretch and/or —OH deformation
9.35	1070	C—O stretch and/or —OH deformation; aliphatic ether of type —CH_2—O—CH_2 (polyethylene glycols and methyl, ethyl, and butyl cellosolves and carbitols all show very broad strong bands in the region 1150–1160 cm^{-1}).
13.15–13.25	760–754	Apparently characteristic of tolylene-2,4-diisocyanate and
14.35	697	its polymers. Not aromatic substitution band; this appears in TDI (but not its polymers) in the normal region (817 cm^{-1}).

The —NCO band at 4.4 μ was used for this work and TODI (3,3′-bitolylene-4,4′-diisocyanate) was used for calibration in the range of 0.001–0.050% by weight. A Wig-L-Bug amalgamator was used to mix the KBr and isocyanate after which 500 mg was pressed, which resulted in a pellet approximately 0.125 cm thick. The exact pellet thickness and weight were then measured and recorded. The baseline technique was used to measure the absorbance of the —NCO band at 2270 cm^{-1}.

The sample absorbance A_0 was related to the pellet thickness b_0 by

$$A_0 = EC_0b_0$$

where E is the molecular extinction coefficient and C is the concentration of —NCO. For a different pellet thickness and concentration

$$A_1 = EC_1b_1$$

and

$$\frac{A_0}{A_1} = \frac{EC_0b_0}{EC_1b_1}, \quad \text{but } C_0 = C_1$$

therefore

$$\frac{A_0}{A_1} = \frac{b_0}{b_1}$$

If A_0 is the corrected absorbance at b_0 (0.125 cm) and A_1 is observed absorbance at pellet thickness b_1, then

$$A_0 = 0.125(A_1/b_1)$$

The calculated microequivalents of —NCO were plotted as a function of absorbance and were found to follow Beer's law up to 0.20 μequiv of —NCO. This same technique was then applied to foam samples after they had been powdered in a Waring blender and ground with an agate mortar and pestle.

While this may not apply to this technique, we have observed calibration curves (equivalents of —NCO as a function of absorbance) of slightly varying slope when solutions of different isocyanates have been prepared. In solution, effects such as hydrogen bonding and steric effects from the isocyanates themselves may account for these observations.

Greth et al. (83) have reported a similar method using the KBr pellet technique to determine unreacted NCO in rigid foams. These investigators found that extreme care must be exercised in preparing the KBr pellet or calibration curves of different slopes are obtained, probably resulting from the —NCO reacting with moisture.

To circumvent this, a standard solution of the isocyanate was prepared using spectro grade chloroform which has been dried over P_2O_5. Equal amounts of standard solutions (50 μl) were then pipetted onto 400 mg of KBr, after which the solvent was removed under vacuum before the disks were pressed.

Approximately 3–13 mg of foam per KBr disk was used depending upon the % NCO expected. The optimum mixing time of powdered foam with the KBr in a Wig-L-Bug amagalmator was found to be 30 sec. A reproducibility of ±4.5% was obtained. This value may partly reflect slight differences in the actual foam samples.

Some interesting results were obtained from this investigation. A sorbitol polyol with a hydroxyl number of 450, catalyst, blowing agent, surfactant, and TDI were prepared by the batch mix process to provide indexes from 95 to 110. After preparation and oven curing at 130°F for 10 min, the foams were placed into polyethylene bags, flushed with dry

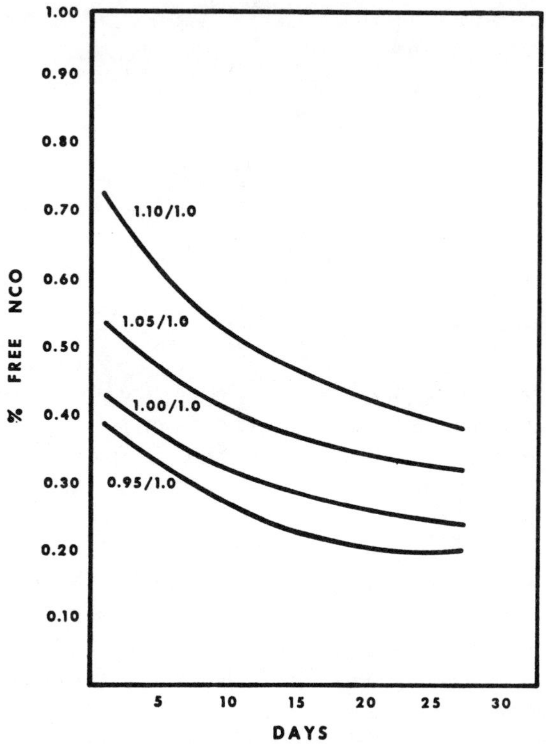

Fig. 6.7. Percent free NCO versus aging for different NCO/OH ratios (83).

nitrogen, and sealed. The free —NCO content was then determined as a function of time. The results are shown in Figure 6.7. These data show that the initial difference of the free —NCO content remained approximately the same during aging and that even the 95 index foam (underindexed) contained free —NCO indicating the presence of excess unreacted hydroxyl groups.

The compressive strengths improved as the foams aged, showing additional reactions and crosslinking. Unexplained was the fact that the underindexed foam gave higher compressive strengths than the 100 and 105 index foams.

TDI-based foams were found to contain less free —NCO than MDI-based foams and the NCO content dropped more rapidly over a four-week aging period. This was explained on the basis of the greater reactivity of TDI. The greatest improvement occurred in the first two weeks. It should be noted that these results are similar to those of Merten et al. (80).

In all cases, these investigators found a considerable amount of free —NCO remains in the foams, available for additional reactions, such as with moisture from the air during normal production and handling. A total variation between 0.15 and 0.92 wt % NCO was found for the various foams.

The effect of polyols were also evaluated in a crude TDI one-shot formulation. By measuring the free —NCO, these investigators were able to determine that the slowest curing polyol accounted for the lower compressive strength of this foam.

This same type of study can also be performed nicely using FMIR and the C—H stretching band to normalize absorbance differences of the NCO group for multiple runs of the same sample caused by contact differences between samples and plates. However, this technique has the limitation that it can be used only for comparing samples that contain the same structural unit in the chain, since different starting materials will produce different CH absorptions which are not comparable among systems.

2. Wet Methods

Kubitz (84) has presented qualitative and quantitative methods for the determination of free —NCO groups in polyurethane foam. The quantitative method is based on the reaction of the residual isocyanate with excess standard *n*-butylamine in tetrahydrofuran and the unreacted *n*-butylamine is determined colorimetrically by malachite green, with which it forms a colorless derivative. The qualitative method depends upon the reaction of isocyanate groups with the colorless secondary amine derived from malachite green and *n*-butylamine to form a product of the characteristic color of malachite green.

Determination of Traces of Isocyanate in Polyurethanes (84)

Reagents

(*1*) Tetrahydrofuran, sodium dried and distilled.

(*2*) Pyridine, reagent grade containing 0.05–0.15% water.

(*3*) Standard solution of *n*-butylamine in tetrahydrofuran; approximately 1.2 μmoles of purified *n*-butylamine/ml of tetrahydrofuran.

(*4*) Malachite green in pyridine, 0.45 g/450 g of pyridine.

(*5*) *n*-Butylamine–malachite green derivative. Sufficient *n*-butylamine is added to 20 g of malachite green to discharge all the color. Extract the white precipitate with diethyl ether in a Soxhlet extractor thimble overnight, evaporate the ether to 50 ml, and chill in ice to bring down the white crystals. Remove the ether and wash with 25 ml of cold ether.

Remove the last traces of ether under vacuum and store in a brown glass bottle.

(*6*) Cumene, Eastman white label cumene (isopropylbenzene).

(*7*) Heavy mineral oil, Nujol.

(*8*) Isocyanate test reagent, 0.20 g of *n*-butylamine–malachite green derivative in 50 ml of cumene. After solution, add 50 ml of mineral oil and mix. Place 0.5–1.0 ml of fresh reagent in small vials.

PROCEDURE

(*a*) *Qualitative*. Place a small piece of polymer into one of the vials containing 0.5–1.0 ml of test reagent and cap tightly. A green color develops in 20–30 sec in the presence of free isocyanate.

(*b*) *Quantitative*. Weigh 0.05 g of polymer for 0.4–1.0% isocyanate and 0.10 g for 0–0.4% isocyanate into a 25 ml glass-stoppered flask.

Pipet 15.0 ml of standard butylamine solution into the flask and extract for ½ hr for 0.6% NCO to 3–4 hr for 1% NCO. Withdraw 5 ml of the unreacted butylamine, through glass wool, and transfer to a 1 cm absorption cell. Pipet 5 ml of malachite green solution into the cell and mix. Read the absorption or transmittance at 625 mμ at the end of 3 min ± 5 sec from the time the malachite green solution began to be added to the cell. Determine the concentration of unreacted *n*-butylamine and then the difference between the initial and final *n*-butylamine concentrations and allow for aliquot size in the final calculations. The unreacted *n*-butylamine is determined from a calibration curve prepared by pipeting mixtures of tetrahydrofuran and standard butylamine to total 5 ml and providing various concentrations of each. Five ml of malachite green in pyridine is added to the cell and mixed. The absorption is read at 625 mμ at 3 min ± 5 sec from the start of mixing as before. A calibration curve is constructed from these data.

The 95% confidence for a duplicate determination of less than 0.60% —NCO in a cured elastomer was reported to be $\pm 0.033\%$. Water, weak amines, ureas, and urethanes did not interfere with the quantitative method. Strong and weak amines which have ionization constants less than 1×10^{-6} did not rcact with malachite green.

The qualitative isocyanate test reagent was found not to be sensitive to oxygen or carbon dioxide. Water and strong mineral acids generate color. The reagent must therefore be kept free from atmospheric moisture and strong mineral acids.

3. THERMAL ANALYSIS

Simha (85) has studied the thermal degradation of polyurethanes using carbon-14 as a tracer and liquid scintillation counting techniques.

Polymers studied were prepared from 2000 mw poly(propylene glycol) and 2,4-tolylenediisocyanate containing carbon-14. The polymers were degraded at 200 and 207°C, passing dry nitrogen through the polymer at a flow rate of 100 ml/min.

In general, for a given number of scissions more carbon-14 was volatilized than was predicted from the total number of urethane bonds for undegraded polymer. Simha proposed two explanations for this: (*1*) the formation of carbodiimide without net scission but loss of CO_2

$$2\sim\!N{=}C{=}O \rightarrow \sim\!N{=}C{=}N\!\sim + CO_2\uparrow$$

and (*2*) the formation of secondary amine without net scission but loss of CO_2

$$\sim\!\underset{\displaystyle\sim N}{\overset{H}{|}}\!\!-\!\!\overset{O}{\overset{\|}{C}}\!\!-\!O\!\sim \;\rightarrow\; \sim\!\overset{H}{\overset{|}{N}}\!\sim + CO_2\uparrow$$

It would be interesting to observe the effect of the sweep gas and whether or not the use of nitrogen, carbon dioxide, or helium would affect the equilibrium conditions. The use of nitrogen may promote carbodiimide formation; carbon dioxide, isocyanate formation; and helium, random chain scission.

Backus et al. (86) studied the thermal decomposition of (*1*) a rigid polyether-based urethane foam, (*2*) a similar foam containing a reactive organophosphorus flame retardant, and (*3*) a flame retardant foam based on a chlorinated polyester resin, using TGA, DTA, and infrared techniques.

TGA results showed that foam *1* in air lost little weight below 250°C, 50% being lost at 250–400°C in two stages. Foam *2* lost weight below 200°C, the loss again occurring in two steps the first of which appeared between 200 and 300°C, approximately 100°C lower. Foam *3* showed similar decomposition characteristics to foam *1*.

It was suggested that the results on foam *2* were attributable to the flame retardant portion of the foam (87) from decomposition at low temperatures yielding Lewis acids which then catalyze formation of solid char rather than flammable products. These results are shown in Figures 6.8–6.10. Similar results were obtained using solid urethane polymers.

The thermal decomposition of urethanes has been reported to occur by: (*1*) dissociation to the isocyanate and alcohol, (*2*) formation of a primary amine, olefin, and carbon dioxide, and (*3*) formation of a secondary amine and carbon dioxide.

Dyer, Newborn, and Wright (88) have reported these simultaneous occurrences during degradation at 300°C, in an inert atmosphere, of the urethane of phenyl isocyanate and 1-phenylethanol.

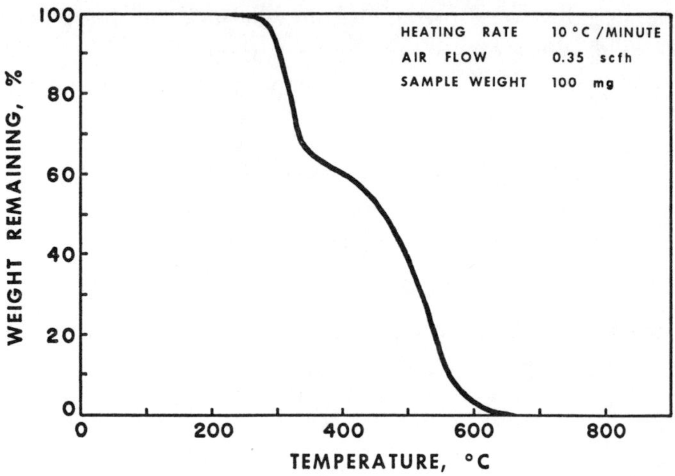

Fig. 6.8. Thermal gravimetric analysis—rigid polyether urethane foam (86).

Backus et al. (86) found the same processes to occur in the case of urethane foams as evidenced by infrared analysis of both chars and volatile products. Foam *1* was found to evolve volatile products at slightly above 200°C and yielded large amounts of both CO_2 and a volatile olefin during the first major stage of weight loss. The olefin was thought to be propene.

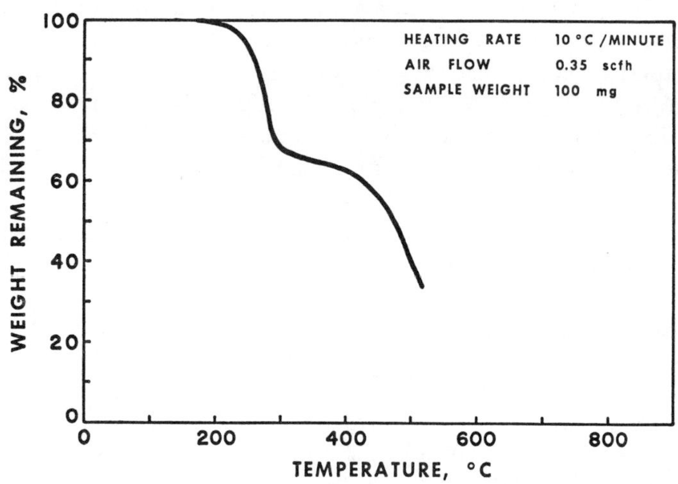

Fig. 6.9. Thermal gravimetric analysis—rigid flame retardant polyether urethane foam (86).

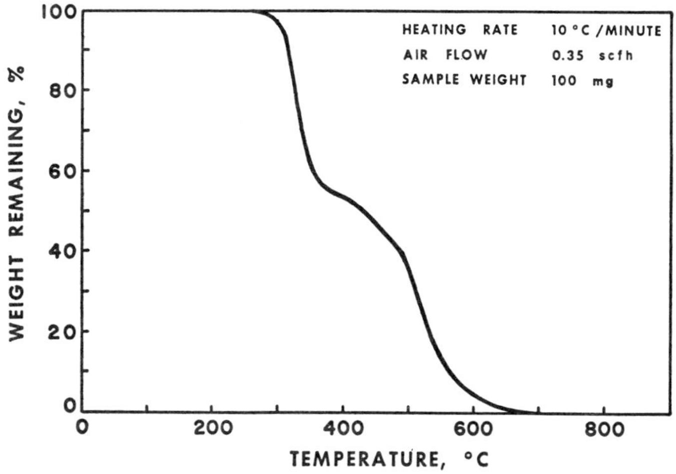

Fig. 6.10. Thermal gravimetric analysis—rigid chlorinated polyester urethane foam (86).

The detection of —NH bonds in the spectra of the 300°C chars of flame retardant polyether foam suggested that degradation of urethane to an amine, olefin, and CO_2 was an important factor. The presence of a different type of carbonyl was suggested to be associated with substituted ureas formed from reactions of urethane degradation products. The infrared spectra of the original foam *1* and the 300°C char are shown in Figures 6.11 and 6.12. The —NH stretching band in Figure 6.12 appears to have a higher frequency, possibly due to free NH as opposed to bonded NH in the original foam spectrum.

DTA thermograms of these foams showed a strong exotherm at approximately the temperature of initial TGA weight loss. Foam *2* exhibited the start of the exotherm at 170°C, approximately 55°C earlier than the other two foams and was attributed to loss of trichlorofluormethane dissolved in the polymer as shown in Figure 6.13. However, this may have been due to partial decomposition of the organophosphorus flame retardant without measurable weight loss, since it was found to degrade at lower temperatures than any of the other raw materials (150–170°C).

Merten, Braun, and Lauerer (89) have recently investigated the effect of high temperature on the structure of urethane foam by infrared using KBr pellet techniques. Several conclusions were reached which were based on the changes observed in the infrared absorption spectra. These investigators found that heating rigid urethane foams not containing urea

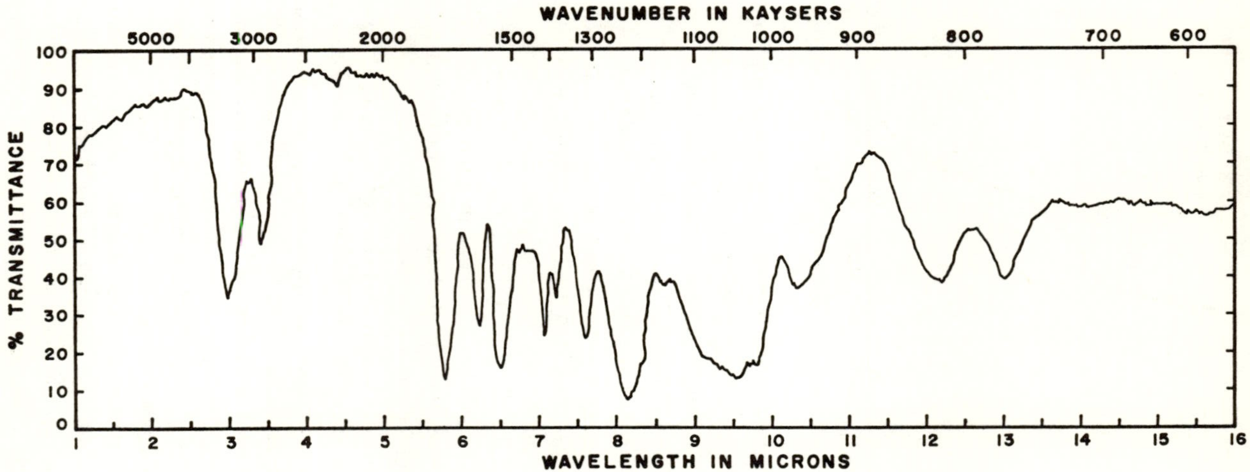

Fig. 6.11. Infrared spectrum of flame retardant polyether urethane foam 2 (86).

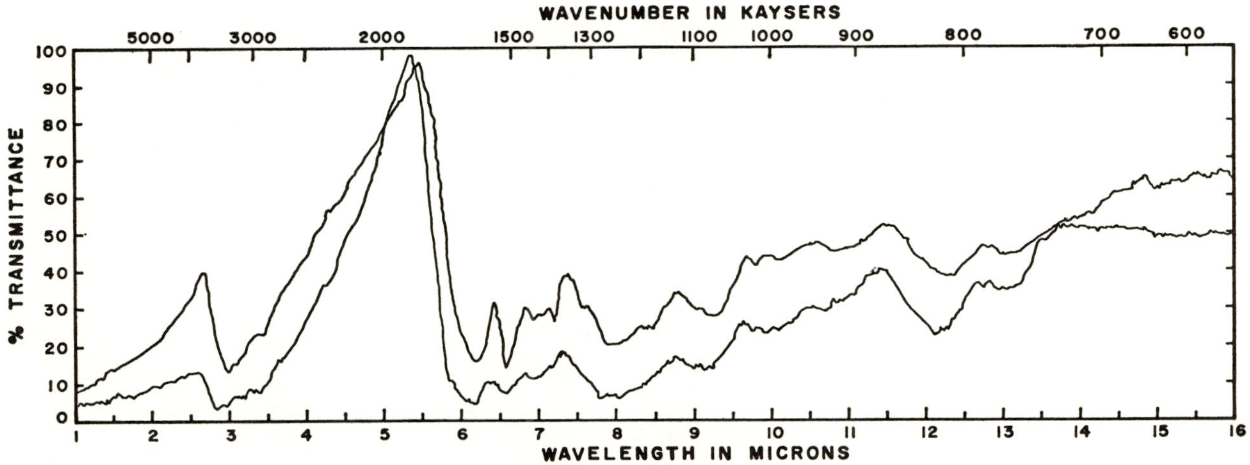

Fig. 6.12. Infrared spectrum of 300°C char of flame retardant polyether urethane foam 2 (86).

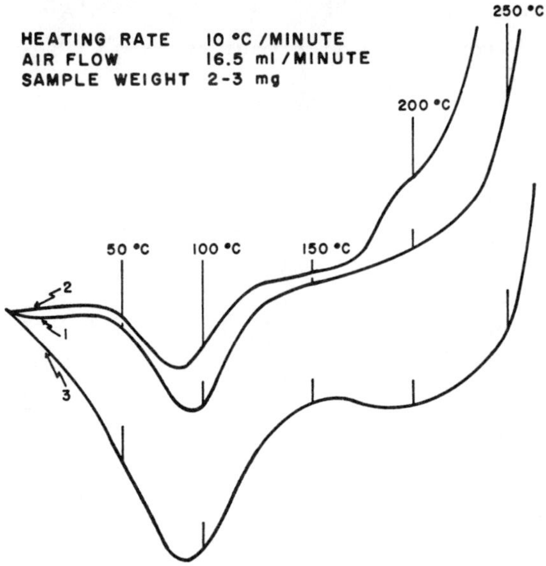

Fig. 6.13. DTA curves of rigid urethane foams in air (86). 1. Polyether foam; 2. flame retardant polyether foam; 3. chlorinated polyether foam.

groups to 200°C for 1 hr transformed all the free NCO to carbodiimide, whereas heating to 260°C for 1 hr caused a general breakdown of the total chemical structure.

The thermal treatment of flexible foams containing urea and biuret groups by heating to 220°C resulted in dissociation of the biuret into urea and isocyanate structures. The NCO was then transformed into carbodiimide groups as evidenced from infrared spectra.

These results are consistent with those of other investigators. The total results (85,86,88,89) of these investigations appear to present a good picture of thermal effects upon urethane foams.

IV. NONCELLULAR POLYURETHANES

A. Preparation (General)

Polyurethane elastomers are prepared generally from a long chain linear diol, a diisocyanate, and a chain-extender such as a diamine, glycol, or water. Both polyesters and polyethers are utilized for the diol portion and *p,p′*-substituted diisocyanates are utilized for the isocyanate portion. These polymers generally display outstanding properties and are being utilized in a wide variety of applications.

The first highly elastic urethanes were developed by Pinten (90) and subsequently improved by Bayer and co-workers (91) to provide elastomers with outstanding properties. Those developed by Bayer and co-workers were prepared from 1,5-naphthalene diisocyanate, poly(ethylene adipate), and chain extenders like water, diamines, or glycols. Polyester urethane elastomers, based on polyester-amides, were prepared in England (92) and liquid casting techniques were introduced by Mueller et al. (93,94).

Recent techniques permit commercial one-shot casting procedures to be used (95) for the production of thermoplastic urethane elastomers based on polyesters (96–100). The latter thermoplastic polyurethanes can be processed by extrusion, injection molding, etc.

Urethane elastomers are prepared generally by first forming a low molecular weight prepolymer, and then building the molecular weight with suitable chain-extending agents. This results in block copolymers which combine relatively flexible or soft segments with a stiff or hard segment. Polyether or polyester diols in the molecular weight range of 1000–2000 are utilized to provide the desired properties.

Hydrogen bonding and van der Waals forces between segments increase the outstanding properties of the polymers and serve as physical crosslinks. Primary crosslinks may be incorporated by using components with a functionality greater than 2 or excess isocyanate groups which are reacted into the polymer network.

Many of the same reactions given in the foam section occur here, the specific steps involved being:

(*1*) Prepolymer Formation:

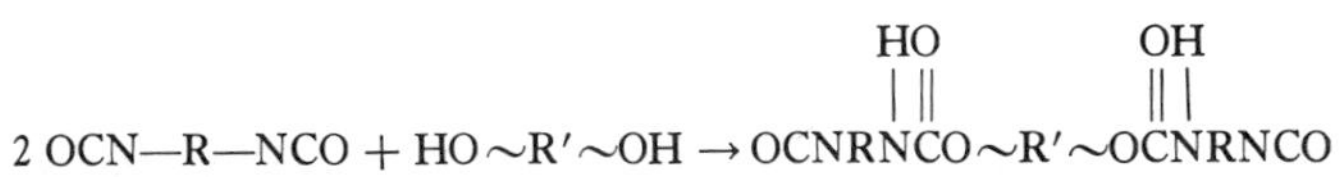

(*2*) Chain Extension:

(*a*) With water

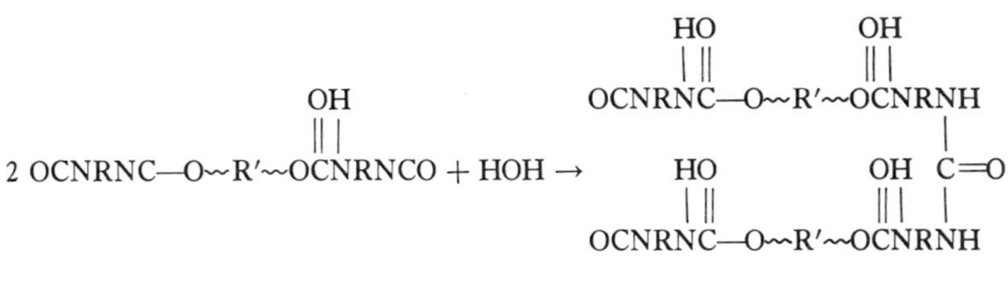

(*b*) With diol

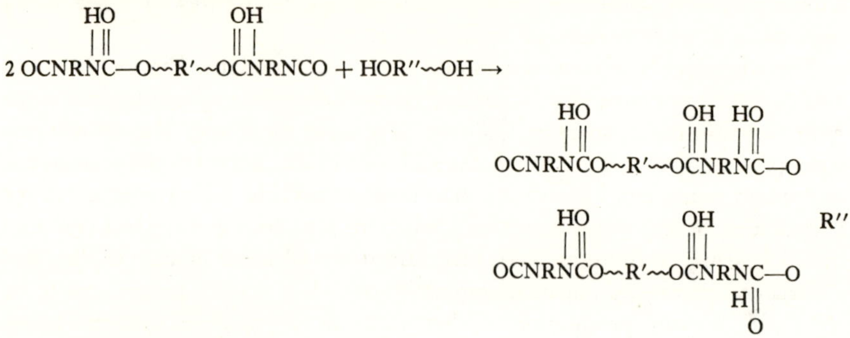

(*c*) With diamines:

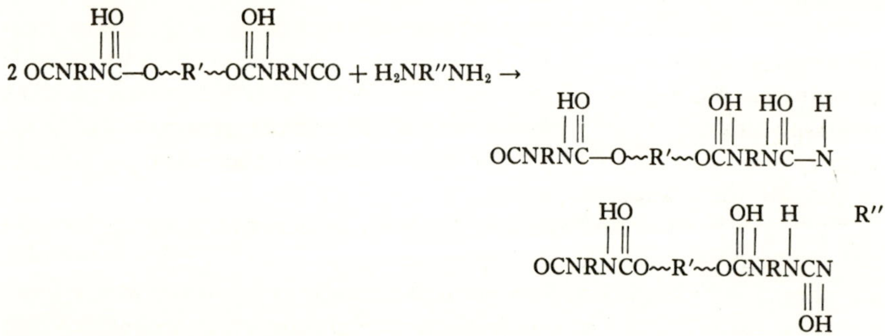

If both urethane and amine groups are present, urea formation will take precedence due to the faster reaction rate.

3. Curing or Crosslinking

Curing or crosslinking of the intermediate molecular weight polymer may be accomplished by introducing triols, and having free NCO groups present and then heating. In the latter case biuret or allophanate crosslinks will be formed by the reactions presented in the foam preparation section.

Additional methods of crosslinking to form high molecular weight polymers have been described in detail (27).

The isocyanates most frequently employed are tolylene diisocyanate 80/20 and pure 2,4 isomer, diphenylmethane diisocyanate as the essentially pure 4,4′ isomer or 90%+ 4,4′ isomer. Adipate polyesters or ε-caprolactone polyesters, linear polyester amides, poly(1,4-oxybutylene) glycols, poly(oxypropylene) glycols, and castor oil are utilized.

B. Cast Elastomers

1. Prepolymer Formation

A minimum viscosity is desired that allows easy handling and good control of the free monomeric isocyanate which is especially important in the case of TDI, due to vapor hazards. In some cases the monomeric isocyanate must be removed after prepolymer preparation.

From the previous reactions it can be seen that the water content of the polyol should be as low as possible, below 0.1%, to prevent the formation of substituted ureas which may react to form biuret branch points and subsequent higher viscosity prepolymers.

2. Curing of Prepolymers

A number of reactants, such as diamines, glycols, amine alcohols, water, and additional resins, have been used to cure prepolymers depending upon the desired properties of the end product.

The presence of acids, bases, and metallic impurities must be closely controlled throughout so that the activity of the system is reproducible for each polymer batch.

The free NCO content must be accurately known so that the final adjustment of NCO/OH ratio will provide the desired molecular weight and properties.

3. One-Shot Preparation

In this type of system the end properties are similar to the prepolymer route providing the exotherm is controlled during the mixing and simultaneous reaction of all components.

If reactants whose active hydrogens vary widely in reactivities are used, selective catalysts must be employed in order to regulate the ratio of urethane groups to polyureas and biurets. Depending upon the system, tin catalysts, such as stannous octoate, to catalyze the NCO/OH reaction have been used (95,100).

4. Thermoplastic Polymers

These elastomers are prepared essentially by a one-shot method in which the degree of reaction and/or crosslinking is closely controlled so that the polymer may be molded or extruded.

Thermoplastic polymers have been prepared by maintaining unreacted isocyanate groups (94,101), by using an excess of hydroxyl component (98,102,103), and by controlling the excess of isocyanate reactant as in the preparation of the Texin polyurethanes (97,104,105).

These thermoplastic types must be kept dry, below 0.1% H_2O; otherwise, hydrolysis during processing may occur, accompanied by both bubbling and loss of properties. It is, therefore, important to be able to analyze with speed and accuracy for the moisture content of these materials.

The properties and examples of each type have been thoroughly covered by Saunders and Frisch (27).

C. Coatings

Bayer and co-workers (2) developed polyester-polyisocyanate polymers which provided coatings possessing flexibility, toughness, excellent electrical properties, and good abrasion and chemical resistance.

One of the earlier problems was the vapor hazard from TDI since these coatings were based on TDI or a derivative of it. The level of TDI in the air should be less than 0.02 ppm. The effects of this exposure and the analytical methods associated with the analysis of air for free TDI will be covered in Chapter 8. This problem has been largely circumvented by utilizing adducts of polyols such as trimethylolpropane or 1,2,6-hexanetriol with TDI, which contain only traces of TDI for use in two-component coatings.

In order to employ a one-component system, Bayer (2,106) and Petersen (107) developed "blocked" isocyanates. The only blocking agent for isocyanates to achieve commercial importance, thus far, is phenol. The ability of blocked isocyanates to split at elevated temperatures led to their successful use in coatings for wire enamels.

Urethane oils, which are one-component systems prepared from isocyanates and drying oils, are also used commercially (2,108–110). Curing is accomplished by air oxidation of the unsaturated portion of the molecule.

Castor oil-diisocyanate coatings (111–116) and both one- and two-component coatings based on castor oil and castor oil derivatives (117–119) have been reported.

1. One-Component Coatings

The initial step in the preparation of a urethane oil is the alcoholysis of an unsaturated glyceride using a polyol such as trimethylolpropane

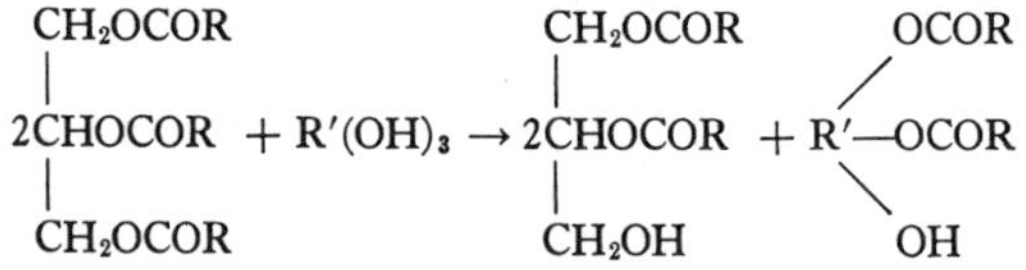

The ratio of glyceride to polyol controls the average number of hydroxyl groups per molecule in the mixed hydroxy esters which result.

Diisocyanate is then added at an NCO/OH ratio of about 1 or less in order to insure that no free diisocyanate is present in the final coating. A solution of this polymer is used as the coating with a catalyst such as cobalt naphthenate.

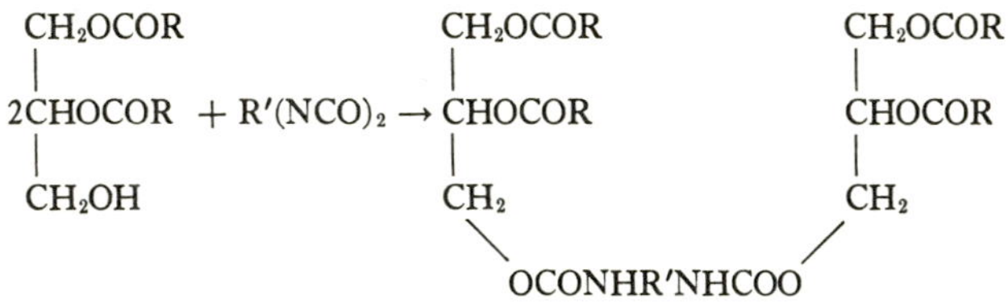

When glycerol linolenate is used as the starting triglyceride, the low molecular weight polyester-urethane will contain four unsaturated fatty acid residues

$$R{=}CHCH_2CH{=}CHCH_2CH{=}CHCH_2CH{=}CH(CH_2)_7$$

These polyunsaturates cure to a high molecular weight polymer by an oxidative mechanism involving the methylene groups adjacent to the double bonds as reported by Farmer et al. (120,121).

a. Isocyanate-Terminated Prepolymers

These coatings are prepared by the reaction of a diisocyanate, usually TDI, and polyols such as polyethers, castor oil, and polyesters. As in other prepolymer preparations it is important to be able to determine the free isocyanate content.

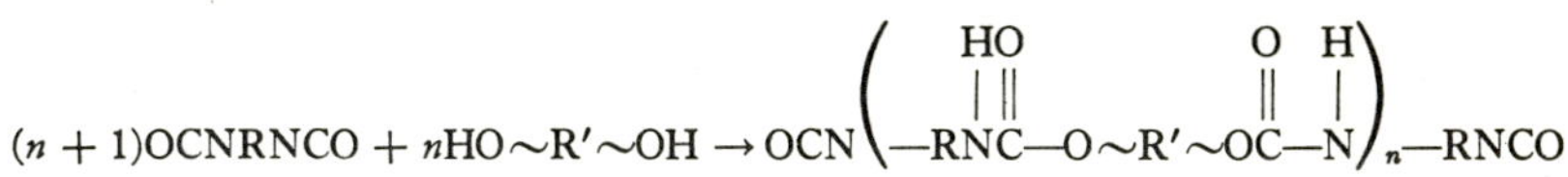

Curing can occur from atmospheric moisture exposure, the generation of carbon dioxide being sufficiently slow so as to allow escape of the gas through the film without bubbling.

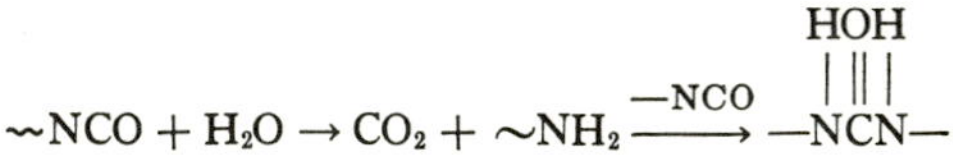

b. Blocked Isocyanates

By "temporarily blocking" the —NCO reactive group with a reactant such as phenol, a stable one-component system containing reactive hydroxyl components can be prepared.

The splitting may occur via two mechanisms, the latter of which Saunders (27) has pointed out to be the more likely.

$$R(NHCOOAr)_2 \xrightleftharpoons{150°C} R(NCO)_2 + 2ArOH$$

or ester interchange

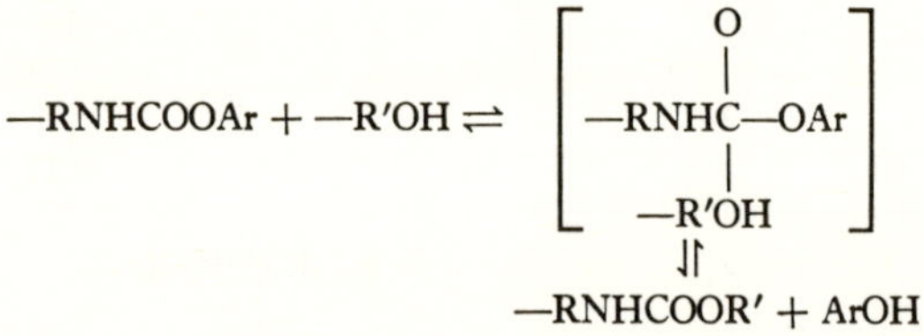

2. Two-Component Systems

Coatings of this type are prepared from isocyanate-terminated prepolymers by reacting with dihydroxyl or polyhydroxyl compounds.

This preparation is usually carried out below 40°C which permits the *para* isocyanate group to react preferentially with each hydroxyl group since it is about eight times as reactive. The *ortho* group is thus free for further reaction. The reaction of trimethylolpropane with three moles of TDI is shown in Scheme III. The polyisocyanate component, usually in the form of a solution (60–75% in an organic solvent) is mixed with the polyol just prior to application.

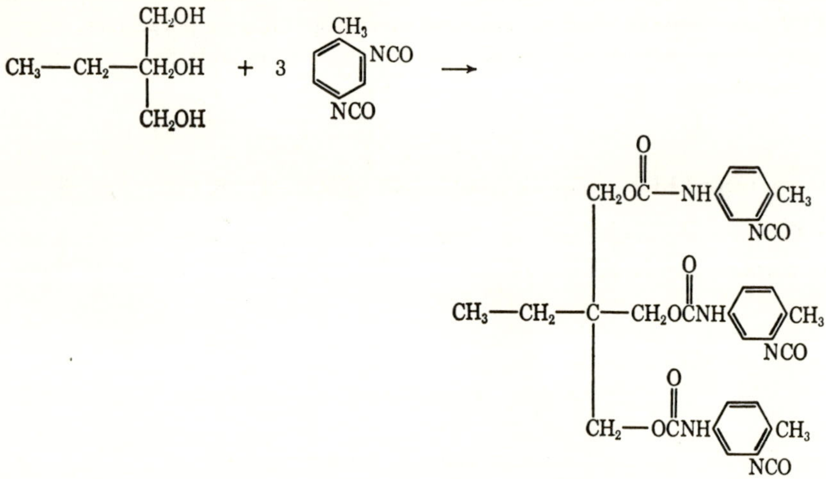

It can be seen that the method reported by Wittendorfer (74) given in the first part of this chapter is very useful for the analysis of polyurethane coatings for trimethylolpropane.

D. Adhesives

Saunders (27) has classified the majority of isocyanate adhesives into three classes: (*1*) polyisocyanates which may be used alone or in mixtures with rubber, (*2*) reaction products (prepolymers) of an excess of polyisocyanate with a polyol, and (*3*) polymeric compounds containing active hydrogen groups which have been modified by reaction with a polyisocyanate.

The major reactions accompanying the preparation and use have been given previously in this chapter and in Chapter 4. The characterization, therefore, and applicable analytical methods given for other types of urethanes can also be used for polyurethane adhesives.

E. Fibers

The chemical structure of the "spandex" fibers is very similar to many of the urethane elastomer systems previously described in that they consist of polyether or polyester soft segments and a polyurea-urethane or polyurethane hard segment.

Saunders and Frisch (27) and Rinke (122) have described the details of the processes and the relationship of structure to properties.

F. Analysis and Characterization

The qualitative and quantitative analytical techniques for cellular polyurethanes are also applicable to elastomeric polyurethanes, except for minor changes in sample handling or technique.

1. Wet Methods

Determination of % NCO and Amine Equivalent (123)

Scope

This method is used to determine the weight of prepolymer which will combine with 1 g-equiv of dibutylamine. The % NCO and/or equivalents per gram can also be calculated.

Principle

If an excess of dibutylamine is added to a prepolymer, the free —NCO groups will react readily with the dibutylamine and the excess subsequently determined by an acidimetric titration, the % NCO, equivalents/gram or $\bar{M}_n$ can be calculated, providing the functionality is known.

$$\text{OCNRNC—O}\sim\text{R}'\sim\text{OCNRNCO} + 2(\text{C}_4\text{H}_9)_2\text{NH} \rightarrow$$
$$(\text{C}_4\text{H}_9)_2\text{NCNRNC—O}\sim\text{R}'\sim\text{OCNRNC—N}(\text{C}_4\text{H}_9)_2$$

The value must be corrected for acidic or basic impurities.

EQUIPMENT

(*a*) Dry methanol.

(*b*) Dry trichlorobenzene.

(*c*) Dry toluene.

(*d*) Dibutylamine reagent (2*N*) in toluene. Add, by graduate, 338 ml di-*N*-butylamine to a 1 liter volumetric flask and dilute to volume with dry toluene.

(*e*) Bromophenol blue indicator. To 0.1 g of bromophenol blue dissolved in 1.5 ml of 1.0*N* NaOH add 1000 ml of distilled water and mix well.

(*f*) Bromcresol green indicator. Titrate 0.100 g of bromcresol green indicator with 1.5 ml of approximately 0.1*N* NaOH until solution has been effected. Transfer to a 100 ml volumetric flask and dilute to volume.

PROCEDURE

(*1*) Weigh by difference, on an analytical balance, 3.5–6.5 g of prepolymer depending upon the expected NCO content into a flask containing 2 to 3 Carborundum boiling chips.

(*2*) Carefully pipet 20 ml of DBA reagent into the flask.

(*3*) Add by dispensing buret or pipet 20 ml of trichlorobenzene carefully washing any sample that has adhered to the sides of the flask down into the main portion.

(*4*) Stopper loosely and heat gently on a hot plate (not exceeding 70°C) until completely dissolved. Remove and let stand 20 min at room temperature while stirring with magnetic stirrer.

(*5*) Add by graduate 100 ml of dry methanol, washing down the sides of the flask.

(*6*) Add 0.5 ml of the 0.1% bromophenol blue indicator and titrate to a yellow endpoint with 1.0*N* HCl.

(*7*) Always run a blank concurrently with the sample following exactly the same procedure but omitting the sample.

CALCULATIONS

$$\text{Amine equivalent} = \frac{\text{Sample weight} \times 1000}{(\text{Ml HCl blank} - \text{Ml HCl sample}) \times N\ \text{HCl}}$$

Report to the nearest 0.1 unit.

The % NCO can be calculated as follows:

$$\%\ \text{NCO} = \frac{0.042 \times 100 \times N\ \text{HCl} \times (\text{ml HCl blank} - \text{ml HCl sample})}{\text{Sample weight}}$$

$$\text{AE} = 4200/\%\ \text{NCO}$$

This procedure is essentially the same as that given in Chapter 2 for TD-80 except for a few changes. Trichlorobenzene is used as the solvent for prepolymer systems because of its greater solvency power. Heating, in this case, is required in order to drive the reaction to quantitative completion in a minimum time period.

Methanol is added before titration to aid solution of all components but mainly to provide a medium that permits ionization to occur readily and subsequent improvement in endpoint detection. If an instrumental titration system is used, this may not be as important depending on the make of electrodes and pH meter.

Determination of Solids in Prepolymer and Prepolymer Coating Solutions (124)

SCOPE

Frequently, it may be necessary to determine the concentration of solids in coating and prepolymer solutions, as a process control prior to formulation and the actual coating operation. The following method is applicable to this determination.

PRINCIPLE

The solids in the solution are determined by evaporating the solvent from a weighed sample and obtaining the weight of the solids remaining.

EQUIPMENT

(*a*) Aluminum weighing dish.
(*b*) Oven equipped with forced air correction heating and cooling.
(*c*) Syringe, 2 cc.
(*d*) Desiccator.
(*e*) Analytical balance.

PROCEDURE

(*1*) Place an aluminum dish in an oven maintained at 105 ± 50°C for ½ hr. Cool in a desiccator and weigh on an analytical balance.

(*2*) Fill a 2 cc syringe with sample, clean the excess sample from the syringe using ethyl acetate and weigh the syringe.

(*3*) Transfer approximately 1 g of sample (0.8–1 cc) from the syringe into the aluminum dish containing 3 cc of ethyl acetate and allow the material to spread out in a thin uniform layer.

(*4*) Reweigh the syringe and determine the weight of sample used.

(*5*) Place the aluminum dish in an oven maintained at 105 ± 50°C for 4 hr.

(*6*) Remove from oven, cool in a desiccator and weigh.

(*7*) Replace the dish in an oven for 1 hr, then cool in the desiccator and reweigh.

(*8*) Repeat Step *7* until the loss in weight is less than 0.003 g.

CALCULATION

$$\% \text{ Solids} = \frac{(\text{Final weight of dish and solids} - \text{Wt. of dish}) \times 100}{\text{Sample weight}}$$

Results are accurate to $\pm 0.05\%$.

Control and knowledge of viscosity is also important during prepolymer preparation and before utilization in a number of systems. One of the simplest, fastest methods by which it can be determined is by the Brookfield viscometer. The full description of this method, precautions, and proper handling of the viscometer is given to every purchaser of a Brookfield viscometer.

2. Gas Chromatographic Methods

In many prepolymer systems, especially those that are utilized for coatings, it is advantageous and/or necessary to be able to determine the concentration of free TDI.

Methods other than gas chromatography have been used for the determination of free TDI in urethane polymers. The method of Remington (125) removes the free TDI and solvents by vacuum distillation and the TDI is then determined colorimetrically after color development with sodium nitrite–Cellosolve reagent, and the absorption is measured at 440 mμ.

While this method has served adequately in the past, some inherent limitations of this procedure are:

(*1*) Variations in recovery due to minute changes in vacuum or temperature.

(*2*) Total time requirements in order to perform multiple analyses.

(*3*) Variations in calibration curves and developed colors on samples as a result of variations in reagents and changes in color intensity as a function of time.

(*4*) The method often gives slightly higher results than other methods. This is especially true when free TDI is present in low concentrations (0.2%).

The following gas chromatographic procedure will permit the determination of TDI in polyurethanes, providing the viscosity is not exceedingly high. This method appears to offer some advantages over more conventional methods in that it is fast, especially where multiple analyses are required, accurate, and reproducible.

The disadvantages are that it is initially more costly to set up because it requires GLC equipment and it is more time consuming if only an occasional sample is to be analyzed.

The Analysis of Free Toluene Diisocyanate in Polyurethane Prepolymer Solutions by Gas–Liquid Chromatography (126)

SCOPE

This method can be used to determine the unreacted or free TDI present in urethane prepolymer as a result of incomplete reaction or the use of nonstoichiometric quantities.

PRINCIPLE

The free TDI in the urethane prepolymer is thermally expelled within the vaporizing chamber of a gas chromatograph but below the decomposition temperature of the parent or matrix material. It is subsequently separated on a column coated with SE-30 from diluent, internal standard, and the vehicle used for the original prepolymer solution. The area produced by the TDI can be directly related to concentration in the original sample.

EQUIPMENT

(*a*) Any good commercial flame ionization gas chromatograph.
(*b*) Copper tubing, ⅛ in. o.d.
(*c*) Silicone gum rubber, SE-30.
(*d*) Analytical balance.
(*e*) Vials, 4 dram.
(*f*) Hamilton 10 ml syringe.
(*g*) Ethyl acetate, urethane grade.
(*h*) Trichlorobenzene, dry.
(*i*) TDI.

CHROMATOGRAPHIC CONDITIONS

Detector: Flame ionization, complete with electrometer.

Column pressure and flow: 11.5 psig nitrogen, restricted to provide a flow of 27.5 ml/min through the column.

Column: 100 ft of ⅛ in. copper tubing coating with a 2½% solution of silicone gum rubber SE-30.

Column temperature: 125°C.

Vaporizer temperature: 160°C.

Combustion mixture: Hydrogen, 20.5 ml/min. Air, 270 ml/min.

CALIBRATION AND PROCEDURE:

(*1*) Prepare standards to cover the desired range by adding known amounts of TDI to approximately 4.5 g portions of prepolymer such as

Mondur CB weighed into vials. One of the standards should have no toluene diisocyanate added to it.

(*2*) To each standard add 0.5 g of trichlorobenzene as an internal standard.

(*3*) To each standard add 10 ml of the *dry* urethane grade ethyl acetate.

(*4*) Prepare samples for analysis by adding 0.5 g of trichlorobenzene to approximately 4.5 g of prepolymer. Dilute with 10 ml of the *dry* urethane grade ethyl acetate.

(*5*) Inject a 1 μl portion of the standards and samples into the chromatograph.

Calculations

(*1*) On each chromatogram, measure the peak height of the trichlorobenzene and the area of the TDI.

Note: On the standards only, subtract from each the area of the TDI obtained on the standard to which no TDI has been added.

(*2*) The ratio of the peak area of the TDI to the peak height of the trichlorobenzene is obtained by the following calculation. The calculation corrects for the amount of internal standard and of prepolymer used.

$$\frac{\text{Peak area of TDI} \times \text{Attenuation of TDI peak} \times \text{Wt of TCB in g} \times 9}{\text{Peak height of TCB} \times \text{Attenuation of TCB} \times (\text{Wt of TDI in g} + \text{wt prep. in g})}$$

This figure is compared with the calibration curve to determine the % TDI. A typical chromatogram using this procedure is shown in Figure 6.14.

Vaporizer temperatures exceeding 175°C caused thermal decomposition of the prepolymer with the release of additional TDI. This was confirmed by differential thermal analysis.

At vaporizer temperatures less than 148°C, broad, tailing peaks occurred without return of the recorder pen to the baseline. At a vaporizer temperature of 160°C, quantitative recovery of TDI added to a prepolymer was found to be complete, ±0.1% at the 0.55% level, without decomposition.

Since TDI is determined on the tail of the TCB peak, peak height ratios yielded results which were neither linear nor reproducible at concentrations less than 0.5% TDI. This is not only because TDI is determined on a descending slope, but is also due to the different releases, with respect to time, of the free TDI from the matrix.

A quick injection of a finely divided spray from a syringe onto a clean vaporizing block yields a higher sharper peak for TDI than a less rapid injection onto a vaporizer that has a nonvolatile residue accumulation.

If nonvolatile residues are allowed to accumulate in the block, nonrepro-

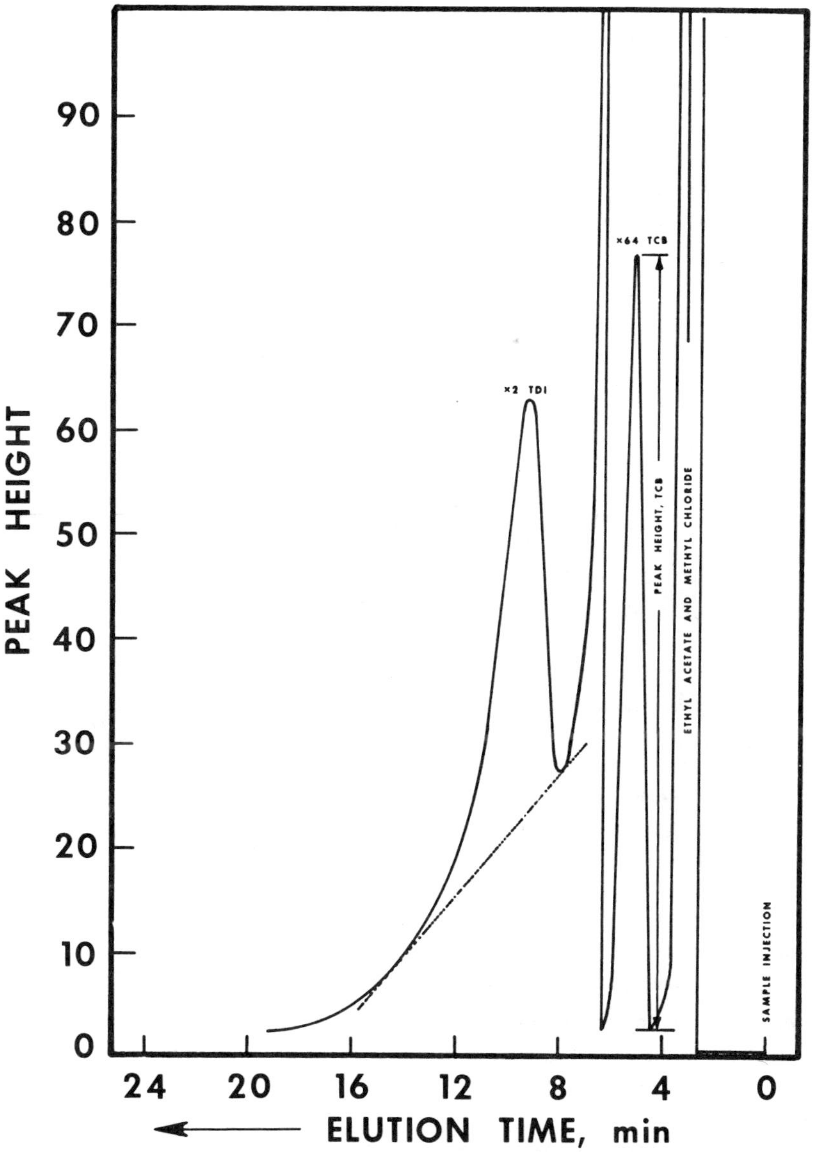

Fig. 6.14. Chromatogram showing baseline drawing for area computation of TDI and resolution of TCB and TDI, 100 ft of column.

ducibility will be experienced. This will be observed by the appearance of a doublet at the apex of the TDI peak due to poor heat transfer through the residue which does not permit instantaneous volatilization. As a result of this, multiple sampling occurs.

Approximately 50 analyses can be run before the injection block must be cleaned out with steel wool. The ⅛ in. o.d. tubular column permits lower column temperatures to be utilized with the retainment of fast elution time and sharp peaks which are good for quantitative work.

This method is similar to the method originally developed by Neubauer et al. (127). Several significant differences exist in the method of Neubauer et al., however. These are: (*1*) a packed column was used (*2*) Free TDI in the range of 2–10% was determined, permitting the utilization of peak height measurements for both TCB and TDI. Below 1% we have found that the employment of peak height measurement for TDI does not provide consistently reproducible results. (*3*) The optimum temperature for complete recovery of TDI, without decomposition, was found to be 150–155°C but a vaporizer temperatures of 175°C ± 2°C was utilized. These investigators reported that more reproducible results were obtained at this temperature and below it vaporization was incomplete. Any error should be compensated for by the similar handling of standards. The incomplete vaporization may be due partly to the fact that approximately five times the amount of sample was used in the Neubauer et al. method as compared with the Mobay method.

3. Infrared Techniques

Similar methods and bands can be used for polyurethane elastomers as have been described in Chapter 4 and in this chapter for cellular materials, for both identification and determination. However, one additional technique that may be of some interest and is useful for determining the concentration of free NCO groups is given below. This technique is relatively simple and fast to use but requires a separate calibration curve for each polymer series containing the same parent compounds. The accuracy is relatively poor, being only ±0.1% absolute, but is nevertheless useful for studying curing rate stability, or the effects of chemical treatment.

Procedure (128)

A small section of the polymer to be analyzed is sectioned with a sharp dissecting knife or razor blade. The section, measuring approximately 5 × 2 mm × 2 mils thick, is taped over the small aperture of a microcell holder, and the holder is positioned in a prealigned beam condenser. The polymer is scanned over the region between 5.5 and 2 μ and the absorbance measured at 4.38 μ for NCO and 3.4 μ for the CH. The ratio of NCO absorbance to CH absorbance is calculated and the corresponding % NCO obtained from a calibration curve.

A calibration curve can be constructed by plotting the NCO/CH ratio vs. known amounts of free NCO. The known amounts of free NCO can be obtained from solution of the polymer, reaction with *n*-dibutylamine and back-titration with acid.

This technique is particularly useful for (*1*) those polymers that cannot be ground and formed readily into a transparent KBr pellet, (*2*) those polymers that are somewhat hard and possess an uneven surface where good contact between an ATR or FMIR plate cannot be achieved, and (*3*) those polymers where the free NCO content at specific distances from the surface is required.

The utilization of the CH stretching vibration absorbance, in a ratio with the NCO asymmetric stretching vibration, alleviates differences in NCO absorbance caused only by variation in film thickness. Techniques in handling the elastomeric sections to be analyzed, which minimize moisture exposure, should be employed. These techniques have been pointed out previously by Greth et al. for the analysis of rigid foams by KBr pellet methods (83).

Spectra of elastomers obtained by this technique are shown in Figure 6.15 (polyester-based elastomer) and Figure 6.16 (polyether-based elastomer).

4. Thermal Methods

Salde and Jenkins (129) have used differential thermal analysis and thermogravimetric analysis to evaluate the thermal properties of several types of polyurethane elastomers. An elastomer prepared from 4,4′-diphenylmethane diisocyanate, a polyether and an aliphatic diamine extender showed, from DTA, two very small endothermic changes, one at 150°C which has not been identified, and one at 205°C that is identical with the softening temperature. The polymer was found to melt at 200°C and exhibited weight loss in two steps respectively, beginning at 280°C and 325°C.

Similar results were obtained on a polymer prepared from 4,4′-diphenylmethane diisocyanate, polyester, and aromatic diamine. Two small endothermic changes were observed, one at 100°C that could be attributable to moisture, and one at 250°C which corresponded to polymer softening. The polymer decomposed in two steps, one beginning at 310°C and the other at 365°C as TGA showed.

These investigators found that polymers prepared from TDI showed a weight loss at 320°C with only one apparent decomposition step in contrast

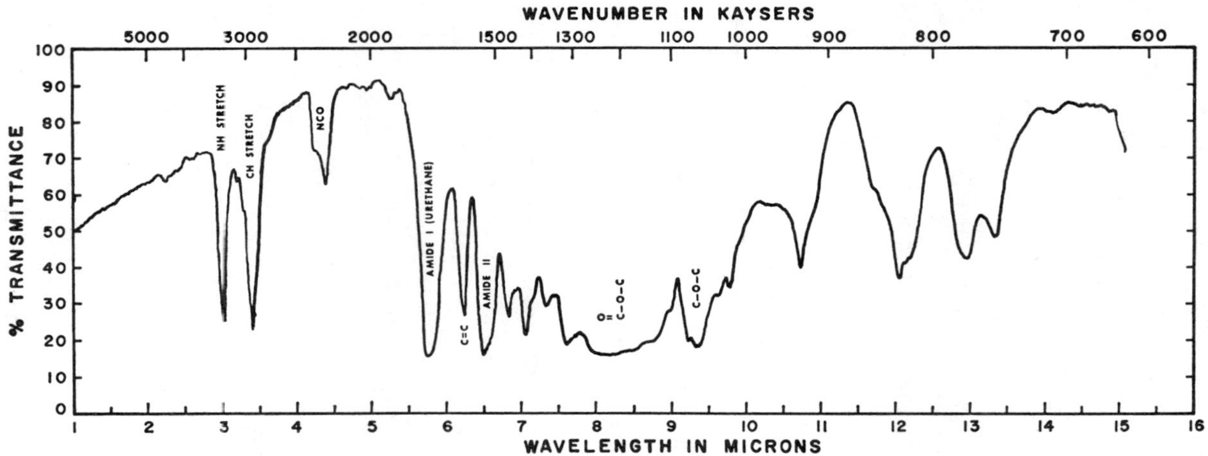

Fig. 6.15. Infrared spectrum of polyester-based polyurethane using microbeam condenser.

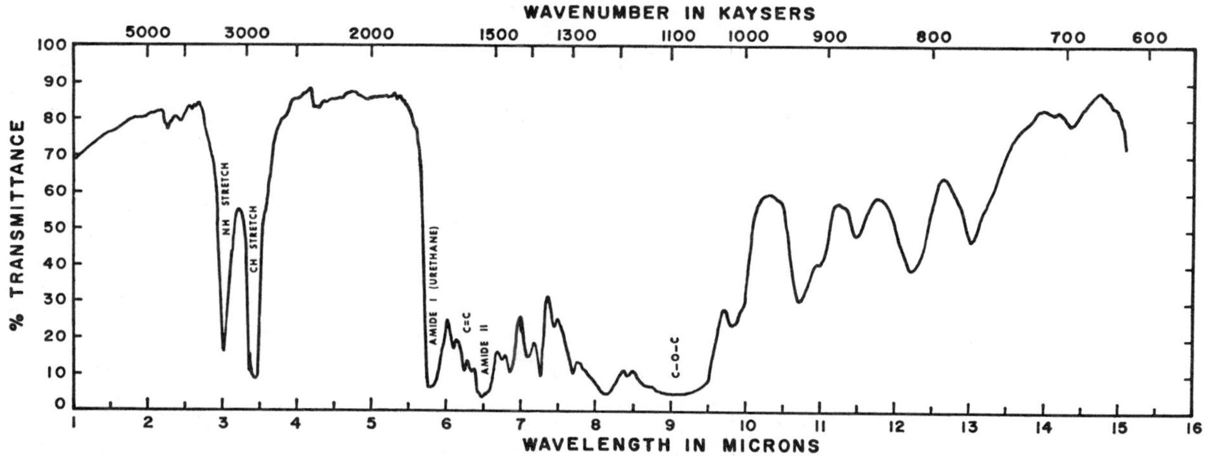

Fig. 6.16. Infrared spectrum of polyether based polyurethane using microbeam condenser.

with the polymers made with MDI. Slightly greater thermal stability of the MDI-based polymers over the TDI-based polymers was found for all cases investigated.

5. Nuclear Magnetic Resonance

Sumi et al. (130) studied the structure of polyurethane elastomers by NMR. These investigators found that urea, urethane, allophanate, and biuret linkages can be discriminated from each other by their different chemical shifts when the measurement is made in polar solvents like *N*,*N*-dimethylacetamide, dimethyl sulfoxide, etc. The results of this study on model compounds were then applied to the analysis of urethane polymer systems: (*1*) a prepolymer made from poly(propylene glycol) and 4,4′-diphenylmethane diisocyanate, (*2*) poly(propylene glycol) and 2,4-toluenediisocyanate, and (*3*) poly(propylene glycol) and mixtures of the 2,4 and 2,6 isomers of toluene diisocyanate.

The MDI-based prepolymer showed only a single NH peak with a chemical shift of 9.47 ppm in DMA which was attributed to the proton of the urethane linkage. All prepolymers prepared from PPG and MDI were found to give the same spectra, irrespective of starting ratios. When water was added to the reaction mixture, the formation of the urea linkage was observed.

The TDI-based prepolymers showed three NH signals which were assigned to the NH protons of 2,4-diurethane (at 9.43 ppm, 8.55 ppm) and a third one at 9.60 ppm to the 4-urethane. Even when 2 moles of 2,4-TDI and 1 mole of PPG were allowed to react, the 2-isocyanate group as well as the 4-isocyanate group were found to form urethane linkages.

Prepolymers, containing mixtures of 2,4- and 2,6-TDI isomers, showed five NH peaks in their NMR spectrum. Three of them were the same as the 2,4 isomer based prepolymers and the remaining two (8.92 ppm and 8.72 ppm) were assigned to the NH protons of 2-urethane.

Relative intensities of the resonance signals were found to be dependent upon the molar ratios of reactants.

6. X-Ray Diffraction

Vasil'ev and Tarakanov (131) studied the effects of temperature on the character of crystallization by X-ray diffraction. The systems studied consisted of polyurethanes prepared from hexamethylenediisocyanate (HMDI) and various glycols, such as ethylene glycol, diethylene glycol, triethylene glycol, and butylene glycol.

Amorphous polymers were prepared by quenching from above the melt temperature and crystalline polymers were prepared by annealing at optimum temperatures to achieve equilibrium crystallization. The diffraction patterns from these materials formed the basis for interpretation of other diffraction patterns.

These investigators concluded that linear polyurethanes crystallize in two stages: (*1*) at low temperatures ($\sim 0°$) there occurs primary crystallization which leads to the formation of two-dimensional ordered structures—pseudocrystals, and (*2*) at higher temperatures secondary crystallization occurs which leads to the three-dimensional ordered structures—ideal crystals.

References

1. Saunders, J. H., and Frisch, K. C., *Polyurethanes, Chemistry and Technology*, Part I, *Chemistry*, Interscience, New York, 1962.
2. Bayer, O. *Angew. Chem.*, **A59,** 257 (1947); *Chem. Abstr.* **42**:6160.
3. Höchtlen, A., *Kunstoffe*, **42,** 303 (1952); *Chem. Abstr.* **47**:2535i.
4. Wilson, C. L. and Hammon, H. G., Fr. Pat. 1,134,076 (to Hudson Foam Plastics Corp.), April 5, 1957.
5. Windemuth, E., Ger. Pat. 929,507, June 27, 1955; *Chem. Abstr.* **52**:3390D.
6. Johnson, P. C., and Packer, E. A., Can. Pat. 565,522, Nov. 4, 1958.
7. Merriman, P., Can. Pat. 569,774, Jan. 27, 1959.
8. Wilson, C. L., and Liberman, R., Can. Pat. 570,900, Feb. 17, 1959.
9. Simon, E., and Thomas, F. W., Can. Pat. 568,167, Dec. 30, 1958.
10. Robitschek, P., U.S. Pat. 3,058,925, Oct. 16, 1962.
11. Barringer, C. M., U.S. Pat. 3,080,329, March 5, 1963.
12. Windemuth, E., Schnell, H., and Bayer, O., U.S. Pat. 2,948,691, August 9, 1960.
13. Loew, G., Belg. Pat. 582,363 (1959).
14. Hood, B. G., Hood, R. S., and Gardella, R. A., Brit. Pat. 914,448, Jan. 2, 1963.
15. Wilson, C. L., Brit. Pat. 916,745, Jan. 30, 1963.
16. Saunders, J. H., Gemeinhardt, P. G., and Sandridge, R. L., Ger. Pat. 1,150,199 (June 12, 1963); *Chem. Abstr.* **59**:12989g.
17. Otey, F. H., Zagoren, B. L., and Mehltretter, C. L., *Ind. Eng. Chem., Prod. Res. Devlop.*, **2**(**4**), 256 (1963).
18. Dioly, R., and Granger, C., Fr. Pat. 1,336,857 (Sept. 6, 1963); *Chem. Abstr.* **60**:3169e.
19. Lyon, C. K., Garrett, V. H., and Goldblatt, L. A., *J. Am. Oil Chemist's Soc.*, **41**(**1**), 23 (1964);
20. Hennig, H. J., Bayer, O., and Braun, G., Ger. Pat. 1,160,173 (Dec. 27, 1963); *Chem. Abstr.* **60**:9443e.
21. Lanham, W. M., U.S. Pat. 3,075,927 (Jan. 29, 1963).
22. Trescher, V., Braun, G., and Dahm, M., Ger. Pat. 1,160,172 (Dec. 27, 1963); *Chem. Abstr.* **60**:9443a.
23. Heiss, H. L., Combs, F. P., Gemeinhardt, P. G., Saunders, J. H., and Hardy, E. E., *Ind. Eng. Chem.*, **51,** 929 (1959).
24. Gmitter, G. T., and Gruber, E. E., *S.P.E. J.*, **13,** 27 (1957).
25. Kane, R. P., Kogan, I. C., and Bradley, H. W., Paper presented before the Division of Rubber Chemistry, Am. Chem. Soc., Cincinnati, Ohio, May, 1958.

26. Scholten, H. G., Schuhmann, J. G., and Tenhoor, R. E., *Ind. Eng. Chem.*, **52**, 613 (1960).
27. Saunders, J. H., and Frisch, K. C., *Polyurethanes, Chemistry and Technology*, Part II, *Technology*, Interscience, New York, 1964.
28. Alzner, B. G., and Frisch, K. C., *Ind. Eng. Chem.*, **51**, 715 (1959).
29. Baker, J. W., and Holdsworth, J. B., *J. Chem. Soc.*, **1947**, 713.
30. Britain, J. W., and Gemeinhardt, P. G., *J. Appl. Polymer Sci.*, **4**, 207 (1960).
31. Hostettler, F., and Cox, E. F., *Ind. Eng. Chem.*, **52**, 609 (1960).
32. Sandridge, R. L., Gemeinhardt, P. G., and Saunders, J. H., Paper presented before the Division of Organic Coatings and Plastics Chemistry, Am. Chem. Soc., Chicago, Illinois, September, 1961, Preprint, vol. 21, (2) 68.
33. Olin Mathieson Chemical Corp., Organic Division, New York TIB No. 00-48U-1263.
34. Weissberg, S. G., Rothman, S., and Wales, M., "Molecular Weights and Sizes," Chapter I, in *Analytical Chemistry of Polymers*, Part II, G. M. Kline, Ed., Interscience, New York, 1962.
35. Beyer, G. L., "Determination of Particle Size and Molecular Weight," in Part I, Chapter V, *Physical Methods of Organic Chemistry—Techniques of Organic Chemistry*, A. Weissberger, Ed. Vol. I, Part I, 3rd Edition, Interscience, New York, 1959.
36. Billmeyer, F. W., Jr., *Textbook of Polymer Science*, Interscience, New York, 1962.
37. Rosenthaler, L., *Z. Anal. Chem.*, **109**, 31 (1937); *Chem. Abstr.* **31**:5297[2].
38. Feigl, F., *Spot Tests in Organic Chemistry*, 6th ed., Elsevier, New York, 1960.
39. Feigl, F., and Jungreis, E., *Mikrochim. Acta*, **6**, 812 (1958); *Chem. Abstr.* **54**:24113e.
40. Davidson, D., *Ind. Eng. Chem.*, *Anal. Ed.*, **12**, 40 (1940).
41. Brauer, G. M. and Horowitz, E., "Systematic Procedures," Chapter I in *Analytical Chemistry of Polymers*, Part III, G. M. Kline, Ed., Interscience, New York, 1962.
42. Pickhardt, W. P., Safranski, L. W., and Mitchell, J., *Anal. Chem.*, **30**, 1298 (1958).
43. Wood, L. A., Madorsky, I., and Paulson, R. A., *J. Res. Natl. Bur. Std.* **64A**, 157 (1960).
44. Belcher, R., Nutten, A. J., and Stephen, W. I., *Mikrochim. Acta*, **1953**, 51, *CA*47:9859c.
45. Scott, W. W., *Standard Methods of Chemical Analysis*, Vol. 2, Fifth Ed., N. H. Furman, Ed., Van Nostrand, Princeton, N.J., 1947.
46. Cogbill, E. and Kirkland, J. J., *Anal. Chem.*, **27**, 1611 (1955).
47. Haslam, J., and Squirrell, D. C. M., *Analyst*, **82**, 511 (1957).
48. Skoda, W., and Schurz, J., *Z. Anal. Chem.*, **162**, 259 (1958); *Chem. Abstr.* **53**:763a.
49. Bartels, U., *Faserforch. u. Textiltech*, **8**, 194 (1957); *Chem. Abstr.* **51**:15994i.
50. Ingram, G., *Mikrochim. Acta*, **1953**, 131; *Chem. Abstr.* **47**:9857i.
51. Alford, W. C., *Anal. Chem.*, **24**, 881 (1952).
52. Mullen, J. D., *Anal. Chim. Acta*, **20**, 16 (1959); *Chem. Abstr.* **53**:22953d.
53. Schöniger, W., *Mikrochim. Acta*, **1955**, 123; *Chem. Abstr.* **49**:8734h.
54. Wickbold, R., *Angew. Chem.*, **64**, 133 (1952); *Chem. Abstr.* **46**:4950h.
55. Wickbold, R., *Angew. Chem.*, **69**, 530 (1957); *Chem. Abstr.* **54**:13948g.
56. Rowe, R. D., *Anal. Chem.*, **37**, 368 (1965).
57. Hinsvark, O. N., and Waltz, R. H., Paper presented at the Pittsburgh Conference on Analytical Chemistry, 1965.
58. Snell, F. D., and Snell, C. T., *Colorimetric Methods of Analysis*, Van Nostrand, Princeton, N.J., 1957.
59. Kolthoff, I. M., and Elving, P. J., *Treatise on Analytical Chemistry*, Part II, Vols. I–XI, Interscience, New York, 1961–1967.
60. Welcher, F. J., *Organic Analytical Reagents*, Vols. I, II, III, and IV, Van Nostrand, Princeton, N.J., 1955.
61. Kolthoff, I. M., and Lingane, J. J., *Polarography*, Vol. II, Interscience, New York, 1952.
62. Swann, M. H., and Esposito, G. G., *Anal. Chem.*, **30**, 107 (1958).

63. Baumann, G. F., and Steingiser, S., *J. Appl. Polymer Sci.*, **1**, 251 (1959).
64. Davison, W. H. T., and Bates, G. R., *Proc. 3rd Rubber Technology Conf., London, England, June* 1954, pp. 281–297; *Chem. Abstr.* **50**:16164i.
65. Corish, P. J., *Anal. Chem.*, **31**, 1298 (1959).
66. Corish, P. J., and Davison, W. H. T., *J. Chem. Soc.*, **1955**, 2431.
67. Davison, W. H. T., and Corish, P. J., *J. Chem. Soc.*, **1955**, 2428.
68. Kappelmeir, C. P. A. (a) *Farben-Ztg* **40,** 1141 (1935); *Chem. Abstr.* **30**:785[4] (b) *Farben-Ztg* **41,** 111 (1936); *Chem. Abstr.* **30**:2526[7] (c) *Paint Oil Chem. Rev.* **99,** No. 12, 20, 22, 24, (1937); *Chem. Abstr.* **31**:7271[7].
69. Parker, E. E., "Polyesters" Chapter XI in *Analytical Chemistry of Polymers*, Part I, G. M. Kline, Ed. Interscience, New York, 1959.
70. Bonar, F. M., and David, D. J., private communication, 1965.
71. Brügel, W., *An Introduction to Infrared Spectroscopy*, Methuen, London, 1962.
72. Stafford, R. W., Francel, R. J., and Shay, J. F., *Anal. Chem.*, **21,** 1454 (1949).
73. Stafford, R. W., Shay, J. F., and Francel, R. J., *Anal. Chem.*, **26,** 656 (1954).
74. Wittendorfer, R. E., *Anal. Chem.*, **36,** 930 (1964).
75. Kappelmeir, C. P. A., Boomen, J. F., and Mostert, J., *Verfkroniek*, **27,** 291 (1954); *Chem. Abstr.* **49**:7889g.
76. Esposito, G. G., and Swann, M. H., *Anal. Chem.*, **33,** 1854 (1961).
77. Percival, D. F., *Anal. Chem.*, **35,** 236 (1963).
78. Schroeder, E., *Plaste Kautschuk*, **9,** 121, 186 (1962); *Chem. Abstr.* **57**:15337i.
79. Brumfield, H. L., "Infrared Spectrophotometric Analysis of Urethane Compounds," Sandia Corporation, Livermore Report SC-R-322, U.S. Atomic Energy Commission TID-4500; *Chem. Abstr.* **60**:15151b.
80. Merten, R., Braun, G., and Lauerer, D., *Kunststoffe*, **55,(4)** 249 (1965); *Chem. Abstr.* **64**:12899e.
81. Niederdellmann, G., Lauerer, D., and Merten, R., Central Research Laboratory, Farbenfabriken Bayer A.G., private communication, November 1965.
82. Murphy, E. B,. and O'Neil, W. A., *SPE Journal*, **18,** 191 (1962).
83. Greth, G. G., Smith, R. G., and Rudkin, G. O., Jr., *J. Cell. Plastics*, **1**(1), 159 (1965).
84. Kubitz, K. A., *Anal. Chem.*, **29,** 814 (1957).
85. Simha, R., *J. Polymer Sci. B.*, **2,** 675 (1964).
86. Backus, J. K., Darr, W. C., Gemeinhardt, P. G., and Saunders, J. H., *J. Cell Plastics*, **1,** 178 (1965).
87. Van Wazer, J. R., *Phosphorus and Its Compounds*, Vol. II, Interscience, New York, 1961.
88. Dyer, E., Newborn, G. E., Jr., and Wright, G. C., "Thermal Degradation of Carbamates," Delaware Chemical Symposium, February 1958.
89. Merten, R., Braun, G., and Lauerer, D., Central Research Laboratory, Farbenfabriken Bayer A.G., private communication, May 1965.
90. Pinten, H., Ger. Pat. Appl. D-90 260 (Dynamid Nobel A.-G., March 1942).
91. Bayer, O., Müller, S., Petersen, S., Piepenbrink, H. F., and Windemuth, E., *Angew Chem.*, **62,** 57 (1950); *Rubber Chem. Technol.*, **23,** 812 (1950); *Chem. Abstr.* **44**:4709c, **45**:8800f.
92. Harper, D. A., Smith, W. F., and White, H. G., *Rubber Chem. Technol.*, **23,** 608 (1950).
93. Müller, E., *Rubber Plastics Age*, **39**(3), 195 (1958).
94. Müller, E., Bayer, O., Petersen, S., Piepenbrink, H. F., Schmidt, F., and Weinbrenner, E., *Angew. Chem.*, **64,** 523 (1952); *Rubber Chem. Technol.*, **26,** 493 (1953); *Chem. Abstr.* **47**:339e, **48**:8576i.
95. Axelrood, S. L., Hamilton, C. W., and Frisch, K. C., *Ind. Eng. Chem.*, **53,** 889 (1961).

96. Pigott, K. A., Archer, W., Gable, C. L., and Steingiser, S., *Mod. Plastics*, **40**(4), 117, December (1962).
97. Pigott, K. A., Britain, J. W., Archer, W., Frye, B. F., Cote, R. J., and Saunders, J. H., *Ind. Eng. Chem. Prod. Res. Develop.*, **1**(1), 28 (1962).
98. Schollenberger, C. S., Scott, H., and Moore, G. R., *Rubber World*, **137**, 549 (1958).
99. Waugaman, C. A., *Mod. Plastics*, **39**(2), 146 (1961).
100. Axelrood, S. L., Smith, L. C., and Frisch, K. C., *Rubber Age*, **96**(2), 233 (1964).
101. Müller, K. E., Schmidt, F. W., Weinbrenner, E., and Piepenbrink, H. F., U.S. Pat. 2,729,618 (to Farbenfabriken Bayer) (Jan. 3, 1956).
102. Schollenberger, C. S., and Dinbergs, K., *SPE Trans.*, **1**(1), 31 (1961).
103. Schollenberger, C. S., Pappas, L. G., Park, J. C., and Vickroy, V. V., Jr., *Rubber World*, **142**(6), 81 (1960).
104. Saunders, J. H., and Pigott, K. A., U.S. Pat. 3,214,411 (to Mobay Chem. Co.) (Oct. 26, 1965).
105. Saunders, J. H., in *The Polymer Chemistry of Synthetic Elastomers*, Interscience, to be published.
106. Bayer, O., *Farbe U. Lack*, **64**, 235 (1958).
107. Petersen, S., *Ann.*, **562**, 205 (1949).
108. Pansing, H. E., *Offic. Dig. Federation Paint Varnish Prod. Clubs*, **30**, No. 396, 37 (1958); *Chem. Abstr.* **53**:20834e.
109. Robinson, E. B., and Waters, R. B., *J. Oil Colour Chemists' Assoc.*, **34**, 361 (1951).
110. Wilson, G. and Stanton, J. M., *Offic. Dig. Federation Paint Varnish Prod. Clubs*, **32**, No. 421, 242, (1960); *Chem. Abstr.* **54**:18982i.
111. Bailey, M. E., *SPE J.*, **14**(2), 41 (1958).
112. Heiss, H. L., Saunders, J. H., Morris, M. R., Davis, B. R., and Hardy, E. E., *Ind. Eng. Chem.*, **46**, 1498 (1954).
113. Metz, H. M., Erlich, A., Smith, M. K., and Patton, T. C., *Paint, Oil, Chem. Rev.*, **121**(8), 6 (1958).
114. Toone, G. C. and Wooster, G. S., *Offic. Dig. Federation Paint Varnish Prod. Clubs*, **32**, No. 421, 230, (1960); *Chem. Abstr.* **54**:18983b.
115. Wells, E. R., *Paint Varnish Prod.*, **51**(3), 41 (1961).
116. Wells, E. R., Hudson, G. A., Saunders, J. H., and Hardy, E. E., *Offic. Dig. Federation Paint Varnish Prod. Clubs*, **31**, No. 416, 1181 (1959).
117. Damusis, A., McClellan, J. M., and Frisch, K. C., *Ind. Eng. Chem.*, **51**, 1386 (1959).
118. Mobay Chemical Co. Bull. TIB 25-C10, "Prepolymer Type Urethane Coatings," May 9, 1959.
119. Remington, W. J. and Athey, R. J., *Offic. Dig. Federation Paint Varnish Prod. Clubs*, **31**, No. 412, 612 (1959).
120. Farmer, E. H., Bloomfield, G. F., Sundralingam, A., and Sutton, D. A., *Trans. Faraday Soc.*, **38**, 348 (1942).
121. Farmer, E. H., and Sutton, D. A., *J. Chem. Soc.*, **1942**, 139.
122. Rinke, H., *Angew. Chem.*, Int. Ed., **1**, 419 (1962).
123. Malec, E. J., and David, D. J., "Mobay Method for The Determination of %NCO and Amine Equivalent," March 1965, Mobay Chemical Co., unpublished.
124. Castilow, A. R., Jones, R. L., and Malec, E. J., "Mobay Method for the Determination of Solids in Prepolymer and Coating Solutions," August 1961, Mobay Chemical Co., unpublished.
125. Remington, W. J., E. I. du Pont de Nemours and Co., Elastomer Chemicals Dept., Bulletin P.B.-5, 1958.

126. Coen, C. S., and David, D. J., "Mobay Method for the Analysis of Free Toleune Diisocyanate in Polyurethane Prepolymer Solutions by Gas Liquid Chromatography," March 1964, Mobay Chemical Co., unpublished.
127. Neubaur, N. R., Skreckoski, G. R., White, R. G., and Kane, A. J., *Anal. Chem.*, **35,** 1647 (1963).
128. David, D. J., and Bonar, F. M., private communication, March 1963.
129. Slade, P. E. Jr., and Jenkins, L. T., *J. Polymer Sci.*, *C*, **6,** 27 (1964).
130. Sumi, M., Chokki, Y., Nakai, Y., Nakabayashi, M., and Kanzawa, T., *Makromol. Chem.*, **78,** 146 (1964); *Chem. Abstr.* **61**:16278.
131. Vasil'ev, B. V., and Tarakanov, O. G., *Vysokomolekul. Soedin.*, **6**(12), 2189 (1964); *Chem. Abstr.* **62**:9302f.

CHAPTER VII

PHYSICAL PROPERTIES OF ISOCYANATES

For the complete analysis and identification of any organic compound, it is necessary to have information concerning its physical properties and suitable derivatives by which it can be characterized. Such data are provided in the accompanying tabulation for 883 isocyanate compounds described in the literature. It is noteworthy that 275 of these compounds were synthesized by Professor Otto Bayer and his co-workers in the laboratories of the former Farbenfabriken Bayer A.G. and are described by Werner Siefken in *Annalen der Chemie*, Volume 562, pages 75–136 (1949). This paper reflects the tremendous pioneering work in isocyanate chemistry made by this organization.

The entire citations given in the subject indexes to *Chemical Abstracts* under the heading of "isocyanic acid, esters" for Volumes 1–57 (1907–1962) were examined for references to literature which described some property of an isocyanate compound. Approximately 2500 abstracts were scanned in this search, of which some 1600 yielded property information. The original literature was consulted wherever possible to obtain all the information available concerning the particular compound. Data on the boiling point, melting point, refractive index, and density with notations concerning infrared or ultraviolet absorption data are given in tabular form. Derivatives suitable for identification purposes are listed under "derivatives." The numbers given under "reference" in the tables correspond to the 558 references given in the bibliography. These are arranged in order according to volume and page numbers in *Chemical Abstracts*.

The compounds have been named as substitution products of a "parent" compound usually with the position of the isocyanate group or groups given next, followed by the position and names of other substituent groups. It is to be noted that almost all of the compounds (excepting the metal isocyanates) are named by *Chemical Abstracts* as ester derivatives of Isocyanic Acid, thus requiring that each complete name include the word ester, and by implication the words "isocyanic acid." Doubtless, this naming system is satisfactory to differentiate this group of compounds from all the other groups of organic compounds, but it does not make for a convenient distinction between isocyanates. We have therefore identified the principal "parent" compound other than the isocyanate group and have arranged them alphabetically according to this parent compound. Thus, phenyl isocyanate becomes Benzene-1-isocyanato in the table and methyl isocyanate becomes methane-1-isocyanato. In cases where there is a choice as to which group is principal, alternate names are given with suitable cross references. Likewise, trivial names are cross-referenced to the principal parent compound name.

There are some exceptions to the alphabetical arrangement evident in the table. The isocyanates containing fluorine are grouped together under "fluoroisocyanates," "perfluoroisocyanates alkyl isocyanates," and "1,1-dihydro perfluoro alkyl isocyanates," since many of them are members of a homologous series. Within these subgroups the compounds are named by their parent compound name, but are not alphabetically arranged. The metal isocyanates are listed under M. We have retained the name "nitraza" as used in the original literature for the group O_2NNH—, which is called "nitramino" by *Chemical Abstracts*. There are several of these compounds listed in which the isocyanate group appears in the side chain.

Likewise, it has seemed expedient to retain the names used in the original publications for those isocyanates containing phosphorus or silicon, and group them under "phosphorus isocyanates" and "silicon isocyanates." Isocyanates containing sulfur are listed under "sulfur isocyanates" with cross references to the parent compound name, where the physical properties of the particular compound will be found.

There are 23 compounds in this table for which the data are so extensive that they could not be tabulated conveniently. Suitable notations for these are made under "other data" referring to separate tables which are arranged with an index following the principal tabulation. It is evident that a considerable amount of information has accumulated on the commercially important isocyanates such as toluene 2,4-diisocyanate, methane-diphenyl-4,4′-diisocyanate, the chloro-phenyl isocyanates, phenyl isocyanate, hexane-1,6-diisocyanate, and cyclohexane-1-isocyanate. Several of the isocyanates have been investigated as reagents for the preparation of well-defined derivatives of the alcohols and amines. Thus, the tables for the nitrophenyl isocyanates, biphenyl-4-isocyanate, and the naphthalene isocyanates contain many urethan and urea derivatives. Azobenzene-4-isocyanate gives many analytically useful derivatives of the amino acids and of primary aliphatic alcohols, as evidenced by Table 7.1.

It is believed that the present compilation includes almost all, if not all, of the isocyanate compounds which have been characterized by some physical properties and described in the literature through the year 1962.

No.	Name	Other Names	Structure
	Abietic Acid Isocyanate	see (T-4) Tetrahydroabietyl isocyanate	
	Abietic Acid Isocyanate	see (D-6) Dehydroabietane-1-isocyanate	
A-1	Acetic Acid Isocyanato	Acetyl isocyanate	CH_3-CO-NCO
A-2	Acetic Acid -isocyanato -butyl ester		OCN-CH_2-CO-O-C_4H_7
A-3	Acetic acid 1,1 diphenyl isocyanate	Diphenyl acetyl isocyanate	$(C_6H_5)_2$CH-CO-NCO
A-4	Acetic acid -isocyanato dodecyl ester	Glycolic acid isocyanato dodecyl ester	OCN-CH_2-CO-O-$C_{12}H_{25}$
A-5	Acetic acid -isocyanato -ethylester	Ethyl isocyanato acetate	OCN-CH_2-CO-O-C_2H_5
A-6	Acetic acid -(-2 isocyanato ethyl)ester	B-Acetyloxy-ethyl isocyanate	CH_3-CO-O-CH_2-CH_2-NCO
A-7	Acetic acid 1-phenyl 1-ethyl -isocyanato	Phenyl ethyl acetyl isocyanate	C_6H_5-CH(C_2H_5)-CO-NCO
A-8	Acetic acid -isocyanato -isobutylester		OCN-CH_2-CO-O-CH_2-CH$(CH_3)_2$
A-9	Acetic acid -isocyanato -isopropylester		OCN-CH_2-CO-O-CH$(CH_3)_2$
A-10	Acetic acid -isocyanato -methylester	Methyl isocyanato acetate	OCN-CH_2-CO-O-CH_3
A-11	Acetic acid (o-phenylisocyanate) ester	2-acetoxy phenyl isocyanate	CH_3-CO-O-C_6H_4-NCO
A-12	Acetic acid -1 phenyl -isocyanate	Phenacetyl isocyanate	C_6H_5-CH_2-CO-NCO
A-13	Acetic acid -(phenyl-2-isocyanato -4,6 dibromo) ester		CH_3-CO-O-C_6H_2(Br_2)(NCO)
A-14	Acetic acid 2-isocyanato 2-phenyl ethyl ester	Isocyanato phenyl glycine ethyl ester	C_6H_5-CH(NCO)-CO-O-C_2H_5
A-15	Acetophenone -4'-isocyanato	p-acetyl-phenyl-isocyanate	CH_3-CO-C_6H_4-NCO
	β-Acetyloxy-ethyl isocyanate	see Acetic acid -2-(-isocyanato ethyl)-ester (A-6)	
	2-Acetoxy phenyl isocyanate	see Acetic acid (o-phenyl isocyanate)ester (A-11)	

Formula	Boiling Point °C.	Boiling Point mm.	Melting Point °C.	Density	Refractive Index °C.	Refractive Index n	Other Data	Derivatives	References
$C_3H_3NO_2$	79 80-80.3	708 737		1.0892 (18°/4°)			Infrared data Ref (371)		(5) (371)
$C_7H_{11}NO_3$	79	6	49.5			1.4301			(537)
$C_{15}H_{11}NO_2$	141-143	2							(546)
$C_{15}H_{27}NO_3$									(141)
$C_5H_7NO_3$	65 66-68 72-74 75-77	10 12 14 16						Phenyl urea MP 111°	(194) (240) (343) (465)
$C_5H_7NO_3$	62-63	8							(251)
$C_{11}H_{11}NO_2$	111-115	11							(108) (153)
$C_7H_{11}NO_3$	76	6	86		25	1.4229			(537)
$C_6H_9NO_3$	64	6	115		25	1.4188			(537)
$C_4H_5NO_3$	73	19			21	1.4286			(526)
$C_9H_7NO_3$	112	12						Ethyl urethane MP 75-76° Methyl urethane MP 100-101°	(32)
$C_9H_7NO_2$	116-120	20							(108)
$C_9H_5Br_2NO_3$			76					Methyl urethane MP 178°	(32)
$C_{11}H_{11}NO_3$	127 152	2 17						Urea from aniline MP 175°	(240)
$C_9H_7NO_2$	139	14	36-37						(441)

No.	Name	Other Names	Structure
	Acetyl isocyanate	see Acetic acid -isocyanate (A-1)	
	p-Acetyl,-phenyl isocyanate	see Acetophenone-4-isocyanato (A-15)	
A-16	Acrylic acid,-isocyanate	Acrylyl isocyanate	CH_2=CH-CO-NCO
A-17	Acrylic acid -(2-isocyanato ethyl), ester		CH_2=CH-CO-O-CH_2CH_2-NCO
	Allylisocyanate	see Propene-3-isocyanate (P-102)	
	n-Amylisocyanate	see Pentane-1-isocyanate (P-6)	
A-18	Anthracene-2-isocyanate		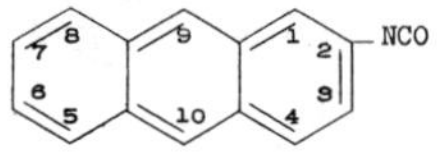
A-19	Anthracene-9-isocyanate		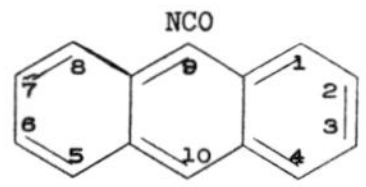
A-20	Anthracene -9,10 bis(methyl isocyanate)		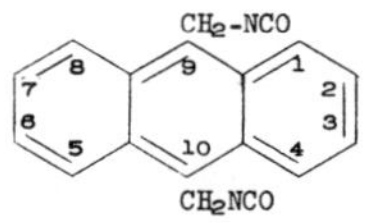
A-21	9,10 Anthraquinone 2,6 diisocyanato		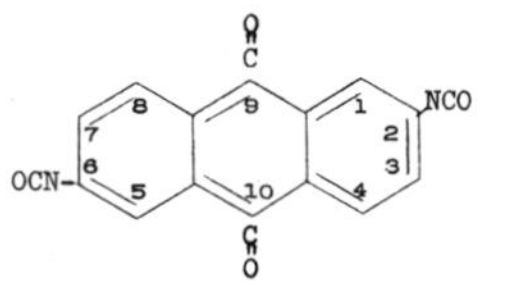
A-22	Azobenzene-4,4'-diisocyanate		OCN-C_6H_4-N=N-C_6H_4-NCO
A-23	Azobenzene, -2-methyl -4,4'diisocyanato		OCN-(CH_3)C_6H_3-N=N-C_6H_4-NCO
A-24	Azobenzene-3-isocyanato	m-Phenylazophenyl isocyanate	C_6H_5-N=N-C_6H_4(NCO)
A-25	Azobenzene-4-isocyanato	p-Phenylazophenyl isocyanate	C_6H_5-N=N-C_6H_4(NCO)

	Formula	Boiling Point °C.	Boiling Point mm.	Melting Point °C.	Density	Refractive Index °C.	Refractive Index n	Other Data	Derivatives	References
	$C_4H_3NO_2$								Phenyl urea MP 147° Benzacryl urethane MP 115°	(330) (395)
	$C_6H_7NO_3$	80-82	10							(394)
	$C_{15}G_9NO$			207.5- 208.0					Methyl urethane MP 231° Ethyl urethane MP 216° Urea from NH_3 MP > 360°	(98)
	$C_{15}H_9NO$			76.0					Ethyl urethane MP 237°	(87)
	$C_{18}H_{12}N_2O_2$			214 215						(305)
	$C_{16}H_6N_2O_4$			250 252						(194)
	$C_{14}H_8N_4O_2$			158- 161					Dimethyl urethane MP 241-243°	(194) (217)
	$C_{15}H_{10}N_4O_2$			123- 125					Dimethyl urethane MP 244-248°	(194)
	$C_{13}H_9N_3O$			45-46						(62)
	$C_{13}H_9N_3O$			97-98				see Table No. 1	Ethyl urethane MP 151° Methyl urethane MP 120-121°	(58) (63) (126) (194) (202) (206) (214) (217) (221) (474)

No.	Name	Other Names	Structure
6	Azobenzene -4-isocyanato -5-methoxy -6-methyl -2'-chloro		ClC_6H_4-N=N-$C_6H_2(CH_3)(CH_3O)(NCO)$
A-27	Azobenzene -4-isocyanato -2-methyl -4'-carbethoxy		$(NCO)(CH_3)C_6H_3$-N=N-C_6H_4-$COOC_2H_5$
A-28	Azobenzene -4'-isocyanato -2'-methyl -3 carbonyl chloride		(COCl)-C_6H_4-N=N-$C_6H_3(CH_3)(NCO)$
A-29	Azobenzene -4-isocyanato -2',3-dimethyl		$(CH_3)C_6H_4$-N=N-$C_6H_3(CH_3)(NCO)$
A-30	Azobenzene 4-isocyanato 4'-ethoxy		$(C_2H_5O)C_6H_4$-N=N-$C_6H_4(NCO)$
A-31	Azobenzene -4-isocyanato -3'-methyl -4'-hydroxy		$(HO)(CH_3)$-C_6H_3-N=N-$C_6H_4(NCO)$
A-32	Azobenzene 1-naphthalene 4,4'diisocyanato		$(OCN)C_6H_4$-N=N-$C_{10}H_6(NCO)$
A-33	Azobenzene -4-isocyanato -2-methyl -4'-methoxy		(CH_3O)-C_6H_4-N=N-C_6H_3-$(CH_3)(NCO)$
A-34	Azobenzene -4-isocyanato -4'-Iodo		(I-C_6H_4-N=N-$C_6H_4(NCO)$

Formula	Boiling Point °C.	Boiling Point mm.	Melting Point °C.	Density	Refractive Index °C.	Refractive Index n	Other Data	Derivatives	References
$C_{15}H_{12}N_3O_2Cl$			153						(63) (194
$C_{17}H_{15}N_3O$			221						(63)
$C_{15}H_{10}ClN_3O_2$			91						(342)
$C_{15}H_{13}N_3O$			59-60					Diurea MP 280-281°	(317)
$C_{15}H_{13}N_3O_2$			141 117						(36)
$C_{14}H_{11}N_3O_2$									(63)
$C_{18}H_{10}N_4O_2$			177-178					Dimethyl urethane MP 210-212°	(194)
$C_{13}H_{13}N_3O_2$			80						(63) (194)
$C_{13}H_8IN_3O$			165					Methyl urethane MP 178-179 Ethyl urethane MP 168-169 Other urethanes	(223 (219)

No.	Name	Other Names	Structure
B-1	1,2 Benanthrene-3-isocyanate		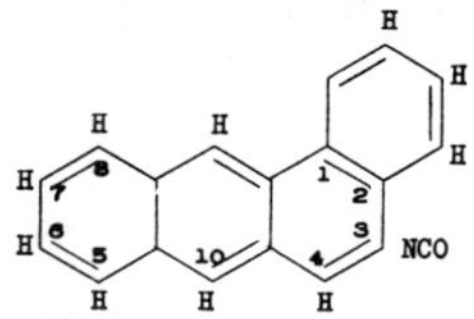
B-2	1,2 Benzanthrene 10-isocyanate		See B-1
B-3	1,2 Benzanthrene-3-isocyanate 9-methyl		see B-1
B-4	1,2 Benzanthrene-3-isocyanato 10-methyl		see B-1
B-5	1,2 Benzanthrene-3-isocyanato 9-10 dimethyl		see B-1
B-6	1,2 Benzanthrene-7-isocyanato 9-10 dimethyl		see B-1
B-7	1,2,5,6,Dibenzanthrene 9-isocyanato		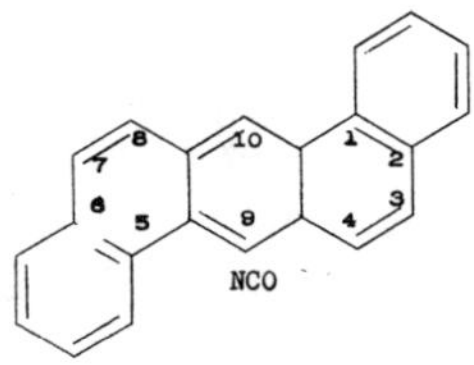
B-8	Benzene 1-isocyanato	Phenylisocyanate	C_6H_5NCO
B-9	Benzene 1-isocyanato 3-bromo		Br-C_6H_4-NCO
B-10	Benzene 1-isocyanato 4-bromo		BrC_6H_4-NCO
B-11	Benzene 4-(bromomethyl) 1-isocy[illegible]ato		$BrCH_2$-C_6H_4-NCO
B-12	Benzene 1-isocyanato 2-butoxy		C_4H_9-O-C_6H_4-NCO

Formula	Boiling Point °C.	Boiling Point mm.	Melting Point °C.	Density	Refractive Index °C.	Refractive Index n	Other Data	Derivatives	References
$C_{19}H_{11}NO$			163-163.5				Carcinogenic inactive (165) UV Absorption (114)	Methyl carbamate MP 203.5-204° Ethyl carbamate MP 211.5-212°	(98) (114) (115) (165) (166) (230) (320)
$C_{19}H_{11}NO$			144-144.5				Carcinogenic inactive (165) UV Absorption (99)(109)(114)	Methyl carbamate MP 227-227.5	(98) (114) (99) (109) (115) (165) (166) (230) (320)
$C_{20}H_{13}NO$			63-64.2					Ethyl carbamate MP 174.2-174.9	(230)
$C_{20}H_{13}NO$			149.5-150				Carcinogenic inactive (165)	Ethyl carbamate MP 201-201.5 Urea from NH_3 MP 348-350°	(115) (165)
$C_{21}H_{15}NO$			115.4-115.8					Ethyl carbamate MP 113.4-114.1	(230)
$C_{21}H_{15}NO$			113.8-114.6						(320)
$C_{23}H_{13}NO$			181-181.5				UV Absorption (157)	Ethyl carbamate MP 224-224.5°	(87) (157)
C_7H_5NO	50 62-64 165	11 20 760	-178	1.0946 (20/4°)	20	1.5363	see Table No. 2	see Table No. 2	
C_7H_4BrNO	127	44					Reactivity with 2-ethylhexanol (520)	(520)	
C_7H_4BrNO			37-41				Infrared spectra (422)	Isopropyl-urethane MP 105° Dimer MP 187-192°	(15) (268) (281) (282) (272) (284) (422) (518)
C_8H_6BrNO	131-135	760	58-60						(353)
$C_{11}H_{13}NO_2$	139	18							(355)

No.	Name	Other Names	Structure
B-13	Benzene 1-isocyanato 3-butoxy		C_4H_9-O-C_6H_4-NCO
B-14	Benzene 1-isocyanato 4-butoxy		C_4H_9-O-C_6H_4-NCO
B-15	Benzene 1-isocyanato 4-n-butyl		C_4H_9-C_6H_4-NCO
B-16	Benzene 1-isocyanato x-t-butyl		$(CH_3)_3C$ C_6H_4-NCO
B-17	Benzene 1-isocyanato 4-cetyl		$C_{16}H_{33}$-C_6H_4-NCO
B-18	Benzene 1-isocyanato 2-chloro	o-chlorophenyl isocyanate	Cl-C_6H_4-NCO
B-19	Benzene 1-isocyanato 3-chloro	m-chlorophenyl isocyanate	Cl-C_6H_4-NCO
B-20	Benzene 1-isocyanato 4-chloro	p-chlorophenyl isocyanate	Cl-C_6H_4-NCO
B-21	Benzene 1-isocyanato 2-(1-chloroethyl)		CH_3-CH(Cl)-C_6H_4-NCO
B-22	Benzene 1-isocyanato 3-(1-chloroethyl)		CH_3-CH(Cl)C_6H_4-NCO
B-23	Benzene 1-isocyanato 3-(chloromethyl)		CH_2Cl-C_6H_4-NCO
B-24	Benzene 1-isocyanato 4-(chloromethyl)		CH_2Cl-C_6H_4-NCO
B-25	Benzene 1-isocyanato 3,5-dibromo		Br_2-C_6H_3-NCO

Formula	Boiling Point °C.	Boiling Point mm.	Melting Point °C.	Density	Refractive Index °C.	Refractive Index n	Other Data	Derivatives	References
$C_{11}H_{13}NO_2$	162- 164	41					Pharmacology and toxicity of urethanes discussed (355)	Carbamate from β-diethyl-amino-ethanol MP 124°	(355)
$C_{11}H_{13}NO_2$	110- 112	4							(355)
$C_{11}H_{13}NO$	121	17					Kinetics of reaction with 2 ethylhexanol (520)		(520)
$C_{11}H_{13}NO$	120	20			20	1.5282			(521)
$C_{23}H_{37}NO$	260- 261	15							(90) (125)
C_7H_4ClNO	 200- 203	10 760						Isopropyl carbamate BP 84° (1 mm) (268) Dimer MP 218-219° (271) Cl^{35} compound (248)	(194) (248) (265) (285) (268) (281) (518)
C_7H_4ClNO	64 83.6 111- 112 200- 203.5	2 10.5 45 760					see Table No. 3	Isopropyl carbamate BP 112-113° (1 mm)	see Table No. 3
C_7H_4ClNO	80.6- 80.9 115- 117 203- 204	9.5 45 760					see Table No. 4	Isopropyl carbamate MP 104° Methylurethane MP 115° Cl^{36} compound (249)	see Table No. 4
C_8H_6ClNO	120- 122	14							(353)
C_8H_6ClNO	116	11							(194)
C_8H_6ClNO	118- 119 116- 118	11 10						Methylurethane MP 88°	(194) (353)
C_8H_6ClNO	115- 117	10	34						(353)
$C_7H_3Br_2NO$	105- 111	0.05 to 0.10	52						(423)

No.	Name	Other Names	Structure
B-26	Benzene 1-isocyanato 2,4-dichloro		Cl_2-C_6H_3-NCO
B-27	Benzene 1-isocyanato 2,5-dichloro		Cl_2-C_6H_3-NCO
B-28	Benzene 1-isocyanato 3,4-dichloro		Cl_2-C_6H_3-NCO
B-29	Benzene 1-isocyanato 3,5-dichloro		Cl_2-C_6H_3-NCO
B-30	Benzene 1-isocyanato 4-diethylamino	Diethylaminophenyl isocyanate	(C_2H_5)N-C_6H_4-NCO
B-31	Benzene 1-isocyanato x-di-isopropyl (mixed isomers)		$[(CH_3)_2CH]_2$-C_6H_3-NCO
B-32	Benzene 1-isocyanato 2,3-dimethyl	o-xylyl-isocyanate	$(CH_3)_2$-C_6H_3-NCO
B-33	Benzene 1-isocyanato 2,4-dimethyl	m-xylyl isocyanato	$(CH_3)_2$-C_6H_3-NCO
B-34	Benzene 1-isocyanato 2,6-dimethyl		$(CH_3)_2$-C_6H_3-NCO
B-35	Benzene 1-isocyanato 3,4-dimethyl		$(CH_3)_2$-C_6H_3-NCO
B-36	Benzene 1-isocyanato 4-dimethylamino	Dimethylamino-phenyl-isocyanate	$(CH_3)_2$N-C_6H_4-NCO
B-37	Benzene 1-methyl isocyanato 4-dimethylamino	Dimethylamino-benzyl-isocyanate	$(CH_3)_2$N-C_6H_4-CH_2NCO
B-38	Benzene 1-isocyanato 2,5-dimethoxy		$(CH_3O)_2$-C_6H_3-NCO

Formula	Boiling Point °C.	Boiling Point mm.	Melting Point °C.	Density	Refractive Index °C.	Refractive Index n	Other Data	Derivatives	References
$C_7H_3Cl_2NO$			57-58.8					Isopropyl carbamate MP 80°	(282) (268)
$C_7H_3Cl_2NO$	83-84 233-234	3 to 4 760	27.4-28.8					Isopropyl Carbamate MP 70 Dimer 233-240° Urea from H_2O MP 289°	(177) (185) (265) (285) (268) (281) (371) (518)
$C_7H_3Cl_2NO$	116.7-118.1 131.8-133.2	1-10 25	45					Methyl urethane MP 111-112° Urea MP 153-154 Urea derivatives (360) (370) (412) (414) (416) (429) (432) (492)	(194) (312) (360) (370) (412) (414) (416) (429) (432) (492) (520)
$C_7H_3Cl_2NO$								Urea derivatives (412)	(412)
$C_{11}H_{14}N_2O$	104	0.01			22	1.5692	Infrared data (526)	HCl salt MP 133-135	(526)
$C_{13}H_{17}NO$	130	16							(521)
C_9H_9NO							Infrared data (422)		(422)
C_9H_9NO	212	760					Reaction with $AlCl_3$ (265) (285)		(265) (285) (518)
C_9H_9NO	87-89 9 91	12 13				1.5360 (20°)			(348) (403)
C_9H_9NO	112.8 to 113.0	25					Infrared (422) kinetics of reaction with 2 ethylhexanol (520)		(422) (520)
$C_9H_{10}N_2O$	90-92	0	39					Urea from water MP 258-259° Phenyl urea MP 207-208° Methyl urethane MP 101-103°	(18)
$C_{10}H_{12}N_2O$	119	4	24.5-25				Sublimes at 110-115° (15 mm)		(441)
$C_9H_9NO_3$	152-154	26					Infrared (371)		(194) (371)

No.	Name	Other Names	Structure
B-39	Benzene 1-isocyanato 2,4-dinitro		$(NO_2)_2$-C_6H_3-NCO
B-40	Benzene 1-isocyanato 4-dodecyl		$C_{12}H_{25}$-C_6H_4-NCO
B-41	Benzene 1-isocyanato 2-ethoxy		C_2H_5-O-C_6H_4-NCO
B-42	Benzene 1-isocyanato 3-ethoxy		C_2H_5-O-C_6H_4-NCO
B-43	Benzene 1-isocyanato 4-ethoxy		C_2H_5-O-C_6H_4-NCO
-	Benzene 1-isocyanato 3-fluoro	see Fluoro isocyanates (F-21)	
-	Benzene 1-isocyanato 4-fluoro	see Fluoro isocyanates (F-22)	
-	Benzene 1-isocyanato 3-trifluoro methyl	see Fluoro isocyanates (F-23)	
B-44	Benzene 1-isocyanato 4-hexadecyloxy	p-Cetyloxy-phenyl isocyanate	$C_{16}H_{33}$-O-C_6H_4-NCO
B-45	Benzene 1-isocyanato 4-iodo		I-C_6H_4-NCO
B-46	Benzene 1-isocyanato x-isopropyl	Cumyl isocyanate	$(CH_3)_2CH$-C_6H_4-NCO
B-47	Benzene 1-isothiocyano		C_6H_5NCS
B-48	Benzene 1-isocyanato 3-isothiocyanato		SCN-C_6H_4-NCO
B-49	Benzene 1-isocyanato 2-methoxy		CH_3O-C_6H_4-NCO

Formula	Boiling Point °C.	Boiling Point mm.	Melting Point °C.	Density	Refractive Index °C.	Refractive Index n	Other Data	Derivatives	References
$C_7H_3N_3O_5$			63					Methyl urethane MP 127° Butyl urethane MP 91°	(93)
$C_{19}H_{29}NO$	230	25							(90) (125)
$C_9H_9NO_2$	106-110	11							(250)
$C_9H_9NO_2$	107-109	6							(250)
$C_9H_2NO_2$	104-106	12							(39) (40) (194) (203) (371) (418) (4)
	113-116	16							
	142	50							
	230-235	760							
$C_{23}H_{37}NO_2$	206-210	3.5	37-39						(125) (141)
C_7H_4INO								Isopropyl carbamate MP 105° Derivatives from aldehydes and ketones (119) from phenols (120) from amines (268)	(118) (119) (120) (268)
$C_{10}H_{11}NO$	110-117	30			20	1.5214 to 1.5218			(521)
C_7H_5NS							DiPole moment 2.85 D		(392)
$C_8H_4ON_2S$	140-142	14	4-6						(194)
$C_8H_7NO_2$	94-96	7					Infrared (422)	Isopropyl Carbamate BP 127-130° (1 mm) $d^{20} = 1.1032$ $N_D^{20} = 1.5$	(194) (250) (422) (268)
	95	10							

No.	Name	Other Names	Structure
B-50	Benzene 1-isocyanato 4-methoxy		$CH_3O-C_6H_4-NCO$
B-51	Benzene 1-isocyanato 4-methoxy		$CH_3O-C_6H_4-NCO$
B-52	Benzene 1-isocyanato 2-nitro		$(NO_2)-C_6H_4-NCO$
B-53	Benzene 1-isocyanato 3-nitro		$(NO_2)-C_6H_4-NCO$
B-54	Benzene 1-isocyanato 4-nitro		$(NO_2)\ C_6H_4\ NCO$
B-55	Benzene 1-isocyanato 2-sulfonic acid, chloride		$NCO-C_6H_4-SO_2Cl$
B-56	Benzene 1-isocyanato 3-sulfonic acid, chloride		$NCO-C_6H_4-SO_2Cl$
B-57	Benzene 1-isocyanato 4-sulfonic acid, chloride		$NCO-C_6H_4-SO_2Cl$
B-58	Benzene 1-isocyanato 3-sulfonic acid, fluoride		$NCO-C_6H_4-SO_2F$
B-59	Benzene 1-isocyanato 2,4,6-tribromo		$Br_3-C_6H_2-NCO$
B-60	Benzene 1-isocyanato 2,4,5, trichloro		$Cl_3-C_6H_2-NCO$
B-61	Benzene 1-isocyanato 2,4,6-trichloro		$Cl_3-C_6H_2\ NCO$
B-62	Benzene 1-isocyanato 3,4,6-trichloro		$Cl_3-C_6H_2-NCO$
B-63	Benzene 1-isocyanato 2,4,6-triiodo		$I_3-C_6H_2-NCO$

Formula	Boiling Point °C.	Boiling Point mm.	Melting Point °C.	Density	Refractive Index °C.	Refractive Index n	Other Data	Derivatives	References
$C_8H_7NO_2$	92-93 94.7 98	7 10 18					Infrared (371) Reactivity (520)		(93) (194) (250) (371) (520)
$C_8H_7NO_2$	82 97.6-97.8 106-110 116-118	2 9.5 16 37					DiPole moment 2.98 (296) Reactivity (520)	Methyl urethane MP 88° Carbodiimide BP 193-194 (0.5 mm.) Dimer MP 171-175°	(40) (93) (194) (296) (432) (505) (524) (520) (281)
$C_7H_4N_2O_3$	135	17	40-41				DiPole Moment 5.22 D (296) Infrared (422)	See Table No. 5	(93) (107) (194) (296) (422)
$C_7H_4N_2O_3$	130-131 136-137	11 12	50-51				DiPole Moment 3.42 D (296) Infrared (422) Reactivity (520)	See Table No. 6	See Table No. 6
$C_7H_4N_2O_3$	137-138 160-162 145	11 18 14	56-57				DiPole Moment (296) Infrared (371) (422) Reactivity (520)	See Table No. 7	See Table No. 7
$C_7H_4ClNO_3S$	132-138	0.7							(354) (430)
$C_7H_4ClNO_3S$	145-150	2							(354)
$C_7H_4ClNO_3S$	132-	2						Adducts (527)	(354) (430) (527)
$C_7H_4FNO_3S$	154-156	31	25					Methyl urethane MP 93°	(194)
$C_7H_2Br_3NO$			92-94					Diurea from water MP 323°	(177) (185)
$C_7H_2Cl_3NO$			61					Isopropyl Carbamate MP 114°	(268) (329)
$C_7H_2Cl_3NO$			64-65					Ethyl urethane MP 118° Methyl urethane MP 132° Phenyl urea MP 295° Diurea MP 295°	(161) (177) (185)
$C_7H_2Cl_3NO$	124-126	11	60-61 53.5						(194) (329)
$C_7H_2I_3NO$			129				Other ureas and urethanes (310)	Methyl urethane MP 209° Ethyl urethane MP 194°	(310)

No.	Name	Other Names	Structure
B-64	Benzene 1-isocyanato 2,4,5-trimethyl		$(CH_3)_3$-C_6H_2-NCO
B-65	Benzene 1-isocyanato 2,4,6-trimethyl	Mesityl isocyanate	$(CH_3)_3$-C_6H_2-NCO
B-66	Benzene 1-isocyanato -2,4,6-trinitro		$(NO_2)_3$-C_6H_2-NCO
B-67	Benzene 1-isocyanato 4-bromo 2-methyl		(Br)(CH_3)-C_6H_3-NCO
B-68	Benzene 1-isocyanato 3-chloro x-isopropyl		(i-C_3H_7)(Cl)-C_6H_3-NCO
B-69	Benzene 1-isocyanato 2-chloro 4-nitro		(Cl)(NO_2)-C_6H_3-NCO
B-70	Benzene 1-isocyanato 4-chloro 3-nitro		(Cl)(NO_2)-C_6H_3-NCO
B-71	Benzene 1-isocyanato 4-chloro 2-nitro		(Cl)(NO_2)-C_6H_3-NCO
B-72	Benzene 1-isocyanato 4-chloro 3-sulfonic acid, chloride		(SO_2Cl)(Cl)-C_6H_3-NCO
	Benzene 1-isocyanato 4-chloro 3-trifluoromethyl	see Fluoro isocyanates (F-24)	
B-73	Benzene 1-isocyanato 3-chlorocarbonyl 5-sulfonic acid, chloride		$(SOCl_2)$(COCl)-C_6H_3-NCO
	Benzene 1-isocyanato 2-(4 chlorophenoxy) 5-trifluoromethyl	see Fluoro isocyanates (F-25)	
B-74	Benzene 1-(propyl-3-isocyanato) 3,4 dichloro		Cl_2-C_6H_3-(C_3H_6NCO)
B-75	Benzene 1-isocyanato 3,5 dinitro 4-dimethylamino		$[(CH_3)_2N-]$ $(NO_2)_2$-C_6H_2-NCO

Formula	Boiling Point °C.	Boiling Point mm.	Melting Point °C.	Density	Refractive Index °C.	Refractive Index n	Other Data	Derivatives	References
$C_{10}H_{11}NO$	100-101	12							(194)
$C_{10}H_{11}NO$	100-102 218-220	10 760							(194) (312) (348) (403)
$C_7H_2N_4O_7$			188-190					Butyl urethane MP 135°	(81) (93)
C_8H_6BrNO								Isopropyl Carbamate	(268)
$C_{10}H_{10}ClNO$	128-134	14			20	1.5410			(521)
$C_7H_3ClN_2O_3$			62					Methyl urethane MP 138-138.5°	(183)
$C_7H_3ClN_2O_3$							Infrared		(371)
$C_7H_3ClN_2O_3$			87						(295)
$C_7H_3Cl_2NO_3S$	140-151	0.4							(354) (430)
$C_8H_3Cl_2NO_4S$	170-175	0.7							(354)
$C_{10}H_9Cl_2NO$	176-178	14						Phenyl urea MP 99°	(194)
$C_9H_8N_4O_5$			101-102					Phenyl urea MP 190-191°	(259)

No.	Name	Other Names	Structure
B-76	Benzene 1-isocyanato 3,4-dimethyl 6-nitro		$(CH_3)_2(NO_2)$-$C_6H_2(NCO)$
B-77	Benzene 1-isocyanato 2,4-dimethyl 5-nitro		$(CH_3)_2(NO_2)$-C_6H_2-NCO
B-78	Benzene 1-isocyanato 4-methyl 3,5-dinitro		$(NO_2)_2(CH_3)$-C_6H_2-NCO
B-79	Benzene 1-isocyanato 4-dodecyl 2-methyl		$(C_{12}H_{25})(CH_3)$-C_6H_3-NCO
B-80	Benzene 1-isocyanato 4-dodecyl 2-nitro		$(C_{12}H_{25})(NO_2)$-C_6H_3-NCO
B-81	Benzene 1-isocyanato 4-dodecyl 3-nitro		$(C_{12}H_{25})(NO_2)$-C_6H_3-NCO
B-82	Benzene 1-isocyanato 5-isopropyl 2-methyl	p-cymene-2-isocyanate caravacryl ester	$[(CH_3)_2CH]$ (CH_3)-C_6H_3-NCO
B-83	Benzene 1-isocyanato 6-isopropyl 3-methyl	p-cymene-3-isocyanate thymylisocyanate	$[(CH_3)_2CH]$ (CH_3)-C_6H_3-NCO
B-84	Benzene 1-isocyanato x-isopropyl 2-methyl	cymyl isocyanate	$[(CH_3)_2CH]$ (CH_3) C_6H_3-NCO
B-85	Benzene 1-isocyanato 5-isothiocyano 2-methyl		$(SCN)(CH_3)$-C_6H_3-NCO
B-86	Benzene 1-isocyanato 4-methyl 3-nitro		$(CH_3)(NO_2)$-C_6H_3-(NCO)
B-87	Benzene 1-isocyanato 2-methyl 4-nitro		$CH_3(NO_2)C_6H_3$-NCO
B-88	Benzene 1-isocyanato 6-methyl 3-sulfonic acid, chloride		$(SO_2Cl)(CH_3)$-C_6H_4-NCO

Formula	Boiling Point °C.	Boiling Point mm.	Melting Point °C.	Density	Refractive Index °C.	Refractive Index n	Other Data	Derivatives	References
$C_9H_8N_2O_3$			87-87.5					Ethyl urethane MP 58°	(79) (276)
$C_9H_8N_2O_3$	110-125	2							(194)
$C_8H_5N_3O_5$								See Table No.8 50 urea derivatives	(100)
$C_{20}H_{31}NO$	192-199	45	24					Phenyl urea MP 135-137°	(125)
$C_{19}H_{28}N_2O_3$								Phenyl urea MP 98°	(125)
$C_{19}H_{28}N_2O_3$								Phenyl urea MP 103°	(125)
$C_{11}H_{13}NO$	76-77	1			22	1.5205			(517)
$C_{11}H_{13}NO$	76-77	1			22	1.5190			(517)
$C_{11}H_{13}NO$	105	14			20	1.5228			(521)
$C_9H_6N_2OS$	164	16.5						Methyl urethane 136-137°	(194)
$C_8H_6N_2O_3$	100-102	0.2							(194)
$C_8H_6N_2O_3$	168 168	23 2	75-78						(183) (194)
$C_8H_6ClNO_3$	163-169	1.5							(354) (430)

No.	Name	Other Names	Structure
B-89	Benzene 1-isocyanato 3-methyl 4-methoxy		$(CH_3O)(CH_3)\text{-}C_6H_3\text{-}NCO$
B-90	Benzene 1-isocyanato x-methyl x-nonyl (mixed isomers)		$(C_9H_{19})(CH_3)\text{-}C_6H_3\text{-}NCO$
B-91	Benzene 1-isocyanato 4-methoxy 2-nitro		$(CH_3O)(NO_2)\text{-}C_6H_3\text{-}NCO$
B-92	Benzene 1-isocyanato 2-methoxy 5-nitro		$(CH_3O)(NO_2)\text{-}C_6H_3\text{-}NCO$
B-93	Benzene 1-isocyanato 3-methoxy 5-nitro		$(CH_3O)(NO_2)\text{-}C_6H_3\text{-}NCO$
B-94	Benzene 1-isocyanato 2-methoxy 4-nitro		$(CH_3O)(NO_2)\text{-}C_6H_3\text{-}NCO$
B-95	Benzene 1-isocyanato 5-nitro 2-propoxy		$(CH_3CH_2\text{-}CH_2\text{-}O)(NO_2)C_6H_3\text{-}NCO$
B-96	Benzene 1-isocyanato 5-nitro 3-propoxy		$(CH_3CH_2CH_2O)(NO_2)C_6H_3\text{-}NCO$
B-97	Benzene 1-isocyanato 2-isopropyl 5-methyl 3-nitro		$\left[(CH_3)_2CH\right](CH_3)(NO_2)\text{-}C_6H_2\text{-}NCO$
B-98	Benzene 1,4 bis(isocyanato-ethyl)		$C_6H_4\text{-}(CH_2CH_2\text{-}NCO)_2$
B-99	Benzene 1,3 bis(isocyanato-methyl)	m-xylylene diisocyanate	$C_6H_4\text{-}(CH_2NCO)_2$
B-100	Benzene 1,4 bis(isocyanato-methyl)	p-xylylene diisocyanate	$C_6H_4(CH_2NCO)_2$
B-101	Benzene 1,3 bis(isocyanato-methyl) 5-t-butyl		$(CH_3)_3C\text{-}C_6H_3\text{-}(CH_2NCO)_2$
B-102	Benzene 1,3 bis(isocyanato-methyl) 4,6 dimethyl		$(CH_3)_2\text{-}C_6H_2\text{-}(CH_2NCO)_2$
B-103	Benzene 1,3 diisocyanato	meta-phenylene diisocyanate	$C_6H_4(NCO)_2$

Formula	Boiling Point °C.	Boiling Point mm.	Melting Point °C.	Density	Refractive Index °C.	Refractive Index n	Other Data	Derivatives	References
C_9H_9NO	119.5	17					Reactivity in terms of Hammett Equation (520)		(520)
$C_{17}H_{25}NO$	122-126	2		0.9675		1.5412			(299) (404)
$C_8H_6N_2O_4$			57.5-58.5						(439)
$C_8H_6N_2O_4$			113						(183)
$C_8H_6N_2O_4$			90-91						(439)
$C_8H_6N_2O_4$			114-116					Methyl urethane MP 146-147°	(183) (385)
$C_{10}H_{10}N_2O_4$			45.6						(439)
$C_{10}H_{10}N_2O_4$									(439)
$C_{11}H_{12}N_2O_3$	130	1	75		55	1.5425			(517)
$C_{12}H_{12}N_2O_2$	142-145	0.1							(194)
$C_{10}H_8N_2O_2$	159-162	12					Infrared (371)	Dimethyl urethane MP 110-111°	(194) (371)
$C_{10}H_8N_2O_2$	172	16	45-46					Dimethyl urethane MP 184°	(194)
$C_{14}H_{16}N_2O_2$	188	2							(359)
$C_{12}H_{12}N_2O_2$	171-173	8			24	1.5407			(503)
$C_8H_4N_2O_2$	102-104 104-106 121 108	8 12 25 15					Kinetics (421)(424) Reactivity (520)	Dimethyl urethane MP 157-157.5	(194) (217) (260) (402) (421) (424) (520)

No.	Name	Other Names	Structure
B-104	Benzene 1,4 diisocyanato	p-phenylene diisocyanate	$C_6H_4(NCO)_2$
B-105	Benzene 2,6 diisocyanato 1-(bromomethyl)		(CH_2Br)-C_6H_3-$(NCO)_2$
B-106	Benzene -2,4-diisocyanato -1-sec. butyl		sec. C_4H_9-C_6H_3-$(NCO)_2$
B-107	Benzene 2,4 diisocyanato 1-tert. butyl		t-C_4H_9-$C_6H_3(NCO)_2$
B-108	Benzene -2,4 diisocyanato 1-chloro		Cl-C_6H_3-$(NCO)_2$
B-109	Benzene 2,4-diisocyanato 1-(chloromethyl)		CH_2Cl-C_6H_3-$(NCO)_2$
B-110	Benzene -diisocyanato -dodecyl		$C_{12}H_{25}$-C_6H_3-$(NCO)_2$
B-111	Benzene 2,4-diisocyanato 1-ethyl		C_2H_5-C_6H_3-$(NCO)_2$
B-112	Benzene 2,4-diisocyanato 1-ethoxy		C_2H_5-O-C_6H_3-$(NCO)_2$
B-113	Benzene -2,4-diisocyanato -1-isobutoxy		iC_4H_9-O-C_6H_3-$(NCO)_2$
B-114	Benzene -2,4-diisocyanato -1-isopropyl		i C_3H_7-C_6H_3-$(NCO)_2$
-	Benzene 2,6-diisocyanate 1-methyl	see Toluene 2,6 diisocyanato (T-21)	
-	Benzene 3,5-diisocyanate 1-methyl	see Toluene 3,5-diisocyanate (T-22)	
B-115	Benzene 2,4-diisocyanato 1-(methane, trichloro)		CCl_3-$C_6H_3(NCO)_2$
B-116	Benzene 2,4-diisocyanato 1-methoxy		CH_3O-C_6H_3-$(NCO)_2$
B-117	Benzene 2,5-diisocyanato 1-methoxy		CH_3O-C_6H_3-$(NCO)_2$

Formula	Boiling Point °C.	Boiling Point mm.	Melting Point °C.	Density	Refractive Index °C.	Refractive Index n	Other Data	Derivatives	References
$C_8H_4N_2O_2$	110-112	12	94-96				See Table No. 9	Dimethyl urethane MP 207°	See Table No. 9
	116-118	14							
	108	10							
$C_9H_5BrN_2O_2$	112-115	0.06							(353)
$C_{12}H_{12}N_2O_2$								Trimer MP 364-367	(375)
$C_{12}H_{12}N_2O_2$	138-140	8						Trimer MP 310-311	(375) (502)
$C_8H_3ClN_2O_2$	122-124	11					Reactivity with 2-ethylhexanol (520)		(194) (520)
	128-129	10							
$C_9H_7ClN_2O_2$	152-155	11							(353)
$C_{20}H_{28}N_2O_2$	159-168	0.9							(502)
	165-175	1.5							
$C_{10}H_8N_2O_2$	124-126	12							(194) (380)
	98-101	4							
$C_{10}H_8N_2O_3$	162-164	16	56						(194)
$C_{12}H_{12}N_2O_3$	180-181	14							(194)
$C_{11}H_{10}N_2O_2$	136-140	12						Dimethyl urethane MP 166-167 Trimer MP 346-350	(194) (375)
$C_9H_3Cl_3N_2O_2$	134-140	0.3							(353)
$C_9H_6N_2O_3$	149	12	75				Reactivity (520)	Urea from water MP 241-245	(194) (327) (520)
$C_9H_6N_2O_3$			89						(194)

No.	Name	Other Names	Structure
B-118	Benzene 2,5-diisocyanato -1-nitro		NO_2-C_6H_3-$(NCO)_2$
B-119	Benzene 2,4-diisocyanato 1-nitro		NO_2-C_6H_3-$(NCO)_2$
B-120	Benzene -,diisocyanato -,nonyl		C_9H_{19}-C_6H_3-$(NCO)_2$
B-121	Benzene -diisocyanato -octyl		C_7H_{17}-C_6H_3 $(NCO)_2$
B-122	Benzene 2,4-diisocyanato 1-propoxy		C_3H_7-O-C_6H_3-$(NCO)_2$
B-123	Benzene 1,3-diisocyanato 4-sulfonic acid, chloride		SO_2Cl-C_6H_3-$(NCO)_2$
B-124	Benzene 1,4-diisocyanato 1-chloro 5-methoxy		(Cl)(CH_3O)-C_6H_2-$(NCO)_2$
B-125	Benzene 2,4-diisocyanato 1,3-dichloro		Cl_2-C_6H_2-$(NCO)_2$
B-126	Benzene 2,5-diisocyanato 1,4-dichloro		Cl_2-C_6H_2-$(NCO)_2$
B-127	Benzene 4,6-diisocyanato 1,3-dichloro		Cl_2-C_6H_2-$(NCO)_2$
B-128	Benzene -diethyl -diisocyanato		$(C_2H_5)_2$-C_6H_2-$(NCO)_2$
B-129	Benzene 2,4-diisocyanato 3,5-diethyl 1-methyl		$(CH_3)(C_2H_5)_2$-C_6H-$(NCO)_2$
B-130	Benzene 3,5-diisocyanato 4,6-diethyl 2-methyl 1-nitro		[ring closure from (1) to (6)] C(NO_2)-C(CH_3)-C(NCO)-C(C_2H_5)-C(NCO)-C$(C_2H_5$ (1) (2) (3) (4) (5) (6)
B-131	Benzene 2,5-diisocyanato 1,4-diethoxy		$(C_2H_5O)_2$-C_6H_2-$(NCO)_2$
B-132	Benzene 4,6-diisocyanato 1,3-diisopropyl		$(i\text{-}C_3H_7)_2$-C_6H_2-$(NCO)_2$
B-133	Benzene -diisocyanato -diisopropyl		$(i\text{-}C_3H_7)_2$-C_6H_2-$(NCO)_2$
B-134	Benzene 2,4-diisocyanato 1,3-dimethyl		$(CH_3)_2$-C_6H_2-$(NCO)_2$

Formula	Boiling Point °C.	Boiling Point mm.	Melting Point °C.	Density	Refractive Index °C.	Refractive Index n	Other Data	Derivatives	References
$C_8H_3N_3O_4$	165-173	4	59-61					Dimethyl urethane MP 157-158	(194)
$C_8H_3N_3O_4$	160-181	14						Dimethyl urethane MP 197-198	(194)
$C_{17}H_{22}N_2O_2$	162-169	3.5							(502)
$C_{16}H_{20}N_2O_2$	227-235	0.9							(502)
	236-240	2.6							
$C_{11}H_{10}N_2O_3$	164-165	15							(194)
$C_8H_3ClN_2O_4S$	164-170	1.2 to 2.5							(354) (430)
$C_9H_5ClN_2O_3$			134						(194)
$C_8H_2Cl_2N_2O_3$	120-126	1.5	64-65					Dimethyl urethane MP 184-185	(194)
$C_8H_2Cl_2N_2O_2$			134-137						(194)
$C_8H_2Cl_2N_2O_2$	140-146	12							(194)
$C_{12}H_{12}N_2O_2$	138-140	11							(194)
$C_{13}H_{14}N_2O_2$							Infrared (371)		(371)
$C_{13}H_{13}N_3O_4$							Infrared (371)		(371)
$C_{12}H_{12}N_2O_4$			128						(194)
$C_{14}H_{16}N_2O_2$	152-156	12	59-61						(521)
$C_{14}H_{16}N_2O_2$	148-160	12							(194)
$C_{10}H_8N_2O_2$	140-144	17							(194)

No.	Name	Other Names	Structure
B-135	Benzene 4,6-diisocyanato 1,3-dimethyl		$(CH_3)_2$-C_6H_2-$(NCO)_2$
B-136	Benzene 2,5-diisocyanato 1,4-dimethyl		$(CH_3)_2$ C_6H_2-$(NCO)_2$
B-137	Benzene 4,6-diisocyanato 1,3-dimethoxy		$(CH_3O)_2$-C_6H_2-$(NCO)_2$
B-138	Benzene 2,5-diisocyanato 1,4-dimethoxy		$(CH_3O)_2$-C_6H_2 $(NCO)_2$
B-139	Benzene 2,6-diisocyanato 4-isopropyl 1-methyl	p-cymene,-2,6-diisocyanate	$(CH_3)(i$-$C_3H_7)$-C_6H_2-$(NCO)_2$
B-140	Benzene 3,5-diisocyanato 4-isopropyl 1-methyl	p-cymene-3,5-diisocyanate	$(CH_3)(i$-$C_3H_7)$-C_6H_2-$(NCO)_2$
B-141	Benzene 2,5-diisocyanato 4-isopropyl 1-methyl	p-cymene-2,5-diisocyanate	$(CH_3)(i$-$C_3H_7)$-C_6H_2-$(NCO)_2$
B-142	Benzene 2,5-diisocyanato 1-methyl 4-methoxy		$(CH_3)(CH_3O)$-C_6H_2-$(NCO)_2$
B-143	Benzene 4,6-diisocyanato 3-methyl 1-nitro		$(CH_3)(NO_2)$ C_6H_2 $(NCO)_2$
B-144	Benzene 2,4-diisocyanato 1,3,5-triisopropyl		$(i$-$C_3H_7)_3$-C_6H-$(NCO)_2$
B-145	Benzene 3,6-diisocyanato 1,2,4,5-tetramethyl	Durene diisocyanate	C(CH_3)-C(CH_3)-C(NCO)-C(CH_3)-C(CH_3)-C(N (1) (2) (3) (4) (5) (6)
B-146	Benzene 1,3,5-triisocyarato	Phenyl-triisocyanate	$C_6H_3(NCO)_3$
B-147	Benzene 2,4,6-triisocyanato 1-chloro		Cl-C_6H_2-$(NCO)_3$
B-148	Benzene 2,4,6-triisocyanato 1-ethyl		C_2H_5-C_6H_2-$(NCO)_3$

Formula	Boiling Point °C.	Boiling Point mm.	Melting Point °C.	Density	Refractive Index °C.	Refractive Index n	Other Data	Derivatives	References
$C_{10}H_8N_2O_2$	139	15	70-71						(194)
$C_{10}H_8N_2O_2$	138-143	14							(194)
$C_{10}H_8N_2O_4$			125					Dimethyl urethane MP 154°	(194)
$C_{10}H_8N_2O_4$			180-181						(194)
$C_{12}H_{12}N_2O_2$	123-126	2	52-57						(517)
$C_{12}H_{12}N_2O_2$	110-112	2							(517)
$C_{12}H_{12}N_2O_2$	125-126	2	46-47						(517)
$C_{10}H_8N_2O_3$			106-107						(194)
$C_9H_5N_3O_4$							Infrared		(371)
$C_{17}H_{22}N_2O_2$	94-95	0.04						Dimethyl urethane MP 264 265°	(550)
$C_{12}H_{12}N_2O_2$			113.5-114						(376) (424)
$C_9H_3N_3O_3$			84-85					Triethyl-urethane MP 191-192° Tri-ureide MP < 350° Corresponding Triamine MP 84°	(198) (204)
$C_9H_2ClN_3O_3$			93.5-94.5					Triethyl urethane MP 154-155° Tri-ureide MP 288-289°	(198)
$C_{11}H_7N_3O_3$			48-49					Triethyl urethane MP 225-226° Tri-ureide MP 289-291° Corresponding triamine MP 73°	(198) (204)

No.	Name	Other Names	Structur
B-149	Benzene 2,4,6-triisocyanato 1-ethoxy		C_2H_5-O-C_6H_2-$(NCO)_3$
B-150	Benzene 2,4,6-triisocyanato 1-methyl		CH_3-C_6H_2-$(NCO)_3$
B-151	Benzene 2,4,6-trisocyanato 1-methoxy		CH_3O-C_6H_2-$(NCO)_3$
B-152	Benzene 2,4,6-triisocyanato 1-phenoxy	S-diphenylether-triisocyanato	C_6H_5-O-C_6H_2-$(NCO)_3$
B-153	Benzene 2,4,6 triisocyanato 1,3,5 trimethyl		C(CH_3)-C(NCO)-C(CH_3)-C(NCO)-C(CH_3)-C((1) (2) (3) (4) (5) (6)
B-154	Benzenesulfonic acid, isocyanate		C_6H_5-SO_2-NCO
B-155	Benzenesulfonic acid, 3-isocyanato-amide, n-methyl, n-phenyl		C_6H_4-(NCO)-SO_2-N(CH_3)C_6H_5
B-156	Benzenesulfonic acid -4-nitro -isocyanato		NO_2-C_6H_4-SO_2-NCO
B-157	Benzenesulfonic acid, 3-isocyanato 4-methoxy (phenyl-4-isocyanato ester)		(CH_3O)(NCO)-C_6H_3-SO_2-O-C_6H_4-NCO
B-158	Benzenesulfonic acid, 3-isocyanato 4-methyl (phenyl-4-isocyanato ester)		(CH_3)(NCO)-C_6H_3-SO_2-O-C_6H_4-NCC
-	Benzhydryl, p,p dimethoxy isocyanate	see Methane 1-isocyanato 1,1-dimethoxy phenyl (M-56)	
-	Benzhydryl, p-methyl isocyanate	see Methane 1 isocyanato 1-p-tolyl 1-phenyl (M-54)	

Formula	Boiling Point °C.	Boiling Point mm.	Melting Point °C.	Density	Refractive Index °C.	Refractive Index n	Other Data	Derivatives	References
$C_{11}H_7N_3O_4$			49-495					Triethyl urethane MP 134 135 Tri-ureide MP 263-264° Corresponding Triamine MP 124-125°	(198)
$C_{10}H_5N_3O_3$	133-137	1-3	78-79					Trimethyl urethane MP 195-200° Tri-ureide MP 207° Triamine MP 121-122°	(194) (198) (204)
$C_{10}H_5N_3O_4$			83-83.5					Triethyl-urethane MP 114-115° Tri-ureide MP 283-284° Triamine MP 117-118°	(198) (204)
$C_{15}H_7N_3O_4$		78-78.5						Tri ureide MP 271-272° Triamine MP 145°	(198)
$C_{12}H_9N_3O_3$	160-180	4-5	93					Triethyl urethane MP 284°	(194)
$C_7H_5NO_3S$								Adducts (527)	(527)
$C_{14}H_{12}N_2O_3S$			74-75						(194)
$C_7H_4N_2O_5S$									(407)
$C_{15}H_{10}N_2O_6S$			104-106						(194)
$C_{15}H_{10}N_2O_5S$			67.5						(194)

No.	Name	Other Names	Structure
-	Benzhydryl, p-methoxy isocyanate	see Methane 1-isocyanato 1-(p-methoxy phenyl) 1-phenyl (M-55)	
-	Benzhydryl p-nitro	see Methane 1-isocyanato 1-(p-nitro phenyl) 1-phenyl (M-54)	
B-159	Benzoic acid-amide, N(methyl isocyanate)	Hippenyl isocyanate	C_6H_5-CO-NH-CH_2NCO
B-160	Benzoic acid -4-bromo -amide N(methyl isocyanate)	p-brom hippenyl isocyanate	Br-C_6H_4-CO-NH-CH_2-NCO
B-161	Benzoic acid -2-isocyanato -1-amide, N,N'dibutyl		OCN-C_6H_4-CO-N$(C_4H_9)_2$
B-162	Benzoic acid -2-isocyanato -1-amide-N-methyl-N'-octadecyl		OCN-C_6H_4-CO-N$(CH_3)(C_{18}H_{37})$
B-163	Benzoic acid 2-isocyanato 1-(morpholinyl, amide)		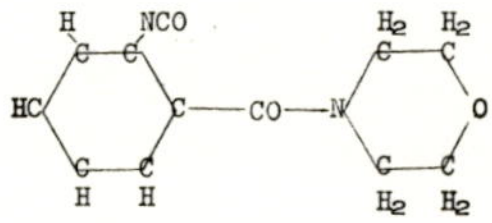
B-164	Benzoic acid -2,4-diisocyanato -1-amide,-N-methyl-N'-octadecyl		$(OCN)_2$-C_6H_3-CO-N$(CH_3)(C_{18}H_{37})$
B-165	Benzoic acid -2,5-diisocyanato -1-amide,-N,N'-dibutyl		$(OCN)_2$-C_6H_3-CO-N$(C_4H_9)_2$
B-166	Benzoic acid, chloride -3-isocyanato		(OCN)-C_6H_4-CO-Cl
B-167	Benzoic acid, chloride -4-isocyanato		(OCN)-C_6H_4-CO-Cl
B-168	Benzoic acid, chloride 3-isocyanato 4 methyl		(OCN)(CH_3)-C_6H_3-CO-Cl
B-169	Benzoic acid, 4-isocyanato dodecyl ester		(OCN)-C_6H_4-CO-O-$C_{12}H_{25}$
B-170	Benzoic acid, 4-isocyanato ethyl ester		(OCN)-C_6H_4-CO-O-C_2H_5
B-171	Benzoic acid 4-isocyanato methyl ester		(OCN)-C_6H_4-CO-O-CH_3

Formula	Boiling Point °C.	Boiling Point mm.	Melting Point °C.	Density	Refractive Index °C.	Refractive Index n	Other Data	Derivatives	References
$C_9H_8N_2O_2$								Diurea MP 246° Ethyl urethane MP 162°	(15) (17)
$C_9H_7BrN_2O_2$			114					Ethyl urethane MP 214° Methyl urethane MP 174°	(15)
$C_{16}H_{22}N_2O_2$			30						(500)
$C_{27}H_{44}N_2O_2$			62-64						(500)
$C_{12}H_{12}N_2O$			150-152						(500)
$C_{28}H_{43}N_3O_3$			57-60						(500)
$C_{17}H_{21}N_3O_3$	192-198	0.2							(500)
$C_8H_4ClNO_2$	60-66	0.01							(342)
	100-105								(396)
$C_8H_4ClNO_2$	54-55	0.07	34-38						(342)
$C_9H_6ClNO_2$	100-108	1.5							(342)
	105-112								
$C_{20}H_{29}NO_3$	203	2.5	40-42					phenyl urea MP 92-94°	(125)
$C_{10}H_9NO_3$			30-32					methyl urethane MP 153-154°	(22) (194)
$C_9H_7NO_3$	122-124	11	49					methyl urethane MP 177-178°	(194) (343)

No.	Name	Other Names	Structure
B-172	Benzoic acid 3,5-diisocyanato methyl ester		$(OCN)_2$-C_6H_3-CO-O-CH_3
B-173	Benzoic acid, chloride 5-isocyanato 3-chlorocarbonyl		OCN-C_6H_3-$(CO\text{-}Cl)_2$
B-174	Benzoic acid -5-isocyanato -2-hydroxy		(OCN)(OH)-C_6H_3-COOH
B-175	Benzoic acid -2-isocyanato nitrile		OCN-C_6H_4-CN
B-176	Benzoic acid 3-isocyanato nitrile		OCN-C_6H_4-CN
B-177	Benzophenone 3,3'-diisocyanato		OCN-C_6H_4-CO-C_6H_4-NCO
B-178	3,4-Benzopyrene-5-isocyanato		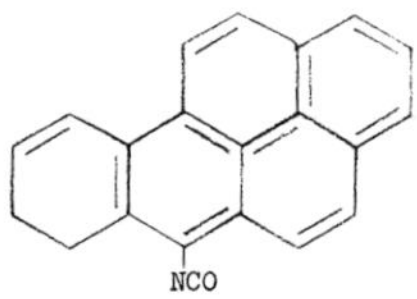
B-179	Benzothiophene -3-(propyl-2-isocyanato-2-methyl)		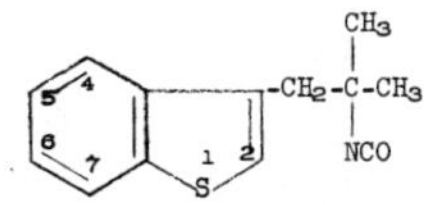
B-180	Benzothiazole 6-methyl 2-(phenyl-4-isocyanate)		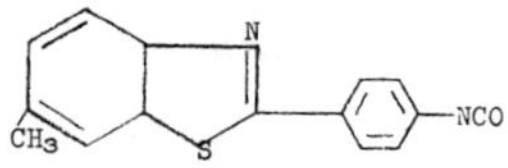
-	Benzotropylium isocyanate	see Norcaradienyl isocyanate (N-41)	
B-181	Benzoyl isocyanate		C_6H_5-CO-NCO
	Benzyl isocyanate	see Methane 1-isocyanato 1-phenyl (M 79)	

Formula	Boiling Point °C.	Boiling Point mm.	Melting Point °C.	Density	Refractive Index °C.	Refractive Index n	Other Data	Derivatives	References
$C_{10}H_6N_2O_4$	140	0.3	68						(504)
$C_9H_3Cl_2NO_3$	140-160	4-6							(342)
$C_8H_5NO_4$								methyl urethane MP 221°	(194)
$C_8H_4N_2O$	124	13	61					Ethyl urethane MP 107° Phenyl urea MP 163°	(16)
$C_8H_4N_2O$	123	12	50-51					Methyl urethane MP 97-98°	(194)
$C_{15}H_8N_2O_3$			118-120					(194)	
$C_{21}H_{11}NO$			183.5-184				UV Spectra (157)	Ethyl urethane MP 249° Urea from NH_3 MP 370°	(115) (157)
$C_{13}H_{13}NOS$	170	17							(201)
$C_{15}H_{10}N_2OS$			143-144					Methyl urethane MP 203-205°	(194)
$C_8H_5NO_2$	73	6						Benzyl carbamate MP 115° Isopropyl carbamate MP 99-100° Benzyl urea MP 165°	(5) (108) (279) (297) (358) (505)
	88-93	10							
	88-96	20							
	202-204	724							

No.	Name	Other Names	Structure
B-182	Bicyclo(2,2,1,)heptane, -3-isocyanato	β-Norbornyl isocyanate	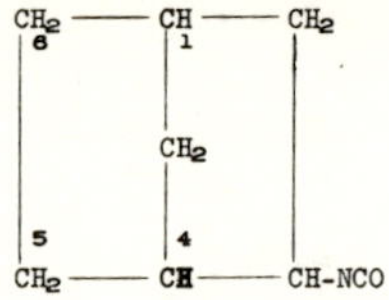
B-183	Bicyclo(2,2,1,)heptane -7-keto -1-isocyanato		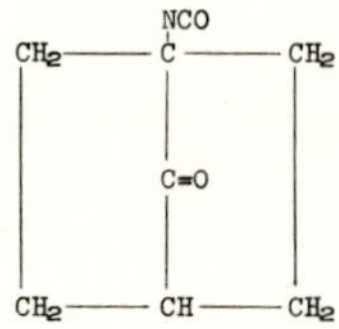
B-184	Bicyclo(2,2,1,)heptane 5-chloro 2-(2-ethyl isocyanate)		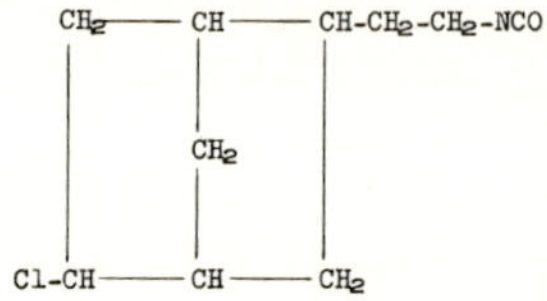
B-185	Bicyclo(2,2,1,)heptane -2-(1-ethyl-isocyanato)		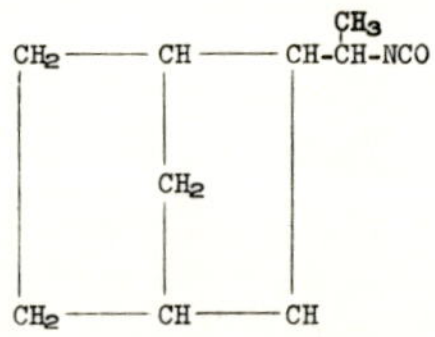
B-186	Bicyclo(2,2,1,) heptane 2-(2-ethyl) isocyanate		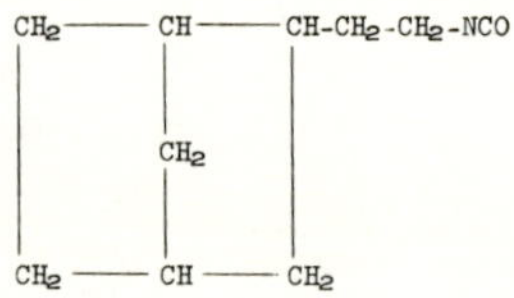
B-187	Bicyclo(2,2,1,)heptene -2-endo-methyl -2-exo-carbonyl isocyanate		(Structure not given)
-	o-Bicyclohexyl-isocyanate	see Cyclohexane 1-cyclohexyl 2-isocyanato	
B-188	Bicyclo(2,2,2,)oct-5-ene -2-isocyanato		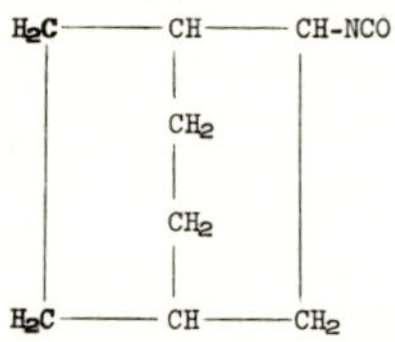
-	Bifluorene	see 9,9'-spirobifluorene -2-isocyanato (F-10)	

Formula	Boiling Point °C.	Boiling Point mm.	Melting Point °C.	Density	Refractive Index °C.	Refractive Index n	Other Data	Derivatives	References
$C_8H_{11}NO$	71-72	9						1) Sym urea MP 257-259 2) Norbornyl Amine BP 156-157° MP 75-80° 3) Norbornyl Amine **-Hydrochloride** MP 260°	(72)
$C_8H_9NO_2$							Infrared Spectra (529)		(529)
$C_{10}H_{14}ClNO$	132-134	5							(510)
$C_{10}H_{15}NO$	78	5							(510)
$C_{10}H_{15}NO$	91-92	5							(510)
	106-109	30							(479)
$C_9H_{11}NO$	101-103	25						Methyl urethane $C_{10}H_{15}NO_2$ MP 107-108°	(384) (31)

No.	Name	Other Names	Structure
B-189	1,1'-Binaphthyl 2,2'-diisocyanato		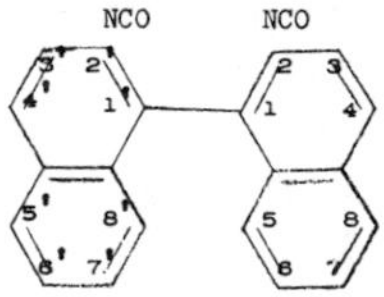 $OCN\text{-}C_{10}H_6\text{-}C_{10}H_6\text{-}NCO$
B-190	Biphenyl 2-isocyanato		$C_6H_5\text{-}C_6H_4\text{-}NCO$
B-191	Biphenyl 4-isocyanato	p-Xenyl -isocyanate	$C_6H_5\text{-}C_6H_4\text{-}NCO$
B-192	Biphenyl -4-isocyanato -4'-iodo		$I\text{-}C_6H_4\text{-}C_6H_4\text{-}NCO$
B-193	Biphenyl -4,4'-di(propyl-3-isocyanato)		$C_6H_4(CH_2CH_2CH_2\text{-}NCO)\text{-}C_6H_4(CH_2CH_2CH_2NCO)$
B-194	Biphenyl -2,4'diisocyanato		$OCN\text{-}C_6H_4\text{-}C_6H_4\text{-}NCO$
B-195	Biphenyl -4,4'-diisocyanato		$OCN\text{-}C_6H_4\text{-}C_6H_4\text{-}NCO$
B-196	Biphenyl -4,4'diisocyanato -2-nitro		$C_6H_3(NCO)(NO_2)\text{-}C_6H_4(NCO)$
B-197	Biphenyl 3,3'-dichloro 4,4'-diisocyanato		$C_6H_3(NCO)(Cl)\text{-}C_6H_3(NCO)(Cl)$
B-198	Biphenyl 3,3'di(chloromethoxy) 4,4'-diisocyanato		$C_6H_3(NCO)(OCH_2Cl)\text{-}C_6H_3(NCO)(OCH_2Cl)$
B-199	Biphenyl 3,3'-diethoxy 4,4'-diisocyanato		$C_6H_3(NCO)(O\text{-}C_2H_5)\text{-}C_6H_3(NCO)(OC_2H_4)$
B-200	Biphenyl -2,2'dimethyl -4,4'-diisocyanato		$C_6H_3(NCO)(CH_3)\text{-}C_6H_3(NCO)(CH_3)$
B-201	Biphenyl -3,3'-dimethyl 4,4'-diisocyanato		$C_6H_3(NCO)(CH_3)\text{-}C_6H_3(NCO)(CH_3)$
B-202	Biphenyl -2,2'dimethoxy -4,4'-diisocyanato		$C_6H_3(NCO)(OCH_3)\text{-}C_6H_3(NCO)(OCH_3)$

Formula	Boiling Point °C.	Boiling Point mm.	Melting Point °C.	Density	Refractive Index °C.	Refractive Index n	Other Data	Derivatives	References
$C_{22}H_{12}N_2O_2$			132-					Dimethyl urethane MP 185 187°	(194)
$C_{13}H_9NO$	92-94 100	1 0.5					Di Pole Moment 2.19 D	Diurea MP 182°	(289) (292) (392)
$C_{13}H_9NO$	137-140 283	2-3 760	55-60				Infrared (386)(422)	See Table No. 10 Trimer MP 374°(386) Semicarbazides (71)	See Table No. 10
$C_{13}H_8INO$			100-101					Methyl urethane MP 191° Ethyl urethane MP 200-200.5° Other urethanes (41)(43)(57)	(41) (35) (43) (57)
$C_{20}H_{20}N_2O_2$	230-250	2-3						Dimethyl urethane MP 156-157°	(194)
$C_{14}H_8N_2O_2$	140-150	(0.5)						Dimethyl urethane MP 171-172°	(194)
$C_{14}H_8N_2O_2$	150-160	0.5	124 107-109					Methyl urethane MP 244° Phenyl urea MP 335°	(194) (217) (260)
$C_{14}H_7N_3O_4$			119-121					Dimethyl urethane	(194)
$C_{14}H_6Cl_2N_2O_2$			165-166						(397)
$C_{16}H_{10}Cl_2N_2O_4$			198-200						(353)
$C_{18}H_{16}N_2O_4$			126				Infrared (371)		(397) (371)
$C_{16}H_{12}N_2O_2$	170-175	0.5 to 0.8		1.1756 (20°)		1.6042 (20°)			(423)
$C_{16}H_{12}N_2O_2$	160-170	0.5	68-70					Dimethyl urethane MP 208°	(194)(217) (397)(421)
$C_{16}H_{12}N_2O_4$			137						(397)

No.	Name	Other Names	Structure
B-203	Biphenyl -3,3'-dimethoxy -4,4'-diisocyanato		$C_6H_3(NCO)(OCH_3)$-$C_6H_3(NCO)(OCH_3)$
B-204	Biphenyl -3,3'diphenoxy -4,4'diisocyanato		$C_6H_3(NCO)(OC_6H_5)$-$C_6H_3(NCO)(OC_6H_5)$
B-205	Biphenyl 2,2',5,5'tetramethoxy 4,4'diisocyanato		$C_6H_2(NCO)(OCH_3)_2$-$C_6H_2(NCO)(OCH_3)_2$
B-206	Biphenyl 2,4,4'triisocyanato		$C_6H_3(NCO)_2$-$C_6H_4(NCO)$
B-207	Bornyl isocyanate		
B-208	Butane 1-isocyanato	Butyl isocyanate	$CH_3CH_2CH_2CH_2$-NCO
B-209	Butane 2-isocyanato	sec. Butyl isocyanate	CH_3-CH_2-CH(NCO)-CH_3
B-210	Butane 2-isocyanato 1-chloro		CH_3-CH_2-CH(NCO)-CH_2Cl
B-211	Butane 1-isocyanato 4-chloro		Cl-CH_2-CH_2-CH_2-CH_2-NCO
-	Butane 1-isocyanato 4-fluoro	See Fluoroisocyanates (F-26)	
B-212	Butane 1-isocyanato 4-isothiocyanato		OCN-$(CH_2)_4$-NCS
B-213	Butane 1-isocyanato 3-methyl	Isopentyl -isocyanate	CH_3-CH(CH_3)-CH_2-CH_2-NCO
B-214	Butane 2-isocyanato 3,3-dimethyl		CH_3-CH(NCO)-C$(CH_3)_3$
B-215	Butane 3-isocyanato 3,4-diphenyl		C_6H_5-CH_2-C(NCO)(C_6H_5)-CH_2-CH_3
B-216	Butane 1-isocyanato 3,3-dinitro		CH_3-C-$(NO_2)_2$-CH_2-CH_2-NCO
-	Butane 1-isocyanato nonafluoro	See Fluoroisocyanates (F-35)	

Formula	Boiling Point °C.	Boiling Point mm.	Melting Point °C.	Density	Refractive Index °C.	Refractive Index n	Other Data	Derivatives	References
$C_{16}H_{12}N_2O_4$	200-210	0.5	121-122				Infrared (371)	Dimethyl urethane MP 215-216°	(194)(217)(371)(397)
$C_{26}H_{16}N_2O_4$			154						(397)
$C_{18}H_{16}N_2O_6$			199-200						(397)
$C_{15}H_7N_3O_3$								Trimethyl urethane MP 204-205°	(194)
$C_{11}H_{17}NO$	114-116	14	69				$[\alpha]_D = 49.14$ (Benzene solution)		(6)
C_5H_9NO	114-116	760		0.880	1.4060 (20°)		Infrared (371)	Urea from NH_3 MP 95-96° Phenyl urea	(194)(207)(235)(267)(262)(324)(352)(371)(505)
C_5H_9NO	101-102	760		0.8804 (20°/4°)	1.3989 (20°)		Infrared (371)	Phenyl urea MP 154°	(194)(207)(371)
C_5H_8ClNO	152-154				1.4469 (17.6)			Phenyl urea MP 128° Benzyl urea MP 106° 2,6-dimethyl-aniline urea MP 190°	(466)
C_5H_8ClNO	70-71	11							(194)
$C_6H_8N_2OS$	96-100	0.5							(451)
$C_6H_{11}NO$	132	760						Urea from water MP 46° Phenyl urea MP 116°	(38)
$C_7H_{13}NO$	40	12					Infrared (371)	Phenyl urea MP 177-178°	(194)(371)
$C_{17}H_{17}NO$	187.5	17						Phenyl urea MP 179°	(42)
$C_5H_7N_3O_5$	106-110	0.001	-1		1.4708 (25°)				(463)(553)

No.	Name	Other Names	Structure
-	Butane 1-isocyanato 1,1-dihydro heptafluoro	See Fluoroisocyanates (F-45)	
B-217	Butane 1-isocyanato 3-nitro 3-methyl		CH_3-C(CH_3)(NO_2)CH_2-CH_2NCO
B-218	Butane 1-isocyanato 1,1-dimethyl 4-phenyl		C_6H_5-CH_2-CH_2-CH_2-C(CH_3)$_2$-NCO
B-219	Butane 2-isocyanato 2-methyl 4-(1-naphthyl)		1-$C_{10}H_7$-CH_2-CH_2-C(CH_3)$_2$-NCO
B-220	Butane 3-isocyanato 3-methyl 4-(1-naphthyl)		1-$C_{10}H_7$-CH_2-C(CH_3)(NCO)-CH_2-CH_3
B-221	Butane 3-isocyanato 3-methyl 4-(2-naphthyl)		2-$C_{10}H_7$-CH_2-C(CH_3)(NCO)-CH_2-CH_3
B-222	Butane 3-isocyanato 3-methyl 4-(1-naphthyl-2-methyl)		1-$C_{10}H_6$(CH_3)-CH_2-C(CH_3)(NCO)-CH_2-CH_3
B-223	Butane 3-isocyanato 3-methyl 4-(1-naphthyl-4-methyl)		1-$C_{10}H_6$(CH_3)-CH_2-C(CH_3)(NCO)-CH_2CH_3
B-224	Butane 3-isocyanato 3-methyl 4-(2-naphthyl-1-tert.butyl)		2-$C_{10}H_6$(C_4H_9)-CH_2-C(CH_3)(NCO)CH_2-CH_3
B-225	Butane 1,4-diisocyanato	Tetramethylene diisocyanate	OCN-(CH_2)$_4$-NCO
B-226	Butane 1,4-diisocyanato 2-methyl		OCN-CH_2-CH(CH_3)-CH_2-CH_2-NCO
B-227	Butane 2,3-diisocyanato 2,3-dimethyl		OCN-C(CH_3)$_2$-C(CH_3)$_2$-NCO
B-228	Dutane 1,4-di(1-propoxy-3-isocyanato)	1,4-Butylene glycol dipropyl ether W,W'	OCN-(CH_2)$_3$-O-(CH_2)$_4$-O-(CH_2)$_3$-NCO
B-229	Butane 1,2,2,triisocyanato		CH_3-CH_2-C(NCO)$_2$-CH_2-NCO
B-230	Butanoic Acid 2-isocyanato 4-(methyl-thio) ethyl ester		CH_3-S-CH_2-CH_2-CH(NCO)-CO-O-C_2H_5

	Formula	Boiling Point °C.	Boiling Point mm.	Melting Point °C.	Density	Refractive Index °C.	Refractive Index n	Other Data	Derivatives	References
	$C_6H_{10}N_2O_3$	68	0.005			1.4520 (25°)				(463)
	$C_{13}H_{17}NO$	142-143 154	12 18			1.5236 (22°)			Corresponding amine BP 123-125° (12 mm)	(158) (406)
	$C_{16}H_{17}NO$	225-230	17							(138)
	$C_{16}H_{17}NO$	235-240	48							(138)
	$C_{16}H_{17}NO$	160-165	2						Corresponding amine and other derivatives (132)	(132)
	$C_{17}H_{19}NO$	155-165	1.5 to 2.0							(132)
	$C_{17}H_{19}NO$	218-220	19							(138)
	$C_{20}H_{25}$-NO	225-230	17							(138)
	$C_6H_8N_2O_2$	102	12						Dimethyl urethane MP 129-130° Diurea MP 172-173°	(194) (391) (401) (454)
	$C_7H_{10}N_2O_2$	98-99	9						Diurea MP 181-182	(194)
	$C_8H_{12}N_2O_2$	135	25							(307)
	$C_{12}H_{20}N_2O_4$	130-135	0.1							(194)
	$C_7H_7N_3O_3$	ND						Explodes on distillation		(23)
	$C_8H_{13}NO_3S$	98	1			1.4735 (25°)				(537)

No.	Name	Other Names	Structure
B-231	Butanoic Acid 2-diphenyl 4-isocyanato nitrile		$OCN-CH_2CH_2-C(C_6H_5)_2-CN$
B-232	1-Butene -2-isocyanato	Ethyl vinyl isocyanate	$CH_2=C(NCO)-CH_2-CH_3$
B-233	2-Butene 1-isocyanato		$CH_3-CH=CH-CH_2-NCO$
B-234	3-Butene 1-isocyanato 3-methyl		$CH_2=C(CH_3)-CH_2-CH_2-NCO$
B-235	2-Butene 1,4-diisocyanato		$OCN-CH_2-CH=CH-CH_2-NCO$
B-236	2-Butenoic Acid, 1-(2-ethyl isocyanato ester)	Crotonic acid, 1-(2-ethyl isocyanato ester)	$CH_3-CH=CH-CO-O-CH_2CH_2NCO$
-	tert-Butyl isocyanate	See Propane 2-isocyanato 2-methyl (P-57)	
-	Butyl Vinyl isocyanate	See N-Hexene -2-isocyanato (H-35)	
-	1,4-Butylene glycol -dipropyl ether W,W'	See Butane 1,4 di)1-propoxy-3-isocyanato) (B-228)	
B-237	Butyric acid 1-isocyanato 2-ethyl		$CH_3-CH_2-CH(C_2H_5)CO-NCO$
B-238	Butyric acid 1-isocyanato 2,2-dimethyl		$CH_3-CH_2-C(CH_3)_2-CO-NCO$
B-239	Butyric acid 1-isocyanato 2,2-diethyl		$CH_3-CH_2-C(C_2H_5)_2-CO-NCO$
B-240	Butyric acid 2-isocyanato ethyl ester		$CH_3-CH_2-CH(NCO)-CO-O-C_2H_5$
B-241	Butyric acid 3-isocyanato ethyl ester		$CH_3-CH(NCO)-CH_2-CO-O-C_2H_5$
B-242	Butyric acid 2-isocyanato 3-methyl ethyl ester		$CH_3-CH(CH_3)-CH(NCO)-CO-O-C_2H_5$

	Formula	Boiling Point °C.	Boiling Point mm.	Melting Point °C.	Density	Refractive Index °C.	Refractive Index n	Other Data	Derivatives	References
	$C_{17}H_{14}N_2O$	180-190	0.1							(433)
	C_5H_7NO	88	760					Infrared (523)	Urea from NH_3 MP 61-62° Phenyl urea MP 109°	(484) (523)
	C_5H_7NO	93	760						Methyl allophanate MP 86.5°C	(399)
	C_6H_9NO									
	$C_6H_6N_2O_2$	ND	-					Unstable	Dimethyl urethane	(194)
	$C_7H_9NO_3$	99-101	9							(394)
	$C_7H_{11}NO_2$	59-61	31							(108)
	$C_7H_{11}NO_2$	65-70	10							(108) (153)
	$C_9H_{15}NO_2$	42-45	2							(546)
	$C_7H_{11}NO_3$	81 64-67	13 4							(240) (381)
	$C_7H_{11}NO_3$	75-80	0.10							(417)
	$C_8H_{13}NO_3$	87 90 94	11 15 20							(240)

No.	Name	Other Names	Structure
C-1	Camphelyl dimethyl isocyanate		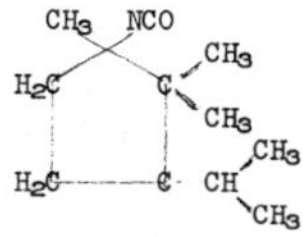
C-2	Camphenyl-1-isocyanate		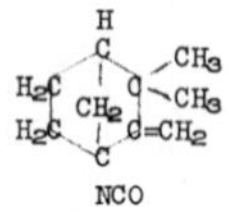
C-3	Apocamphane-7-isocyanate (ℓ-⇑)		
C-4	Apocamphor-7 yl-isocyanate (d trans ⇑)		
—	Caproicacid-2-isocyanato ethyl ester	see (DL) Hexanoic acid 2-isocyanato ethyl ester (H-30)	
-	Caproic acid 6-isocyanato methyl ester	see Hexanoic acid 6-isocyanato methyl ester (H-33)	
-	Caproic acid α butyl, hydroxy ethyl ester	see Hexanoic acid -2 isocyanato -2 butyl (H31) ethyl ester	
C-5	Carbamic acid -N,N-di(4-phenyl isocyanate) -chloride		$(C_6H_4NCO)_2$-N-CO-Cl
C-6	Carbamic acid-N(1-phenyl-2-methyl-5-isocyanato) -methyl ester	Mono-methyl carbamate of 2,4 toluene diisocyanate	$C_6H_3(CH_3)(NCO)$-NHCO-O-CH_3
C-7	Carbamic acid-N(1-phenyl-2-methyl-5-isocyanato) -ethyl ester	Mono-ethyl carbamate of 2,4-toluene diisocyanate	$C_6H_3(CH_3)(NCO)$-NH-CO-O-C_2H_5
-	4-Carbododecyl -phenyl -isocyanate	see Benzoic acid 4-isocyanato dodecyl ester (B-169)	
-	4-Carboethoxy -phenyl -isocyanate	see Benzoic acid 4-isocyanato ethyl ester (B-170)	
-	4-Carbomethoxy -phenyl -isocyanate	see Benzoic acid 4-isocyanato methyl ester (B-171)	

Formula	Boiling Point °C.	mm.	Melting Point °C.	Density	Refractive Index °C.	n	Other Data	Derivatives	References
$C_{12}H_{21}NO$	109-112	14		0.9470 (20°)			$[\alpha]^{24}_{D} = +16.5°$ in alcohol	Di-urea M.P. 154°	(65)
$C_{11}H_{15}NO$			234						(97)
	193-198	7							(368)
	94-96						$[\alpha]^{11}_{D} = 30.1°$ 2% in EtOH $\alpha^{21}_{D} = 28.76°$		(314) (211) (171)
	225	1	77						(401)
$C_{15}H_{8}ClN_{3}O_{3}$									
$C_{10}H_{10}N_{2}O_{3}$			87					phenyl urea M.P. 190°	(194)
$C_{11}H_{12}N_{2}O_{3}$			72					phenyl urea M.P. 189°	(194)

No.	Name	Other Names	Structure
C-8	Carbazole 3-isocyanato 9-ethyl		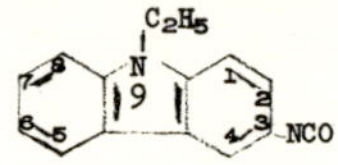
C-9	Carbazole 3,6-diisocyanato 9 ethyl		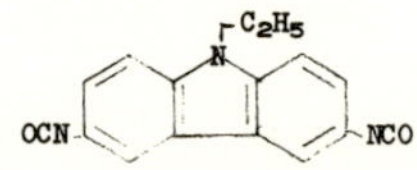
C-10	Carbazole -9(3-propyl isocyanato)		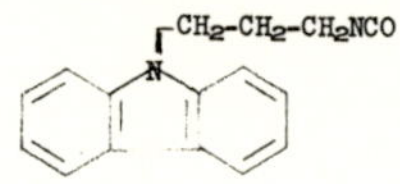
-	Caravacryl ester	see Benzene, -1-isocyanato, -2-methyl-5-isopropyl (B-82)	
-	Chaulmogryl acid	see Homohydrocarpyl Isocyanic acid (H-37)	
C-11	Cholest-5-ene- -3α-yl-isocyanate		See reference
C-12	Cholest-5-ene- -3β-yl-isocyanate		See reference
C-13	Chrysene 2-isocyanato		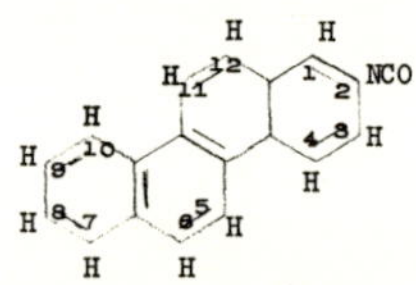
C-14	Chrysene -12,-isocyanato		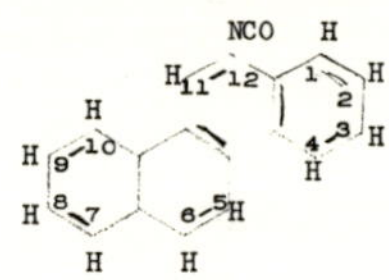
C-15	Chrysene 2,8-diisocyanate		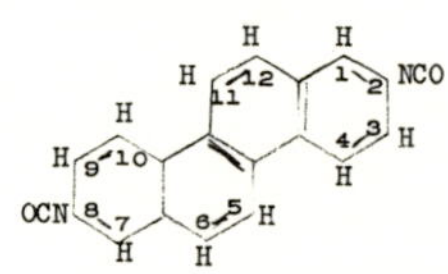
C-16	Coumarin -6-isocyanato		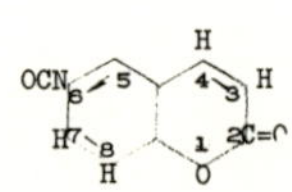

	Formula	Boiling Point °C.	Boiling Point mm.	Melting Point °C.	Density	Refractive Index °C.	Refractive Index n	Other Data	Derivatives	References
	$C_{15}H_{12}N_2O$	172-178	0.1	48					methyl urethane M.P. 118-120°	(194)
	$C_{16}H_{11}N_3O_2$			122-123						(194)
	$C_{16}H_{14}N_2O$	240-242	7						methyl urethane M.P. 107-109°	(194)
										(326) (303)
										(303) (326)
	$C_{19}H_{11}NO$							Infrared data		(371)
	$C_{19}H_{11}NO$			155-156					Methyl urethane M.P. 204°	(194)
	$C_{20}H_{10}N_2O_2$			274						(194)
	$C_{10}H_5NO_3$			163					Ethyl urethane M.P. 156° phenyl urea M.P. 225-226°	(52)

No.	Name	Other Names	Structure
C-17	Coumarin -6-isocyanato -4,-7-dimethyl		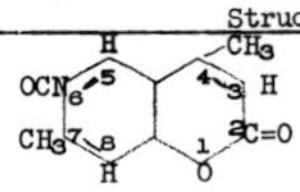
C-18	Coumarin -7-isocyanato -3-phenyl		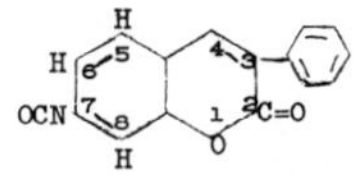
C-19	Coumarin -7-isocyanato -3-(p-methyl phenyl)		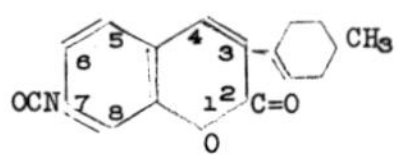
C-20	Coumarin -7-isocyanato -3-(p-chloro-phenyl)		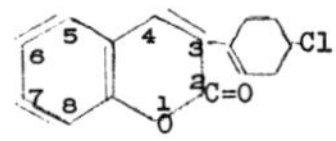
-	Crotonic acid 2-isocyanato ethyl ester	see 2-Butenoic acid -(2-isocyanato ethyl)ester (B-236)	
-	Cumyl isocyanate	see Benzene 1-isocyanato x-isopropyl (B-46)	
-	Cyano ethyl isocyanate	see Ethane 1-cyano 2-isocyanato (E-8)	
-	Cyano pentyl isocyanate	see Pentane 1-cyano 5-isocyanato (P-9)	
-	Cyano propyl isocyanate	see Propane 1-cyano 3-isocyanato (P-60)	
-	o-Cyano phenyl isocyanate	see Benzoic acid 2-isocyanato nitrile (B-175)	
-	m-Cyano phenyl isocyanate	see Benzoic acid 3-isocyanato nitrile (B-176)	

Formula	Boiling Point °C.	Boiling Point mm.	Melting Point °C.	Density	Refractive Index °C.	Refractive Index n	Other Data	Derivatives	References
$C_{12}H_9NO_3$			215					Ethyl urethane M.P. 196° Urea from NH_3 M.P. 250° Diurea M.P. 328-330° phenyl urea M.P. 265-266°	(52)
$C_{16}H_9NO_3$			233				Fluorescent compounds	Methyl urea M.P. 320° Ethyl urea M.P. 315° Propyl urea M.P. 233° Dimethyl urea M.P. 247°	(389) (472) (485) (548)
$C_{17}H_{11}NO_3$			214						(389) (485) (548)
$C_{16}H_8ClNO_3$			225						(389) (485) (548)

No.	Name	Other Names	Structure
C-21	Cyclobutane 1,3-diisocyanato 2,4-diphenyl	α-truxillic isocyanate	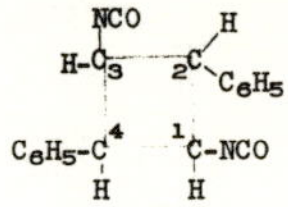
C-22	Cycloheptadiene- -6-isocyanato -(dibenzo [ac] -2,3,9,10- tetramethoxy)		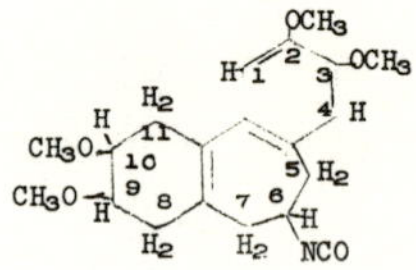
C-23	1,3,5-Cycloheptatrienyl -1-isocyanato		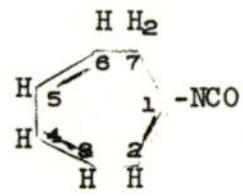
C-24	1,3,6-Cycloheptatrienyl -1-isocyanato		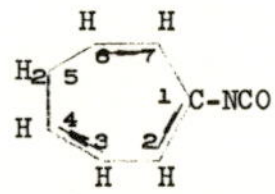
C-25	2,4,6-Cycloheptatrienyl -1-isocyanato		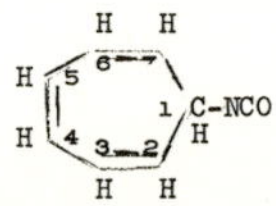
C-26	Cyclohexane 1-isocyanato		CH_2-CH_2-CH_2-CH_2-CH_2-CH-NCO
C-27	Cyclohexane- -hexachloro -1-isocyanato		CHCl-CHCl-CHCl-CHCl-CHCl-CCl(NCO)
C-28	Cyclohexane- -1-isocyanato -2-hydroxy		OCN-CH-CH(OH)-CH_2-CH_2-CH_2-CH_2
C-29	Cyclohexane -1-isocyanato 2-iodo		CH_2-CH_2-CH_2-CH_2-CHI-CH(NCO)
C-30	Cyclohexane -1-isocyanato 4-isothiocyano		CH_2-CH_2-CH(NCS)-CH_2-CH_2-CH(NCO)

Formula	Boiling Point °C.	Boiling Point mm.	Melting Point °C.	Density	Refractive Index °C.	Refractive Index n	Other Data	Derivatives	References
$C_{18}H_{14}N_2O_2$	ND		ND				Infrared data	Ethyl diurethane M.P. 157-159°	(333)
$C_{20}H_{25}NO_5$	Not iso-lated								(374)
C_8H_7NO	61	6						Urea from NH_3 M.P. 140° Phenyl urea M.P. 183-184°	(388)
C_8H_7NO	70.5-72	10						Urea from NH_3 M.P. 109 Phenyl urea M.P. 184	(388)
C_8H_7NO	67	6			1.5455 (20°)		Infrared data	Urea from NH_3 M.P. 126-127° Phenyl urea M.P. 185°	(388) (339)
$C_7H_{11}NO$	54 63-68 165-168	10 18 760		0.9359	25	1.4514	Infrared (371)	See Table No. 11	See Table No. 11
$C_7H_5Cl_6NO$	decomp	0.1					Infrared		(491)
$C_7H_{11}NO_2$	decomp						Decomposes violently	Ethyl urethane M.P. 105°	(212)
$C_7H_{10}INO$	94	2	80 (FP)					Urea from NH_3 M.P. 155°	(47)
$C_8H_{10}N_2OS$	116-120	0.8							(451)

No.	Name	Other Names	Structure
C-31	Cyclohexane -1-(phenyl-4-isocyanate)		C_6H_{11}-C_6H_4-NCO
C-32	Cyclohexane -1-(phenyl-3-methyl-4-isocyanate)		C_6H_{11}-$C_6H_3(CH_3)(NCO)$
C-33	Cyclohexane -1-cyclohexyl 2-isocyanato	o-Bicyclohexyl isocyanate	C_6H_{11}-CH-CH(NCO)-CH_2-CH_2-CH_2
C-34	Cyclohexane 1,3-diisocyanato		$C_6H_{10}(NCO)_2$
C-35	Cyclohexane 1,4-diisocyanato		$C_6H_{10}(NCO)_2$
C-36	Cyclohexane 2,4-diisocyanato 1-ethyl		$C_6H_9(C_2H_5)(NCO)_2$
C-37	Cyclohexane 2,4-diisocyanato 1-methyl		$C_6H_9(CH_3)(NCO_2)$
C-38	Cyclohexane 1-(phenyl-4-isocyanato) 4-isocyanato		OCN-C_6H_{10}-C_6H_4-NCO
C-39	Cyclohexane 1,1-di(phenyl-4-isocyanato)		C_6H_{10}-$(C_6H_4NCO)_2$
C-40	Cyclohexane 1,2-di(methyl isocyanato)		$C_6H_{10}(CH_2NCO)_2$
C-41	Cyclohexane 1,4-di(methyl isocyanato)		$C_6H_{10}(CH_2NCO)_2$
C-42	Cyclohexane 1-methyl isocyanate 2-(propyl-3-isocyanate) 3,5-dimethyl		$C_6H_8(CH_3)_2(CH_2NCO)(CH_2CH_2CH_2NCO)$
C-43	1-Cyclohexene 1-isocyanato		OCN-C=CH-CH_2-CH_2-CH_2-CH_2 1 2 3 4 5 6
C-44	1-Cyclohexene 1-isocyanato 2-methyl		OCN-C=C(CH_3)-CH_2-CH_2-CH_2-CH_2 1 2 3 4 5 6

Formula	Boiling Point °C.	Boiling Point mm.	Melting Point °C.	Density	Refractive Index °C.	Refractive Index n	Other Data	Derivatives	References
$C_{13}H_{15}NO$	135.5 128-130	6 4			20	1.5444			(194) (521)
$C_{14}H_{17}NO$	138-142	4							(194)
$C_{13}H_{21}NO$	89-90	0.5-1.0						Di-urea M.P. 225-228°	(136)
$C_8H_{10}N_2O_2$	120-130	14							(194)
$C_8H_{10}N_2O_2$	85-90 122-124 136	0.5 12 15						Dimethyl urethane cis M.P. 139-140° trans M.P. 264°	(194) (391) (418) (451)
$C_{10}H_{14}N_2O_2$	140-144	19							(194)
$C_9H_{12}N_2O_2$	127-129	12					Infrared cis, trans		(194) (371)
$C_{14}H_{14}N_2O_2$	155-160	0.7	38-44					Dimethyl urethane M.P. 216°	(194)
$C_{20}H_{18}N_2O_2$							Infrared data		(371)
$C_{10}H_{14}N_2O_2$	165-168	23							(194)
$C_{10}H_{14}N_2O_2$	154-156	11						Dimethyl urethane M.P. 163°	(194)
$C_{14}H_{22}N_2O_2$	168-172	4							(194)
C_7H_9NO	61-63	18						Phenyl urea M.P. 197°	(80) (170)
$C_8H_{11}NO$	73-74	18						Phenyl urea M.P. 170°	(80)

No.	Name	Other Names	Structure
C-45	4-Cyclohexene 1-(methyl isocyanate) 2-(propyl-3-isocyanate) 4,5 di-methyl		CH_2-C(CH_3)=C(CH_3)-CH_2-CH(CH_2NCO)-CH— 3 4 5 6 1 2 —($CH_2CH_2CH_2$NCO)
-	Cyclohexyl- benzyl isocyanate	see Methane 1-isocyanato 1-cyclohexyl 1-phenyl	
C-46	10-$\triangle^2$ Cyclopentenyl decyl isocyanate		
C-47	Cyclopropane 1-isocyanato 3-(3,4-dimethoxyphenyl)		$C_6H_3(CH_3O)_2$-CH-CH_2-CH-NCO
-	Cyclopropane- -phenanthrene	see Norcardiene dibenzo isocyanate (N-40)	
-	Cyclotetramethylene sulfone 3-isocyanato	see Sulfone cyclotetramethylene -3-isocyanato S-53	
-	p-Cymene 2-isocyanate	see Benzene 1-isocyanato 2-methyl 5-isopropyl (B-82)	
-	p-Cymene 3-isocyanato	see Benzene 1-isocyanato 3-methyl 6-isopropyl (B-83)	
-	p-Cymene 2,5-diisocyanato	see Benzene 1-methyl 4-isopropyl -2,5-diisocyanato (B-141)	
-	p-Cymene 2,6-diisocyanato	see Benzene 1-methyl 4-isopropyl 2,6-diisocyanato (B-139)	
-	p-Cymene 3,5-diisocyanato	see Benzene 1-methyl 4-isopropyl 3,5-diisocyanato (B-140)	
-	Cymyl isocyanate	see Benzene 1-isocyanato 2-methyl x-isopropyl (B-84)	

Formula	Boiling Point °C.	Boiling Point mm.	Melting Point °C.	Density	Refractive Index °C.	Refractive Index n	Other Data	Derivatives	References
	163-172	4							(226)
									(55)
$C_{12}H_{13}NO_3$									(255)

No.	Name	Other Names	Structure
D-1	Decalin -2-isocyanato	Naphthalene -decahydro -2-isocyanato	$C_{10}H_{17}NCO$
-	Deca,-3,8,diene tricyclo	see Tricyclodecca,3,8-diene 4-isocyanato (T-28)	
D-2	Decane 1-isocyanato		CH_3-$(CH_2)_9$-NCO
D-3	Decane 1,10-diisocyanato		OCN-$(CH_2)_{10}$-NCO
-	Decane 1-isocyanato heneicosanefluoro	see Fluoro isocyanates (F-41)	
-	Decane -1-isocyanato -1,1-dihydro -nonadecafluoro	see Fluoro isocyanates (F-51)	
D-4	Decanedioic acid 2,9-diisocyanato dimethyl ester	Dimethyl diisocyanato sebacate	CH_3-O-CO-CH(NCO)-$(CH_2)_6$-CH(NCO)-CO-O-CH_3
D-5	9-Decene 1-isocyanato		CH_2=CH-$(CH_2)_7$-CH_2-NCO
D-6	Dehydroabietane -1-isocyanato		See reference
-	1,2,5,6-Dibenzanthrene 9-isocyanato	see Benzanthrene (B-7)	
D-7	Dibenzofuran 2,7-diisocyanato		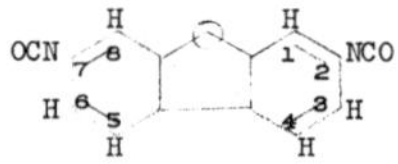
D-8	Dibenzothiophene 2,7-diisocyanato diox ide	Benzene sulfone 4,4'diisocyanato	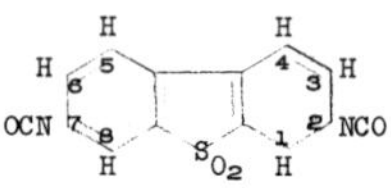
-	Diethylene glycol diphenyl 2,2'diisocyanato	see Ethane(oxy-bis) 2-(phenoxy-2-isocyanato) (E-27)	
-	Diphenyl acetyl isocyanate	see Acetic acid -1,1-diphenyl -isocyanate (A-3)	

Formula	Boiling Point °C.	Boiling Point mm.	Melting Point °C.	Density	Refractive Index °C.	Refractive Index n	Other Data	Derivatives	References
$C_{11}H_{17}NO$	116-117	12							(194)
$C_{11}H_{21}NO$	85-89	14							(450)
$C_{12}H_{20}N_2O_2$	134 128 154-157	0.54 0.8 4.5						Dimethyl urethane MP 115° Di-urea MP 189-189.5	(127) (194) (454)
$C_{14}H_{20}N_2O_6$	192-194	11							(381)
$C_{11}H_{19}NO$	56-58	0.1							(505)
$C_{20}H_{27}NO$	50-60	(0.01)					Infrared data α = 1.74° C = 2.97 benzene $[\alpha]_D^{24}$ = + 58.6°		(278)
$C_{14}H_6N_2O_3$			164						(519)
$C_{14}H_6N_2O_4S$			288-290						(194)

No.	Name	Other Names	Structure
-	Diphenyl methane sulfone 4,4'diisocyanato	see Thioxanthene -2,7-diisocyanato dioxide (T-14)	
-	S-Diphenyl ether triisocyanato	see Benzene 1-phenoxy -2,4,6-triisocyanato (B-152)	
-	Diphenyl, methyl isocyanate	see Methane -1-isocyanato -1,1-diphenyl (M-52)	
-	Diphenyl, p-tolyl, methyl isocyanate	see Methane 1-isocyanato 1,1-diphenyl 1-p-tolyl (M-58)	
D-9	Disulfide, diphenyl 4,4'diisocyanato		$OCN-C_6H_4-S-S-C_6H_4NCO$
D-10	Disulfide, diphenyl 6,6'diisocyanato 3,3'dimethyl		$OCN(CH_3)C_6H_3-S-S-C_6H_3(CH_3)NCO$
D-11	Disulfide, diphenyl 5,5'diisocyanato 4,4'dimethyl		$OCN(CH_3)C_6H_3-S-S-C_6H_3-(CH_3)NCO$
D-12	Disulfide, diphenyl 5,5'diisocyanato 2,2'dimethyl		$OCN(CH_3)C_6H_3-S-S-C_6H_3(CH_3)NCO$
D-13	Disulfide, diphenyl 3,3'diisocyanato 4,4'dimethoxy		$OCN(CH_3O)C_6H_3-S-S-C_6H_3(CH_3O)NCO$
D-14	Disulfide, diphenyl 4,4'diisocyanato 3,3'dimethoxy		$OCN(CH_3O)C_6H_3-S-S-C_6H_3(CH_3O)NCO$
D-15	Dodecane -1-isocyanato	Lauryl isocyanate	$CH_3-(CH_2)_{11}-NCO$
D-16	Dodecane- -1,-12-diisocyanato	Dodecamethylene diisocyanate	$OCN-(CH_2)_{12}-NCO$
D-17	Dodecane -2-isocyanato -2-methyl		$CH_3-(CH_2)_8-C(NCO)(CH_3)-CH_2-CH_3$
-	Durene diisocyanate	see Benzene -1,2,4,5-tetramethyl -3,6-diisocyanato (B-145)	

	Formula	Boiling Point °C.	Boiling Point mm.	Melting Point °C.	Density	Refractive Index °C.	Refractive Index n	Other Data	Derivatives	References
	$C_{14}H_8N_2O_2S_2$			58-60						(194)
	$C_{16}H_{12}N_2O_2S_2$			74-76						(194)
	$C_{16}H_{12}N_2O_2S_2$			88						(194)
	$C_{16}H_{12}N_2O_2S_2$			77-78						(194)
	$C_{16}H_{12}N_2O_4S_2$			66-67						(194)
	$C_{16}H_{12}N_2O_4S_2$			101						(194)
	$C_{13}H_{25}NO$	106-110 109-114 140-145	1 3 11							(85) (207) (127) (450) (130) (168) (194)
	$C_{14}H_{24}N_2O_2$	135 168-169	0.8 3						Dimethylurethane MP 117-118° Di-urea MP 201-204	(194) (454)
	$C_{14}H_{27}NO$	90-115	2							(242)

No.	Name	Other Names	Structure
E-1	Estrone 3-hydroxy 17-isocyanato		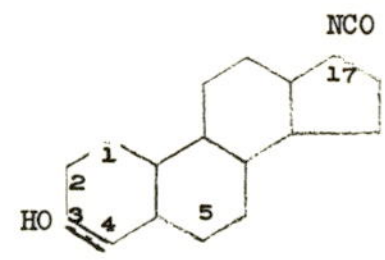
E-2	Ethane 1-isocyanato	Ethyl isocyanate	CH_3-CH_2-NCO
E-3	Ethane, 1,2 bis, (phenyl sulfonyl- 3-isocyanato- 4-methoxy)		$(CH_3O)(NCO)C_6H_3$-SO_2-CH_2-CH_2-SO_2— —$C_6H_3(CH_3O)$NCO
E-4	Ethane, 1,2 bis (phenyl thio- -3-isocyanato- 4-methoxy)		$(CH_3O)(NCO)C_6H_3$-S-CH_2-CH_2-S-$C_6H_3(CH_3O)$N
E-5	Ethane,-1,2-diamino bis N,N'(phenyl- sulfonyl-4-isocyanato)		$(NCO)C_6H_4$-SO_2-NH-CH_2-CH_2-NH-$SO_2C_6H_4$(NCO
E-6	Ethane-1,2-diamino bis N,N'(phenyl- sulfonyl-3-isocyanato- -4-methoxy)		$(CH_3O)(NCO)C_6H_3$-SO_2-NH-CH_2-CH_2-NH-SO_2— —$C_6H_3(CH_3O)$NCO
E-7	Ethane 1-isocyanato 2-chloro		$ClCH_2$-CH_2-NCO
E-8	Ethane 1-cyano 2-isocyanato		NC-CH_2-CH_2-NCO
-	Ethane 1-isocyanato 2-fluoro	see Fluoro isocyanates (F-27)	
-	Ethane 1-isocyanato 1,1 dihydra 2-trifluoro	see Fluoro isocyanates (F-43)	
-	Ethane 1-isocyanato -pentafluoro	see Fluoro isocyanates (F-33)	
E-9	Ethane 1-isocyanato 1-nitro	1-Nitroethyl isocyanate	CH_3-$CH(NO_2)$-NCO
E-10	Ethane 1-isocyanato 2-(isobutoxy-methoxy)		CH_3-$CH(CH_3)$-CH_2-O-CH_2-O-CH_2-CH_2-NCO
E-11	Ethane 1-isocyanato 2-methoxy		CH_3-O-CH_2-CH_2-NCO

Formula	Boiling Point °C.	Boiling Point mm.	Melting Point °C.	Density	Refractive Index °C.	Refractive Index n	Other Data	Derivatives	References
	ND		ND					Estrone-Protein conjugates See references	(419) (508)
C_3H_5NO	60	760		0.9039 (20°/4°)	1.3801 (20°)		See Table No. 12	See Table No. 12	See Table No. 12
$C_{18}H_{16}N_2O_8S_2$			265-267						(194)
$C_{18}H_{16}N_2O_4S_2$			118						(194)
$C_{16}H_{14}N_4O_6S_2$			178						(194)
$C_{18}H_{18}N_4O_8S_2$			192						(194)
C_3H_4ClNO	135 42 38-39	760 16 11						Di-urea MP 128-129° Phenyl urea MP 124°	(194) (84) (85) (130)
$C_4H_4N_2O$	104	12						Dicyanethyl urea MP 148°	(194)
$C_3H_4N_2O_3$	69	32				1.4317 25°	Unstable		(393)
$C_8H_{15}NO_3$	80-82	7							(168)
$C_4H_7NO_2$	123-124	760							(24)

No.	Name	Other Names	Structure
E-12	Ethane 1-isocyanato 2-phenyl	β-phenylethyl- -isocyanate	C_6H_5-CH_2-CH_2-NCO
E-13	Ethane 1-isocyanato 2-(p-n-butyl phenyl)		n-C_4H_9-C_6H_4-CH_2-CH_2-NCO
E-14	Ethane -1-isocyanato -2-(3,4-dimethoxy phenyl)		$(CH_3O)_2$-C_6H_3-CH_2-CH_2-NCO
E-15	Ethane -1-isocyanato -2-(2,3,4-trimethoxy phenyl)		$(CH_3O)_3$-C_6H_2-CH_2CH_2-NCO
E-16	Ethane -1-isocyanato -2-(3,4,5-trimethoxy phenyl)		$(CH_3O)_3$-C_6H_2-CH_2-CH_2-NCO
E-17	Ethane(trimer) 1-isocyanato 2-hydroxy 2-(3-methoxy phenyl)		$[CH_3O$-C_6H_4-CH(OH)-CH_2-NCO$]_3$
E-18	Ethane 1-isocyanato 1,1-diethyl 2-phenyl		C_6H_5-CH_2-C$(C_2H_5)_2$-NCO
E-19	Ethane 1-isocyanato 1-methyl 1-ethyl 2-phenyl		C_6H_5-CH_2-C$(CH_3)(C_2H_5)$-NCO
E-20	Ethane -1-isocyanato -1-phenyl -2,2-diphenyl	1,2,2,triphenyl ethyl isocyanate	$(C_6H_5)_2$-CH-CH(C_6H_5)-NCO
E-21	Ethane -1-isocyanato -1-methyl -2-(p-methoxy phenyl)		CH_3O-C_6H_5-CH_2-CH(CH_3)-NCO
E-22	Ethane -1-ethoxy -2-(phenoxy-2-isocyanato)		C_2H_5O-CH_2CH_2-O-C_6H_4NCO
E-23	Ethane- -1-ethoxy -2-(phenoxy-3-isocyanato)		C_2H_5O-CH_2CH_2-O-C_6H_4NCO
E-24	Ethane -1-ethoxy -2-(phenoxy-4-isocyanato)		C_2H_5-O-CH_2CH_2-O-C_6H_4NCO
E-25	Ethane -1-methoxy -2-(phenoxy-2-isocyanato)		CH_3O-CH_2CH_2-O-C_6H_4NCO

Formula	Boiling Point °C.	Boiling Point mm.	Melting Point °C.	Density	Refractive Index °C.	Refractive Index n	Other Data	Derivatives	References
C_9H_9NO	98-100 109-111 112-114	10 13 15						Phenyl urea MP 148° Urea from H_2O MP 138° Di-urea, MP 112° Ethyl carbamate MP 36° BP 145-148° (13 mm)	(11) (85) (130) (194) (241)
$C_{13}H_{17}NO$	153-156	16			1.5061 (24°)				(246) (273)
$C_{11}H_{13}NO_3$	142 174	0.8 16						Urea from NH_3 MP 157° Corresponding amine BP 153° (12 mm)	(86)
$C_{12}H_{15}NO_4$								Corresponding amine BP 160-180° (1 mm)	(151)
$C_{12}H_{15}NO_4$									(146)
$(C_{10}H_{11}NO_3)_3$			202-203						(222)
$C_{13}H_{17}NO$	142-145	18							(135) (190)
$C_{12}H_{15}NO$	137-138	16							(135) (190)
$C_{21}H_{17}NO$			122-123						(229)
$C_{11}H_{13}NO_2$	112-113 155-159	3 23			1.5168 (23°)				(246) (273)
$C_{11}H_{13}NO_3$	90-92	0.21						Corresponding amine BP 93° (0.05 mm)	(447) (442)
$C_{11}H_{13}NO_3$	123-125	0.06						Corresponding amine BP 111° (0.05 mm)	(447) (442)
$C_{11}H_{13}NO_3$	96	0.10						Corresponding amine BP 101° (0.10 mm)	(447) (442)
$C_{10}H_{11}NO_3$	82	0.07						Corresponding amine BP 110° (0.16 mm)	(447) (442)

No.	Name	Other Names	Structure
E-26	Ethane -1-methoxy -2-(phenoxy-3-isocyanato)		$CH_3O-CH_2CH_2-O-C_6H_4NCO$
E-27	Ethane, oxy bis 2-(phenoxy-2-isocyanato)	Diethylene glycol diphenyl 2,2'diisocyanate	$C_6H_4(NCO)-O-CH_2CH_2-OCH_2CH_2-O-C_6H_4NCO$
E-28	Ethane 1,2-diisocyanato	Ethylene diisocyanate	$OCN-CH_2CH_2-NCO$
E-29	Ethane -1-isocyanato -1,1-diimethyl -2-(p-isocyanato phenyl)		$OCN-C_6H_4-CH_2-C(CH_3)_2-NCO$
E-30	Ethane -1-phenyl -2-(phenyl-2 ,4 diisocyanato		$C_6H_5-CH_2-CH_2-C_6H_3(NCO)_2$
E-31	Ethane 1,2-di(phenoxy-2-isocyanato)		$OCN-C_6H_4-O-CH_2CH_2-O-C_6H_4(NCO)$
E-32	Ethane -1-isocyanato -2-(p-phenyl isocyanato)		$OCN-C_6H_4-CH_2CH_2-NCO$
E-33	Ethane -1-isocyanato -1,1-dimethyl -2,-(4-isopropyl phenyl)		$i-C_3H_7-C_6H_4-CH_2-C(CH_3)_2NCO$
E-34	Ethane -1-isocyanato -1,1-dimethyl -2-(4-methoxy phenyl)		$CH_3O-C_6H_4-CH_2-C(CH_3)_2-NCO$
E-35	Ethane -1-isocyanato -1,1-dimethyl -2-(4-propyl phenyl)		$C_3H_7-C_6H_4-CH_2-C(CH_3)_2NCO$
E-36	Ethane -1-isocyanato -1,1-dimethyl -2-(3-methyl cyclohexyl)		$(CH_3)-C_6H_{10}-CH_2-C(CH_3)_2-NCO$
E-37	Ethane -1-isocyanato -1,1-dimethyl -2-(4-methyl cyclohexyl)		$(CH_3)C_6H_{10}-CH_2-C(CH_3)_2NCO$
E-38	Ethane -1-isocyanato -1,1-dimethyl -2-(4-cyclohexyl phenyl)		$(C_6H_{11})-C_6H_4-CH_2-C(CH_3)_2-NCO$
E-39	Ethane -1-isocyanato -1,1-dimethyl -2-phenyl		$C_6H_5-CH_2-C(CH_3)_2-NCO$
E-40	Ethane -1-isocyanato -1,1-dimethyl -2-(4-tert-butyl phenyl)		$(CH_3)_3-C-C_6H_4-CH_2-C(CH_3)_2-NCO$
E-41	Ethane -1-isocyanato -1,1-dimethyl -2,-(2,4-dimethyl phenyl)		$(CH_3)_2-C_6H_3-CH_2-C(CH_3)_2-NCO$

Formula	Boiling Point °C.	Boiling Point mm.	Melting Point °C.	Density	Refractive Index °C.	Refractive Index n	Other Data	Derivatives	References
$C_{10}H_{11}NO_3$	91	0.11						Corresponding amine BP 112° (0.11 mm)	(447) (442)
$C_{18}H_{16}N_2O_5$									(194)
$C_4H_4N_2O_2$	75-76	4							(194)
$C_{12}H_{12}N_2O_2$	145-146	8						Dimer MP 168-169° Urea MP 180-181°	(307) (318)
$C_{16}H_{12}N_2O_2$	223-227	20						Dimethyl urethane MP 154-155°	(194)
$C_{16}H_{12}N_2O_4$									(194)
$C_{10}H_8N_2O_2$	156	15							(194)
$C_{14}H_{19}NO$	160	22				1.5095 (24°)			(406)
$C_{12}H_{15}NO_2$	142	15							(190) (135) (382)
$C_{14}H_{19}NO$	145-147	13				1.5108 (22°)			(480)
$C_{12}H_{21}NO$	113-114	15							(182)
$C_{12}H_{21}NO$	120	20							(182)
$C_{17}H_{23}NO$	145-150	2						Corresponding amine BP 157-159° (3 mm) MP 45°	(159)
$C_{11}H_{13}NO$	225 142-145 112-115 98-100	760 20 20 15							(124) (135) (190)
$C_{15}H_{21}NO$	160	17							(135) (190)
$C_{13}H_{17}NO$	151-152	20				1.5158 (22°)			(480)

No.	Name	Other Names	Structure
E-42	Ethane -1-isocyanato -1,1-dimethyl -2(2,5-dimethyl phenyl)		$(CH_3)_2$-C_6H_3-CH_2-C$(CH_3)_2$-NCO
E-43	Ethane -1-isocyanato -1,1-dimethyl -2-(3,4-dimethyl phenyl)		$(CH_3)_2$-C_6H_3-CH_2-C$(CH_3)_2$-NCO
E-44	Ethane -1-isocyanato -1,1-dimethyl -2-(2,6-dimethyl-4-tert.butyl-phenyl)		[$(CH_3)_3$C]-$(CH_3)_2$$C_6H_2$-$CH_2$-C$(CH_3)_2$-NCO
E-45	Ethane -1-isocyanato -1,1-dimethyl -2(2,4,6-trimethyl phenyl)		$(CH_3)_3$-C_6H_2-CH_2-C$(CH_3)_2$-NCO
E-46	Ethane -1-isocyanato -1,1-dimethyl -2-(4 ethyl phenyl)		C_2H_5-C_6H_4-CH_2-C$(CH_3)_2$-NCO
E-47	Ethene, -isocyanate	Vinyl isocyanate	CH_2=CH-NCO
E-48	Ethene, -1,2-diisocyanato	trans, Vinylene diisocyanate	OCN-CH=CH-NCO
E-49	Ethene 2-phenyl 1-isocyanato	Styryl isocyanate	C_6H_5-CH=CH-NCO
E-50	Ethene 1-phenyl 1-isocyanato	Phenyl vinyl isocyanate	CH_2=C(C_6H_5)-NCO
E-51	Ethene 1-(phenyl,4-methyl,-3-isocyanate)		CH_2=CH-$C_6H_3$$(CH_3)$NCO
E-52	Ethene 1-methyl 1-(phenyl-4-isocyanato)		CH_2=C(CH_3)-C_6H_4NCO
E-53	Ethene 1-phenyl isocyanato (meta, para)	Vinyl phenyl isocyanate	CH_2=CH-C_6H_4NCO
E-54	Ethene -1-cyclohexyl -1-isocyanato		CH_2=C(C_6H_{11})NCO
-	Ethene (trans) 1,1'diphenyl 2-isocyanato	see Stilbene (trans) 2-isocyanato (S-35)	
-	Ethene (trans) 1,1'diphenyl 4-isocyanato	see Stilbene (trans) 4-isocyanato (S-36)	
-	Ethene 1,1'diphenyl	see Stilbene 4,4'-diisocyanato (S-39)	

Formula	Boiling Point °C.	Boiling Point mm.	Melting Point °C.	Density	Refractive Index °C.	Refractive Index n	Other Data	Derivatives	References
$C_{13}H_{17}NO$	149-151	20			1.5172 (22°)				(480)
$C_{13}H_{17}NO$	147-148	13			1.5200 (21°)				(406)
$C_{17}H_{25}NO$			54°						(135) (190)
$C_{14}H_{19}NO$	146-149	13							(480)
$C_{13}H_{17}NO$	134-135	13			1.5096 (22°)				(406)
C_3H_3O	38-39 36	760 700		0.9388 (20°/4°)	25° 20°	1.4167 1.4188		See references (133)(269) (332)(377) (533)	(133)(144) (269)(332) (365)(377) (494)(523) (533)
$C_4H_2N_2O_2$	150	750	67-69					Diethyl urethane MP 218-220°	(287)
C_9H_7NO	107 110	12 18		1.101 (17°)			Kinetics of reaction with methanol (494) Infrared (523)	Methyl carbamate MP 126°	(9) (494) (523)
C_9H_7NO	61 99	5 25			15° 25°	1.5598 1.5554	Infrared (523)	Phenyl urea MP 142° Urea from NH_3 MP 131°	(80) (458) (523)
$C_{10}H_9NO$	52-54	0.1						Phenyl urea MP 195-196°	(189)
$C_{10}H_9NO$	65-73	2-3							(266)
C_9H_7NO	39-41	0.05						Phenyl urea MP 170-171°	(189)
$C_9H_{13}NO$	77	14			25°	1.4770	Infrared (523)	Urea from NH_3 MP 108° Phenyl urea MP 147-148°	(484) (523)

No.	Name	Other Names	Structure
-	Ethene 1,1'diphenyl 4-isocyanato 4'-dimethyl amino	see Stilbene 4-isocyanato 4'-dimethylamino (S-37)	
-	Ethene 1,1'diphenyl 4-isocyanato 2-methyl 4'-dimethyl amino	see Stilbene 4-isocyanato 2-methyl 4'-dimethylamino (S-38)	
E-55	Ether benzyl, methyl 3-isocyanato		$OCN-C_6H_4-CH_2-O-CH_3$
E-56	Ether benzyl, phenyl 2'-isocyanato		$C_6H_5-CH_2-O-C_6H_4NCO$
E-57	Ether benzyl, phenyl 3'-isocyanato		$C_6H_5-CH_2-O-C_6H_4NCO$
E-58	Ether benzyl, phenyl 4'-isocyanato		$C_6H_5CH_2-O-C_6H_4NCO$
E-59	Ether cyclohexyl propyl 3'-isocyanato		$C_6H_{11}-O-CH_2CH_2CH_2-NCO$
E-60	Ether cyclohexyl propyl 4-isooctyl 3'-isocyanato		$i-C_8H_{17}-C_6H_{10}-O-(CH_2)_3-NCO$
E-61	Ether diphenyl 2'-isocyanato		$C_6H_5-O-C_6H_4NCO$
E-62	Ether diphenyl 3'-isocyanato		$C_6H_5-O-C_6H_4NCO$
E-63	Ether diphenyl 4'-isocyanato		$C_6H_5-O-C_6H_4NCO$
E-64	Ether diphenyl 2,4'diisocyanato		$NCO-C_6H_4-O-C_6H_4-NCO$
E-65	Ether diphenyl 4,4'diisocyanato		$NCO-C_6H_4-O-C_6H_4-NCO$
E-66	Ether diphenyl 2',4',6'triisocyanato		$C_6H_5-O-C_6H_2(NCO)_3$

Formula	Boiling Point		Melting Point	Density	Refractive Index		Other Data	Derivatives	References
	°C.	mm.	°C.		°C.	n			
$C_9H_9NO_2$	109	11						Urea MP 122-124°	(194)
$C_{14}H_{11}NO_2$	115	0.5					Corresponding amino and nitro compound given in abstract.	Diethyl amino ethanol carbamate, MP 104-105° LD_{50}= 235 mg/kg	(387)
$C_{14}H_{11}NO_2$	130-134	?					Corresponding amino and nitro compound given in abstract.	Diethyl amino ethanol carbamate, MP 127-128° LD_{50}= 600 mg/kg	(387)
$C_{14}H_{11}NO_2$	145-146	0.4					Corresponding amino and nitro compound given in abstract.	Diethyl amino ethanol carbamate, MP 130-131° LD_{50}= 500 mg/kg	(387)
$C_{10}H_{17}NO_2$	113-114	11						Phenyl urea MP 100°	(194)
$C_{18}H_{33}NO_2$	140-152	0.10 to 0.20							(194)
$C_{13}H_9NO_2$	162-164	14					Corresponding amino and nitro compound given in abstract.	Diethyl aminoethanol carbamate, MP 174.5° LD_{50}= 250 mg/kg	(387)
$C_{13}H_9NO_2$	118-120	1					Corresponding amino and nitro compound given in abstract.	Diethyl aminoethanol carbamate, MP 132° LD_{50}= 500 mg/kg	(387)
$C_{13}H_9NO_2$	110-114	1					Corresponding amino and nitro compound given in abstract.	Diethyl aminoethanol carbamate MP 136-137° LD_{50}= 660 mg/kg	(387)
$C_{14}H_8N_2O_3$			60-62						(194)
$C_{14}H_8N_2O_3$			66-68						(194)
$C_{15}H_7N_3O_4$			78-78.5					Triureide from NH_3 MP 271-272°	(198)

No.	Name	Other Names	Structure
E-67	Ether diphenyl 2,4-diisocyanato 2'-isocyanato		$C_6H_3(NCO)_2$-O-C_6H_4NCO
E-68	Ether diphenyl 2,4-diisocyanato 4'-isocyanato		$C_6H_3(NCO)_2$-O-C_6H_4NCO
E-69	Ether diphenyl 2,4-diisocyanato 3'-isocyanato 4'-methyl		C_6H_3(NCO)2-O-$C_6H_3(CH_3)$(NCO)
E-70	Ether diphenyl 2,4-diisocyanato 3'-isocyanato 4'-methoxy		$C_6H_3(NCO)_2$-O-$C_6H_3(CH_3O)$(NCO)
E-71	Ether -diphenyl -2,4,4'triisocyanato -3'chloro		$C_6H_3(NCO)_2$-O-C_6H_3Cl(NCO)
E-72	Ether -diphenyl -2'-isocyanato -4,4'dichloro		C_6H_4Cl-O-C_6H_3Cl(NCO)
E-73	Ether -phenethyl -phenyl -2'-isocyanato		$C_6H_5CH_2CH_2$-O-C_6H_4NCO
E-74	Ether -phenethyl -phenyl -3'-isocyanato		$C_6H_5CH_2CH_2$-O-C_6H_4NCO
E-75	Ether -phenethyl -phenyl -4'-isocyanato		$C_6H_5CH_2CH_2$-O-C_6H_4NCO
-	Ethyl isocyanate	see Ethane, isocyanato (E-2)	
-	Ethyl isocyanato acetate	see Acetic acid isocyanato, ethyl ester (A-5)	
-	Ethyl vinyl isocyanate	see 1-Butene -2-isocyanato (B-232)	
-	Ethylene diisocyanate	see Ethane 1,2-diisocyanato (E-23)	
-	Ethylene glycol diphenyl ether 2,2'-diisocyanate	see Ethane,-1,2-di(phenoxy- 2-isocyanato) (E-31)	
-	Ethoxy methyl isocyanate	see Methane 1-isocyanato 1-ethoxy (M-33)	
-	Ethoxy propyl -isocyanate	see Propane 1-isocyanato 3-ethoxy (P-67)	

Formula	Boiling Point °C.	Boiling Point mm.	Melting Point °C.	Density	Refractive Index °C.	Refractive Index n	Other Data	Derivatives	References
$C_{15}H_7N_3O_4$	170- 178	0.5	62.5- 63.5						(367)
$C_{15}H_7N_3O_4$	168- 181 180- 187	0.30 0.65	69.5- 70.5						(367)
$C_{16}H_9N_3O_4$	180 189	0.65 0.75	78- 79.5						(367)
$C_{16}H_9N_3O_5$	193- 199	0.35	91- 93						(367)
$C_{15}H_6ClN_3O_4$	166 174	0.10 0.15	72-74						(367)
$C_{13}H_7Cl_2NO_2$	150- 170	0.5 to 1.5	46-47						(194)
$C_{15}H_{13}NO_2$	136	1					Corresponding amino and nitro compound given in abstract.		(387)
$C_{15}H_{13}NO_2$	146- 149	1					Corresponding amino and nitro compound given in abstract.	Diethyl amino-ethanol carbamate MP 119° LD_{50}= 205 mg/kg	(387)
$C_{15}H_{13}NO_2$	154	0.3					Corresponding amino and nitro compound given in abstract.	Diethyl amino ethanol carbamate, MP 152° LD_{50}= 770 mg/kg	(387)

No.	Name	Other Names	Structure
F-1	Fenchyl-(D-1) isocyanato		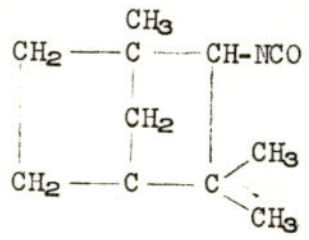
F-2	Fenchelyl isocyanato		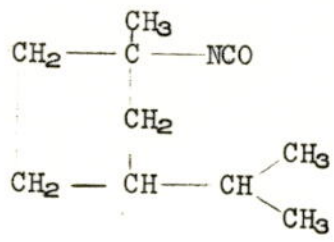
F-3	Ferrocenyl isocyanato		C_5H_5-Fe-C_5H_4-NCO
F-4	Ferrocenyl methyl isocyanato		C_5H_5-Fe-C_5H_4-CH_2NCO
F-5	Fluorene -1-isocyanato		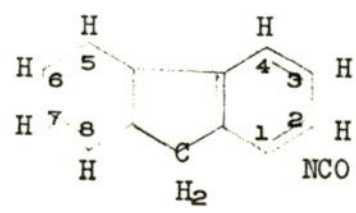
F-6	Fluorene -2-isocyanato		
F-7	Fluorene -2,7-diisocyanato		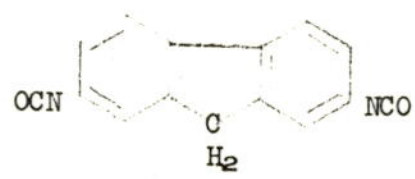
F-8	Fluorene -2-acetamido -7-isocyanato		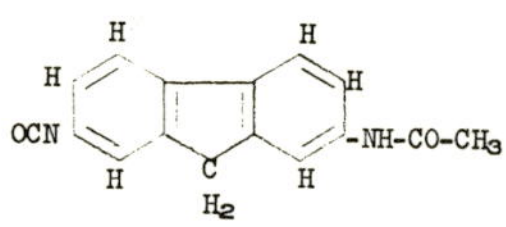
F-9	9-Fluorenone -1-isocyanato		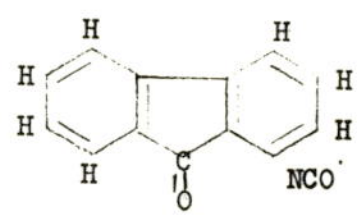

Formula	Boiling Point °C.	Boiling Point mm.	Melting Point °C.	Density	Refractive Index °C.	Refractive Index n	Other Data	Derivatives	References
$C_{11}H_{17}NO$				0.9860 (19.7°)			$[\alpha]_D^{19.7} = +19.47$		(14)
$C_{10}H_{17}NO$	201			0.9210 (19°)	1.4461 (19°)		$[\alpha]_D^{19} = 3.04$	Urea from NH_3 MP 129-130° Fenchelyl amine, BP 173°, d(21°) 0.832	(558)
$C_{11}H_9NOFe$	70-80	0.1	36-39					Methyl urethane MP 145-147° Ethyl urethane MP 90-92°	(443) (398)
$C_{12}H_{11}NOFe$	110-120	0.1						Amino methyl Ferrocenyl C_5H_5-Fe-C_5H_4-CH_2NH_2 MP 30-35°	(398)
$C_{14}H_9NO$			77					Ethyl urethane MP 132-132.6° 1-Amino fluorene MP 124-124.6°	(264) (187)
$C_{14}H_9NO$			69-70					Urea from aniline MP 205° Numerous carbamates (174)	(174) (94)
$C_{15}H_8N_2O_2$			133						(194)
$C_{16}H_{12}N_2O_2$			216-220						(243)
$C_{14}H_7NO_2$			144-146					Di-urea from water MP 264°	(210)

No.	Name	Other Names	Structure

F-10 9,9'-Spirobifluorene
-2-isocyanato

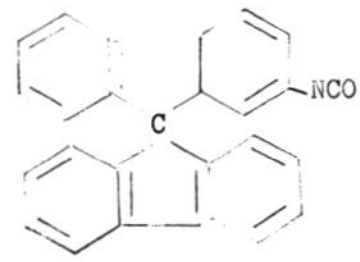

F-11 Fluorescein
-isocyanato

F-12 Furan-
-2-isocyanato

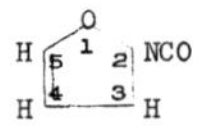

F-13 Furan-
-3-isocyanato

F-14 Furan-
-2-isocyanato
-5-Bromo

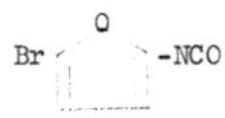

F-15 Furan,-
-2-isocyanato
-5-methyl

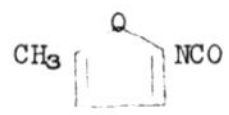

F-16 Furan,-
-3-isocyanato
-2-methyl

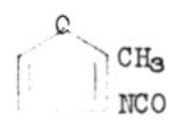

F-17 Furan,-
-3-isocyanato
2,4-dimethyl

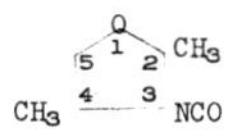

F-18 Furan-
-3-isocyanato
-2,5-dimethyl

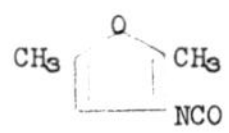

F-19 Furan
2,5 bis(2-isocyanato ethyl)

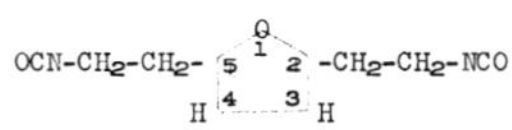

F-20 Furan,-
tetrahydro
-2,5 bis(-2-isocyanato ethyl)

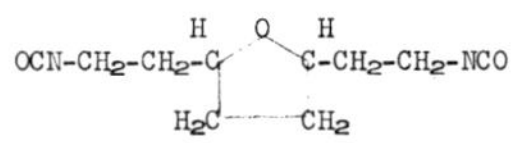

Formula	Boiling Point °C.	Boiling Point mm.	Melting Point °C.	Density	Refractive Index °C.	Refractive Index n	Other Data	Derivatives	References
$C_{26}H_{15}NO$			182-184					9,9-Spirobi-fluorene -azide MP 170-171° 2-Nitro MP 235-236° 2-Amino MP 227-229	(215)
							Used with proteins as a means of identifying antigens.		(199) (122) (313) (325) (522) (528)
$C_5H_3NO_2$	111-112	760						Urea from water MP 190°	(89)
	54	40							
$C_5H_3NO_2$							Infrared and UV absorption		(345)
$C_5H_2BrNO_2$	75	35							(64)
$C_6H_5NO_2$	53-54	35						Sym urea MP 182° Methyl ure-thane, MP 64-66°	(64)
$C_6H_5NO_2$	42	15						Sym urea MP 227° Methyl Urethane BP 128° (14 mm)	(64)
$C_7H_7NO_2$	74-75	31						Sym urea MP 285°	(64)
$C_7H_7NO_2$	68	20						Sym urea MP 247° Methyl urethane BP 107° (2.5 mm)	(64)
$C_{10}H_{10}N_2O_3$	126-128	4							(436)
$C_{10}H_{14}N_2O_3$	143	5							(436)

No.	Name	Other Names	Structure
	Fluoroisocyanates		

Many of the following compounds are members of homologous series. In order to preserve the data in systematic order, they are not arranged alphabetically.

No.	Name	Other Names	Structure
F-21	Benzene- 1-isocyanato 3-fluoro		$F-C_6H_4NCO$
F-22	Benzene -1-isocyanato -4-fluoro		$F-C_6H_4NCO$
F-23	Benzene 1-isocyanato 3-trifluoromethyl		$CF_3-C_6H_4NCO$
-	Benzene 1-isocyanato 3-sulfonic acid, fluoride	see Benzene, 1-isocyanato 3-sulfonic acid fluoride (B-58)	
F-24	Benzene 1-isocyanato 4-chloro 3-trifluoromethyl		$(CF_3)(Cl)-C_6H_3-NCO$
F-25	Benzene 1-isocyanato 2-(4-chlorophenoxy) 5-trifluoromethyl		$(CF_3)(NCO)-C_6H_3-(O-C_6H_4Cl)$
F-26	Butane 1-isocyanato 4-fluoro		$CH_2F-CH_2CH_2CH_2-NCO$
F-27	Ethane 1-isocyanato 2-fluoro		CH_2F-CH_2-NCO
F-28	Methane 1-isocyanato 1,1-difluoro 1-ethoxy		$C_2H_5O-CF_2-NCO$
F-29	Methane 1-isocyanato 1,1-difluoro 1-methoxy		CH_3O-CF_2-NCO
F-30	Methane 1-isocyanato trifluoro		CF_3-NCO
F-31	Propane 1-isocyanato 3-fluoro		$CH_2F-CH_2-CH_2-NCO$
F-32	Propane 2-isocyanato 1,3-hexafluoro		$CF_3-CH(NCO)-CF_3$
-	Sulfuryl fluoride -isocyanate	see Sulfuryl fluoride, isocyanate (S-63)	

	Formula	Boiling Point °C.	Boiling Point mm.	Melting Point °C.	Density	Refractive Index °C.	Refractive Index n	Other Data	Derivatives	References
	C_7H_4FNO	66-68.5	29							(247)
	C_7H_4FNO								Several ureas tested for anti-viral and anti-bacterial properties (412)	(412)
	$C_8H_4F_3NO$	54 104	11 83					Reactivity with 2-ethyl-hexanol (520)		(247) (520)
	$C_8H_3ClF_3NO$	86-90	14							(194)
	$C_{14}H_7ClF_3NO_2$	105-108	0.08							(288)
	C_5H_8FNO	72-72.5	42						Phenyl urea MP 117.5-118°	(364)
	C_3H_4FNO	100-101	760						Phenyl urea MP 144-145°	(364)
	$C_4H_5F_2NO_2$	RT	760							(425)
	$C_3H_3F_2NO_2$	33-34 43-44	760 760							(425) (506)
	C_2F_3NO	36	760					See Table No. 13		See Table No. 13
	C_4H_6FNO	126	760						Phenyl urea MP 105-105.5°	(364)
	C_4HF_6NO	Not characterized							Phenyl urea MP 211°	(475)

No.	Name	Other Names	Structure
	PERFLUORO ALKYL ISOCYANATES		
F-33	Ethane- -1-isocyanato -Pentafluoro		CF_3-CF_2-NCO
F-34	Propane- -1-isocyanato -heptafluoro		CF_3-$(CF_2)_2$NCO
F-35	Butane -1-isocyanato -nonafluoro		CF_3-$(CF_2)_3$NCO
F-36	Pentane 1-isocyanato undecafluoro		$CF_3(CF_2)_4$NCO
F-37	Hexane 1-isocyanato tridecafluoro		$CF_3(CF_2)_5$NCO
F-38	Heptane 1-isocyanato pentadecafluoro		$CF_3(CF_2)_6$NCO
F-39	Octane- -1-isocyanato -heptadecafluoro		$CF_3(CF_2)_7$NCO
F-40	Nonane 1-isocyanato nonadecafluoro		$CF_3(CF_2)_8$NCO
F-41	Decane- 1-isocyanato heneicosanefluoro		CF_3-$(CF_2)_9$NCO
F-42	Undecane- -1-isocyanato -tricosanefluoro		$CF_3(CF_2)_{10}$NCO

Formula	Boiling Point °C.	Boiling Point mm.	Melting Point °C.	Density	Refractive Index °C.	Refractive Index n	Other Data	Derivatives	References
C_3F_5NO	-9 -8.5 to -5	742 760							(244) (383) (483)
C_4F_7NO	25 24-26	740 759						Ethyl urethane BP 55-58°(9 mm) BP 66-69 (16.5 mm) $N_D^{20°}$ = 1.3403 $d^{25°}$ = 1.458	(244) (336) (373) (357) (363) (383)
C_5F_9NO	52-53	753		1.568 (20°)	25°	1.279			(244) (383)
$C_6F_{11}NO$	75-78	735		1.696 (20°)	25°	1.290			(244) (383)
$C_7F_{13}NO$	99	740							(244) (383)
$C_8F_{15}NO$	119	740						Ethyl carbamate BP 71-74° (1.5 mm) N_D^{20} = 1.3291	(244) (383)
$C_9F_{17}NO$	140	740							(244) (383)
$C_{10}F_{19}NO$	160-161	743							(244) (383)
$C_{11}F_{21}NO$	180	740							(244) (383)
$C_{12}F_{23}NO$	200	740							(244) (383)

No.	Name	Other Names	Structure
	1,1-DIHYDRO PERFLUORO ALKYL ISOCYANATES		
F-43	Ethane 1-isocyanato 2-trifluoro 1,1 dihydra		$CF_3\text{-}CH_2NCO$
F-44	Propane,- 1-isocyanato 1,1-dihydra pentafluoro		$CF_3\text{-}CF_2\text{-}CH_2\text{-}NCO$
F-45	Butane 1-isocyanato 1,1-dihydra Heptafluoro		$CF_3\text{-}(CF_2)_2CH_2\text{-}NCO$
F-46	Pentane 1-isocyanato 1,1-dihydra nonafluoro		$CF_3\text{-}(CF_2)_3\text{-}CH_2\text{-}NCO$
F-47	Hexane 1-isocyanato 1,1-dihydra undecafluoro		$CF_3\text{-}(CF_2)_4CH_2\text{-}NCO$
F-48	Heptane 1-isocyanato 1, 1-dihydra tridecafluoro		$CF_3(CF_2)_5CH_2NCO$
F-49	Octane 1-isocyanato 1,1-dihydra pentadecafluoro		$CF_3(CF_2)_6CH_2NCO$
F-50	Nonane -1-isocyanato -1,1-dihydra -heptadecafluoro		$CF_3(CF_2)_7CH_2NCO$
F-51	Decane -1-isocyanato -1,1-dihydra -nonadecafluoro		$CF_3(CF_2)_8CH_2NCO$

Formula	Boiling Point °C.	Boiling Point mm.	Melting Point °C.	Density	Refractive Index °C.	Refractive Index n	Other Data	Derivatives	References
$C_3H_2F_3NO$	53-56	747		1.387 (27°)	27°	1.3486		Urea from water MP 157°	(256) (304)
$C_4H_2F_5NO$	72	757		1.484 (20°)	20°	1.3350		Urea from water MP 166-167°	(256) (304)
$C_5H_2F_7NO$	90	737	-78	1.512 (20°)	20°	1.3152		Urea from water MP 121°	(256) (304)
$C_6H_2F_9NO$	110	740							(256) (304)
$C_7H_2F_{11}NO$	127	740							(256) (304)
$C_8H_2F_{13}NO$	147	740							(256) (304)
$C_9H_2F_{15}NO$	167	740							(256) (304)
$C_{10}H_2F_{17}NO$	184	740							(256) (304)
$C_{11}H_2F_{19}NO$	203 93 80	746 14 1	46-47					Urea MP 144-146° Phenyl urea MP 137-138°	(256) (304)

No.	Name	Other Names	Structure
-	Glutaric acid -2-isocyanato -diethyl ester	see Pentanedioic acid -2-isocyanato diethyl ester (P-20)	
-	Glutaric acid 2-isocyanato dimethyl ester	see Pentanedioic acid -2-isocyanato dimethyl ester (P-19)	
-	Glycolic acid isocyanato dodecyl ester	see Acetic acid isocyanato dodecyl ester (A-4)	
G-1	Guanyl isocyanate		$H_2N-\underset{\underset{NH}{\|\|}}{C}-NCO$

	Formula	Boiling Point °C.	Boiling Point mm.	Melting Point °C.	Density	Refractive Index °C.	Refractive Index n	Other Data	Derivatives	References
	$C_2H_3N_3O$									(68)

No.	Name	Other Names	Structure
H-1	Hendecane 1-isocyanato	Undecyl isocyanate	$CH_3-(CH_2)_{10}-NCO$
H-2	Heptadecane -1-isocyanato	Stearyl isocyanate	$CH_3-(CH_2)_{16}-NCO$
-	Heptadecyl (1,1-dimethyl) isocyanate	see Octadecane -2-isocyanato -2-methyl (O-12)	
H-3	Heptane 1-isocyanato	Heptyl isocyanate	$CH_3-(CH_2)_6-NCO$
H-4	Heptane 4-isocyanato		$CH_3-(CH_2)_2-CH(NCO)-(CH_2)_2-CH_3$
H-5	Heptane 1-isocyanato 7-chloro		$Cl-CH_2-(CH_2)_5-CH_2-NCO$
H-6	Heptane 1,7-diisocyanato	Heptamethylene diisocyanate	$OCN-(CH_2)_7-NCO$
H-7	Heptane -4-isocyanato -4-methyl		$CH_3-(CH_2)_2-C(NCO)(CH_3)-(CH_2)_2-CH_3$
-	Heptane 1-isocyanato pentadecafluoro	see Fluoro isocyanates (F-38)	
H-8	Heptane 4-isocyanato 4-n-propyl		$CH_3-(CH_2)_2-C(NCO)(C_3H_7)-(CH_2)_2-CH_3$
-	Heptane -1-isocyanato -1,1-dihydra -tridecafluoro	see Fluoro isocyanates (F-48)	
H-9	3-Heptene -4-isocyanato		$CH_3-CH_2-CH=C(NCO)-CH_2-CH_2-CH_3$
H-10	3-Heptene -2,2,6,6,-tetramethyl -4-isocyanato		$(CH_3)_3-C-CH=C(NCO)-CH_2-C(CH_3)_3$
H-11	Hexadecane -1-isocyanato		$CH_3-(CH_2)_{15}-NCO$
H-12	Hexadecane,- -1,16-diisocyanato	Hexadecamethylene diisocyanate	$OCN-(CH_2)_{16}-NCO$
H-13	Hexadecane,- -2-isocyanato -2-methyl	1,-1,dimethyl- -1-pentadecyl isocyanate	$CH_3-(CH_2)_{13}-C(NCO)(CH_3)-CH_3$
H-14	Hexane 1-isocyanato		$CH_3-(CH_2)_5-NCO$

Formula	Boiling Point °C.	Boiling Point mm.	Melting Point °C.	Density	Refractive Index °C.	Refractive Index n	Other Data	Derivatives	References
$C_{12}H_{23}NO$	86-88 103	1 3							(142) (505)
$C_{18}H_{35}NO$	151 170-175 192-203 200-203	0.4 2 4 15	15-16				Infrared (371)		(91) (371) (400) (450)
$C_8H_{15}NO$	87.5-88.5	28						1,3-diheptyl-urea, MP 91° Phenyl urea MP 64-65°	(324) (505)
$C_8H_{15}NO$	171 58-59	760 13							(13) (194)
$C_8H_{14}ClNO$	111	13							(194)
$C_9H_{14}N_2O_2$	103-104 140-142	6 14						Dimethyl urethane MP 97.5-98°	(160) (181) (194) (239)
$C_9H_{17}NO$	76	18						Heptane-4-amino -4-methyl BP 145°(760)	(163)
$C_{11}H_{21}NO$	65-66	2							(200)
$C_8H_{13}NO$	55-56	17						Phenyl urea MP 102°	(80)
$C_{12}H_{21}NO$	85-87	8				1.4485			(277)
$C_{17}H_{33}NO$	186-188	14							(194)
$C_{18}H_{32}N_2O_2$	185	0.33						N,N'Hexa-decamethylene urea MP 180-181	(454)
$C_{18}H_{35}NO$	173-175	4-5							(134)
$C_7H_{13}NO$	162-163	760							(505)

No.	Name	Other Names	Structure
H-15	Hexane -2-isocyanato, -2-n-butyl -1-(1-piperidyl)		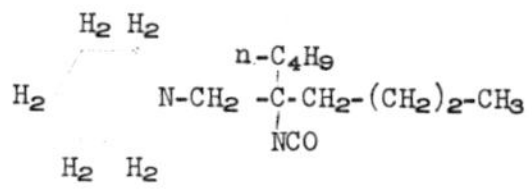
H-16	Hexane 1-isocyanato 6-chloro		$ClCH_2-(CH_2)_4-CH_2-NCO$
H-17	Hexane -1-isocyanato -2-ethyl	2-ethyl-hexyl, isocyanato	$CH_3-(CH_2)_3-CH(C_2H_5)-CH_2-NCO$
H-18	Hexane -1-isocyanato -6-isothiocyano		$OCN-(CH_2)_6-NCS$
H-19	Hexane 3-isocyanato 3-methyl		$CH_3-(CH_2)_2-C(NCO)(CH_3)-CH_2-CH_3$
H-20	Hexane -3-isocyanato -5-methyl		$CH_3-CH(CH_3)-CH_2-CH(NCO)-CH_2-CH_3$
-	Hexane, 1-isocyanato tridecafluoro	see Fluoro isocyanates (F-37)	
-	Hexane, 1-isocyanato 1,1-dihydra undecafluoro	see Fluoro isocyanates (F-47)	
H-21	Hexane 1,6-diisocyanato	Hexamethylene diisocyanate	$OCN-(CH_2)_6-NCO$
H-22	Hexane 1,6-diisocyanato 3-butoxy		$OCN-(CH_2)_2-CH(OC_4H_9)-(CH_2)_3-NCO$
H-23	Hexane 1,6-diisocyanato 2,5-diethyl		$OCNCH_2-CH(C_2H_5)-(CH_2)_2-CH(C_2H_5)CH_2NCO$
H-24	Hexane 1,6-diisocyanato 3-methoxy		$OCN-(CH_2)_2-CH(OCH_3)-(CH_2)_3-NCO$
H-25	Hexanoic acid, 6-isocyanato chloride		$(NCO)(CH_2)_5-COCl$
H-26	Hexanoic acid -2-butyl isocyanate	dibutyl acetyl isocyanate	$CH_3-(CH_2)_3-CH(C_4H_9)CO-NCO$

Formula	Boiling Point °C.	Boiling Point mm.	Melting Point °C.	Density	Refractive Index °C.	Refractive Index n	Other Data	Derivatives	References
$C_{16}H_{30}N_2O$	146-148	6						Carbamate derivatives (227)	(200) (227)
$C_7H_{12}ClNO$	108	12					Infrared (371)		(194) (371)
$C_8H_{17}NO$	75-79	15							(194)
$C_8H_{12}N_2OS$	110	0.5							(451)
$C_8H_{15}NO$	160-161	760						Sym diurea MP 182-183° Hexane-3-methyl -3-amino BP 131-132°	(37)
$C_8H_{15}NO$	165	750						Phenyl urea MP 100° Ethyl urethane BP 128°(28 mm)	(38)
$C_8H_{12}N_2O_2$				See Table No. 14					See Table No. 14
$C_{12}H_{20}N_2O_3$	142-155	2							(194)
$C_{12}H_{20}N_2O_2$	95-96	2							(428)
$C_9H_{14}N_2O_3$	142-145	12						p-chlorophenyl urea MP 217-218°	(194)
$C_7H_{10}ClNO_2$	132-140	24							(396)
$C_{11}H_{19}NO_2$	68-73	12							(108) (153)

No.	Name	Other Names	Structure
H-27	Hexanoic acid, 2-ethyl, isocyanate		$CH_3-(CH_2)_3-CH(C_2H_5)-CO-NCO$
H-28	Hexanoic acid, 2-ethyl 3-methyl, -isocyanate		$CH_3-(CH_2)_2-CH(CH_3)-CH(C_2H_5)-CO-NCO$
H-29	Hexanoic acid 2-ethyl 5-methyl isocyanate		$CH_3-CH(CH_3)-(CH_2)_2-CH(C_2H_5)-CO-NCO$
H-30	Hexanoic acid (DL, 2-isocyanato -ethyl ester	Caproic acid -2-isocyanato -ethyl ester	$CH_3-(CH_2)_3-CH(NCO)-CO-O-C_2H_5$
H-31	Hexanoic acid -2-isocyanato -2-butyl ethyl ester		$CH_3-(CH_2)_3-C(C_4H_9)(NCO)-CO-O-C_2H_5$
H-32	Hexanoic acid -3-isocyanato -5-methyl -ethyl ester		$CH_3-CH(CH_3)-CH_2-CH(NCO)CH_2-CO-O-C_2H_5$
H-33	Hexanoic acid 6-isocyanato methyl ester	Caproic acid 6-isocyanato- -methyl ester	$OCN-(CH_2)_5-CO-O-CH_3$
H-34	Hexanoic acid -6-isocyanato -4,4-dinitro, methyl ester		$OCN-(CH_2)_2-C(NO_2)_2-CH_2-CO-O-CH_3$
H-35	1-Hexene -2-isocyanato	1-butyl vinyl isocyanate	$CH_2=C(NCO)-CH_2-CH_2-CH_2-CH_3$
H-36	1-Hexene 5-(phenyl isocyanato) (o,m,p)	5-Hexenyl- -phenyl isocyanate	$CH_2=CH-CH_2-CH_2-CH(C_6H_4NCO)-CH_3$
-	1-Hexylheptyl isocyanate	see Tridecane-7-isocyanate (T-29)	
-	1-Hexyl vinyl isocyanate	see 1-octene-2-isocyanate (O-10)	
-	Hippenyl isocyanate	see Benzoic acid, amide, -N(methyl isocyanate) (B-159)	
-	Hippenyl isocyanate p-bromo	see Benzoic acid, (4-bromo) amide-N-methyl isocyanate (B-160)	
-	Histamine isocyanate	see Imidazole, 4-ethyl isocyanate (I-1)	
H-37	Homohydrocarpyl isocyanate		

Formula	Boiling Point °C.	Boiling Point mm.	Melting Point °C.	Density	Refractive Index °C.	Refractive Index n	Other Data	Derivatives	References
$C_9H_{15}NO_2$	78-85	20							(108) (153)
$C_{10}H_{17}NO_2$	58-60	5							(546)
$C_{10}H_{17}NO_2$	100-105	30							(108) (153)
$C_9H_{15}NO_3$	104	14						Urea from NH_3 MP 165°	(240)
$C_{13}H_{23}NO_3$	107 133-135	4.5 12.7						Phenyl urea MP 151-152	(128)
$C_{10}H_{17}NO_3$	112-113	18			Optical Rotatory Power α_D^{22}= -6.5° ± 1°(C=4.77 - CCl_4) α_D^{25}= -5.9 ± 1°(C= 2.31 C_6H_6)			Urea from NH_3 MP 60.5-63° α_D^{22}=-32° ± 0.5° (C = 2.06 in EtOH)	(470)
$C_8H_{13}NO_3$	68 123	0.1 15			25°	1.4378			(194) (537)
$C_8H_{11}N_3O_7$	140	0.15			25°	1.4793			(463) (553)
$C_7H_{11}NO$	66	60			25°	1.4335	Infrared data Reaction rate	Urea from NH_3 MP 66-67° Phenyl urea MP 95-96°	(484) (523) (494)
$C_{13}H_{15}NO$	147-148	16			20°	1.5338			(521)
							Prepared from chaulmogryl acid		(55)

No.	Name	Other Names	Structure
I-1	Imidazole -4-ethyl isocyanate	Histamine isocyanate	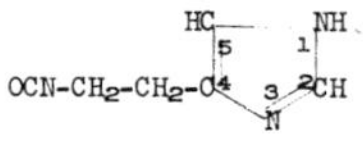
I-2	Imidazole -4-(ethyl acetamide- isocyanato)		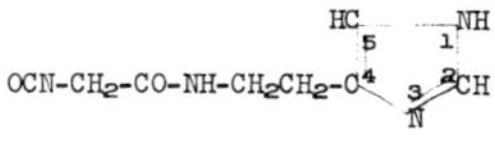
I-3	Indan 5-isocyanato 1,3,3,6-tetramethyl 1-(4-methyl-3-isocyanato- phenyl)		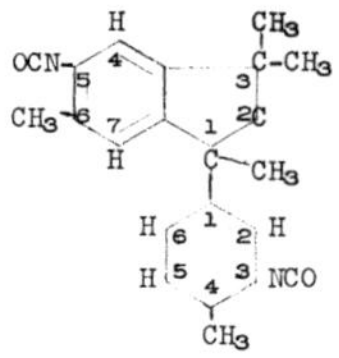
I-4	Indan -5(Propyl-2-isocyanato-2-methyl)		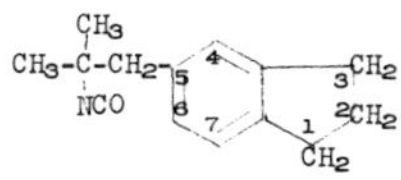
I-5	Indan 6-isocyanato 1,3,3,-trimethyl 1-(4-isocyanato phenyl)		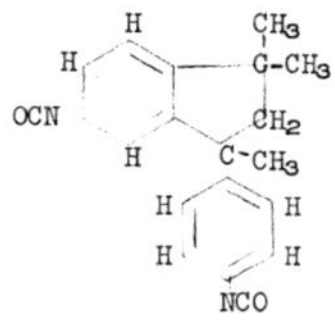
-	Isobutyl isocyanate	see Propane 1-isocyanato 2-methyl (P-56)	
I-6	Isocyanic acid		HNCO
-	Isohexyl isocyanate	see Pentane -1-isocyanato -4-methyl (P-10)	
-	Isopropenyl isocyanate	see Propene -2-isocyanato (P-104)	
-	p-Isopropenyl phenyl isocyanate	see Ethene 1-methyl 1-(phenyl-4-isocyanate) (E-52)	
-	Isopropyl isocyanate	see Propane -2-isocyanato (P-54)	
I-7	Isopropylidene aminooxymethyl isocyanate		CH_3-C=N-O-CH_2-NCO CH_3

Formula	Boiling Point °C.	Boiling Point mm.	Melting Point °C.	Density	Refractive Index °C.	Refractive Index n	Other Data	Derivatives	References
$C_6H_7N_3O$								Histamine or Imidazole-4-ethyl amine	(150)
									(150)
$C_{22}H_{22}N_2O_2$	168	0.2	77-79						(435)
$C_{14}H_{17}NO$	141	0.6							(180)
$C_{20}H_{18}N_2O_2$			90-93						(435)
CHNO							See Table No. 16		See Table No. 16
$C_5H_8N_2O_2$	45	12							(542)

No.	Name	Other Names	Structure
L-1	Lauryl isocyanate	see Dodecane 1-isocyanate (D-15)	
L-2	Leucic Acid 2-isobutyl 2-isocyanato ethyl ester	see Pentanoic acid 2-isocyanato 2-isobutyl 4-methyl ethyl ester (P-25)	
M-1	p-Menthane -1,8-diisocyanato		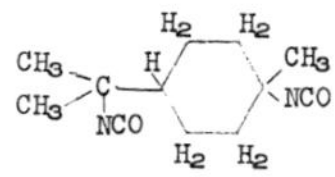
M-2	1,p-carbomenthoxy phenyl isocyanate		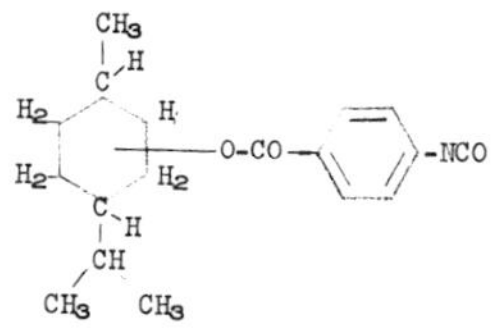
M-3	l-menthyl isocyanate		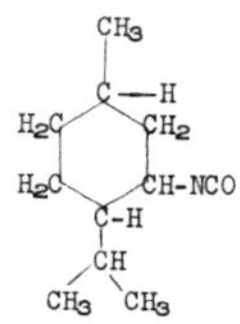
M-4	(±) Menthyl -isocyanate		
M-5	d-Carvomenthyl isocyanate		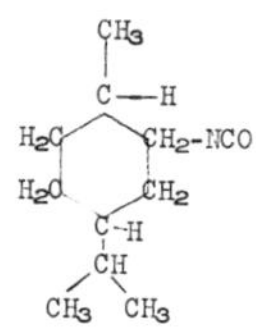
M-6	+ Isomenthyl isocyanate		
M-7	+ Neomenthyl isocyanate		
-	Mesityl isocyanate	see Benzene 1-isocyanato 2,4,6-trimethyl (B-65)	

	Formula	Boiling Point °C.	Boiling Point mm.	Melting Point °C.	Density	Refractive Index °C.	Refractive Index n	Other Data	Derivatives	References
	$C_{12}H_{18}N_2O_2$	90-100	760				1.4713			(316)
	$C_{18}H_{23}NO_3$			58-61				$[\alpha]_D^{25} = -74.25°$ (C = 0.4943 in benzene)		(245)
	$C_{11}H_{19}NO$	108-110 98-100 77-80	10-13 15 1		0.9355 (18°/4°)		1.4605 (22°)	Optical Rotation see Ref.	Methyl carbamate MP 53° Ethyl carbamate MP 54.5-55.5°	(6) (8) (14)
		120-123	16						Ethyl carbamate MP 93-94°	(238)
	$C_{11}H_{19}NO$			206-207						(67)
		116-117	35							(238)
		108-110	16							(238)

No.	Name	Other Names	Structure
M-8	Barium isocyanate		$Ba(OCN)_2$
M-9	Cadmium isocyanate		$2Cd(OCN)_2 \cdot 5H_2O$
M-10	Cuprous isocyanate		$Cu(OCN)$
M-11	Cupric isocyanate		$Cu(OCN)_2$
M-12	Chromyl isocyanate		$CrO_2(NCO)_2$
M-13	Ferrous isocyanate	see Ferrocenyl isocyanate	
M-14	Germanium isocyanate		$Ge(NCO)_4$
M-15	Germanium-diisopropyl diisocyanate		$(i\text{-}C_3H_7)_2Ge(NCO)_2$
M-16	Germanium-ethyl-mono--isocyanate		$(C_2H_5)_3Ge(NCO)$
M-17	Germanium ethyl-di--isocyanate		$(C_2H_5)_2Ge(NCO)_2$
M-18	Germanium ethyl-tri--isocyanate		$C_2H_5\text{-}Ge(NCO)_3$
M-19	Germanium-trichloro -isocyanate		$GeCl_3(NCO)$
M-20	Lithium -isocyanate		$Li(NCO)$
M-21	Mercury -isocyanate		$Hg(NCO)_2$
M-22	Mercurous -isocyanate		$Hg_2(NCO)_2$
M-23	Mercury -isocyanate -potassium chloride		$Hg(NCO)_2\text{-}KCl$
M-24	Mercury- -isocyanate -potassium -cyanate		$3\ Hg(NCO)_2\text{-}2\ KOCN$

R(Ml) Molar Refractivity
Δ H - Molar heat of vaporization
Δ H/T = Troutons Constant

Vapor Pressure

$\log_{10} p = A - \frac{B}{T}$

Formula	Boiling Point °C.	mm.	Melting Point °C.	Density	Refractive Index °C.	n	R (ML)	ΔH kCal	ΔH/Tβ	Vapor Pressure A	B	References
	Solubility: in water (20°) = 10.6 gms/100 ml											(405)
	Solubility: in water (20°C) 9.1 x 10^{-3} molar											(405)
												(405)
												(405)
	Stable only as a deep red solution in CCl_4											(131)
	203.7 to 204.2	760	-8 ± 0.5	1.7714 (20°) 1.760 (30°)	20 25	1.4858 1.4793	-	12.6	26.4	8.678	2757	(129) (131) (192)
	239 72-73	760 1		1.225 $\frac{20°}{4}$	20	1.464	54.7	-	-	-	-	(322)
	200.4 ± 0.5 109.4	760 48	-26.4	1.514 (20°)	20	1.4519	47.3	11.0	23.1	7.9424	2396	(192)
	226 135	760 52	-32	1.330 (20°)	20	1.4619	44.4	11.9	23.8	8.0861	2597	(192)
	225.4 139	760 52	-31	1.534 (20°)	20	1.4739	41.7	12.7	25.3	8.4195	2760	(192)
	112	760										(154)
												(449)
	Solubility: In ether (20°) - 1.3 gms/100 ml In water (0°) 3.7 gms/100 ml In water (20°) 7.5 gms/100 ml											(76) (405)
												(405)
												(76)
												(76)

No.	Name	Other Names	Structure
M-25	Mercury isocyanate -potassium cyanate -potassium acetate		$[2\ Hg(NCO)_2]$ KOCN-KOOCCH$_3$
M-26	Potassium- -copper -isocyanate		$K\ Cu_3(NCO)_4$
M-27	Potassium- copper -isocyanate		$K\ Cu(NCO)_3$
M-28	Rubidium copper isocyanate		$RbCu_3(NCO)_4$
M-29	Silver isocyanate potassium cyanate		2 AgNCO·KOCN
M-30	Sodium copper isocyanate		$NaCu_4(OCN)_5$

R(Ml) Molar Refractivity
Δ H - Molar heat of vaporization
Δ H/T = Troutons Constant

Vapor Pressure

$\log_{10} p = A - \frac{B}{T}$

	Formula	Boiling Point °C.	mm.	Melting Point °C.	Density	Refractive Index °C.	n	R (ML)	ΔH kCal	ΔH/T β	Vapor Pressure A	B	References
													(76)
													(405)
													(405)
													(405)
													(76)
													(405)

No.	Name	Other Names	Structure
-	Methacrylic acid isocyanate	see Propenoic acid -2-methyl -isocyanate (P-109)	
-	Methacrylic acid -2-isocyanato ethyl ester	see Propenoic acid -2-methyl -2(isocyanato ethyl ester) (P-110)	
M-31	Methane 1-isocyanato	Methyl isocyanate	CH_3NCO
M-32	Methane diisocyanato	Methylene diisocyanate	$OCN-CH_2-NCO$
M-33	Methane 1-isocyanato 1-ethoxy	Ethoxy methyl -isocyanate	$CH_3-CH_2-O-CH_2-NCO$
M-34	Methane 1-isocyanato 1-methoxy	Methoxy methyl -isocyanate	CH_3-O-CH_2-NCO
M-35	Methane 1-isocyanato 1-nitro	Nitromethyl -isocyanate	O_2N-CH_2-NCO
M-36	Methane 1-isocyanato 1-octadecyloxy		$CH_3(CH_2)_{17}-O-CH_2-NCO$
M-37	Methane sulfonic acid -isocyanate		CH_3-SO_2-NCO
-	Methane 1-isocyanato trifluoro	see Fluoro isocyanates (F-30)	
-	Methane 1-isocyanato 1-difluoro 1-ethoxy	see Fluoro isocyanates (F-28)	
-	Methane 1-isocyanato 1-difluoro 1-methoxy	see Fluoro isocyanates (F-29)	
M-38	Methane 1-isocyanato 1-bromo 1-phenyl		$C_6H_5-CH(Br)-NCO$
M-39	Methane 1-isocyanato 1-(p-bromophenyl) 1-phenyl		$(C_6H_5)(C_6H_4Br)-CH-NCO$
M-40	Methane 1,1-dimethyl 1,1-di(3-chloro,4-isocyanato-phenyl)		$(NCO)(Cl)C_6H_3-C(CH_3)_2-C_6H_3(Cl)(NCO)$
M-41	Methane 1-isocyanato 1-(p-chlorophenyl) 1-phenyl		$(C_6H_5-)(C_6H_4Cl)CH-NCO$
M-42	Methane 1-isocyanato 1-cyclohexyl 1-phenyl		$(C_6H_{11})(C_6H_5)CH-NCO$

Formula	Boiling Point °C.	Boiling Point mm.	Melting Point °C.	Density	Refractive Index °C.	Refractive Index n	Other Data	Derivatives	References
C_2H_3NO	37-39	760		0.960 (20°/4°)	27	1.3630	See Table No. 16	Phenyl urea MP 148-150°	See Table No. 16
$C_3H_2N_2O_2$	37-38	14						Phenyl urea MP 238-240°	(231)
$C_4H_7NO_2$	105-106	760							(24)
$C_3H_5NO_2$	87.5	760							(24)
$C_2H_2N_2O_3$							Infrared (393)		(393)
$C_{20}H_{39}NO_2$									(141)
$C_2H_3NO_3S$	73-75	12							(5)
C_8H_6BrNO	114-116	18							(75)
$C_{14}H_{10}BrNO$	173.5	4			23	1.6027			(280)
$C_{17}H_{12}Cl_2N_2O_2$			100-101						(194)
$C_{14}H_{10}ClNO$	164	3			23	1.5863			(280)
$C_{14}H_{17}NO$	111	0.8							(280)

No.	Name	Other Names	Structure
M-43	Methane 1-cyclohexyl-4-isocyanato 1-phenyl-4-isocyanato		$C_6H_{10}(NCO)-CH_2-C_6H_4NCO$
M-44	Methane 1-cyclohexyl-1-isocyanato 1-phenyl-4-isocyanato		$C_6H_{10}(NCO)-CH_2-C_6H_4(NCO)$
M-45	Methane -1-cyclohexyl -1,1-di(phenyl-4-isocyanato)		$C_6H_{11}-CH-(C_6H_4NCO)_2$
M-46	Methane 1-cyclohexyl-4-isocyanato 1-cyclohexyl		$C_6H_{11}-CH_2-C_6H_{10}(NCO)$
M-47	Methane 1,1-di(cyclohexyl-4-isocyanato)		$C_6H_{10}(NCO)-CH_2-C_6H_{10}(NCO)$
M-48	Methane- -1,1-di(cyclohexyl-2-methyl-4-isocyanato)		$C_6H_9(CH_3)(NCO)-CH_2-C_6H_9(CH_3)(NCO)$
M-49	Methane -1,1-di(cyclohexyl-3,5-dimethyl-4-isocyanato)		$C_6H_8(CH_3)_2(NCO)-CH_2-C_6H_8(CH_3)_2(NCO)$
M-50	Methane -1-methyl -1,1-di(cyclohexyl-4-isocyanato)		$C_6H_{10}(NCO)-CH(CH_3)-C_6H_{10}(NCO)$
M-51	Methane 1,1-dimethyl 1,1-di(cyclohexyl-4-isocyanato)		$C_6H_{10}(NCO)-C(CH_3)_2-C_6H_{10}(NCO)$
M-52	Methane -1-isocyanato -1,1-diphenyl	Diphenyl methyl isocyanate	$(C_6H_5)_2CH-NCO$
M-53	Methane 1-isocyanato 1-(p-nitrophenyl) 1-phenyl	p-nitro,benz- -hydryl isocyanate	$(C_6H_5)(C_6H_4NO_2-)CH-NCO$
M-54	Methane 1-isocyanato 1-p-tolyl 1-phenyl	p-methyl benzhydryl isocyanate	$(C_6H_5)(C_6H_4CH_3)CHNCO$
M-55	Methane 1-isocyanato 1-(p-methoxy phenyl) 1-phenyl	p-methoxy benzhydryl isocyanate	$(C_6H_5)(CH_3O-C_6H_4)CH-NCO$
M-56	Methane 1-isocyanato 1,1-dimethoxy phenyl	p,p'dimethoxy benzhydryl isocyanate	$(CH_3O-C_6H_4)_2-CH-NCO$
M-57	Methane 1-isocyanato 1,1,1-triphenyl	Trityl isocyanate Triphenyl methyl isocyanate	$(C_6H_5)_3-C-NCO$
M-58	Methane 1-isocyanato 1,1-diphenyl 1-p-tolyl	Diphenyl,p-tolyl methyl isocyanate	$(C_6H_5)_2(CH_3-C_6H_4)-C-NCO$
M-59	Methane 1-isocyanato 1-(p-phenyl isocyanate)	4-Phenyl isocyanate-methyl isocyanate	$OCN-C_6H_4-CH_2NCO$

Formula	Boiling Point °C.	Boiling Point mm.	Melting Point °C.	Density	Refractive Index °C.	Refractive Index n	Other Data	Derivatives	References
$C_{15}H_{16}N_2O_2$	152-156	0.3							(194)
$C_{15}H_{16}N_2O_2$	168-169	3						Dimer MP 185-186° Urea MP 185-186°	(307) (318)
$C_{21}H_{20}N_2O_2$	195-200	0.1						Dimethyl urethane MP 173°	(194)
$C_{14}H_{23}NO$	120-123	0.25							(194)
$C_{15}H_{22}N_2O_2$	165-180 179	0.5 0.9					Infrared data (2)		(194) (371) (418)
$C_{17}H_{26}N_2O_2$	200-212	5							(194)
$C_{19}H_{30}N_2O_2$	205-212	1						Dimethyl urethane MP 204-207°	(194)
$C_{16}H_{24}N_2O_2$	198-208	4						Dimethyl urethane MP 182-183°	(194)
$C_{17}H_{26}N_2O_2$	208-230	8							(194)
$C_{14}H_{11}NO$	148	4							(102)
$C_{14}H_{10}N_2O_3$	160	0.2	56-57						(280)
$C_{15}H_{13}NO$	178	16							(280)
$C_{15}H_{13}NO_2$	148	2	52	1.137 (23°)	23	1.5773			(280)
$C_{16}H_{15}NO$			59-61						(280)
$C_{20}H_{15}NO$			98					Urea from aniline MP 242°	(188) (20)
$C_{21}H_{17}NO$								Urea from aniline MP 213-215°	(25)
$C_8H_6N_2O_2$	152	19							(194)

No.	Name	Other Names	Structure
M-60	Methane 1-isocyanato 1-methyl 1-(m-Phenyl-isocyanate)		$C_6H_4(NCO)$-$CH(CH_3)$-NCO
M-61	Methane- -diphenyl,- -4,4'diisocyanate "MDI"	see Table No. 17	$C_6H_4(NCO)$-CH_2-$C_6H_4(NCO)$
M-62	Methane, -diphenyl,- -4,4'diisocyanato -3,3'dimethyl		$C_6H_3(CH_3)(NCO)$-CH_2-$C_6H_3(CH_3)(NCO)$
M-63	Methane -diphenyl -4,4'diisocyanato -2,2'dimethyl		$C_6H_3(CH_3)(NCO)$-CH_2-$C_6H_3(CH_3)(NCO)$
M-64	Methane,- -diphenyl -4,4'diisocyanato -2,2',5,5' tetramethyl		$C_6H_2(CH_3)_2(NCO)$-CH_2-$C_6H_2(CH_3)_2(NCO)$
M-65	Methane, -diphenyl, -4,4'diisocyanato, -3,3'dimethyl, -5,5'dimethoxy		$C_6H_2(CH_3)(OCH_3)(NCO)$-CH_2-$C_6H_2(CH_3)$- $(OCH_3)(NCO)$
M-66	Methane,- -diphenyl, -4,4'diisocyanato, -3,3'-dimethoxy,		$C_6H_3(OCH_3)(NCO)$-CH_2-$C_6H_3(OCH_3)(NCO)$
M-67	Methane- -diphenyl -3,3'diisocyanato -4,4'dimethoxy		$C_6H_3(OCH_3)(NCO)$-CH_2-$C_6H_3(OCH_3)(NCO)$
M-68	Methane -diphenyl -3,3'diisocyanato -4,4'diethoxy		$C_6H_3(OC_2H_5)(NCO)$-CH_2-$C_6H_3(OC_2H_5)(NCO)$
M-69	Methane -dimethyl -diphenyl -4,4'diisocyanato		$C_6H_4(NCO)$-$C(CH_3)_2$-$C_6H_4(NCO)$
M-70	Methane -diphenyl -4-isocyanato		C_6H_5-CH_2-C_6H_4NCO
M-71	Methane- -diphenyl -4,6,4'triisocyanato -3-methyl		$C_6H_2(CH_3)(NCO)_2$-CH_2-$C_6H_4(NCO)$
M-72	Methane -diphenyl -2,4,4'triisocyanato		$C_6H_3(NCO)_2$-CH_2-$C_6H_4(NCO)$
M-73	Methane- -diphenyl -2,2'5,5'tetraisocyanato -4,4'dimethyl		$C_6H_2(CH_3)(NCO)_2$-CH_2-$C_6H_2(CH_3)(NCO)_2$
M-74	Methane 1-phenyl-4-isocyanato 1-triphenyl		$(C_6H_5)_3$-C-C_6H_4-NCO

Formula	Boiling Point °C.	Boiling Point mm.	Melting Point °C.	Density	Refractive Index °C.	Refractive Index n	Other Data	Derivatives	References
$C_{10}H_8N_2O_2$	134 100-111	11 2.6						Dimethyl urethane MP 112-113°	(194) (505)
$C_{15}H_{10}N_2O_2$	170	1	43				see Table No. 17	see Table No. 17	see Table No. 17
$C_{17}H_{14}N_2O_2$							Kinetics of reaction with 1-butanol		(424)
$C_{17}H_{14}N_2O_2$	208-211	5	65-67						(194)
$C_{19}H_{18}N_2O_2$	200-240	1.2	127.6						(194)
$C_{19}H_{18}N_2O_4$			128-129						(194)
$C_{17}H_{14}N_2O_4$			97-98						(194)
$C_{17}H_{14}N_2O_4$			90-93						(194)
$C_{19}H_{18}N_2O_4$			74						(194)
	212	5-6	91.5				Infrared data (371)		(194) (371) (448)
$C_{14}H_{11}NO$	120	0.07						Methyl urethane MP 78-79° Sym urea MP 158-159°	(194)
$C_{17}H_{11}N_3O_3$	224	4	75-77					Trimethyl urethane MP 167°	(194)
$C_{16}H_9N_3O_3$	Not distillable								(194)
			154-156						(194)
$C_{26}H_{19}NO$			168-169						(174)

No.	Name	Other Names	Structure
M-75	Methane- -triphenyl- -4-nitro -4'4''diisocyanato		$C_6H_4(NO_2)\text{-}CH\text{-}(C_6H_4NCO)_2$
M-76	Methane,- -triphenyl- -3-nitro -4'4''diisocyanato		$C_6H_4(NO_2)\text{-}CH\text{-}(C_6H_4NCO)_2$
M-77	Methane,- -triphenyl- -2',5',2'',5'',tetramethyl -4',4''diisocyanato		$C_6H_5\text{-}CH\text{-}[C_6H_2(CH_3)_2(NCO)]_2$
M-78	Methane -triphenyl- -4,4',4''triisocyanato		$CH(C_6H_4NCO)_3$
M-79	Methane, 1-isocyanato 1-phenyl	Benzyl isocyanate	$C_6H_5\text{-}CH_2\text{-}NCO$
M-80	Methane 1-isocyanato 1-(p-methoxy-phenyl)		$CH_3O\text{-}C_6H_4\text{-}CH_2NCO$
M-81	Methane (1) -1-isocyanato 1-methyl 1-phenyl		$C_6H_5\text{-}CH(CH_3)(NCO)$
M-82	Methane- -1-isocyanato -1-methyl -1-phenyl		$C_6H_5\text{-}CH(CH_3)(NCO)$
M-83	Methane- 1-isocyanato 1,1-dimethyl 1-phenyl		$C_6H_5\text{-}C(CH_3)_2(NCO)$
M-84	Methane 1-isocyanato 1-ethyl 1-phenyl		$C_6H_5\text{-}CH(C_2H_5)NCO$
M-85	Methane 1-isocyanato 1,1-diethyl 1-phenyl		$C_6H_5\text{-}C(C_2H_5)_2(NCO)$

Formula	Boiling Point °C.	Boiling Point mm.	Melting Point °C.	Density	Refractive Index °C.	Refractive Index n	Other Data	Derivatives	References
$C_{21}H_{13}N_3O_4$								Phenyl urea MP 224-226	(194)
$C_{21}H_{13}N_3O_4$								Phenyl urea MP 145° (Sintered)	(194)
$C_{25}H_{22}N_2O$			116- 118						(194)
$C_{22}H_{13}N_3O_3$			89-96				Infrared (371)		(194) (253) (260) (371)
C_8H_7NO	162- 163 193- 194 103 82-84 88	760 670 17 10 5-6					Infrared (361)(371)	Urea from aniline MP 167-168° Urea from - chloro-ethyl- amine MP 105-106° 1,1-Benzyl-3- propyl urea MP 88° 1,1-Benzyl-3- phenyl urea MP 124° Benzyl allo- phanate MP 187-188°	(91) (130) (152) (191) (194) (233) (324) (361) (371) (505) (524) (241)
$C_9H_9NO_2$	102- 104	1							(172)
$C_9H_9NO_2$	82-83	12- 14					$[\alpha]_D^{24} = -2$ (benzene)	Urea from NH_3 MP 121-122° $[\alpha]_D^{25°} = +48.8°$	(116)
C_9H_9NO	90-94 96	13 18						Ethyl carba- mate, MP 37° BP 173°/23 mm Ref. 5	(102) (241) (194) (309)
$C_{10}H_{11}NO$	50-52	0.16			22°	1.5038		Urea from NH_3 MP 191° Urea from aniline MP 192-194°	(193)
$C_{10}H_{11}NO$	105- 106	17							(309)
$C_{12}H_{15}NO$	115	13						Urea from NH_3 MP 197°	(42)

No.	Name	Other Names	Structure
-	Methyl isocyanate	see Methane 1-isocyanato (M-31)	
-	Methyl isocyanato acetate	see Acetic acid -isocyanato- -methyl ester (A-10)	
-	Methyl -3-isocyanato -propionate	see Propanoic acid 3-isocyanato methyl ester (P-94)	
-	Methylene diisocyanate	see Methane diisocyanate (M-32)	
-	Morpholine	see Benzoic acid -2-isocyanato -1-morpholinyl amide (B-163)	
-	Myristyl isocyanate	see Tetradecane 1-isocyanate (T-1)	
-	Naphthalene -azo-phenyl 4,4'-diisocyanate	see Azobenzene 1-naphthyl 4,4'diisocyanato (A-32)	

	Formula	Boiling Point °C.	Boiling Point mm.	Melting Point °C.	Density	Refractive Index °C.	Refractive Index n	Other Data	Derivatives	References

No.	Name	Other Names	Structure
N-1	Naphthalene 1,4-bis(isocyanato methyl)		$C_{10}H_6(CH_2NCO)_2$
N-2	Naphthalene 1,5-bis(isocyanato methyl)		$C_{10}H_6(CH_2NCO)_2$
-	Naphthalene decahydro -2-isocyanato	see Decalin 2-isocyanato (D-1)	
N-3	Naphthalene 1,4-diisocyanato		$C_{10}H_6(NCO)_2$
N-4	Naphthalene 1,5-diisocyanato		$C_{10}H_6(NCO)_2$
N-5	Naphthalene 2,6-diisocyanato		$C_{10}H_6(NCO)_2$
N-6	Naphthalene 2,7-diisocyanato		$C_{10}H_6(NCO)_2$
N-7	Naphthalene 1-isocyanato		$C_{10}H_7(NCO)$
N-8	Naphthalene -2-isocyanato		$C_{10}H_7NCO$
N-9	Naphthalene 1-isocyanato 4-chloro		$C_{10}H_6(Cl)NCO$
N-10	Naphthalene 1-isocyanato 8-chloro		$C_{10}H_6(Cl)NCO$
N-11	Naphthalene 2-isocyanato 1-chloro		$C_{10}H_6(Cl)NCO$
N-12	Naphthalene 2-isocyanato 5-8 dibromo		$C_{10}H_5(Br)_2NCO$
N-13	Naphthalene 1-isocyanato 7,8-dihydro		$C_{10}H_8NCO$
N-14	Naphthalene 1-isocyanato 3-hydroxy		$C_{10}H_6(OH)NCO$
N-15	Naphthalene 1-isocyanato x-isopropyl		$C_{10}H_6(iC_3H_7)NCO$
N-16	Naphthalene 1-isocyanato 2-methyl		$C_{10}H_6(CH_3)NCO$

Formula	Boiling Point °C.	Boiling Point mm.	Melting Point °C.	Density	Refractive Index °C.	Refractive Index n	Other Data	Derivatives	References
$C_{14}H_{10}N_2O_2$	183-184	1						Dimethyl urethane MP 167°	(194)
$C_{14}H_{10}N_2O_2$			88-89					Dimethyl urethane MP 215-216°	(194)
$C_{12}H_6N_2O_2$			67-70					Dimethyl urethane MP 220-222°	(194)
$C_{12}H_6N_2O_2$	190	8	129-131				Heat of combustion 1317.0 kCal/mole (555) Heat capacity 0.287 Cal/gram/°C (555) Infrared data (254)	Dimethyl urethane MP 245-246	(194) (218) (254) (260) (555)
$C_{12}H_6N_2O_2$			152-154					Dimethyl urethane MP 244-246	(194)
$C_{12}H_6N_2O_2$			152-155						(194)
$C_{11}H_7NO$	269-270 140-142 120	760 12 2		1.1774 (20°/4°)			See Table No. 18	See Table No. 18	See Table No. 18
$C_{11}H_7NO$	144	14	56				See Table No. 19	See Table No. 19	(82) (254) (96) (281) (194) (386)
$C_{11}H_6ClNO$	116.5-117	1.5	44-45						(258)
$C_{11}H_6ClNO$	125-127	0.1						Methyl urethane MP 119-120°	(194)
$C_{11}H_6ClNO$	120	0.3	52-53					Methyl urethane MP 115-116°	(194)
$C_{11}H_5Br_2NO$								(110)	(110)
$C_{11}H_9NO$	137-138	16						Methyl urethane MP 86-87° Phenyl urea 210-212°	(194)
$C_{11}H_7NO_2$			205°						(28)
$C_{14}H_{13}NO$	152-161	3							(521)
$C_{12}H_9NO$	161-162	14	29						

No.	Name	Other Names	Structure
N-17	Naphthalene 1-isocyanato 5-nitro		$C_{10}H_6(NO_2)NCO$
N-18	Naphthalene 1-isocyanato 4-sulfonic acid, chloride		$C_{10}H_6(SO_2Cl)(NCO)$
N-19	Naphthalene 1-isocyanato 6-sulfonic acid, chloride		$C_{10}H_6(SO_2Cl)(NCO)$
N-20	Naphthalene 1-isocyanato 7-sulfonic acid, chloride		$C_{10}H_6(SO_2Cl)(NCO)$
-	Naphthalene 1,2,3,4-tetrahydro 1-isocyanato	see Tetralin 1-isocyanato (T-5)	
-	Naphthalene 1,2,3,4-tetrahydro 2-isocyanato	see Tetralin 2-isocyanato (T-6)	
-	Naphthalene 1,2,3,4-tetrahydro 1,5-diisocyanato	see Tetralin 1,5-diisocyanato (T-7)	
-	Naphthalene 5,6,7,8-tetrahydro 1-isocyanato 2-ethoxy	see Tetralin 1-isocyanato 2-ethoxy (T-8)	
N-21	Naphthalene 1,3,7-triisocyanato		$C_{10}H_5(NCO)_3$
-	Naphthyl see also the following: Binaphthyl 2,2'diisocyanate (B-189) Butane 2-isocyanato 2-methyl 4-(1-naphthyl) (B-219) Butane 3-isocyanato 3-methyl 4-(1-naphthyl) (B-220) Butane 3-isocyanato 3-methyl 4(2-naphthyl) (B-221) Butane 3-isocyanato 3-methyl 4-(1-naphthyl-2-methyl) (B-222)		

	Formula	Boiling Point °C.	Boiling Point mm.	Melting Point °C.	Density	Refractive Index °C.	Refractive Index n	Other Data	Derivatives	References
	$C_{11}H_6N_2O_3$			121-122					Methyl urethane MP 170°	(194)
	$C_{11}H_6ClNO_3S$	168-174	1.8							(354) (430)
	$C_{11}H_6ClNO_3S$	184-187 210-215	0.04 3							(354) (430)
	$C_{11}H_6ClNO_3S$	213-219 198-222	0.6 0.85							(354) (430)
	$C_{13}H_5N_3O_3$			162-163					Trimethyl urethane MP 217°	(194)

No.	Name	Other Names	Structure
	Butane 3-isocyanato 3-methyl 4-(1-naphthyl-4-methyl) (B-223)		
	Butane 3-isocyanato 3-methyl 4-(2-naphthyl-1-tert.butyl) (B-224)		
	Pentane 3-isocyanato 3-methyl 5-(1-naphthyl) (P-12)		
	Propane 2-isocyanato 2-methyl 3-(1-naphthyl) (P-91)		
	Propane 2-isocyanato 2-methyl 3-(2-naphthyl) (P-92)		
	Propane 2-isocyanato 2-methyl 3-(1-naphthyl-2-methyl) (P-93)		
N-22	2-Nitraza 1,4-butane diisocyanate		OCN-CH$_2$-N(NO$_2$)-CH$_2$-CH$_2$-NCO
N-23	3-Nitraza 1,5-pentane diisocyanate		OCN-CH$_2$-CH$_2$-N(NO$_2$)-CH$_2$-CH$_2$-NCO
N-24	2,5-Dinitraza -1,6-hexane diisocyanate		OCN-CH$_2$-N(NO$_2$)-CH$_2$-CH$_2$-N(NO$_2$)-CH$_2$-NCO
N-25	3,6-Dinitraza 1,8-octane diisocyanate		OCN-CH$_2$-CH$_2$-N(NO$_2$)-CH$_2$-CH$_2$-N(NO$_2$)-CH$_2$-CH$_2$-NCO
N-26	3,7-Dinitraza 1,9-nonane diisocyanate		OCN-CH$_2$-CH$_2$-N(NO$_2$)-CH$_2$-CH$_2$-CH$_2$-N(NO$_2$)-CH$_2$-CH$_2$-NCO
N-27	3,3,5,7,7,Penta- nitro nitraza -5-nonane 1,9-diisocyanate		OCN-(CH$_2$)$_2$-C(NO$_2$)$_2$-CH$_2$-N(NO$_2$)-CH$_2$-C(NO$_2$)$_2$-(CH$_2$)$_2$-NCO
N-28	3,6,9-Trinitraza -1-11-undecane diisocyanate		OCN-CH$_2$-CH$_2$-N(NO$_2$)-(CH$_2$)$_2$-N(NO$_2$)-(CH$_2$)$_2$-N(NO$_2$)-(CH$_2$)$_2$-NC

Formula	Boiling Point °C.	Boiling Point mm.	Melting Point °C.	Density	Refractive Index °C.	Refractive Index n	Other Data	Derivatives	References
$C_5H_6N_4O_4$			1-2						(495)
$C_6H_8N_4O_4$			25-26						(495)
$C_6H_8N_6O_6$			95-98						(495)
$C_8H_{12}N_6O_6$			91.5-93 (97.3% purity)						(495)
$C_9H_{14}N_6O_6$			35						(495)
$C_{10}H_{12}N_8O_{12}$			92-95						(514)
$C_{10}H_{16}N_8O_8$			90-95						(495)

No.	Name	Other Names	Structure
N-29	3,6,6,9-tetranitro 3,9-dinitraza 1,11-undecane diisocyanate		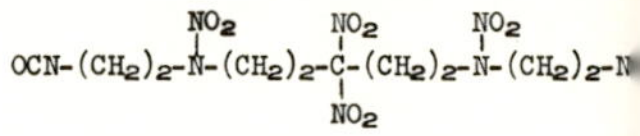
N-30	3,10-Dinitraza 1,12-dodecane diisocyanate		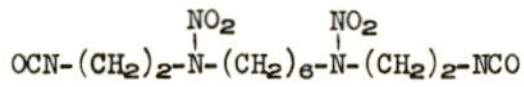
N-31	3-Nitraza, -butyl, -isocyanate		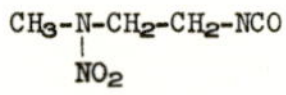
N-32	2,4,4-trinitro -2-aza-amyl -isocyanate		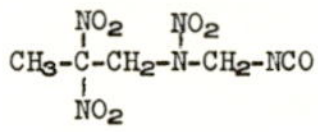
N-33	Nonadecane -2-isocyanato		$CH_3\text{-}(CH_2)_{16}\text{-}CH(NCO)\text{-}CH_3$
N-34	Nonane -5-diethylamino methyl -5-isocyanato		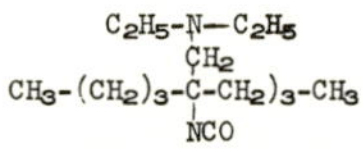
N-35	Nonane -1,9-diisocyanato		$OCN\text{-}(CH_2)_9\text{-}NCO$
N-36	Nonane,- -5-isocyanato -5-ethyl		$CH_3\text{-}(CH_2)_3\text{-}C(NCO)(C_2H_5)\text{-}(CH_2)_3\text{-}CH_3$
N-37	Nonane -5-isocyanato -5-n-butyl		$CH_3\text{-}(CH_2)_3\text{-}C(NCO)(C_4H_9)\text{-}(CH_2)_3\text{-}CH_3$
N-38	Nonane -3,3,5,5,7,7,hexanitro -1,9-diisocyanato		$OCN\text{-}(CH_2)_2\text{-}C(NO_2)_2\text{-}CH_2\text{-}C(NO_2)_2\text{—}CH_2\text{-}C(NO_2)_2\text{-}(CH_2)_2NCO$

Formula	Boiling Point °C.	Boiling Point mm.	Melting Point °C.	Density	Refractive Index °C.	Refractive Index n	Other Data	Derivatives	References
$C_{11}H_{16}N_8O_{10}$			88-90						(514)
$C_{12}H_{20}N_6O_6$			75-76						(495)
$C_4H_7N_3O_3$			25-26						(334) (539)
$C_5H_7N_5O_7$			102-103				Explodes at RT		(539)
$C_{20}H_{39}NO$	140-160	0.1							(194)
$C_{15}H_{30}N_2O$	126	4							(200)
$C_{11}H_{18}N_2O_2$	121 128	0.5 2.5						Dimethyl Urethane MP 102-105° Di-urea MP 177.5 - 177.8°	(194) (454)
$C_{12}H_{23}NO$								Phenyl urea MP 94° Diethylamino ethanol carbamate BP 164-168° (2.5 mm) N_D^{25}= 1.4568	(42) (227)
$C_{14}H_{27}NO$	103-105	1							(194)
$C_{11}H_{12}N_8O_{14}$			126-128						(515)

No.	Name	Other Names	Structure
-	Nonane 1-isocyanato 1,1-dihydro heptafluoro	see Fluoro isocyanates (F-50)	
-	Nonane 1-isocyanato Nonadecafluoro	see Fluoro isocyanates (F-40)	
N-39	1-Nonene -2,9-diisocyanato		$CH_2=C(NCO)-(CH_2)_7-NCO$
-	Norbornyl isocyanate	see Bicyclo[2,2,1] heptane 3-isocyanato	
N-40	Norcaradiene dibenzo -isocyanate	Benzotropylium -isocyanate (Attempted - Does not crystallize)	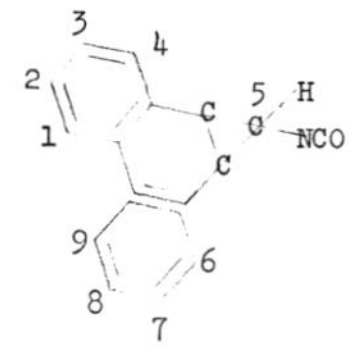
N-41	Norcaradienyl isocyanate	Tropylium isocyanate	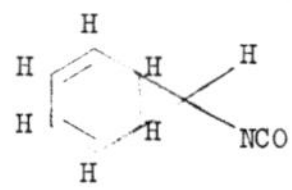 Norcaradienyl isocyanate
N-42	6-nor-cholestan -3 β-acetoxy -5,6-seco -5-oxo -7-isocyanato		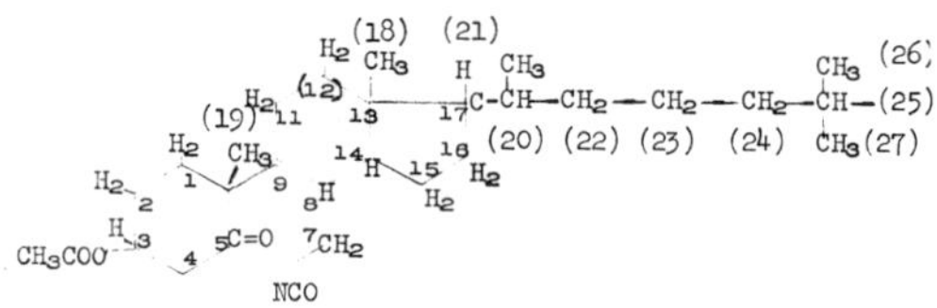

Formula	Boiling Point °C.	Boiling Point mm.	Melting Point °C.	Density	Refractive Index °C.	Refractive Index n	Other Data	Derivatives	References
$C_{11}H_{16}N_2O_2$	100	2			25	1.4668	Infrared data (523)		(523) (535)
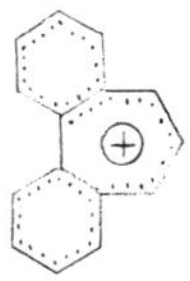 Tropylium cation			78						(462) (440) (351)
C_8H_7NO 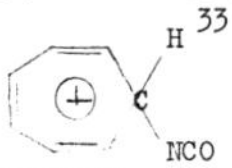 Tropylium isocyanate	80-85 33	13 0.25	198				Norcoradienyl isocyanate (solid) Tropylium isocyanate (liquid)	Di-urea MP 145° Phenyl urea MP 184°	(275) (339) (350) (388)
$C_{29}H_{47}NO_4$			88- 88.5						(488)

No.	Name	Other Names	Structure
O-1	Octane 1-isocyanato		$CH_3(CH_2)_7$-NCO
O-2	Octane 2-isocyanato	2-Octyl isocyanate	CH_3-$(CH_2)_5$-CH(NCO)-CH_3
O-3	Octane -2 Methyl -2 Isocyanato	1,1 Dimethyl Heptyl isocyanate	CH_3-$(CH_2)_5$-C$(CH_3)_2$-NCO
O-4	Octane, -4 isocyanato -2,7 dimethyl	isocybutyl isohexyl isocyanate	$(CH_3)_2$-CH-CH_2-CH(NCO)-$(CH_2)_2$-CH$(CH_3)_2$
O-5	Octane, 2-isocyanato -2-Methyl-4-Ethyl	1,1 Dimethyl-3 Ethyl Heptyl isocyanate	CH_3-$(CH_2)_3$-CH(C_2H_5)-CH_2-C$(CH_3)_2$-NCO
O-6	Octane 1,8 diisocyanato	Octa methylene diisocyanate	OCN-$(CH_2)_8$-NCO
O-7	Octane- -1,6 diisocyanate	Octamethylene ester	OCN-$(CH_2)_5$-CH(NCO)-CH_2-CH_3
O-8	Octane- 3,6 diisocyanato	1,4 Diethyl- tetramethylene ester	CH_3-CH_2-CH(NCO)-$(CH_2)_2$-CH(NCO)CH_2-CH_3
O-9	Octane, -2-ethyl -1,8 diisocyanato	2-ethyloctamethylene ester	OCN-CH_2-CH(C_2H_5)-$(CH_2)_6$-NCO
	Octane 1-isocyanato 1,1-dihydra pentadecafluoro	see Fluoro isocyanates (F-49)	
	Octane -1-isocyanato -heptadecafluoro	see Fluro isocyanates (F-39)	

Formula	Boiling Point °C.	Boiling Point mm.	Melting Point °C.	Density	Refractive Index °C.	Refractive Index n	Other Data	Derivatives	References
$C_9H_{17}NO$	94-95	18		0.8768 (20/4)	20	1.4314			(207) (450) (487)
	114-115	50			22	1.4290			
	200-204	760							
$C_9H_{17}NO$	68	5		0.8260 (25/25)	25	1.4230			(331)
$C_{10}H_{19}NO$	85-86	11						Octane, 2-amino 2-methyl BP 165-170° (760 mm)	(162)
$C_{11}H_{21}NO$	105-115	25						Urea from H_2O MP 228-230° Phenyl urea MP 190° Corresponding Amine BP 186-190° (744 mm)	(28)
	210-220	760							
$C_{12}H_{23}NO$	104-106	19						(134)	
$C_{10}H_{16}N_2O_4$	98-104	0.1						Dimethyl urethane MP 111-112° N,N'octa-methylene urea MP 165°	(160) (184) (194) (428)
	126-128	2							
	146-148	11							
$C_{10}H_{16}N_2O_2$	85-88	0.1						Methyl urethane MP 85.5-86°	(454) (428)
$C_{10}H_{16}N_2O_2$	66-70	0.5						Methyl diurethane MP 150-154°	(428)
$C_{12}H_{20}N_2O_2$	110-112	0.1						Dimethyl urethane MP 51.5-53°	(428)

No.	Name	Other Names	Structure
O-10	1-Octene 2-isocyanato	1-Hexyl vinyl	$CH_2{=}C(NCO)-(CH_2)_5-CH_3$
O-11	Octadecane 1-isocyanato		$CH_3-(CH_2)_{17}-NCO$
O-12	Octadecane -2-isocyanato -2-methyl	1,1 Dimethyl- heptadecyl isocyanate	$CH_3-(CH_2)_{15}-C(NCO)(CH_3)_2$
-	Oleyl isocyanate	see 9-octadecene 1-isocyanate (O-14)	
-	Octadecyloxy -methyl -isocyanate	see Methane 1-isocyanato 1-octadecyloxy (M-36)	
-	n-octyloxo -propyl -isocyanate	see Propane 1-isocyanato 3-n-octyloxo (P 74)	
O-13	12-Oleane -3 β-acetoxy -11-oxonor 20-β-isocyanato		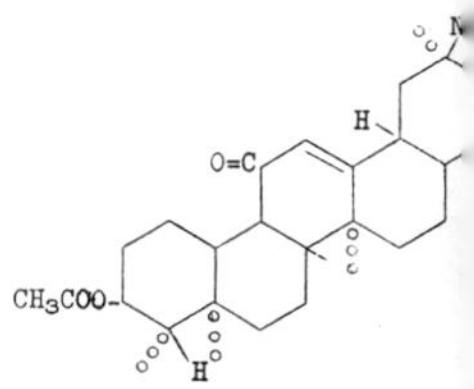
O-14	9-Octadecene -1-isocyanato	Oleyl isocyanate	$CH_3-(CH_2)_7-CH{=}CH-(CH_2)_8-NCO$
O-15	1,2,4 oxadiazole -5-isocyanato -3 phenyl		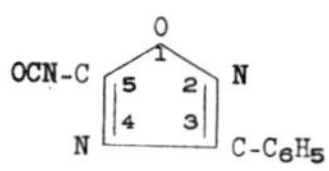
O-16	Oxazole -2-phenyl -4-isocyanato		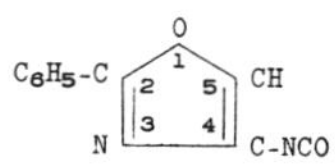

Formula	Boiling Point °C.	Boiling Point mm.	Melting Point °C.	Density	Refractive Index °C.	Refractive Index n	Other Data	Derivatives	References
$C_9H_{15}NO$	60	8			25	1.4408	Infra-red data (523)	Urea from NH_2 MP 45° Phenyl urea MP 91-92°	(484) (494) (523)
$C_{19}H_{37}NO$	135.5-137 150-170 172-173 190-210 210-220	0.2 3 5 10 20					Infra-red (254)	Octadecylurea from NH_3 MP 101° Dioctadecylurea MP 105-106°	(85) (127) (130) (141) (194) (218) (254) (418) (505) (556)
$C_{20}H_{39}NO$	165	0.9							(134)
$C_{32}H_{47}NO_4$			293-294				$[\alpha]_D^{20} = +164.0°$ C = 0.86		(530)
$C_{19}H_{35}NO$	162-163 181-183.5 135-140	1 4 0.05							(274) (127) (194)
$C_9H_5N_3O_2$			60						(544)
$C_{10}H_6N_2O_2$	116	0.8	194				Infra-red data	Cyclohexylcarbamate MP 87-88°	(498)

No.	Name	Other Names	Structure
P-1	Pentadecane 1-isocyanato	Palmityl isocyanate	$CH_3-(CH_2)_{14}NCO$
P-2	Pentadecane 4-isocyanato		$CH_3-(CH_2)_2-CH(NCO)-CH_2)_{10}-CH_3$
P-3	Pentadecane -8-isocyanato -8-methyl		$CH_3-(CH_2)_6-C(NCO)(CH_3)-(CH_2)_6-CH_3$
P-4	Pentadecane -8-isocyanato -8-n-butyl		$CH_3-(CH_2)_6-C(NCO)(C_4H_9)-(CH_2)_6-CH_3$
P-5	Pentadecane 1-15 diisocyanato		$OCN-(CH_2)_{15}-NCO$
P-6	Pentane 1-isocyanato	n-amyl isocyanate	$CH_3-(CH_2)_4-NCO$
P-7	Pentane -3-isocyanato		$CH_3-CH_2-CH(NCO)-CH_2-CH_3$
P-8	Pentane 1-isocyanato 5-chloro		$Cl-(CH_2)_5-NCO$
P-9	Pentane -1-cyano -5-isocyanato		$NC-(CH_2)_5-NCO$
P-10	Pentane 1-isocyanato -4-methyl	Isohexyl isocyanate	$(CH_3)_2-CH-(CH_2)_3-NCO$
P-11	Pentane 2-isocyanato 2-methyl		$CH_3-CH_2-CH_2-C(CH_3)(NCO)-CH_3$
P-12	Pentane 3-isocyanato 3-methyl 5 (1-naphthyl)		$i-C_{10}H_7-CH_2-CH_2-C(CH_3)(NCO)-CH_2-CH_3$
P-13	Pentane -2-isocyanato -2,4,4-trimethyl		$CH_3-C(CH_3)_2-CH_2-C(CH_3)(NCO)-CH_3$
-	Pentane 1-isocyanato 1,1-dihydro nonafluoro	see Fluoro isocyanates (F-46)	
-	Pentane 1-isocyanato undecafluoro	see Fluoro isocyanates (F-36)	
P-14	Pentane 1,5 diisocyanato		$OCN-(CH_2)_5-NCO$
P-15	Pentane -2-methyl -2,4-diisocyanato		$OCN-C(CH_3)_2-CH_2-CH(CH_3)-NCO$

Formula	Boiling Point °C.	Boiling Point mm.	Melting Point °C.	Density	Refractive Index °C.	Refractive Index n	Other Data	Derivatives	References
$C_{16}H_{31}NO$	159-162	3							(450)
$C_{16}H_{31}NO$	150-170	12							(194)
$C_{17}H_{35}NO$	130-132	2							(200)
$C_{20}H_{39}NO$	176-177	6.5							(200)
$C_{17}H_{30}N_2O_2$	157-160	0.20						N,N+ di-urea MP 185-186°	(454)
$C_6H_{11}NO$	136-137 48-50	760 14					Infra-red (371)	Ethyl urethane MP 218° Urea from NH_3 MP 89-91° Phenyl urea MP 92-93°	(2) (194) (207) (324) (371) (505) (524)
$C_6H_{11}NO$	125-126	760					Infra-red (371)	Phenyl urea MP 144-145°	(194) (371)
$C_6H_{10}ClNO$	100	13							(194)
$C_7H_{10}N_2O$	133	11					Infra-red (371)		(194) (371)
$C_7H_{13}NO$	50 46	760 14						di-isohexyl urea BP 122-123° MP 44°	(38) (194)
$C_7H_{13}NO$	136-137	760						Diurea MP 155°	(37) (42)
$C_{17}H_{19}NO$	223-227	15							(138)
$C_9H_{17}NO$	64-66 170-172	30 760		0.8588 (25/25)	20 25	1.429 1.4243			(242) (267) (308) (331)
$C_7H_{10}N_2O_2$	116 122-125	11 15						Diurea MP 233-234°	(12) (194) (270) (454)
$C_8H_{12}N_2O_2$	90-92	10		1.019 (20.5°)	20	1.4445		Methyl urethane MP 81°	(461)

No.	Name	Other Names	Structure
P-16	Pentane -2,2 dimethyl -1,5 diisocyanato		$OCN-CH_2-C(CH_3)_2-CH_2)_3-NCO$
P-17	Pentane -3,3-dinitro -1,5-diisocyanato		$OCN-CH_2-CH_2-C(NO_2)_2-CH_2-CH_2-NCO$
P 18	Pentane -2,2,4,trimethyl -1,5-diisocyanato		$OCN-CH_2-C(CH_3)_2-CH_2-CH(CH_3)-CH_2-NCO$
P-19	Pentanedioic acid, 2-isocyanato Dimethylester	Glutaric acid 2-isocyanato dimethylester	$CH_3-O-CO-(CH_2)_2-CH(NCO)-CO-O-CH_3$
P-20	Pentanedioic acid, -2-isocyanato diethylester	Glutaric acid -2-isocyanato diethylester	$C_2H_5-O-CO-(CH_2)_2-CH(NCO)-CO-O-C_2H_5$
P-21	(DL) Pentanoic acid, -2-Isocyanato ethylester	DL Valeric acid 2-Isocyanato ethylester	$CH_3-CH_2-CH_2-CH(NCO)-CO-O-C_2H_5$
P-22	(DL) Pentanoic acid -2-isocyanato -4-methyl ethylester	DL Valeric acid -2-isocyanato -4-methyl ethylester	$(CH_3)_2-CH-CH_2-CH(NCO)-CO-O-C_2H_5$
P-23	(L) Pentanoic acid, -2-isocyanato -4-methyl ethyl ester	L-Valeric acid -2-isocyanato -4-methyl ethyl ester	$(CH_3)_2CH-CH_2-CH(NCO)-CO-O-C_2H_5$
P-24	(DL) Pentanoic acid, -2-isocyanato -3-methyl ethylester	DL-Valeric acid, -2-isocyanato -3-methyl ethylester	$CH_3-CH_2-CH(CH_3)-CH(NCO)-CO-O-C_2H_5$
P-25	Pentanoic acid -2-isocyanato -2-isobutyl -4-methyl ethylester	Leucic acid, 2-isobutyl 2-isocyanato ethylester	$(CH_3)_2CH\ CH_2\ C(isoC_4H_9)(NCO)\ CO-O-C_2H$
P-26	Pentanoic acid -5-isocyanato cholesterol ester	Valeric acid, δ-hydroxy,-isocyanato cholesterol ester	$OCN-CH_2-(CH_2)_3-CO-O-C_{27}H_{45}$
P-27	Pentanoic acid 2-diphenyl 4-isocyanato nitrile	3,3 diphenyl,- -3 cyano, -1-methyl -1-propyl -isocyanate	$CH_3CH(NCO)CH_2C(C_6H_5)_2CN$
P-28	Pentanoic acid -2-ethyl -3-methyl -isocyanato		$CH_3-CH_2-CH(CH_3)-CH(C_2H_5)-CO-NCO$
P-29	1-Pentene 1-isocyanato		$CH_3-CH_2-CH_2-CH_2=CH-NCO$
P-30	1-Pentene 2,5 diisocyanato		$CH_2=C(NCO)-(CH_2)_3-NCO$

Formula	Boiling Point °C.	Boiling Point mm.	Melting Point °C.	Density	Refractive Index °C.	Refractive Index n	Other Data	Derivatives	References
$C_9H_{14}N_2O_2$	120-122	12							(194)
	132-134	16							
$C_7H_8N_4O_6$			39.5					Phenylurea MP 186-187°	(220)
$C_{10}H_{16}N_2O_2$	106	5						Diurea MP 159-162°	(194)
$C_8H_{11}NO_5$	103	1.5	118		25	1.4442			(537)
$C_{10}H_{15}NO_5$	116	2	96		25	1.4390			(537)
$C_8H_{13}NO_3$	94	14							(240)
$C_9H_{15}NO_3$	97	11							(240)
	104.5	15							
$C_9H_{15}NO_3$	93	8	113		25	1.4273	$\alpha_D^{20} = -22.40$		(240) (537)
	104.5	15							
$C_9H_{15}NO_3$	94	11							(240)
$C_{13}H_{23}NO_3$	95	3.9						Urea from NH_3 MP 159.5-160.5	(128)
$C_{33}H_{53}NO_3$									(141)
$C_{18}H_{16}N_2O$			111-114						(433) (444) (347)
$C_9H_{15}NO_2$	55-56	11							(108) (153)
C_6H_9NO	133	760					Infrared data (523) Reaction rate (494)		(523) (494)
$C_7H_8N_2O_2$	65	3			25	1.4707			(523) (535)

No.	Name	Other Names	Structure
P-31	4-Pentenoic acid 2-ethyl isocyanato		CH_2=CH-CH_2-CH(C_2H_5)-CO-NCO
P-32	4-Pentenoic acid -2,2-dimethyl isocyanato		CH_2=CH CH_2 C(CH_3)$_2$-CO NCO
-	Phenacetyl isocyanate	see Acetic acid 1-phenyl isocyanate (A-12)	
-	p-phenyl azophenyl isocyanate	see Azobenzene 4-isocyanato (A-25)	
-	Phenylethyl acetyl isocyanato	see Acetic acid 1-phenyl 1-ethyl isocyanato (A-7)	
P-33	Phenanthrene -3-isocyanato		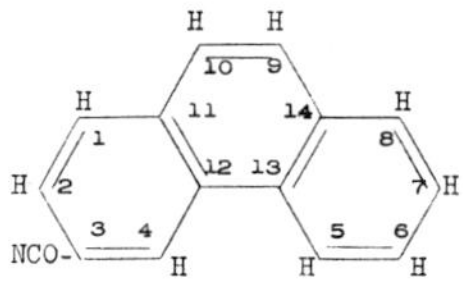
P-34	Phenanthrene -4-isocyanato		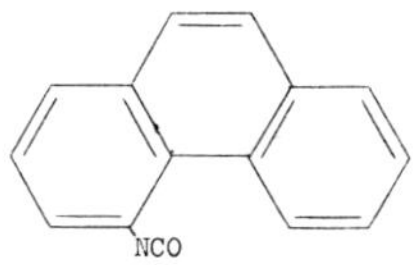
	Phenanthrene 5-isocyanato	see Norcaradienyl isocyanate (N-41)	

PHOSPHOROUS ISOCYANATES

No.	Name	Other Names	Structure
P-35	Benzyl phosphonyl diisocyanate		C_6H_5-CH_2-PO-(NCO)$_2$
P-36	Diethyl- (benzyl isocyanate) phosphonate		(C_2H_5O)$_2$-PO-CH(NCO)-C_6H_5
P-37	Diethyl- -phosphonyl -isocyanate		(C_2H_5O)$_2$-PO-NCO

Formula	Boiling Point °C.	Boiling Point mm.	Melting Point °C.	Density	Refractive Index °C.	Refractive Index n	Other Data	Derivatives	References
$C_8H_{11}NO_2$	83-85	35							(108) (153)
$C_8H_{11}NO_2$	87-89	40							(546)
$C_{15}H_9NO$	198-204	4	48					Methyl urethane MP 140-142°	(194)
			68.6-69						(349)
$C_9H_7N_2O_3P$	145	1-2						Urea from Aniline MP 188-189° Methyl urethane MP 179-181°	(411)
$C_{12}H_{18}NO_4P$	131-134	2		1.1825 (20°)	20	1.5052		Urea from NH_3	(328) (344)
$C_5H_{10}NO_4P$	60-65	760							(411)

No.	Name	Other Names	Structure
P-38	Diethyl- -(2 propyl isocyanato) phosphonate		$(C_2H_5O)_2$ PO-C(CH_3)NCO
P-39	Ethyl phosphonyl diisocyanate		C_2H_5-PO-(NCO)$_2$
P-40	Isopropyl phosphonyl diisocyanate		$(CH_3)_2$CH-PO-(NCO)$_2$
P-41	Phenyl phosphonyl isocyanate		C_6H_5-O-PO-(NCO)$_2$
P-42	Phenyl diisocyanate phosphine		C_6H_5-P-(NCO)$_2$

Formula	Boiling Point °C.	Boiling Point mm.	Melting Point °C.	Density	Refractive Index °C.	Refractive Index n	Other Data	Derivatives	References
$C_8H_{16}NO_4P$	75-77	3		1.1014 (20°)	20	1.4349	Molar Refractivity 52.4	Urea from NH_3 MP 139°	(328)
								Urea from Aniline MP 137-138° Urea from p-toluidine MP 157-158° Ethyl urethane (344)	(344)
$C_4H_5NO_3P$	61-65	0.65-1.0						Diphenyl urea MP 204-205° Methyl urethane MP 142-144°	(411)
	65	1							
	69	1.5							
$C_5H_6N_2O_3P$	61-67	1.0-1.75						Methyl urethane MP 163-165° Urea from aniline MP 192-193°	(411)
$C_8H_5N_2O_4P$							Polymerizes	Urea from aniline MP 192-193°	(411)
$C_8H_5N_2O_2P$	118-122	3						Urea from aniline MP 175 178° Methyl urethane MP 161 163°	(411)

No.	Name	Other Names	Structure
P-43	Phosphorous triisocyanate		$P(NCO)_3$
P-44	Phosphoryl triisocyanate		$PO(NCO)_3$
P-45	Triphenyl-phosphine-imine-isocyanate		$[(C_6H_5)_3P\text{-}NH_2]\,NCO$
P-46	Chloromethyl phosphonyl diisocyanate		$ClCH_2\text{-}PO\text{-}(NCO)_2$
P-47	Phosphorous chloro diisocyanate		$PCl(NCO)_2$
P-48	Phosphorous dichloro isocyanate		$PCl_2(NCO)$
P-49	Phosphorous fluoro diisocyanate		$PF(NCO)_2$
P-50	Phosphorous difluoro isocyanate		$PF_2(NCO)$
P-51	Phosphine diethyl (ethyl, isocyanato-1-methyl) sulfide		$(C_2H_5)_2\text{-}PS\text{-}C(CH_3)NCO$
P-52	Thiophosphoryl triisocyanate		$PS(NCO)_3$

R(Ml) Molar Refractivity
Δ H - Molar heat of vaporization
Δ H/T = Troutons Constant

Vapor Pressure

$\log_{10} p = A - \frac{B}{T}$

Formula	Boiling Point °C.	mm.	Melting Point °C.	Density	Refractive Index °C.	n	R (ML)	ΔH kCal	ΔH/Tβ	Vapor Pressure A	B	References
$C_3N_3O_3P$	169.3	760	-2	1.439 (20°)	25	1.5320	-	11.9	26.8	8.7455	2595	(103) (449)
	60.5-63.0	10		1.515 (20°)								
$C_3N_3O_4P$	193.1	760	5.0	1.570 (20°)	20	1.4804	-	13.41	28.8	9.1682	2931	(121) (145)
$C_{19}H_{17}N_2OP$			120									(469)
$C_3H_2ClN_2O_3P$	76-82	0.6			Urea from aniline - MP 199-200° Dimethyl urethane - MP 181-183°							(411)
$C_2ClN_2O_2P$	135	760		1.500 (25°)	25	1.5239						(449)
	54	25										
CCl_2NOP	104.4	760	-99	1.512 (25°)	26	1.511						(145) (449)
$C_2FN_2O_2P$	98.7	760	FP -55	1.475	20	1.4678	-	9.25	24.8	8.3210	2022.7	(175)
CF_2NOP	123		FP ca. -108	1.444 (20°)	20	1.3700						(175)
$C_8H_{16}NOPS$	71-74	3		1.1102 (20°)	20	1.4778						(328)
					Urea from NH_3 - MP 104-106° Di-urea - MP 72- 73°							
$C_3N_3O_3PS$	215	760	8.8	1.538		1.5116		16.0	33	10.032	3492	(131)
	150	60										
	135	30										

No.	Name	Other Names	Structure
	Piperidyl	**see Hexane** 2-isocyanato 2-n-butyl 1-(1-piperidyl (H-15)	
P-53	Propane 1-isocyanato	n-propyl isocyanate	CH_3-CH_2-CH_2-NCO
P-54	Propane 2-isocyanato	Isopropyl isocyanate	CH_3-CH(CH_3)-NCO
P-55	Propane -1,3 diisocyanato	Trimethylene diisocyanate	OCN-CH_2-CH_2-CH_2-NCO
P-56	Propane 1-isocyanato 2-methyl	Isobutyl -isocyanate	CH_3-CH(CH_3)-CH_2-NCO
P-57	Propane 2-isocyanato 2-methyl	tert-Butyl isocyanate	$(CH_3)_3$-C-NCO
P-58	Propane 1-isocyanato 1,2 dimethyl		CH_3-CH(CH_3)-CH(CH_3)-NCO
P-59	Propane 1-isocyanato 3-dimethylamino		$(CH_3)_2$-N-CH_2-CH_2-CH_2-NCO
P-60	Propane 1-cyano 3 isocyanato	1-Cyanpropyl isocyanate	NC-CH_2-CH_2-CH_2-NCO
P-61	Propane 1-chloro 2 isocyanato		CH_3-CH(NCO)CH_2-Cl
P-62	Propane 1-isocyanato 2 chloro		CH_3-CH(Cl)-CH_2-NCO
P-63	Propane 1-isocyanato 3-(4 chloro butoxy)		Cl$(CH_2)_4$-O-$(CH_2)_3$-NCO

Formula	Boiling Point °C.	Boiling Point mm.	Melting Point °C.	Density	Refractive Index °C.	Refractive Index n	Other Data	Derivatives	References
C_4H_7NO	86-88	760						Dipropyl urea MP 105° (324)	(26) (194) (207) (224) (324) (505)
C_4H_7NO	70-75	760		0.8669 (20/4)	20 25	1.3886 1.3744	Raman data (60)	Diisopropylurea MP 192° Urea from NH_3 MP 154-156° Phenyl urea MP 156°	(3) (10) (38) (60) (207) (324) (331) (505)
$C_5H_6N_2O_2$	71 86-87 97	4 14 21							(194) (213)
C_5H_9NO	104-105	760					Infra-red (371)	Phenyl urea MP 151-152° 1,3 Diisobutyl urea MP 135°	(7) (194) (168) (207) (324) (371) (505) (547)
C_5H_9NO	84-88	760		0.8359 (25/25)	-	1.3824		Phenyl urea MP 153° or 167-168° 1,3 ditert-. Butyl urea 242° or 248°	(137) (194) (267) (308) (324) (331)
$C_6H_{11}NO$	118-120	760					Infra-red (371)	Phenyl urea MP 143-144°	(194) (371)
$C_6H_{12}N_2O$	50-60	13						Phenyl urea MP 93°	(490)
$C_5H_6N_2O$	103	10					Infra-red (371)	Urea MP 142°	(194) (371)
C_4H_6ClNO	141-142				18.2	1.4472		α Naphthyl urea MP 176° Phenyl urea MP 132°	(466)
C_4H_6ClNO	131-132				20	1.4496		α Naphthyl urea MP 144-145° Benzyl urea MP 214-215° Phenyl urea MP 143°	(466)
$C_8H_{14}ClNO_2$	76	0.10						(194)	

No.	Name	Other Names	Structure
P-64	Propane 1-isocyanato 3-chloro		$ClCH_2-CH_2-CH_2-NCO$
P-65	Propane 2-isocyanato 2-chloro		$CH_3-C(Cl)(NCO)-CH_3$
-	Propane 1-isocyanato 3-fluoro	see Fluoro isocyanates (F-31)	
-	Propane 2-isocyanato 1,3 hexafluoro	see Fluoro isocyanates (F-32)	
-	Propane 1-isocyanato heptafluoro	see Fluoro isocyanates (F-34)	
-	Propane 1-isocyanato 1,1 dihydro pentafluoro	see Fluoro isocyanates (F-44)	
P-66	Propane 1-isocyanato 3-methoxy		$CH_3-O-CH_2\ CH_2-CH_2-NCO$
P-67	Propane 1-isocyanato 3-ethoxy		$CH_3-CH_2-O-CH_2-CH_2-CH_2-NCO$
P-68	Propane 1-isocyanato 3-propoxy		$CH_3-CH_2-CH_2-O-CH_2-CH_2-CH_2-NCO$
P-69	Propane 1-isocyanato 3-isopropoxy		$CH_3-CH(CH_3)-O-(CH_2)_3-NCO$
P-70	Propane 1-isocyanato 3-isobutoxy		$CH_3-CH(CH_3)-CH_2-O-(CH_2)_3-NCO$
P-71	Propane 1-isocyanato 3 butoxy		$CH_3-(CH_2)_3-O-(CH_2)_3-NCO$
P 72	Propane 1 isocyanato 3-sec. butoxy		$CH_3-CH_2-CH(CH_3)-O-(CH_2)_3-NCO$
P-73	Propane 1-isocyanato 3-(2 butoxy-ethoxy)		$CH_3-(CH_2)_3-O-(CH_2)_2-O-(CH_2)_3-NCO$
P-74	Propane 1-isocyanato 3-octyloxy		$CH_3-(CH_2)_7-O-(CH_2)_3-NCO$
P-75	Propane 1-isocyanato 3-iso-octyloxy		$CH_3-CH(CH_3)-(CH_2)_5-O-(CH_2)_3-NCO$
P-76	Propane 1-isocyanato 3-decyloxy		$CH_3-(CH_2)_9-O-(CH_2)_3-NCO$
P-77	Propane 2-isocyanato 2-methyl 3(vinyloxy)		$CH_2=CH-O-CH_2-C(CH_3)_2-NCO$

Formula	Boiling Point °C.	Boiling Point mm.	Melting Point °C.	Density	Refractive Index °C.	Refractive Index n	Other Data	Derivatives	References
C_4H_8ClNO	54.7	16						Phenyl urea MP 127°	(194)
C_4H_8ClNO	26-28	110							(557)
$C_5H_9NO_2$								Phenyl urea MP 79-80°	(194)
$C_6H_{11}NO_2$									(194)
$C_7H_{13}NO_2$	60-61	11							(194)
$C_7H_{13}NO_2$									(194)
$C_8H_{15}NO_2$	78	14						Phenyl urea MP 80°	(194)
$C_8H_{15}NO_2$								Phenyl urea MP 59-60°	(194)
$C_8H_{15}NO_2$								Phenyl urea MP 53-54°	(194)
$C_{10}H_{19}NO_3$	123-124	11							(194)
$C_{12}H_{23}NO_2$	135	11						Phenyl urea MP 55-56°	(194)
$C_{12}H_{23}NO_2$	125-128	11							(194)
$C_{14}H_{27}NO_2$	160-162	11						Phenyl urea MP 58-59°	(194)
$C_7H_{11}NO_2$	62	25							

No.	Name	Other Names	Structure
P-78	Propane 1-isocyanato 3(1-propoxy-3-isocyanato)		$OCN-(CH_2)_3-O-(CH_2)_3-NCO$
P-79	Propane 2-isocyanato 2-nitro		$CH_3-C(NO_2)(NCO)-CH_3$
P-80	Propane 3,3,3 trinitro 1-isocyanato		$(NO_2)_3\ C-CH_2-CH_2\ NCO$
P-81	Propane 1-isocyanato 3-phenyl		$C_6H_5-CH_2-CH_2\ CH_2-NCO$
P-82	Propane 1 isocyanato 1-methyl 3-phenyl		$C_6H_5-CH_2-CH_2-CH(CH_3)-NCO$
P-83	Propane 2-isocyanato 1,3-diphenyl		$C_6H_5-CH_2-CH(NCO)-CH_2-C_6H_5$
P-84	Propane 1-isocyanato 1,1-dimethyl 3-cyclohexyl		$C_6H_{11}-CH_2-CH_2-C(CH_3)-NCO$
P-85	Propane 1-isocyanato 1,1-dimethyl 3-(3-methyl cyclohexyl)		$(CH_3)C_6H_{10}-CH_2-CH_2-C(CH_3)_2-NCO$
P-86	Propane 1-isocyanato 3-(3 phenyl isocyanato)		$OCN-C_6H_4-CH_2-CH_2-CH_2-NCO$
P-87	Propane 1-isocyanato 3 (4-phenyl isocyanato)		$OCN-C_6H_4-CH_2-CH_2\ CH_2-NCO$
P-88	Propane 1-isocyanato 1-methyl 3-(4-phenyl isocyanato)		$OCN-C_6H_4-CH_2-CH_2-CH(CH_3)-NCO$
P-89	Propane 1-isocyanato 1-methyl 3-(3-phenyl isocyanato)		$OCN-C_6H_4-CH_2-CH_2-CH(CH_3)-NCO$
P-90	Propane 1,3(phenoxy-4-isocyanato)		$C_6H_5(NCO)-O-CH_2-CH_2-CH_2-O-C_6H_5(NCO)$
P-91	Propane -2-isocyanato -2-methyl -3-(1-naphthyl)		$1-C_{10}H_7-CH_2-C(CH_3)_2-NCO$
P-92	Propane 2-isocyanato 2-methyl 3-(2-naphthyl)		$2-C_{10}H_7-CH_2-C(CH_3)_2-NCO$
P-93	Propane 2-isocyanato 2-methyl 3-(1-naphthyl-2-methyl)		$1-C_{10}H_6(CH_3)-CH_2-C(CH_3)_2-NCO$
P-94	Propanoic acid 3-isocyanato methyl ester		$OCN-CH_2-CH_2-CO-O-CH_3$

Formula	Boiling Point °C.	Boiling Point mm.	Melting Point °C.	Density	Refractive Index °C.	Refractive Index n	Other Data	Derivatives	References
$C_8H_{12}N_2O_3$	137	14						Diurea MP 141°	(194)
$C_4H_6N_2O_3$									(393)
$C_4H_4N_4O_7$	73-75	0.001		1.5266 (25°)	25	1.4805		Ethyl urethane MP 66°	(536) (553)
$C_{10}H_{11}NO$	115	12							(194)
$C_{11}H_{13}NO$	117	12							(194)
$C_{16}H_{15}NO$	197	15-18						Urea from NH_3 MP 138°	(246) (273)
$C_{12}H_{21}NO$	132	20							(182)
$C_{13}H_{23}NO$	133-135	15							(182)
$C_{11}H_{10}N_2O_2$	118-120	0.5							(194)
$C_{11}H_{10}N_2O_2$	100-115	0.5							(194)
$C_{12}H_{12}N_2O_2$	142-145	3.5							(194)
$C_{12}H_{12}N_2O_2$	146-149	2							(194)
$C_{17}H_{14}N_2O_4$			103-105						(194)
$C_{15}H_{15}NO$	176-180	14							(138)
$C_{15}H_{15}NO$	155-157	1							(132)
$C_{16}H_{17}NO$	162-164	1							(132)
$C_5H_7NO_3$	61 82-84	6 18	88		25	1.4312			(417) (537)

No.	Name	Other Names	Structure
P-95	Propanoic acid 3 isocyanato ethyl ester		$OCN-CH_2-CH_2-CO-O-C_2H_5$
P-96	Propanoic acid -2-isocyanato -2-methyl ethyl ester		$OCN-C(CH_3)_2-CO-O-C_2H_5$
P-97	Propanoic acid -isocyanato 3-bromo		$Br-CH_2-CH_2-CO-NCO$
P-98	Propanoic acid 3-chloro 2-(ethyl isocyanate) ester		$ClCH_2CH_2-CO-O-CH_2CH_2-NCO$
P-99	Propanoic acid 3-isocyanato 3-phenyl methyl ester		$C_6H_5-CH(NCO)-CH_2-CO-O-CH_3$
P-100	Propanoic acid 3-isocyanato 3-phenyl ethyl ester		$C_6H_5-CH(NCO)-CH_2-CO-O-C_2H_5$
P-101	Propanoic acid 2 isocyanato 3-phenyl ethyl ester		$C_6H_5-CH_2-CH(NCO)-CO-O-C_2H_5$
P-102	Propene -3-isocyanato	Allyl isocyanate	$CH_2=CH\ CH_2-NCO$
P-103	Propene -1-isocyanato	Propenyl isocyanate	$CH_3-CH=CH-NCO$
P-104	Propene -2-isocyanato	Isopropenyl isocyanate	$CH_2=C(CH_3)-NCO$
P-105	Propene -2-methyl -1 isocyanato		$CH_3-C(CH_3)=CH-NCO$
P-106	Propene 2,3 diisocyanato		$OCN-CH_2-C(NCO)=CH_2$

Formula	Boiling Point °C.	Boiling Point mm.	Melting Point °C.	Density	Refractive Index °C.	Refractive Index n	Other Data	Derivatives	References
$C_6H_9NO_3$	69	3	116		25	1.4272		Urea from NH_3 MP 101°	(240) (417)
	94-95	20						Phenyl urea MP 174°	(537)
$C_7H_{11}NO_3$	61.5	12						5,5 dimethyl-hydantoin from Conc. NH_4OH MP 175°	(240)
$C_4H_4BrNO_2$	68-69	10			27	1.4915	Infra-red (356)	Methyl carbamate MP 138-139°	(356) (408)
					29	1.4900			(467)
	54-55	2						Ethyl carbamate MP 111-113° Phenyl urea MP 183°	(482) (366)
$C_6H_8ClNO_3$	136-137	16							(394)
$C_{11}H_{11}NO_3$	152-153	12							(549)
$C_{12}H_{13}NO_3$	110-115	0.05							(417)
$C_{12}H_{13}NO_3$	120-121	1.5							(240) (525)
	152	10							
	163	16							
	173	26							
C_4H_5NO	87-89	760						Phenyl urea MP 105-106°	(194) (207) (235) (505)
C_4H_5NO	81	760					Reaction rate (494)	Phenyl urea MP 150-151°	(224) (377) (494)
	37-42	260					Infra-red (523)	p-ethoxy phenyl urea MP 195-196°	(523)
C_4H_5NO	37-38	302		0.8776 (25/4)	25	1.4020	Reaction rate (494)	α Naphthyl urea MP 228°	(133) (224)
					20	1.4170	Infra-red (523)	Urea from NH_3 MP 74-75	(332)
	42	390			20	1.4160			(338) (365)
	62-63	757							(377) (494) (523)
C_5H_7NO	40-41	105						p-tolyurea MP 188-189° p-Ethoxyphenyl urea MP 142-143° Phenyl urea MP 186-187°	(377)
$C_5H_4N_2O_2$	98-103	30							(538)

No.	Name	Other Names	Structure
P-107	1-Propene -1-phenyl -1-isocyanato		CH_3-CH=C(C_6H_5)NCO
P-108	Propenoic acid -2 chloro -isocyanato		CH_2=C(Cl)-CO-NCO
P-109	Propenoic acid -2-methyl -isocyanato	Methacrylic acid -isocyanate	CH_2=C(CH_3)-CO-NCO
P-110	Propenoic -2-methyl -2-(isocyanato ethyl) ester	Methacrylic acid -2-(isocyanato ethyl) ester	CH_2=C(CH_3)-CO-O-CH_2CH_2-NCO
P-111	Propyne -3-isocyanato	Propargyl isocyanate	CH≡C-CH_2-NCO
P-112	Pyran -tetrahydro -2-isocyanato -3-bromo		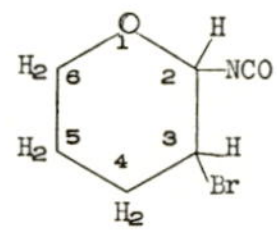
P-113	Pyrazine 2,5 diisocyanato		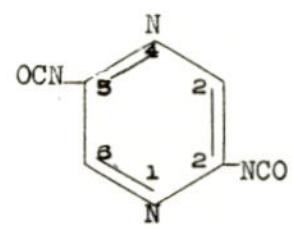
P-114	5-Pyrazoline, -3-methyl -1-phenyl -4-(azobenzene-4-isocyanato)		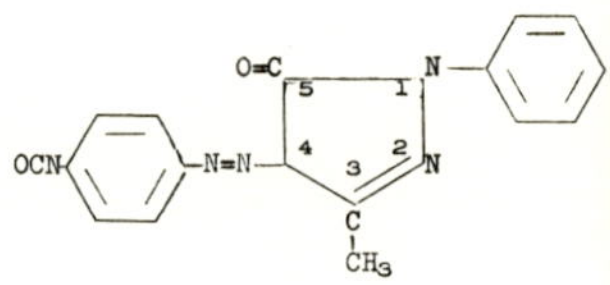
P-115	Pyrene -3,8-diisocyanato		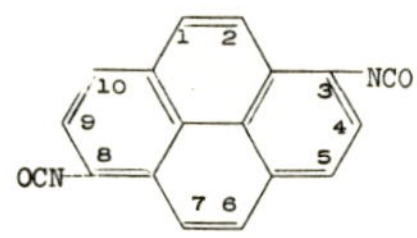
P-116	Pyrene -3-isocyanato		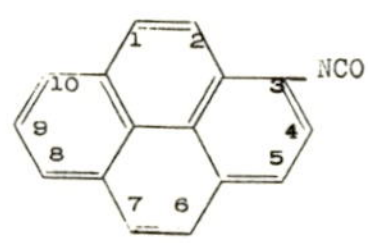

Formula	Boiling Point °C.	Boiling Point mm.	Melting Point °C.	Density	Refractive Index °C.	Refractive Index n	Other Data	Derivatives	References
$C_{10}H_9NO$	110-112	18						Phenyl urea MP 185°	(80)
$C_4H_2ClNO_2$								Benzyl urea MP 115°	(330)
$C_5H_5NO_2$								Phenyl urea MP 129°	(330)
$C_7H_9NO_3$	87-89	10							(394) (300)
C_4H_3NO	25-27	80						Urea from H_2O MP 187-188° Urea from NH_3 MP 128-129° Phenyl urea MP 129-130°	(420)
	31-32	140							
$C_6H_8BrNO_2$	100-102	2						Urea from NH_3 MP 147-148° Diurea MP 149-150° Ethyl carbamate MP 98-99°	(298)
$C_6H_2N_4O_2$			250					Ethyl carbamate	(88)
$C_{17}H_{13}N_5O_2$			162						(63)
$C_{18}H_8N_2O_2$			234					Dimethyl urethane MP 264°	(194)
$C_{17}H_9NO$			92					Methyl urethane MP 203°	(194)

No.	Name	Other Names	Structure
P-117	4-Pyrimidone 2-ethylthio 5-methyl isocyanato		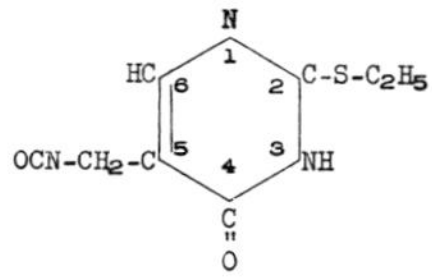
P-118	Pyrrole 2-(carboxy, methyl ester) 3-bromo 4-methyl 5-isocyanato		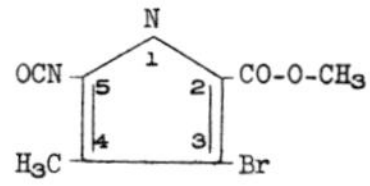
P-119	Pyrrole -2,4 dimethyl -3-acetyl -5-isocyanato		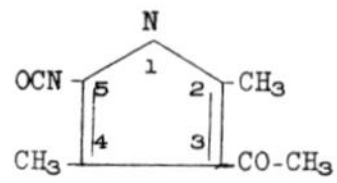
P-120	Pyrrole 1-methyl 2,5 bis(isocyanato ethyl)		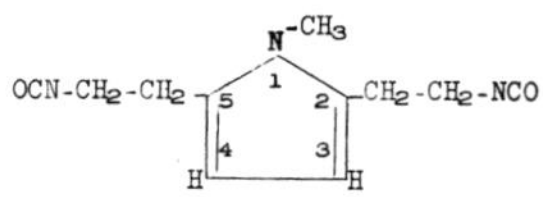
P-121	2-Pyrrolidone -3-isocyanato -1-tosyl		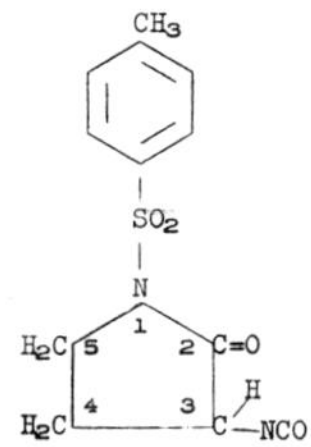

Formula	Boiling Point °C.	Boiling Point mm.	Melting Point °C.	Density	Refractive Index °C.	Refractive Index n	Other Data	Derivatives	References
$C_8H_8N_3O_2S$			189-191					Sym urea MP 270-272°	(83)
$C_8H_7BrN_2O_3$			> 260					Ethyl urethane MP 165°	(73)
$C_9H_{10}N_2O_2$			301					Corresponding amine MP 235° Benzyl urethane MP 163°	(73)
$C_{11}H_{13}N_3O_2$									(436)
$C_{12}H_{12}N_2O_4S$			101-102						(456)

No.	Name	Other Names	Structure
Q-1	Quinoline 4-isocyanato 6-Hydroxy		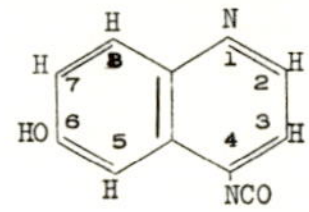
Q-2	Quinoline 4-isocyanato 6-methoxy		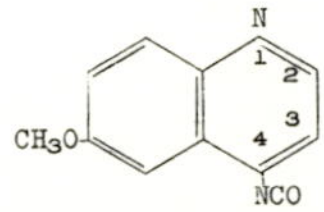
Q-3	Quinoline 4-isocyanato 2-phenyl		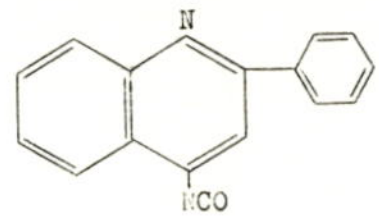
Q-4	Quinoline 4-isocyanato 2-phenyl 6-methyl		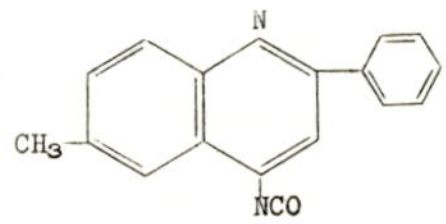
Q-5	Quinoline 4-isocyanato 2-phenyl 6-methoxy		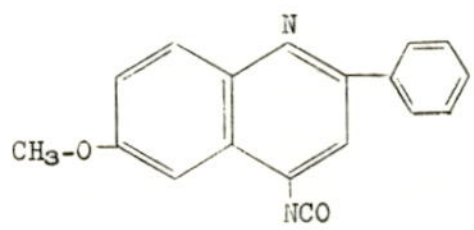
Q-6	Quinoline 4-isocyanato 2-phenyl 6-ethoxy		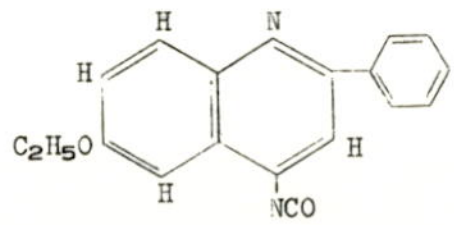
Q-7	Quinoline 4-isocyanato 2-phenyl 7-chloro		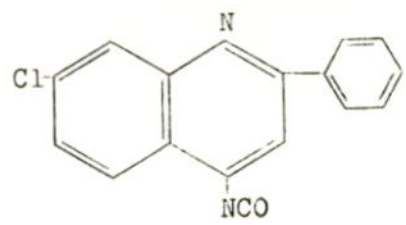
Q-8	Quinoline 4-isocyanato 2-phenyl 8-methyl		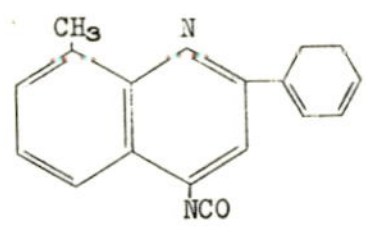
Q-9	Quinoline 4-isocyanato 2-p-tolyl		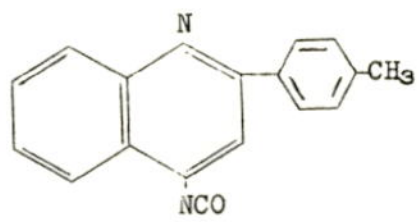
Q-10	Quinoline 3-isocyanato 2-phenyl		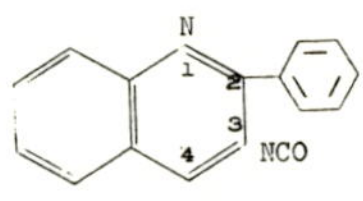

Formula	Boiling Point °C.	Boiling Point mm.	Melting Point °C.	Density	Refractive Index °C.	Refractive Index n	Other Data	Derivatives	References
$C_{10}H_6N_2O_2$			178						(44)
$C_{11}H_8N_2O_2$			171						(44)
$C_{16}H_{10}N_2O$			231						(30)
$C_{17}H_{12}N_2O$			246-247					Methyl urethane MP 178°	(54) (123)
$C_{17}H_{12}N_2O_2$			221						(51)
$C_{18}H_{14}N_2O_2$			213						(51)
$C_{16}H_9ClN_2O$			106.5					Ethyl urethane MP 148.5°	(164)
$C_{17}H_{12}N_2O$			261					Ethyl urethane MP 134°	(54)
$C_{17}H_{12}N_2O_2$			206					Ethyl urethane MP 98°	(53)
$C_{16}H_{10}N_2O$			262					Ethyl urethane MP 119°	(53)

No.	Name	Other Names	Structure
	SILICON ISOCYANATES		
S-1	n-Butyl Triisocyanato silane		nC_4H_9-Si(NCO)$_3$
S-2	Cyclohexyl isocyanato silane		C_6H_{11}-SiH_2(NCO)
S-3	Cyclohexyl triisocyanato silane		C_6H_{11}-Si-(NCO)$_3$
S-4	Dibutyl diisocyanato silane		(n-C_4H_9)$_2$Si(NCO)$_2$
S-5	Diethoxy diisocyanato silane		(C_2H_5O)$_2$-Si-(NCO)$_2$
S-6	Diethyl diisocyanato silane		(C_2H_5)$_2$Si-(NCO)$_2$
S-7	Dimethoxy diisocyanato silane		(CH_3O)$_2$-Si-(NCO)$_2$
S-8	Dimethyl,-acetoxymethyl isocyanato silane		CH_3-CO-O-CH_2-Si(CH_3)$_2$NCO
S-9	Dimethyl diisocyanato silane		(CH_3)$_2$-Si-(NCO)$_2$
S-10	Dimethyl--di(3 isocyanatophenoxy) silane		(CH_3)$_2$-Si-(O-C_6H_4NCO)$_2$
S-11	Diphenyl -diisocyanato silane		(C_6H_5)$_2$-Si-(NCO)$_2$
S-12	Ethoxy triisocyanato silane		(C_2H_5O)-Si-(NCO)$_3$
S-13	Ethyl triisocyanato silane		(C_2H_5)-Si-(NCO)$_3$
S-14	n-Heptyl--triisocyanato silane		CH_3-(CH_2)$_6$-Si(NCO)$_3$
S-15	Isopropyl triisocyanato silane		(CH_3)$_2$CH-Si(NCO)$_3$
S-16	Methoxy triisocyanato silane		CH_3O-Si(NCO)$_3$

R(Ml) Molar Refractivity
Δ H - Molar heat of vaporization
Δ H/T = Troutons Constant

Vapor Pressure

$\log_{10} p = A - \frac{B}{T}$

Formula	Boiling Point °C.	Boiling Point mm.	Melting Point °C.	Density	Refractive Index °C.	Refractive Index n	R (ML)	ΔH kCal	$\frac{\Delta H}{T\beta}$	Vapor Pressure A	Vapor Pressure B	References
$C_7H_8N_3O_3Si$	215.5 115	760 23	-37°	1.141 (20°)	20	1.4479	48.53	12.5	25.6	8.483	2737	(178) (216
$C_7H_{13}NOSi$	163.7 82-83	760 29		0.990 (20°)	20	1.4668	43.40					(476)
$C_9H_{11}N_3O_3Si$	253 109- 110	760 1		1.208 (20°)	20	1.4787	55.65					(476)
$C_{10}H_{18}N_2O_2Si$	133	25										(216)
$C_8H_{10}N_2O_4Si$	175.4	760		1.108 (20°)	20	1.4046	44.71	11.0	24.6	8.2577	2411	(179)
$C_8H_{10}N_2O_2Si$	176.7 112- 113	775 101		1.0223 (20°)	20°	1.4348	43.4					(205)
$C_4H_6N_2O_4Si$	152.1	760		1.208 (20°)	20°	1.4028						(143)
$C_6H_{11}NO_3Si$	51	12		-	20	1.4272						(554)
$C_4H_6N_2O_2Si$	139.2 72-75	760 77	-31.2	1.076 (20°)	20° Infra-red data Ref. 6	1.4221	33.6	10.6	24.1	8.154	2174	(178) (459) (216) (471) (225) (551)
$C_{16}H_{14}N_2O_4Si$	150- 155	0.05										(512)
$C_{14}H_{10}N_2O_2Si$	319.6	760	22.9	1.188 (20°)	20	1.5675	73.3	15.7	26.5	8.6683	3421	(179)
$C_5H_5N_3O_4Si$	179.6			1.236 (20°)	20	1.4251	41.06	11.5	25.4	8.4374	2515	(179)
$C_5H_5N_3O_3Si$	183.5 124-5	757 103		1.2191 (20°)	20	1.4468	40.1					(205)
$C_{10}H_{15}N_3O_3Si$	258 111- 113	760 10		1.077 (20°)	20	1.4516	63.6					(540)
$C_6H_7N_3O_3Si$	192 78	760 16		1.1626 (20°)	20	1.4444	45.2					(205)
$C_4H_3N_3O_4Si$	168.4	760		1.313 (20°)	20°	1.4287						(143)

No.	Name	Other Names	Structure
S-17	Methyl phenyl diisocyanato silane		CH_3 / C_6H_5 > $Si(NCO)_2$
S-18	Methyl triisocyanato silane		$CH_3-Si(NCO)_3$
S-19	Phenyl triisocyanato silane		$C_6H_5-Si-(NCO)_3$
S-20	n-Propyl triisocyanato silane		$n-C_3H_7-Si(NCO)_3$
S-21	Silicon tetraisocyanate		$Si(NCO)_4$
S-22	Triethoxy isocyanato silane		$(C_2H_5O)_3-Si-NCO$
S-23	Triethyl isocyanato silane		$(C_2H_5)_3-Si-NCO$
S-24	Trimethoxy isocyanato silane		$(CH_3O)_3-Si-NCO$
S-25	Trimethyl isocyanato silane		$(CH_3)_3-Si-(NCO)$
S-26	Trimethyl (4-isocyanatophenyl) silane		$(CH_3)_3-Si-C_6H_4NCO$
S-27	Triphenyl isocyanato silane		$(C_6H_5)_3-Si-NCO$
S-28	Vinyl triisocyanato silane		$CH_2=CH-Si-(NCO)_3$
S-29	Ethyl (1,2 dibromo) triisocyanato silane		$CH_2Br-CH_2Br-Si-(NCO)_3$
S-30	Ethyl (1,2 dichloro) triisocyanato silane		$CH_2Cl-CH_2Cl-Si-(NCO)_3$
S-31	Silicon trichloro monoisocyanate		$Si-Cl_3-(NCO)$
S-32	Silicon dichloro diisocyanate		$SiCl_2(NCO)_2$
S-33	Silicon monochloro triisocyanate		$SiCl(NCO)_3$

R(Ml) Molar Refractivity
Δ H - Molar heat of vaporization
Δ H/T = Troutons Constant

Vapor Pressure
$\log_{10} p = A - \frac{B}{T}$

Formula	Boiling Point °C.	mm.	Melting Point °C.	Density	Refractive Index °C.	n	R (ML)	ΔH kCal	ΔH/Tβ	Vapor Pressure A	B	References
$C_9H_8N_2O_2Si$	123-124	20		1.1533 (20°)	20	1.5177						(459)
$C_4H_3N_3O_3Si$	170.8	760	2.7	1.267 (20°)	20	1.4430	35.39	11.2	25.3	8.415	2457	(179)
$C_9H_5N_3O_3Si$	251.9	760		1.273 (20°)	20	1.5210	55.45	13.9	26.4	8.6663	3038	(179)
$C_6H_7N_3O_3Si$	198.2 101.5-102	700 30		1.1726 (20°)	20	1.4462	44.9					(205)
$C_4N_4O_4Si$	185.6 86	760 13	26	1.434 (25°) 1.442 (20°)	20	1.4610	37.29	12.9	26.5	9.0198	2816	(103) (179) (140) (205) (143) (449)
$C_7H_{15}NO_4Si$	172.4	760		1.015 (20°)	20	1.3922	48.2	10.9	24.4	8.2179	2380	(179)
$C_7H_{15}NOSi$	165.1 99-100	760 101		0.8895 (20°)	20	1.4295	45.6					(205)
$C_4H_9NO_4Si$	137.6	760		1.123 (20°)	20	1.3839						(143)
C_4H_9NOSi	91	760	-49	0.867 (20°)	20 Infra-red data 551	1.3960	31.92	8.3	22.7	7.8446	1807	(179) (505) (225) (551) (471)
$C_{10}H_{13}NOSi$	87-89	5	Derivatives:	Methyl urethane - MP 71.7° Di-urea from water - MP 257-258° Additional derivatives (452)								(452)
$C_{19}H_{15}NOSi$	372	760	95					16.1	24.9	8.3234	3511	(179) (209)
$C_5H_3N_3O_3Si$	176.3	760		1.251 (20°)	20	1.4587	39.44					(513)
$C_5H_3Br_2N_3O_3Si$	264	760		1.918 (20°)	20	1.5357						(513)
$C_5H_3Cl_2N_3O_3Si$	248	760		1.481 (20°)	20	1.4970	49.81					(513)
CCl_3NOSi	86.8	760	-69	1.445 (20°)	20	1.4262						(140)
$C_2Cl_2N_2O_2Si$	117.8	760	-80	1.437 (20°)	20	1.4380						(140)
$C_3ClN_3O_3Si$	152	760	-35	1.437 (25°)	20	1.4507						(140)

No.	Name	Other Names	Structure
S-34	Silicon triisocyanate thiocyanate		$Si(NCO)_3SCN$
-	Spirobifluorene	see Fluoro isocyanates	
-	Stearyl isocyanate	see Heptadecyl isocyanate	
-	Steroid isocyanate	see Norcholestane "N"	
-	Steroid isocyanate	see Oleane "O"	
-	Steroid isocyanate	see Estrone "E"	
-	Steroid isocyanate	see Cholest-5-ene	
-			

R(Ml) Molar Refractivity
Δ H - Molar heat of vaporization
Δ H/T = Troutons Constant

Vapor Pressure

$\log_{10} p = A - \frac{B}{T}$

Formula	Boiling Point °C.	Boiling Point mm.	Melting Point °C.	Density	Refractive Index °C.	Refractive Index n	R (ML)	ΔH kCal	ΔH/Tβ	Vapor Pressure A	Vapor Pressure B	References
$C_4N_4O_3SiS$	126	28										(154)

No.	Name	Other Names	Structure
S-35	Stilbene (trans) 2-isocyanato	Ethene 1,2- diphenyl 2-isocyanato	$C_6H_4(NCO)CH{=}CHC_6H_5$
S-36	Stilbene (trans) 4-isocyanato	Ethene 1,2- diphenyl 4-isocyanato	$C_6H_4(NCO)CH{=}CHC_6H_5$
S-37	Stilbene 4-isocyanato 4'-dimethylamino	Ethene 1,2- diphenyl 4-isocyanato 4'-dimethylamino	$C_6H_4(NCO)CH{=}CHC_6H_4N(CH_3)_2$
S-38	Stilbene 4-isocyanato 2 methyl 4'-dimethylamino	Ethene 1,2- diphenyl 4-isocyanato 4'-dimethylamino	$C_6H_4(CH_3)(NCO)CH{=}CHC_6H_4N(CH_3)_2$
S-39	Stilbene 4,4' diisocyanato	Ethene 1,2- diphenyl 4,4' diisocyanate	$C_6H_4(NCO)CH{=}CHC_6H_4(NCO)$
-	Styryl isocyanate	see Ethane 2-phenyl 1-isocyanato	
S-40	Succinic acid 2-isocyanato dimethyl ester	Dimethyl isocyanato succinate	$CH_3\text{-O-CO-CH(NCO)-}CH_2\text{-CO-O-}CH_3$
S-41	Succinic acid 2-isocyanato diethylester	Diethyl isocyanato succinate	$C_2H_5\text{-O-CO-CH(NCO)-}CH_2\text{-CO-O-}C_2H_5$
-	SULFUR CONTAINING ISOCYANATES	see Benzene 1-isocyanato 3-isothiocyano (B-48)	
-		see Benzene 1-isothiocyano (B-47)	
-		see Benzene 1-isocyanato 5-isothiocyano 2-methyl (B-85)	
-		see Benzene 1,3 diisocyanato 4 sulfonic acid chloride (B-123)	
-		see Benzene 1 isocyanato 2-sulfonic acid chloride (B-55)	
-		see Benzene 1-isocyanato 3-sulfonic acid chloride (B-56)	

Formula	Boiling Point °C.	Boiling Point mm.	Melting Point °C.	Density	Refractive Index °C.	Refractive Index n	Other Data	Derivatives	References
$C_{15}H_{11}NO$	155	0.9	96					Urea from water MP 228°	(486)
$C_{15}H_{11}NO$	205	1	113					Urea from H_2O MP 300° Urethanes fluoresce in U.V. light	(486) (543)
$C_{17}H_{16}N_2O$								(243)	
$C_{18}H_{18}N_2O$			(a) 113.5-114.9 (b) 112-114.2						(243)
$C_{16}H_{10}N_2O_2$			149					Urethanes fluoresce in U.V. light	(486) (543)
$C_7H_9NO_5$	97	1	135		25	1.4445			(537)
$C_9H_{13}NO_5$	81 132	0.5 8	80		25	1.4357			(537)

No.	Name	Other Names	Structure
		see Benzene 1-isocyanato 4-sulfonic acid chloride (B-57)	
-		see Benzene 1-isocyanato 3-sulfonic acid chloride 4-chloro (B-72)	
-		see Benzene 1-isocyanato 3-sulfonic acid chloride 6-methyl (B-88)	
-		see Benzene 1-isocyanato 5-sulfonic acid chloride 3-chlorocarbonyl (B-73)	
-		see Benzene 1-isocyanato 3-sulfonic acid fluoride (B-58)	
-		see Benzene sulfonic acid 3-isocyanato amide N-methyl,-N-Phenyl (B-155)	
-		see Benzene sulfonic acid -isocyanate (B-154)	
-		see Benzene sulfonic acid -3-isocyanato -4-methyl -phenyl-4-isocyanato ester (B-158)	
-		see Benzene sulfonic acid 3-isocyanato -4-methoxy -phenyl-4-isocyanato ester (B-157)	
-		see Benzene sulfonic acid -isocyanate -4,-nitro (B-156)	
-		see Benzo(b)thiophene -3-(propyl,-2 isocyanato-2-methyl) (B-179)	
-		see Benzotriazole -6-methyl -2(phenyl-4-isocyanato) (B-180)	
-		see Butane 1-isocyanato 4-isothiocyano (B-212)	

Formula	Boiling Point °C.	Boiling Point mm.	Melting Point °C.	Density	Refractive Index °C.	Refractive Index n	Other Data	Derivatives	References
-					see Butanoic acid 2-isocyanato 4-(methylthio), ethyl ester (B-230)				
-					see Cyclohexane -1-isocyanato -1-isothiocyano (C-30)				
-					see Dibenzothiophene -2,7-diisocyanato -dioxide (D-8)				
-					see Disulfide diphenyl 4,4' diisocyanato (D-9)				
-					see Disulfide diphenyl 6,6' diisocyanato 3,3 dimethyl (D-10)				
-					see Disulfide diphenyl 5,5' diisocyanato 4,4' dimethyl (D-11)				
-					see Disulfide diphenyl 5,5' diisocyanato 2,2' dimethyl (D-12)				
-					see Disulfide diphenyl 3,3' diisocyanato 4,4' dimethoxy (D-13)				
-					see Disulfide diphenyl 4,4' diisocyanato 3,3' dimethoxy (D-14)				
-					see Ethane 1,2 diamino bis N,N'(phenyl sulfonyl-4 isocyanato) (E-5)				
-					see Ethane 1,2 diamino bis N,N'(phenyl sulfonyl-3-isocyanato-4-methoxy) (E-6)				
-					see Ethane 1,2 diamino bis(phenyl sulfonyl-3-isocyanato-4-methoxy) (E-3)				
-					see Ethane 1,2 diamino bis(phenylthio-3-isocyanato-4-methoxy) (E-4)				
-					see Methane sulfonic acid -isocyanate (M-37)				

No.	Name	Other Names	Structure
-		see Naphthalene -1-isocyanato -4-sulfonic acid chloride (N-18)	
-		see Naphthalene -1-isocyanato -6-sulfonic acid -chloride (N-19)	
-		see Naphthalene 1-isocyanato 7-sulfonic acid chloride (N-20)	
-		see Pyrimidone -2-ethylthio -5-methyl isocyanate (P-117)	
S-42	Sulfide-butyl phenyl 3'-isocyanato		$OCN\text{-}C_6H_4\text{-}S\text{-}C_4H_9$
S-43	Sulfide dibenzyl 4,4' diisocyanato 3,3'-dimethoxy		$CH_3\text{-}O\text{-}(NCO)C_6H_3\text{-}CH_2\text{-}S\text{-}CH_2\text{-}C_6H_3(NCO)OC$
S-44	Sulfide diethyl 2,2'diisocyanato		$OCN\ CH_2\text{-}CH_2\ S\text{-}CH_2CH_2NCO$
S-45	Sulfide dihexyl 6,6' diisocyanato		$OCN\text{-}(CH_2)_6\text{-}S\text{-}(CH_2)_6\text{-}NCO$
S-46	Sulfide diphenyl 4-isocyanato		$OCN\text{-}C_6H_4\text{-}S\text{-}C_6H_5$
S-47	Sulfide diphenyl 2,4 diisocyanato		$C_6H_5\text{-}S\text{-}C_6H_3(NCO)_2$
S-48	Sulfide diphenyl 2,4,4' triisocyanato		$(OCN)_2\text{-}C_6H_3\text{-}S\text{-}C_6H_4(NCO)$
S-49	Sulfide diphenyl 2,4,2' triisocyanato		$(OCN)_2\text{-}C_6H_3\text{-}S\text{-}C_6H_4(NCO)$
S-50	Sulfide dipropyl 3,3' diisocyanato		$(OCN)\text{-}(CH_2)_3\text{-}S\text{-}(CH_2)_3\text{-}NCO$
S-51	Sulfide dodecyl phenyl 4' isocyanato		$OCN\text{-}C_6H_4\text{-}S\text{-}C_{12}H_{25}$
S-52	Sulfide ethyl, phenyl 3'-isocyanato		$OCN\text{-}C_6H_4\text{-}S\text{-}C_2H_5$

	Formula	Boiling Point °C.	Boiling Point mm.	Melting Point °C.	Density	Refractive Index °C.	Refractive Index n	Other Data	Derivatives	References
	$C_{11}H_{13}NOS$									(112)
	$C_{18}H_{16}N_2O_4S$									(194)
	$C_6H_8N_2O_2S$	100-120	0.5-1.0						Di-urea MP 210°	(194)
	$C_{14}H_{24}N_2O_2S$	180-185	0.7-0.8						Dimethyl urethane MP 87°	(194)
	$C_{13}H_9NOS$	156	4	25						(441)
	$C_{14}H_8N_2O_2S$			73-74						(194)
	$C_{15}H_7N_3O_3S$	172	0.3	79-80						(67)
		188	0.35							
	$C_{15}H_7N_3O_3S$	184	0.35	63.5-65						(367)
	$C_8H_{12}N_2O_2S$	118	0.3-0.4							(194)
	$C_{19}H_{29}NOS$	204	3	39-40					Urea from aniline MP 128-132°	(125)
	C_9H_9NOS	127	10							(112)

No.	Name	Other Names	Structure
S-53	Sulfone cyclotetra methylene -3-isocyanato		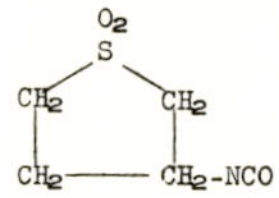
S-54	Sulfone dibenzyl 4,4' diisocyanato 3,3' dimethoxy		$(CH_3O)(NCO)$-C_6H_3-CH_2-SO_2-CH_2-$C_6H_3(NCO)($
S-55	Sulfone diphenyl 4-isocyanato		C_6H_5-SO_2-$C_6H_4(NCO)$
S-56	Sulfone diphenyl 4,4' diisocyanato		$(NCO)C_6H_4$-SO_2-$C_6H_4(NCO)$
S-57	Sulfone diphenyl 2,4 diisocyanato 4'methyl		CH_3-C_6H_4-SO_2-$C_6H_3(NCO)_2$
S-58	Sulfone diphenyl 3,3' diisocyanato 4,4' dimethoxy		$(CH_3O)(NCO)C_6H_3$ SO_2-$C_6H_3(NCO)(CH_3O$
S-59	Sulfone methyl, phenyl 4' isocyanato		CH_3-SO_2-$C_6H_4(NCO)$
S-60	Sulfuric acid diisocyanate		$SO_2(NCO)_2$
S-61	Sulfuric acid monochloride monoisocyanate		OCN-SO_2-Cl
S-62	Sulfuryl, pyro chloride isocyanato		OCN-SO_2-O-SO_2-Cl
S-63	Sulfuryl fluoride isocyanate		OCN-SO_2-F
		see Thiophene -2-(3-propyloxyphenyl) -3-(N-benzoic acid amide -4-isocyanato (T 10)	
		see Thiophene -2-carboxylic acid methyl ester (T-11)	

Formula	Boiling Point °C.	Boiling Point mm.	Melting Point °C.	Density	Refractive Index °C.	Refractive Index n	Other Data	Derivatives	References
$C_5H_7NO_3S$								Phenyl urea MP 190°	(194)
$C_{18}H_{16}N_2O_6S$									(194)
$C_{13}H_9NO_3S$			126-127						(441)
$C_{14}H_8N_2O_4S$			149 156-158 154					Dimethyl urethane MP 262-264°	(113) (155) (194)
$C_{15}H_{10}N_2O_4S$			180						(194)
$C_{16}H_{12}N_2O_6S$			165						(194)
$C_8H_7NO_3S$			96-97						(441)
$C_2N_2O_4S$	139	760		1.588				Diethyl urethane MP 169°	(409)
$CClNO_3S$	106-107 52	760 94		1.6264 (20°)	27	1.4435			(311) (341) (335) (434) (455) (457)
$CClNO_6S_2$	67-68	12	-32	1.792 (20°)	27	1.447			(335) (372)
$CFNO_3S$	62-63	760							(489) (509)

No.	Name	Other Names	Structure
-		see Thiophene -2,5-di(2 ethyl isocyanate) (T-12)	
-		see Thiophene -2-isocyanato -3-methyl (T-13)	
-		see 4-Toluene sulfonic acid -isocyanate (T-24)	
-		see 4-Toluene sulfonic acid 3-isocyanato amide, N-(4-isocyanato-3-methyl-phenyl) (T-25)	
-		see 4-Toluene sulfonic acid methyl isocyanate (T-26)	
-		see Thiapyrium isocyanate (T-9)	

	Formula	Boiling Point °C.	Boiling Point mm.	Melting Point °C.	Density	Refractive Index °C.	Refractive Index n	Other Data	Derivatives	References

No.	Name	Other Names	Structure
T-1	Tetradecane 1-isocyanato	Myristyl -isocyanate	$CH_3-(CH_2)_{13}-NCO$
T-2	Tridecane-1,13,- diisocyanato	Tridecamethylene diisocyanate	$OCN-(CH_2)_{13}-NCO$
T-3	Tetradecane- -1,14-diisocyanato	Tetradeca methylene diisocyanate	$OCN-(CH_2)_{14}-NCO$
T-4	Tetrahydroabietyl isocyanate		
T 5	Tetralin 1 isocyanato	Naphthalene 1,2,3,4 tetrahydro 1-isocyanate	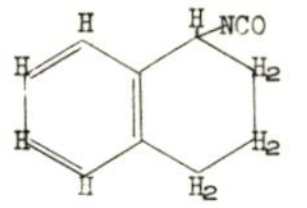
T-6	Tetralin 2-isocyanato	Naphthalene 1,2,3,4 tetrahydro 2-isocyanate	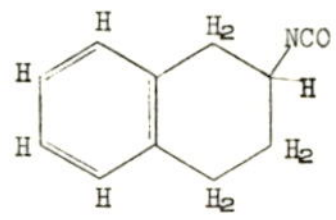
T-7	Tetralin 1,5 diisocyanato cis,-trans	Naphthalene 1,2,3,4 tetrahydro 1,5 diisocyanate	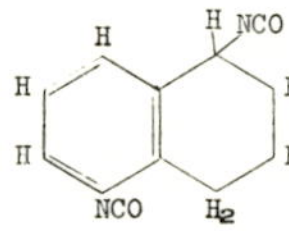
T-8	Tetralin 5-isocyanato 6-ethoxy	Naphthalene 5,6,7,8 tetrahydro 1-isocyanato 2-ethoxy	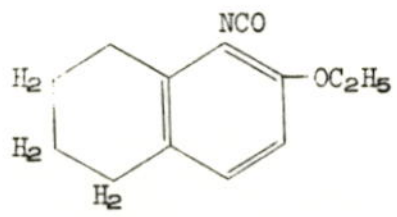
	Tetramethylene diisocyanate	see Butane 1,4 diisocyanate	
T-9	Thiapyrylium -isocyanate		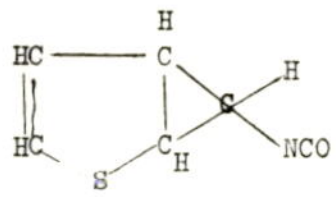
		Prepared from thiophene through the hydrazide. The isocyanate is converted with HBr followed by NaI to give the pyrlium cation with I^- as the anion. Compare with tropylium cation.	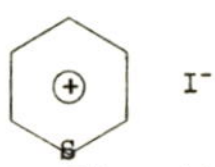 Thiapyrylium cation with Iodine anion
T-10	Thiophene 2-(3-propyloxyphenyl) 3-(N-benzoic acid, amide 4-isocyanato		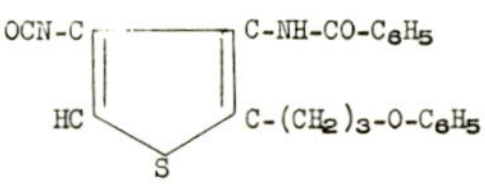

Formula	Boiling Point °C.	Boiling Point mm.	Melting Point °C.	Density	Refractive Index °C.	Refractive Index n	Other Data	Derivatives	References
$C_{15}H_{29}NO$	165-170	14							(194) (450)
	128-133	3							
$C_{15}H_{26}N_2O_2$	155-160	0.04						N,N' Trideca-methylene urea MP 189-190°	(454)
$C_{16}H_{28}N_2O_2$	175	0.59						N,N' Tetradeca-methylene urea MP 195 196°	(454)
	199-202	3-4	59-61						(291)
$C_{11}H_{11}NO$	134-135	14						Methyl urethane MP 62-63° Phenyl urea MP 193-194°	(194)
$C_{11}H_{11}NO$	134-136	10						Phenyl urea MP 169-170	(194) (130)
	117-118	0.04							(194)
	127-130	1-2							
$C_{13}H_{15}NO_2$							Infrared Data (371)		(371)
C_6H_5NOS									(493) (339)
$C_{21}H_{18}N_2O_3S$			195						(156)

No.	Name	Other Names	Structure
T-11	Thiophene 2-carboxylic acid, methyl ester 3-isocyanato		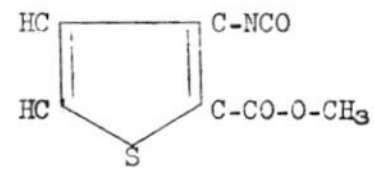
T-12	Thiophene 2,5 di(2-ethyl isocyanate)		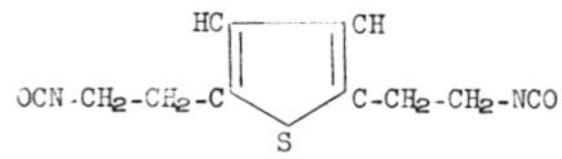
T-13	Thiophene 2 isocyanato 3-methyl		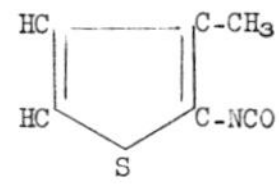
T-14	Thioxanthene -2,7 diisocyanato -dioxide	Diphenyl methane sulfone 4,4' diisocyanate	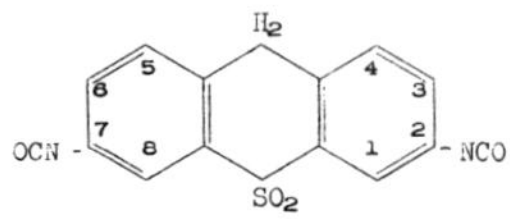
T-15	1-Thujyl carbimide		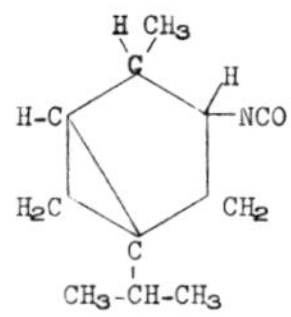
-	Thymine isocyanate	see Uracil (U-5)	
-	Thymyl isocyanate	see Benzene 1-isocyanato 3-methyl 6-isopropyl (B-83)	
T-16	Toluene 2-isocyanato	o Tolyl isocyanate	CH_3 C_6H_4 NCO
T-17	Toluene 4-isocyanato	p-Tolyl, -isocyanate	CH_3-C_6H_4-NCO
T-18	Toluene 3-isocyanato	m-Tolyl isocyanate	CH_3-C_6H_4-NCO
T-19	Toluene 2,4-diisocyanato	4-Methyl-m- phenylene diisocyanate	CH_3-C_6H_3-$(NCO)_2$

Formula	Boiling Point °C.	Boiling Point mm.	Melting Point °C.	Density	Refractive Index °C.	Refractive Index n	Other Data	Derivatives	References
$C_7H_5NO_3S$								N,N' bis-urea MP 200-202°	(252)
$C_{10}H_{10}N_2O_2S$	155-157	5							(436)
C_6H_5NOS								Urea from aniline MP 203 204°	(252)
$C_{15}H_8N_2O_4S$			166-168					Dimethyl urethane 235-237°	(194)
$C_{11}H_{17}NO$			141-142						(95)
C_8H_7NO	63.2-63.4	9.5					see Table No. 20		see Table No. 20
	184-186	760							
C_8H_7NO	67.6 67.8	10					see Table No. 21		see Table No. 21
	75	15							
	187-189	760							
C_8H_7NO	65.7-66.3	4.5					see Table No. 22	see Table No. 22	see Table No. 22
	51	3-5							
	187	760							
$C_9H_6N_2O_2$	124-126	18	21				see Table No. 23	see Table No. 23	see Table No. 23
	147-148	40							

No.	Name	Other Names	Structure
T-20	Toluene 2,5 diisocyanato	3-Methyl-p- phenylene diisocyanate	$CH_3(C_6H_3)(NCO)_2$
T-21	Toluene 2,6-diisocyanato	Benzene, 1 methyl 2,6 diisocyanate	$CH_3(C_6H_3)(NCO)_2$
T-22	Toluene 3,5-diisocyanato	Benzene-1-methyl -3,5 diisocyanate	$CH_3(C_6H_3)(NCO)_2$
T-23	Toluene 4-isocyanato 2-chloro		CH_3-Cl-C_6H_3-NCO
T-24	4-Toluene sulfonic acid -isocyanate		CH_3-C_6H_4-SO_2-NCO
T-25	4-Toluene sulfonic acid 3-isocyanato, amide, N-(4-isocyanato -3-methyl-phenyl)		$CH_3(NCO)C_6H_3$-SO_2-NH-$C_6H_3(NCO)CH_3$
T-26	4-Toluene sulfonic acid -methyl isocyanate		CH_3-C_6H_4-SO_2-CH_2NCO
T-27	1,3,5 Triazine -2,4 diphenyl -6-isocyanato		C(NCO)=N=C(C_6H_5)=N-C(C_6H_5)=N (ring: 1 to 2) 2 3 4 5 6 1
T-28	Tricyclo [5,2,1,0,2,6] deca-3,8,-diene 4-isocyanate		see Reference
T-29	Tridecane,- -7-isocyanate	1-Hexyl heptyl isocyanate	CH_3-$(CH_2)_5$-CH(NCO)-$(CH_2)_5$-CH_3
-	Trimethylene diisocyanate	see Propane 1,3 diisocyanate (P-55)	
T-30	Triptycyl isocyanate		
-	Trityl isocyanate	see Methane -1-isocyanato 1,1,1 triphenyl (M-57)	
-	Tropylium isocyanate	see Norcardienyl isocyanate (N-41)	
-	Tropylium dibenzo isocyanate	see Norcaradiene dibenzo isocyanate (N-40)	
-	α Truxillic isocyanate	see Cyclobutane 1,3 diisocyanate 2,4 diphenyl (C-21)	

Formula	Boiling Point °C.	Boiling Point mm.	Melting Point °C.	Density	Refractive Index °C.	Refractive Index n	Other Data	Derivatives	References
$C_9H_6N_2O_2$	138-	15	39				FP 36-37° Infrared (371)	Dimethyl urethane MP 158°	(194) (260) (371) (402)
$C_9H_6N_2O_2$	129-133	18					Infrared data (371) Kinetics (424) Vapor-liquid Equilibria (437)		(194) (371) (424) (437)
$C_9H_6N_2O_2$	133.5-134.5	17					Infrared data (371)		(194) (371)
C_8H_6ClNO	121-122.5	26					Reactivity with 2-ethyl-hexanol (520)		(520)
$C_8H_7NO_3S$	114-116	0.5						Adducts (527)	(407) (426) (473) (507) (527)
$C_{16}H_{13}N_3O_4S$			191-198						(194)
$C_9H_9NO_3S$								Di-urea MP 183°	(293)
$C_{16}H_{10}N_4O$							Not isolated very reactive identified spectroscop-ically		(534)
$C_{11}H_{13}NO$									(511)
$C_{14}H_{27}NO$	139-141	12							(194)
			248-249					Triptazide MP 247-248° Triptycylamine MP 221-221.5°	(286)

No.	Name	Other Names	Structure
U-1	Undecane -1,11 diisocyanato	Undecamethylene diisocyanate Hendecamethylene diisocyanate	$OCN-(CH_2)_{11}-NCO$
U-2	Undecane -6-isocyanato 6-n butyl	1-Butyl-1-pentyl- -hexyl isocyanate	$CH_3-(CH_2)_4-C(NCO)(C_4H_9)(CH_2)_4CH_3$
U-3	Undecane -6-isocyanato -6-pentyl	1,1 Diphenyl hexyl isocyanate	$CH_3-(CH_2)_4-C(NCO)(C_5H_{11})-(CH_2)_4-CH_3$
-	Undecane 1-isocyanato tricosanefluoro	see Fluoro isocyanates (F-42)	
U-4	Undecanoic acid -11-isocyanato ethyl ester	Ethyl-11-isocyanato hendecanoate	$OCN(CH_2)_{10}-CO-O-C_2H_5$
-	Undecyl isocyanate	see Hendecane 1-isocyanato (H-1)	
U-5	Uracil -5-(Methyl-isocyanato)	Thymine,- -isocyanate	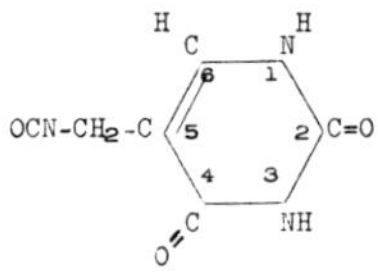
-	DL Valeric acid 2-isocyanato ethyl ester	see (DL) Pentanoic acid, -2-isocyanato ethyl ester (P-21)	
-	DL Valeric acid -2-isocyanato -4-methyl ethyl ester	see (DL) Pentanoic acid, -2-isocyanato -4-methyl ethyl ester (P-22)	
-	L-Valeric acid -2-isocyanato -4-methyl ethyl ester	see (L) Pentanoic acid, -2-isocyanato -4-methyl ethyl ester (P-23)	
-	DL-Valeric acid, -2-isocyanato -3-methyl ethyl ester	see (DL) Pentanoic acid, -2-isocyanato -3-methyl ethyl ester (P-24)	
-	Valeric acid, δ-Hydroxy, -isocyanato cholesterol ester	see Pentanoic acid -5-isocyanato cholesterol ester (P-26)	

Formula	Boiling Point °C.	Boiling Point mm.	Melting Point °C.	Density	Refractive Index °C.	Refractive Index n	Other Data	Derivatives	References
$C_{13}H_{22}N_2O_2$	124 151-152 164-165	0.06 3 3						Dimethyl urethane MP 97-98° N,N'-Hendeca methylene-urea MP 192-194°	(160) (194) (454)
$C_{16}H_{31}NO$	120-121	2							(200)
$C_{17}H_{33}NO$	126-127	1						Carbamate from diethylamino ethanol BP 180-183° (2.5 mm) N_D^{25}=1.4564	(200) (228)
$C_{14}H_{25}NO_3$	149	2	100		25	1.4450			(537)
$C_6H_5N_3O_3$			273-275					Urea (sym) MP 315° Ethyl urethane MP 256-257° Benzyl urethane MP 261-263°	(83)

No.	Name	Other Names	Structure
-	DL-Isocyanato valine, ethyl ester	see Butyric acid (DL), -2-isocyanato -3-methyl ethyl ester (B-242)	
-	Vinyl isocyanate	see Ethene isocyanate (E-47)	
-	(trans) Vinylene diisocyanate	see Ethene -1,2 diisocyanate (E-48)	
-	1,1 dimethyl-,1,2-vinyloxy ethyl isocyanate	see Propane,- 2-isocyanato 2-methyl 3-(vinyloxy) (P-77)	
-	Cyclohexyl vinyl isocyanate	see Ethene,- -1-cyclohexyl,- -1-isocyanato (E-54)	
-	Vinyl phenyl isocyanate	see Ethene 1-(phenyl isocyanate,m,p) (E-53)	
-	2-Methyl-5-vinyl phenyl isocyanate	see Ethene 1-(phenyl,4 methyl,-3-isocyanate) (E-51)	
-	Phenyl vinyl isocyanate	see Ethene 1-phenyl 1-isocyanato (E-50)	
	o-Xylyl isocyanate	see Benzene 1-isocyanato 2,3,-dimethyl (B-32)	
-	m-Xylyl isocyanate	see Benzene 1-isocyanato 2,4 dimethyl (B-33)	
-	m-Xylylene diisocyanate	see Benzene 1,3-bis(isocyanato-methyl) (B-99)	
-	p-Xylylene diisocyanate	see Benzene 1,4-bis(isocyanato-methyl) (B-100)	
-	p-Xenyl isocyanate	see Biphenyl 4-isocyanato (B-191)	

INDEX TO TABLES

TABLE 7.1

Compound A-25: Azobenzene-4-isocyanato; *p*-phenyl azo phenyl isocyanate

(*1*) *Properties*

MP, °C	Ref.
97–98	58, 63, 194
94–95	126
100	214
97.5–98	217
95.5–96	221

(*2*) *Derivatives*

Urethanes (221)

The following colored (usually orange to red) urethanes (*p*-phenylazophenylcarbamic acid esters) were obtained by heating the isocyanate with various alcohols in C_6H_6 and recrystallizing from acetone–petroleum ether.

	MP, °C
Methyl	119.5
Ethyl	151.5
Propyl	144
Butyl	128.5–129
Isobutyl	164
Amyl	111.5–112
Isoamyl	108.5–.09
Hexyl	105–106
Heptyl	110
Octyl	110.5–111
Nonyl	111
Decyl	107.5–108
Hendecyl	110.5
Dodecyl	111.5–112
Tridecyl	108 –108.5
Tetradecyl	100.5–101
Pentadecyl	99 –99.5
Hexadecyl	102.5–103
Heptadecyl	98.5– 99
Octadecyl	103 –103.5
Isopropyl	172 –173

Urethanes (206)

	MP, °C
Urethane from MeOH—Me-4-phenylazocarbanilate	122
from EtOH—Et-4-phenylazocarbanilate	153
from PrOH—Pr-4-phenylazocarbanilate	146
from IsoPrOH—isoPr-4-phenylazocarbanilate	174
from 2-MePrOH—2 MePr-4-phenylazocarbanilate	131

TABLE 7.1 (*Continued*)

Urea (206)

4,4′-Bis(phenylazo)carbanilide 274

$C_6H_5—N=N—C_6H_4—NH—CO—NH—C_6H_4—N=N—C_6H_5$

Derivatives from amino acids → phenyl azo anilino formyl amino acids.

Type formula

$$p\text{-PhN}=\text{N}—\text{C}_6\text{H}_4\text{NH}—\text{CO}—\text{NH}—\underset{\displaystyle \text{R}}{\underset{|}{\text{CH}}}—\text{COOH.}$$

Amino acid (202)	MP, °C
Glycine	206
Sarcosine	143
L(+)-Alanine	194
DL-Alanine	203
L(−)-Phenylalanine	174
DL-Serine	202
DL-Valine	191
L-(−)-Leucine	185
L(+)-Isoleucine	190
L(−)-Tyrosine	191
DL-Methionine	165
L(−)-Aspartic Acid	219
L(+)-Glutamic Acid	184
L(−)-Histidine	191
L(−)-Tryptophan	200
DL-Proline	187
L(−)-Hydroxy Proline	201
L(±)-Cystine	188
L(+)-Lysine	222
L(+)-Ornithine	224
L(+)-Arginine	210

TABLE 7.1 (*Continued*)

Ethyl esters of amino acids reacted with C_6H_5—N=N—C_6H_4NCO give colored derivatives. Several pairs of the derivatives were separated by chromatography. The absorbent used was active alumina. The solvents were AcOEt and C_6H_6; AcOBu and C_6H_6 or AcOEt and petroleum ether in equal volume; as eluting agents, other mixtures of similar solvents were selected.

Ethyl ester of amino acid	*p*-C_6H_5—N=N—C_6H_4NCO derivative MP, °C
Glycine	164 –164.5
DL-Alanine	144 –145
DL-α-Amino isobutyric acid	153 –154
DL-α-Amino butyric	189
DL Valine	163 –164
L-Leucine	111 –112
DL-Phenyl glycine	179 –181
DL-Phenylalanine	168.5–169.5
DL-Aspartic Acid	112 –113
DL-α-Amino adipic acid	126 –128

(*3*) *Reactions*

—

(*4*) *References*

58	126	202	214	221
63	194	206	217	473

NOTE: The numerous derivatives of phenylazophenyl isocyanate given in this table were reported by Musuyama (202) and by Musuyama and Hamada (221), who separated the highly colored compounds from the amino acids and aliphatic primary alcohols by chromatographic absorption.

Zeile and Oetzel (206) developed microanalytical procedures for the separation and determination of the amino acids by partitioning the derivatives in suitable solvent pairs, and studied their ultraviolet absorption spectra.

TABLE 7.2
COMPOUND B-8: Benzene-1-isocyanato; phenyl isocyanate

(*1*) *Properties*

Boiling point			Absolute density			Refractive index	
°C	mm	Ref.	°C	*d*	Ref.	*n*	Ref.
165	760	91	0.0	1.1152	21	$1.5365^{20°}_{He}$	34
163	760	1	10.0	1.1049	21		
161.7	752	82	20.0	1.0943	21	1.5363^{20}_{D}	82
62–64	20	236	30.0	1.0840	21		
60–63	18	520	40.0	1.0736	21		
50	11	552	50.0	1.0630	21		
55–57	10	194	15.	1.092	1		
			20.	1.0946	82		
			11.6	1.1010	34		

Viscosity, cp			Vapor pressure, mm Hg		
°C	*n*	Ref.	°C	*p*	Ref.
0.0	1.326	21	75.0	36.3	21
11.0	1.111	21	80.0	45.3	21
16.0	1.024	21	85.0	55.9	21
20.0	0.963	21	90.0	68.2	21
25.0	0.899	21	95.0	83.9	21
30.0	0.834	21			
35.0	0.779	21			
40.0	0.739	21			

Density, $d = -0.001044t + 1.1152(21)t$ = temperature, °C
Vapor pressure = $P = 157 - 4.76t + 0.042t^2$ (21)
Viscosity, $n = 1.326 - 0.02163t + 0.000174t^2$ (21)

	Ref.
Infrared absorption	254, 361, 371
Raman spectra	49, 45
Ultraviolet spectra	78, 176
Dipole moment 2.23 D	61
2.28 D	77
2.29 D	392
Heat of combustion 821 kcal/mole	555
Calculation of latent heat and entropy	445

TABLE 7.2 (*Continued*)

(2) *Derivatives*
Urethane (Carbamates)

Alcohol or phenol used	MP, °C		
	Ref. 48	Ref. 46	Ref. 139
Methyl	47	47	—
Ethyl	51	52	—
n-Propyl	57	58	—
Isopropyl	90	90	—
n-Butyl	55	oil	—
Isobutyl	—	80	—
Isoamyl	—	55	—
Heptyl	—	68	—
Octanol-2	—	114	—
Ethylene glycol	—	157	—
Cyclohexanol	82	—	—
Phenol	126	—	126.5
o-Cresol (2-Methyl Phenol)	144	—	142.5
m-Cresol (3-Methyl Phenol)	125	—	124.5
p-Cresol (4-Methyl Phenol)	115	—	113
Benzyl alcohol	78	—	—
3,4-Dimethyl Phenol (*o*-4-Xylenol)	120	—	119.3
3,5-Dimethyl Phenol (*m*-5-Xylenol)	148	—	149.5
2,5-Dimethyl Phenol (*p*-Xylenol)	162	—	166
2,4-Dimethyl Phenol (*m*-4-Xylenol)	112	—	103
2,6-Dimethyl Phenol (*m*-2-Xylenol)	133	—	—
2,4,6-Trimethyl Phenol (mesitol)	—	—	143
2,4,5-Trimethyl Phenol (pseudocumenol)	110	—	—
5-Methyl-2-isopropyl phenol (thymol)	106	—	—
2-Methyl-5-isopropyl phenol (carvacrol)	135	—	—
α-Naphthol	178	—	—
o-Dihydroxy benzene (catechol)	169	—	—
m-Dihydroxy benzene (resorcinol)	164	—	—
p-Dihydroxy benzene (quinol)	224	—	—
3,4-Dihydroxy toluene (homocatechol)	166	—	—
3,5-Dihydroxy toluene (orcinol)	154	—	—

Ureas	MP, °C	Ref.
sym Diphenyl urea	147	194
1,1 Dicyclohexyl-3-phenyl urea	174–175	324
1,1 Diisopropyl-3-phenyl urea	118	324
1,1 Dibenzyl-3-phenyl urea	124	324
4-Methyl-3-nitro carbanilide	218–219	323
2-Methyl-5-nitro carbanilide	218–219	323
N-Phenyl-*N'*-(5-chloro-2-pyridyl)urea	214	412
N-Phenyl-*N'*-(5-nitro-2-pyridyl)urea	238	412
N-Phenyl-*N'*-(5-nitro-2-thiazolyl)urea	259	412

TABLE 7.2 (*Continued*)

	Dimer	175	281
	Trimer	285	386, 378
	Diphenyl carbodiimide, BP, 107–121° (0.3 mm)		432
	BP, 110–120° (0.5 mm)		432
(3)	*Reactions*		
	Preparations phenyl urethane with P_2O_5 or PCl_5		84
	Reaction with $AlCl_3$		265
	Catalyst for the reaction of PhNCO with alcohols		496
	Kinetics of reaction with 2-ethylhexanol		40, 520
	Kinetics of reaction with methanol		494
	Kinetics of reaction with 1-Butanol		424
	Thermal dissociation		324

(4) *References*

1	61	176	281	386	496
21	77	194	285	392	505
34	78	197	323	412	520
45	82	207	324	424	521
46	84	236	361	432	524
48	91	254	371	445	552
49	139	265	378	494	555

TABLE 7.3

Compound B-19: Benzene, 1-isocyanato-3-chloro; *m*-chlorophenyl isocyanate

(*1*) *Properties*

	Ref.
Infrared data	371
Quadrapole coupling frequency, Cl^{35}	248

(*2*) *Derivatives*

		Ref.
Isopropyl carbamate	BP, 112–113°C (1 mm)	268
	$d^{20} = 1.1913$	
	$n_D^{20} = 1.5402$	
Ethyl urethane	BP, 95°/0.05 mm	545
Triisocyanurate	MP, 256°	545
Dimer	MP, 171–172°	281
Dimer	MP, 177°	518
Dimer	MP, 227–230°	545
Dimer	MP, 153–154°	126

(*3*) *Reactions*

m-Chlorophenylisocyanate with Cl^{35}	248
with $AlCl_3$	265, 285
Reactivity with 2-ethyl hexanol	520
Homopolymerization	505

(*4*) *References*

126	265	281	518
194	268	371	520
248	285	505	545

TABLE 7.4
Compound B-20: Benzene, 1-isocyanato, 4-chloro; *p*-chlorophenyl isocyanate

(*1*) *Properties*
Boiling range

°C	mm	Ref.	MP, °C	Ref.
65	2–5	505	31–32	194
80.6–80.9	9.5	194	30–32	282
93.5–95	10	520	32	294
107–108	41	524	32.5	518
115–117	45	370		
203–204	760	265		

Infrared data 371, 422
Dipole moment, 0.84 D (61)

(*2*) *Derivatives*
Urethanes
Methyl, MP 115° (194)
Isopropyl, MP 104° (268)

Ureas (412)	MP	Formula
N′-(*p*-Chlorophenyl)-*N*-(5-chloro-2-pyridyl)	236	$C_{12}H_9ON_3Cl_2$
N′-(*p*-Chlorophenyl)-*N*-(5-nitro-2-pyridyl)	285	$C_{12}H_9O_3N_4Cl$
N′-(*p*-Chlorophenyl)-*N*-(5-nitro-2-thiazolyl)	284	$C_{20}H_7O_3N_4ClS$
N′-(*p*-Chlorophenyl)-*N*-(2-pyrimidyl)	238	$C_{11}H_9ON_4Cl$
N′-(*p*-Chlorophenyl)-*N*-(4,6-dimethyl-2-pyrimidyl)	216	$C_{13}H_{23}ON_4Cl$
N′-(*p*-Chlorophenyl)-*N*-(2,4-dimethyl-6-pyrimidyl)	249	$C_{13}H_{13}ON_4Cl$

Dimer, MP 172–174° (281)

p—Cl^{36}—C_6H_4NCO (249)

(*3*) *Reactions*
Reactivity with 2-ethyl hexanol (520)
Reaction with $AlCl_3$ (265, 285)
Homopolymerization (505)

(*4*) *References*

61	265	282	370	422	520
194	268	285	371	505	524
249	281	294	412	518	

TABLE 7.5
Compound B-52: Benzene, 1-isocyanato-2-nitro; *ortho*-nitrophenyl isocyanate

(*1*) *Properties*

BP, °C	mm	Ref.	MP, °C	Ref.
135	17	194	40–41	194
137	18	93	40	93

Dipole moment, 5.22 D (296)
Infrared data (422)

(*2*) *Derivatives*
Urethanes (107)

Phenols tested	MP, °C
Phenol	96–98
o-Cresol	113–114
m-Cresol	85–86
p-Cresol	97–98
1,2,4-Xylenol	117–119
1,4,5-Xylenol	90–91
1,3,4-Xylenol	99–101
α-Naphthol	130
β-Naphthol	143
o-Nitrophenol	124
m-Nitrophenol	158
p-Nitrophenol	175
o-Chlorophenol	109–110
m-Chlorophenol	96–97
p-Chlorophenol	126–127
o-Bromophenol	122
m-Bromophenol	91–92
p-Bromophenol	129–130
o-Iodophenol	150–151
m-Iodophenol	98–100
p-Iodophenol	133–135
2,4-Dichlorophenol	123
2,4-Dibromohpenol	121–122
2,4,6-Trichlorophenol	153–155
2,4,6-Tribromophenol	172–174
Guaiacol	136–138
Resorcinol monomethyl ether	99–100
Hydroquinone monomethyl ether	156

(*3*) *Reactions*
—

(*4*) *References*
93, 107, 194, 296, 422

TABLE 7.6

Compound B-53: Benzene-1-isocyanato-3-nitro; *m*-nitrophenyl isocyanate

(*1*) *Properties*

BP, °C	mm	Ref.	MP, °C	Ref.
130–131	11	194	50–51	194
136–137	12	520		

Dipole Moment, 3.42 D (296)

Infrared data (422)

(*2*) *Derivatives*

Urethanes

Isopropyl, MP 191° (268)

Phenols tested (107)	MP, °C
Phenol	125–126
o-Cresol	129–130
m-Cresol	109
p-Cresol	141
1,2,4-Xylenol	130–131
1,4,5-Xylenol	129
1,3,4-Xylenol	118–119
α-Naphthol	143
o-Nitrophenol	141
m-Nitrophenol	163–164
p-Nitrophenol	197–198
o-Chlorophenol	116
m-Chlorophenol	117–118
p-Chlorophenol	136
o-Bromophenol	138–139
m-Bromophenol	132–133
p-Bromophenol	140–141
o-Iodophenol	143
m-Iodophenol	163
p-Iodophenol	152–153
2,4-Dichlorophenol	154
2,4-Dibromophenol	136
2,4,6-Trichlorophenol	169–171
2,4,6-Tribromophenol	201
Guaiacol	142–143
Resorcinol monomethyl ether	97–102
Hydroquinone monomethyl ether	125
β-Naphthol	152–153

Dimer, MP, 182–184° (281)

(*3*) *Reactions*

Reactivity with 2-ethyl hexanol (520)

(*4*) *References*

101	194	281	422
107	268	296	520

TABLE 7.7

Compound B-54: Benzene-1-isocyanato-4-nitro; *p*-nitrophenyl isocyanate

(*1*) *Properties*

BP, °C	mm	Ref.	MP, °C	Ref.
160–162	18	66	57	46, 50, 194
137–138	11	194	56–57	66
145	14	520		

Dipole moment, 3.42 D (296)

Infrared data (371, 422)

(*2*) *Derivatives*

Dimer, MP 208° (281)

p-Nitrophenyl ureas (106)

Amines tested	MP, °C
Aniline	220
o-Toluidine	201
m-Toluidine	205 d.
p-Toluidine	259
m-Xylidine	215
α-Naphthylamine	236
β-Naphthylamine	274
p-Aminodiphenyl	235–236
o-Nitroaniline	256
m-Nitroaniline	285
p-Nitroaniline	301–303 (sublimed)
o-Chloroaniline	233
m-Chloroaniline	271
p-Chloroaniline	263
o-Bromoaniline	228
m-Bromoaniline	263
p-Bromoaniline	279–280
o-Iodoaniline	224
m-Iodoaniline	272
p-Iodoaniline	288
o-Aminophenol	212
p-Aminophenol	235 d.
o-Anisidine	191
p-Anisidine	229 d.
o-Phenetidine	178–179
p-Phenetidine	202 d.
Ethyl anthranilate	186
Ethyl *m*-aminobenzoate	195–196
Ethyl *p*-aminobenzoate	254–255
2-Nitro-4-aminotoluene	260 d.
3-Chloro-4-aminotoluene	209–210
3-Bromo 4-aminotoluene	204–205
4-Nitro-2-aminotoluene	261–262
3-Nitro-2-aminotoluene	278 d.
5-Nitro-2-aminotoluene	246–247
5-Chloro-2-aminotoluene	264
5-Bromo-2-aminotoluene	268–269
5-Iodo-2-aminotoluene	264
4-Nitro-3-aminotoluene	263–264
6-Nitro-3-aminotoluene	283–284
6-Chloro-3-aminotoluene	246
6-Bromo-3-aminotoluene	248–249
6-Iodo-3-aminotoluene	239
Monomethylaniline	123
Diphenylamine	175
Acetamide	295–296
Benzamide	260
Acetanilide	254–255
Aminocyclohexane	169–170

p-Nitrophenyl urethanes

Alcohols tested	MP, °C	Ref.
Methyl	179.5	46
Ethyl	129.	46

TABLE 7.7 (*Continued*)

(2) *Derivatives* (*Continued*)

Alcohols tested	MP, °C	Ref.
Propyl	115	46
Isopropyl	116	46
Isopropyl	107	268
Butyl	95.5	46
Sec-Butyl	75	46
Isobutyl	80	46
Amyl	86	46
Isoamyl	97.5	46
Hexyl	103	46
Heptyl	102	46
Ethylene Glycol	135.5	46
Nonyl	104	59
Undecyl	99.5	59
Cetyl	117–118	59
Chlosterol	204–205	

Phenols tested (107)	MP, °C
Phenol	167
o-Cresol	162
m-Cresol	143
p-Cresol	214
1,2,4-Xylenol	168
1,4,5-Xylenol	153
1,3,4-Xylenol	171
α-Naphthol	189
β-Naphthol	182
o-Nitrophenol	190
m-Nitrophenol	225
p-Nitrophenol	232
o-Chlorophenol	143
m-Chlorophenol	144
p-Chlorophenol	196
o-Bromophenol	140
m-Bromophenol	137
p-Bromophenol	198
o-Iodophenol	144 d.
m-Iodophenol	167 d.
p-Iodophenol	213
2,4-Dichlorophenol	205
2,4-Dibromophenol	186
2,4,6-Trichlorophenol	166 d.
2,4,6-Tribromophenol	198 d.
Guaiacol	157

TABLE 7.7 (*Continued*)

(2) *Derivatives* (*Continued*)

Phenols tested	MP, °C
Resorcinol monomethyl ether	130
Hydroquinone monomethyl ether	189

(3) *Reactions*
Reactivity with 2-ethyl hexanol (520)

(4) *References*

46	66	268	422
50	106	281	520
56	107	296	
59	194	371	

TABLE 7.8

Compound B-78: Benzene-1-isocyanato-3,5-dinitro-4-methyl; 3,5-dinitro-4-methyl phenyl isocyanate

(*1*) *Properties*

The derivatives described below were prepared *in situ* from 3,5-dinitro-4-methyl benzazide (without separation of the isocyanate) by reaction with the various amines. The properties of the isocyanate were not described.

(*2*) *Derivatives*

Properties of 3,5-dinitro-4-methyl phenyl ureas

Amines tested	MP, °C	Amines tested	MP, °C
Aniline	221	Ethyl anthranilate	204–205 d.
o-Toluidine	231	Ethyl *m*-aminobenzoate	209
m-Toluidine	220	Ethyl *p*-aminobenzoate	258 d.
p-Toluidine	233 d.	Aminocyclohexane	201
m-Xylidine	233–234 d.	Diphenylamine	206–207
p-Aminodiphenyl	233	Acetamide	237–238
α-Naphthylamine	260 d.	Monomethylaniline	179–180
β-Naphthylamine	253–254 d.	Benzylamine	181
o-Nitroaniline	258–260	2-Nitro-4-aminotoluene	286–287 d.
m-Nitroaniline	278–279	3-Nitro-4-aminotoluene	198
p-Nitroaniline	245 d.	3-Bromo-4-aminotoluene	246 d.
o-Chloroaniline	254 d.	3-Chloro-4-aminotoluene	237–238 d.
m-Chloroaniline	239–240	3-Iodo-4-aminotoluene	246 d.
p-Chloroaniline	242–243	6-Nitro-3-aminotoluene	279–280 d.
o-Bromoaniline	257 d.	4-Nitro-3-aminotoluene	234–235 d.
m-Bromoaniline	235	6-Bromo-3-aminotoluene	259 d.
p-Bromoaniline	232–233 d.	6-Chloro-3-aminotoluene	270–271 d.
o-Iodoaniline	264–265 d.	6-Iodo-3-aminotoluene	281–282 d.
m-Iodoaniline	246–247 d.	3-Nitro-2-aminotoluene	252–253 d.
p-Iodoaniline	260–261 d.	4-Nitro-2-aminotoluene	256–257 d.
o-Aminophenol	164 d.	5-Nitro-2-aminotoluene	247–248 d.
p-Aminophenol	239 d.	5-Bromo-2-aminotoluene	240
o-Anisidine	182–183	5-Chloro-2-aminotoluene	228–229
p-Anisidine	201–202	5-Iodo-2-aminotoluene	254–255 d.
o-Phenetidine	223 d.		
p-Phenetidine	211–212		

(*3*) *Reactions*

—

(*4*) *Reference:* 100

TABLE 7.9

Compound E-2: Benzene 1,4-diisocyanato; *p*-phenylene diisocyanate

(*1*) *Properties*

BP, °C	mm	Ref.	MP, °C	Ref.
116–118	14	260	94–96	194
111–112	12	555	94–95	555
110–112	12	194	93–94	402
108	10	520	92	217

Infrared data (371)
Crystallographic data (232)
Heat of combustion, 883.3 kcal/mole (555)
Heat capacity, 0.316 cal/g/°C (555)

(*2*) *Derivatives*
Dimethyl urethane MP, 209–210° (194)
MP, 207°C (402)

(*3*) *Reactions*
Kinetics of the reaction with 1-butanol (424)
Reactivity with 2-ethyl hexanol (520)

(*4*) *References*

194	346	520
217	371	555
232	402	
260	424	

TABLE 7.10

Compound B-191: Biphenyl,4-isocyanato; isocyanic acid, biphenyl ester (CA name); *p*-xenyl isocyanate; *p*-xenyl carbimide; diphenyl isocyanate; 4-biphenyl isocyanate; *p*-phenyl phenyl isocyanate; 4-biphenylyl isocyanate

(*1*) *Properties*
Infrared data (422, 386)

(*2*) *Derivatives*
Trimer, MP 374° (386)
Semicarbazides (71)
Carbamates

Alcohol used	MP, °C	
	Ref. 48	Ref. 69
Methyl	127	124
Ethyl	119	119
n-Propyl	129	130
Isopropyl	138	137
n-Butyl	109	109
n-Amyl	99	102
Hexyl	—	98
Heptyl	—	105
Octyl	—	110
Nonyl	—	115
Decyl	—	111
Undecyl	—	106
Dodecyl	—	113
Phenol	173	174
o-Cresol (*o*-Me-phenol)	151	—
m-Cresol (*m*-Me-phenol)	161	—
p-Cresol (*p*-Me-phenol)	198	199
Benzyl alcohol	156	158
Phenylethyl *β*	—	151
α-Naphthyl	190	—
β-Naphthol	—	—
Cyclohexanol	166	140
3,4-Dimethyl Phenol (*o*-4-Xylenol)	183	—
2,6-Dimethyl Phenol (*m*-2-Xylenol)	198	—
2,4-Dimethyl Phenol (*m*-4-Xylenol)	184	—
3,5-Dimethyl Phenol (*m*-5-Xylenol)	150	—
2,5-Dimethyl Phenol (*p*-Xylenol)	162	—
2,4,5-Trimethyl Phenol (Pseudo cumenol)	196	—
5-Methyl-2-Isopropyl phenol (thymol)	194	194
2-Methyl-5-Isopropyl phenol (carvacrol)	166	—
o-Dihydroxy benzene (catechol)	decomp.	—
m-Dihydroxy benzene (resorcinol)	decomp.	—
p-Dihydroxy benzene (quinol)	decomp.	—
3,4-Dihydroxy toluene (homocatchol)	193	—
3,5-Dihydroxy toluene (orcinol)	196	—
1-Menthyl	—	157

TABLE 7.10 (*Continued*)

Ureas

Amine used	MP, °C (70)
Methyl	186
Ethyl	210
n-Propyl	195
Isopropyl	—
n-Butyl	153
n-Amyl	152
Hexyl	—
Heptyl	146
Aniline	240 d.
o-Tolyl	225
m-Tolyl	212 d.
p-Tolyl	246
α-Naphthyl	238
β-Naphthyl	255
p-Methoxy-aniline	232
m-Nitro phenyl	227 d.
Dimethyl	175
Diethyl	136
Dipropyl	124
Piperidine	136
4-Biphenyl	312 (321, Ref. 9)
Ammonia	210
Methyl phenyl	136
Diphenyl	174 d.

(*3*) *Reactions*
Reaction with $AlCl_3$ (265, 285)
Reactivity with 2-ethyl hexanol (520)

(*4*) *References*

48	74	386
69	174	422
70	265	
71	285	

TABLE 7.11

Compound C-26: Cyclohexane-1-isocyanato; cyclohexyl isocyanate

(*1*) *Properties*

BP, °C	mm	Ref.
54	10	186
54	11	85, 130, 194
48–52	15	152
65	15	532
55–63	17	250
63–65	18	331
166	746	186
175	750	19
165–168	760	207
170–180	760	91

Density 0.9359 (°C) (331)
Refractive index, 25°: 1.4514 (331)
Infrared data (371)

(*2*) *Derivatives*

Ureas	MP, °C	Ref.
1,3-Dicyclohexyl urea	227	324
$C_6H_{11}NH—CO—NH—C_6H_{11}$	230–232	438
	229–230	19
	228	130
1-Cyclohexyl urea	196–197	130
$C_6H_{11}—NH—CO—NH_2$	192–194	186
N-Phenyl-*N′*-cyclohexyl urea	182	19
$C_6H_{11}—NH—CO—NH—C_6H_5$	174.5	324
	185–186	532
1,1-Dicyclohexyl-3-*n*-Propyl urea	146	324
$(C_6H_{11})_2—N—CO—NH(C_3H_7)$		
Urethanes		
Ethyl urethane	57	19
$C_6H_{11}—NH—CO—O—C_2H_5$	58–59	186
	60	241

Amine

Cyclohexyl amine ($C_6H_{11}NH_2$), BP 131–133 (760 mm) (250)

(*3*) *Reactions*

Homopolymerization (505)
Thermal dissociation (324)

(*4*) *References*

19	186	241	438
85	191	250	505
91	194	324	532
130	197	331	
152	207	371	

TABLE 7.12
Compound E-2: Ethane-1-isocyanato; ethyl isocyanate

(*1*) *Properties*

BP, °C	mm	Ref.	Density	Ref.	Refractive index	Ref.
58–61	760	207	$d_4^{14.4} = 0.9063$	34		
59.5–60	760	505				
59.8	759	82	$d_4^{15.7} = 0.965$	34	$n_D^{20} = 1.3808$	82
60.	760	7, 194				
60.1	760	494				
60–63	760	207	$d_4^{20} = 0.903$	34		
			$d_4^{20} = 0.9031$	82		

Heat of combustion, 424.5 kcal/mol (555)
Raman data (60)
UV absorption spectra (78)
NMR data (319)
Dipole moment, 2.81 D (77, 82)

(*2*) *Derivatives*
None

(*3*) *Reactions*
Reaction rate with methanol (494)
Homopolymerization (505)
Flash photolysis (410)

(*4*) *References*

7	77	207	410
26	78	262	494
34	82	319	505
60	194	352	555

TABLE 7.13

Compound F-30: Methane, 1-isocyanato, trifluoro; trifluoro methyl isocyanate; CF_3NCO

(*1*) *Properties*

BP,°C	mm	Ref.
−36°	760	336, 446
−35°	760	390

Infrared data (336, 483)
Latent heat of vaporization = 5380 cal/mole (336)
Vapor pressure:
Log P_{mm} = 7.838 = 1176/T (336)
Range −78° to −45°
Trouton's constant = 22.7 (336)

(*2*) *Derivatives*

Ethyl *N*-trifluoromethyl carbamate, MP 28.5–29° (336)
$CF_3NCO + HCl(l) \rightarrow CH_3NHCOCl$; $n_D^{23} = 1.3545$; $d^{23} = 1.5551$
$CF_3NCO + HBr(l) \rightarrow CF_3NHCOBr$; $n_D^{23} = 1.3869$; $d^{23} = 1.9489$

(*3*) *Reactions*

"Attempts to prepare cyanogen fluoride by a reaction analogous to the Hoffman reaction, starting with trifluoroacetyl chloride CF_3COCl, namely $CF_3COCl \rightarrow CFCONH_2 \rightarrow CF_3NCO \rightarrow CF_3NH_2 \rightarrow FCN$ stopped at CF_3NCO, which is surprisingly stable" (257)

(*4*) *References*

257	357	415
306	363	446
336	390	483

TABLE 7.14

Compound H-21: Hexane, 1-6-diisocyanato; hexamethylene diisocyanato; isocyanic acid, hexamethylene ester (CA name)

(*1*) *Properties*

BP, °C	mm	Ref.
150–152	30	148, 169
143–144	23	148, 169
125–130	18	148, 169
130–132	14	194
130–135	14	260
130	12	477
130	11	555
120–125	10	227
124	10	218
108–111	5	227
105–109	5	391
92–96	1	227

Density and viscosity vs. temperature (477)

Temp., °C	Density, d	Viscosity, η
20	1.0465	24.1
30	1.0378	20.0
40	1.0290	17.2
50	1.0202	14.9
60	1.0115	13.1
70	1.0025	11.7
80	0.9935	10.5

$d_t = 1.0465 - 9.3(t - 20°) \times 10^{-5}$
$\eta = 0.1562e^{2950/RT}$

Vapor pressure vs. temperature (477)

Temp., °C	P_{mm}
130	10
140	17
150	26
160	39
170	53
180	70.5

$\log P = 8.428 - 2978/T$

TABLE 7.14 (*Continued*)

Refractive index 20°C (477)

	λ	*n*
G Line	435.8	1.46395
E Line	546.8	1.45532
D Line	589.3	1.45319
C Line	656.3	1.45067

Heat of evaporation at 130–180°C = 13,800 cal/mole (calc.) (477).
E_{ev} = Energy of evaporation at 130–180°C = 13,100 cal/mole (477).
E_{vis} = Energy of viscous flow = 2950 cal/mole (calc.) (477).
E_{ev}/E_{vis} = 4.4: indicates this is an associated liquid (477).
Infrared absorption data (254).

(*2*) *Derivatives*

	MP, °C	Ref.
Dimethyl urethane	113–114	194
Dimethyl urethane	112	195
Diethyl urethane	79–81	195
Dipropyl urethane	95–96.5	195
Di-*n*-butyl urethane	88.5	195
Dicyclohexyl urethane	112–115	195
Hexamethylene diurea from excess NH_3	200–202	195
Diisobutyl urea	192	195

(*3*) *Reactions*

Kinetics of the reaction with 1-butanol (424)

(*4*) *References*

148	194	254	477
152	195	260	555
169	218	391	
191	227	424	

TABLE 7.15
Compound I-6: Isocyanic acid

(*1*) *Properties*

Constitution	(92, 104, 111)
Crystal structure	(283, 301)
Dissociation energy	(78)
Electron diffraction	(111)
Force constants	(237, 261)
Flash photolysis	(410, 453)
Free radical	(410, 460, 497, 499, 501)
Fused system HNCO + H_2O	(92)
Infrared absorption	(208)
Microwave absorption	(208, 460)
Molecular structure	(111, 208, 234)
Polymerization	(173)
Thermodynamics	(234, 271)
UV absorption	(78)
X-ray	(283, 301)

(*2*) *Derivatives*
Carbamyl fluoride and bromide (104)
Cyanates and isocyanates (234)

(*3*) *Reactions*
—

(*4*) *References*

78	111	234	271	410	497
92	173	237	283	453	499
104	208	261	301	460	501

TABLE 7.16

Compound M-31: Methane, 1-isocyanato; methyl isocyanate

(1) *Properties*

Boiling point (760 mm)

°C	Ref.
42–45	26
37.4–37.8	33
40–50	91
36–39	168
37–39	194, 207
42–77	224
38	236
44–45	427

Density, *d*	*t/t*	Ref.
0.9744	14.9/4	34
0.9673	15.7/4	34
0.960(3)	20/4	34
0.9230	27/4	427

Refractive index (34)

n = 1.37182–1.36946
n_{He} = 1.37400–1.37168
n_e = 1.37915–1.37662
n_y = 1.38365 (14.9°C)
n_y = 1.38087 (15.7°C)
n_{He} = 1.3717 (20°C)
n_{He} = 1.3697 (20°C)

Heat of combustion	269.4 kcal/mol (555)
Molar refractivity, found	13.73, calc. 13.68 (427)
Raman spectra	(60, 105)
Ultraviolet spectra	(78)
Infrared spectra	(105)
Near ultraviolet spectra	(117)
Far infrared spectra	(541)
Electron diffraction	(111)
Thermochemical data	(555)
Structure	(111)
Photochemical decomposition	(117)
Preparation	(26, 149, 168, 224)

(2) *Derivatives*

Phenyl urea, MP 148–150° (194)

(3) *Reactions*

—

(4) *References*

26	91	168	224
33	105	194	236
34	111	196	427
60	117	197	541
78	149	207	555

TABLE 7.17
Compound M-61: Methane, diphenyl, 4,4′-diisocyanato

Alternate names[a]:

(1)	MDI	
(2)	Isocyanic acid, methylene diphenyl ester	(CA Name)
(3)	Methylene diphenyl diisocyanate	(147)
(4)	*p*-Methylene diphenyl diisocyanate	(167)
(5)	Diphenylmethane, 4,4′-diisocyanate	(194)(362)(421)
(6)	4,4′-Diphenylmethane diisocyanate	(217)(418)(424)
(7)	Methylene bis(*p*-phenyl isocyanate)	(254)
(8)	*p,p*-Diisocyanato, diphenyl methane	(321)(337)
(9)	Bis(4-isocyanato phenyl methane)	(371)
(10)	Methane, bis(4-isocyanato phenyl)	(Note)
(11)	4,4′-Diisocyanato ditan	(Note)

[a] NOTE: Compare with methane derivatives listed in *Handbook of Chemistry and Physics*, 46th ed., Chemical Rubber Publishing Co., Cleveland, O.

(*1*) *Properties*
Infrared data (254, 371)
Activation energy, 10.5 kcla/mol (421)

(*2*) *Derivatives*
Diphenylmethane 4,4′ diurethanes (362)

Derivative	Recrystallization solvent	MP, °C
Methyl	Methanol	190
Ethyl	Ethanol	135
n-Propyl	Ligroin and ethanol	118
Isopropyl	Ligroin and ethanol	153
n-Butyl	Ligroin and ethanol	121
tert-Butyl	Ethylene chloride; and ethanol	275
n-Amyl	Ligroin and ethanol	111 d.
Isoamyl	Ligroin and ethanol	105
Cinnamyl	Ethylene chloride and ethanol	169
Benzyl	Ethylene chloride and ethanol	197
Menthol	Ethylene chloride and ethanol	172
Borneol	Ligroin and ethanol	197
Cholesterol	Ethylene chloride and ethanol	188–193
Phenol	Ethylene chloride and ethanol	187
Chlorobutanol	Ethylene chloride and ethanol	192–196
Thymol	Ethylene chloride and ethanol	162
Carvacrol	Ethylene chloride and ethanol	162

Dimethyl urethane, MP 183–183.5°C (194, 217)

(*3*) *Reactions*
Kinetics of the reaction with 1-butanol (424)
Kinetics of the reaction with diethylene glycol adipate polyester (421)

(*4*) *References*

147	254	362	421
167	321	371	424
194	337	418	448
217			

TABLE 7.18

Compound N-7: Naphthalene-1-isocyanato; 1-naphthyl-isocyanate; α-naphthyl isocyanate

(*1*) *Properties*
Dipole moment, 2.30 D (82), 2.32 D (392)
Dielectric relaxation and microwave absorption (392)
Infrared spectra (254, 371)
Raman spectra (49)

(*2*) *Derivatives*
Urethanes (27, 29)

Alcohol	MP, °C	Formula
Methyl	124	$C_{12}H_{11}O_2N$
Benzyl	134.5	$C_{18}H_{15}O_2N$
Cinnamyl	114	$C_{20}H_{17}O_2N$
Phenylethyl	119	$C_{19}H_{17}O_2N$
Lauryl	80	$C_{23}H_{33}O_2N$
n-Amyl	68	$C_{16}H_{19}O_2N$
Furfuryl	129–130	$C_{16}H_{13}O_3N$
m-Xylyl	116	$C_{19}H_{17}O_2N$
o-Methoxybenzyl	135–136	$C_{19}H_{17}O_3N$
Ethylene glycol	176	$C_{24}H_{20}O_4N_2$
Trimethylene glycol	164	$C_{25}H_{22}O_4N_2$
Glycerol	191–192	$C_{36}H_{29}O_6N_3$
Ethylene-bromohydrin	86– 87	$C_{13}H_{12}O_2NBr$
Trimethylene-bromohydrin	73– 74	$C_{14}H_{14}O_2NBr$
Ethylene-chlorohydrin	101	$C_{13}H_{12}O_2NCl$
Trimethylene-chlorohydrin	76	$C_{14}H_{14}O_2NCl$
Phenylmethyl carbinol	106	$C_{19}H_{17}O_2N$
Phenylethyl carbinol	102	$C_{20}H_{19}O_2N$
Menthol	119	$C_{21}H_{27}O_2N$
Borneol	127	$C_{21}H_{25}O_2N$
Isoborneol	130	$C_{21}H_{25}O_2N$
Cholesterol	160	$C_{38}H_{53}O_2N$
Benzoin	140	$C_{25}H_{19}O_3N$
Diphenyl carbinol	135–136	$C_{24}H_{19}O_2N$
Cyclohexanol	128–129	$C_{17}H_{19}O_2N$
2-Methylcyclohexanol	154–155	$C_{18}H_{21}O_2N$
3-Methylcyclohexanol	122	$C_{18}H_{21}O_2N$
4-Methylcyclohexanol	159–160	$C_{18}H_{21}O_2N$
Methylhexyl carbinol	63– 64	$C_{19}H_{25}O_2N$
Diethyl carbinol	71– 72	$C_{16}H_{19}O_2N$
Triphenyl carbinol	No reaction	
Diethylmethyl carbinol	No reaction	
Citronellol	No reaction	
Phenol	132–133	$C_{17}H_{13}O_2N$
o-Cresol	141–142	$C_{18}H_{15}O_2N$
m-Cresol	127–128	$C_{18}H_{15}O_2N$

TABLE 7.18 (*Continued*)

Urethanes Alcohol	MP, °C	Formula
p-Cresol	146	$C_{18}H_{15}O_2N$
Thymol	160	$C_{21}H_{21}O_2N$
Carvacrol	116	$C_{21}H_{21}O_2N$
o-Nitrophenol	112–113	$C_{17}H_{12}O_4N_2$
m-Nitrophenol	167	$C_{17}H_{12}O_4N_2$
p-Nitrophenol	150–151	$C_{17}H_{12}O_4N_2$
o-Chlorophenol	120	$C_{17}H_{12}O_2NCl$
m-Chlorophenol	157–158	$C_{17}H_{12}O_2NCl$
p-Chlorophenol	165–166	$C_{17}H_{12}O_2NCl$
o-Bromophenol	128–129	$C_{17}H_{12}O_2NBr$
p-Bromophenol	168–169	$C_{17}H_{12}O_2NBr$
2,4,6-Tribromophenol	153	$C_{17}H_{10}O_2NBr_3$
2-Chloro-5-hydroxytoluene	153–154	$C_{18}H_{14}O_2NCl$
4-Hydroxy-1,2-dimethylbenzene	141–142	$C_{29}H_{17}O_2N$
4-Hydroxy-1,3-dimethylbenzene	134–135	$C_{19}H_{17}O_2N$
2-Hydroxy-1,4-dimethylbenzene	172–173	$C_{19}H_{17}O_2N$
Resorcinol-monomethyl ether	128–129	$C_{18}H_{15}O_3N$
Guiacol	118	$C_{18}H_{15}O_3N$
Eugenol	122	$C_{21}H_{19}O_3N$
Isoeugenol	149–150	$C_{21}H_{19}O_3N$
Orcinol	160	$C_{29}H_{22}O_4N_2$
Resorcinol	No reaction	
Hydroquinol	No reaction	
Catechol	No reaction	
Pyrogallol	No reaction	
α-Naphthol	152	$C_{21}H_{15}O_2N$
β-Naphthol	156–157	$C_{21}H_{15}O_2N$
1-Nitro-2-naphthol	128–129	$C_{21}H_{14}O_4N_2$
o-Aminophenol	201	$C_{17}H_{14}O_2N_2$

Urethanes (46, 50)

	MP, °C
Ethyl	79
n-Propyl	105
Isopropyl	78
n-Butyl	71
sec-Butyl	97
Isobutyl	104
n-Amyl	68
Isoamyl	67
Hexyl	59
Heptyl	62
Octanol-2	63

TABLE 7.18 (*Continued*)

Ureas (29)

Amine used	MP, °C	Formula
Methylamine	196–197	$C_{12}H_{12}ON_2$
Ethylamine	199–200	$C_{13}H_{14}ON_2$
Dimethylamine	158–159	$C_{13}H_{14}ON_2$
Diethylamine	127–128	$C_{15}H_{18}ON_2$
Benzylamine	202–203	$C_{18}H_{16}ON_2$
Isoamylamine	131–132	$C_{16}H_{20}ON_2$
Diisoamylamine	94– 95	$C_{21}H_{30}ON_2$
Diisobutylamine	118–119	$C_{19}H_{26}ON_2$
Di-*n*-propylamine	92– 93	$C_{17}H_{22}ON_2$
Acetone-oxime	68– 69	$C_{14}H_{14}O_2N_2$
Benzald-oxime	149	$C_{18}H_{14}O_2N_2$
Benzophenone-oxime	154–155	$C_{24}H_{18}O_2N_2$
Acetamide	211–212	$C_{13}H_{12}O_2N_2$
Acetanilide	116–117	$C_{19}H_{16}O_2N_2$

Dinaphthyl urea, MP 298°C (386)

Trimer, MP 335°C (386)

(*3*) *Reactions*
Homopolymerization (505)

(*4*) *References*

1	46	82	258	392
27	49	194	371	505
29	50	254	386	

TABLE 7.19

Compound N-8: Naphthalene-2-isocyanato; 2-naphthyl isocyanate; β-naphthyl isocyanate

(*1*) *Properties*
Dipole moment = 2.34 D (82)
Infrared spectra (254)
Infrared spectra of trimer (386)

(*2*) *Derivatives*
Urethanes (96)

Phenols tested	MP, °C	Appearance of crystals
Phenol	149	Colorless plates from ligroin
o-Cresol	127–129	Colorless needles from ligroin
m-Cresol	123	Small colorless needles from ligroin
p-Cresol	159	Colorless plates from ligroin
1,2,4-Xylenol (3,4-dimethylphenol)	148–149	Colorless plates from ligroin

TABLE 7.19 (*Continued*)

Phenols tested	MP,°C	Appearance of crystals
1,4,5-Xylenol (2,5-dimethylphenol)	143–145	Colorless plates from ligroin
1,3,4-Xylenol (2,4-dimethylphenol)	140	Colorless needles from ligroin
Thymol	140–141	Colorless prisms from ligroin
Isothymol	89– 91	Colorless prisms from ligroin and stout needles from 50% alcohol
α-Naphthol	174–175	Colorless prisms from ligroin
β-Naphthol	202–203	Colorless prisms from ligroin
o-Nitrophenol	120–121	Very pale yellow prisms from ligroin
m-Nitrophenol	124	Pale yellow prisms from ligroin
p-Nitrophenol	172–173	Very pale yellow needles from ligroin
o-Chlorophenol	136–137	Colorless plates from ligroin
m-Chlorophenol	116–117	Large colorless prisms from ligroin
p-Chlorophenol	169–170	Colorless plates and prisms from ligroin
o-Bromophenol	128	Colorless prisms from ligroin
m-Bromophenol	118–119	Colorless prisms from ligroin
p-Bromophenol	175–176	Colorless prisms from ligroin
o-Iodophenol	150–152	Colorless rods and prisms from ligroin
m-Iodophenol	148	Colorless plates from ligroin
p-Iodophenol	189	Colorless plates from ligroin
2,4-Dichlorophenol	166	Colorless plates from ligroin
2,4-Dibromophenol	150–151	Colorless prisms from ligroin
2,4,6-Trichlorophenol	161–162	Colorless prisms from ligroin
2,4,6-Tribromophenol	181–183	Colorless prisms from ligroin
Methyl salicylate	Charred and dec. at 290	Colorless plates and prisms from ligroin and methyl alcohol
Guaiacol	108–109	Colorless plates from ligroin
Resorcinol monomethylether	93– 95	Colorless prisms from ligroin and 95% alcohol
Hydroquinone monomethylether	167	Colorless needles and prisms from ligroin

Methyl urethane, MP 113°C (194)
Dimer, MP 194–199°C (281)
Trimer, MP 344°C (386)
Di-2-naphthyl urea, MP 310°C (386)

(*3*) *Reactions*

—

(*4*) *References*

82, 96, 194, 254, 281, 386

TABLE 7.20

Compound T-16: Toluene-2-isocyanato; *o*-tolyl isocyanate

(*1*) *Properties*

Infrared data (371, 422)

(*2*) *Derivatives*

Urethane from isopropyl alcohol, MP 62° (268)

(*3*) *Reactions*

Kinetics of the reaction with 1-butanol (424)

Reaction with $AlCl_3$ (265, 285)

(*4*) *References*

91	268	422
194	285	424
265	371	518

TABLE 7.21

Compound T-17: Toluene-4-isocyanato; *p*-tolyl-isocyanate

(*1*) *Properties*

Infrared data	(371)(422)
Dielectric Relaxation and Microwave absorption	(392)
Dipole moment 2.68 D	(392)

(*2*) *Derivatives*

Urethane from isopropyl alcohol, MP 52°C. (268)

Dimer, MP 186–187°C (518)

Di(*p*-tolyl) carbodiimide, MP 56.5–57.5°C, BP 165°C/2.4 mm (432)

(*3*) *Reactions*

Kinetics of the reaction with 2-ethyl hexanol (520)

Reactivity in terms of the Hammett equation (520)

Reaction with $AlCl_3$ (265, 285)

Kinetics of reaction with 1-butanol (424)

(*4*) *References*

(91)	(266)	(392)	(518)
(93)	(268)	(422)	(520)
(194)	(285)	(424)	(524)
(265)	(371)	(432)	

TABLE 7.22

Compound T-18: Toluene-3-isocyanato; *m*-tolyl isocyanate

(*1*) *Properties*
Infrared data (371, 422)

(*2*) *Derivatives*
Urethane from isopropyl alcohol, MP 90° (268)

(*3*) *Reactions*
Kinetics of the reaction with 2-ethylhexanol (520)
Reactivity in terms of Hammett equation (520)
Reaction with $AlCl_3$ (265, 285)
Homopolymerization (505)

(*4*) *References*

194	285	505
265	371	518
268	422	520

TABLE 7.23

Compound T-19: Toluene 2,4-diisocyanato; 4-methyl-*m*-phenylenediisocyanate

(*1*) *Properties*

	Refs.
Infrared data	254, 371
Density $d_t = d_{25}[1.-8.31 \times 10^{-5}(t-25)]$ for range 25–90°	464
Refractive index; 10.97% in monochlorobenzene, $n_D^{25} = 1.5385$	
99.60% in monochlorobenzene, $n_D^{25} = 1.5658$	
Viscosity, $\eta = 0.0995e^{-3330/RT}$	464
Vapor pressure of saturated vapors at 100–155° is given by $\log P = -(2342/T + 7.4006)$	464
Vapor–liquid equilibria for mixtures of 2,4- and 2,6-toluenediisocyanates	437
Vapor–liquid equilibria of binary solutions of 2,4-toluenediisocyanate in monochlorobenzene and trichlorobenzene	478
Heat of combustion, 1012.2 kcal/mole	555
Heat capacity, 0.395 cal/g/°C	555
Energy of evaporation is given by $E_{ev} = RT + 9700$ cal/mole	464
Dielectric relaxation and dipole moment	392
Microwave absorption	392
Dipole moment, 2.52 D	392

TABLE 7.23 (*Continued*)

(2) *Derivatives*

Dicarbamates	MP, °C	Ref.
Methyl	170–171	194, 362 379, 402 413
Ethyl	135–136.5	302, 379 413
n-Propyl	104	413
Isopropyl	134–136	302, 413 379
n-Butyl	80– 81	379, 413
Isobutyl	120	413
sec-Butyl	93	413
tert-Butyl	208	362
n-Amyl	89	413
Isoamyl	76	413
Neopentyl	132	413
n-Hexyl	83	413
n-Heptyl	82	413
n-Octyl	69– 70	413, 379
n-Nonyl	72	413
n-Decyl	72– 73	413
n-Undecyl	61	413
n-Dodecyl	87	413, 302
n-Tetradecyl	90	413
n-Hexadecyl	93– 94	413
n-Octadecyl	99.5–100	413
Cyclohexyl	158	413
Benzyl	101–102	413, 362
β-Phenylethyl	129.5–130	413
Cinnamyl	117	362
Geraniol	89	362
Menthol	174–179	362
Borneol	217–220	362
Chlosterol	184–187	362
Linalool	228–232	362
Phenol	144	362
Chlorobutanol	189–194	362
Thymol	190	362
Carvacrol	159	362
m-Chlorophenol	120.5	362
o-Chlorophenol	160	362
Trichlorophenol	162	362

TABLE 7.23 (*Continued*)

Carbanilates, R	MP, °C	Ref.	Structure
Methyl	94 –96	379	
Ethyl	51 –59	379	
Butyl	BP 144 at 0.2 mm	379	
Octyl	40 –42	379	
Hexadecyl	70.5–73.5	379	

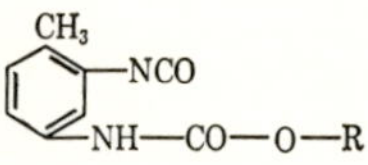

3-Amino-4-methyl carbanilide (323)
MP 200–201°

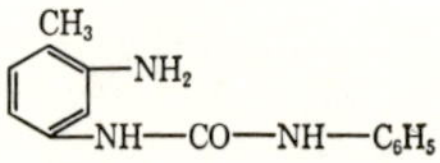

Dimer, MP 158° (281)

Ureas (481)

(*a*) Di(3-isocyanato-4-methylphenyl)urea
MP 176–179°

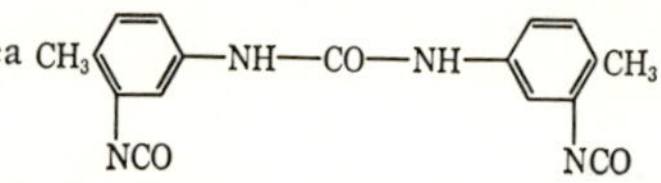

(*b*) Dimethyl,4,4′ -Dimethylcarbanido-4,4′ dicarbamate
MP, 220–220.5°

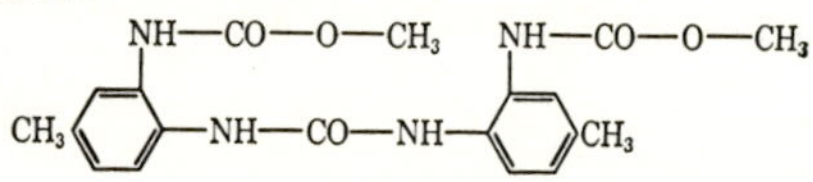

(*3*) *Reactions*

Rate constants and activation energy of the reaction with diethylene glycol adipate polyesters (421)
Kinetics of the reaction with 1-butanol (424)
Reactivity in terms of the Hammett equation (520)
Homopolymerization (505)
Relative reactivity of NCO groups (323)

(*4*) *References*

194	362	413	478
254	371	421	481
281	392	424	505
302	379	437	520
323	402	464	555
327			

References

Asterisk (*) signifies data from abstract

1. A. W. Hofmann, *Ber.* **3,** 653–658 (1870).
2. E. Custer, *Ber.* **12,** 1328–1334 (1879).
3. A. W. Hofmann, *Ber.* **15,** 752–762 (1882).

4. Chininfabriken Zimmer and Co., G. m b, H. German Pat. 133,760 (1901) (See *Chem. Zentr.* 1902, II 553).*
5. O. C. Billeter, *Ber.* **36,** 3213–3218 (1903).
6. A. Neville and R. H. Pickard, *J. Chem. Soc.* **85,** 685–690 (1904).
7. 2:1697[6] R. Anschutz, *Ann.* **359,** 202–216, (1908).
8. 3:424[5] C. Vallee, *Ann. Chim Phys.* (7) **15,** 331–434, November, (1898).
9. 3:2122[1] M. O. Forster, *J. Chem. Soc.* **95,** 433–439 (1909).
(558) 4:761 O. Wallach, *Ann.* **369,** 63–103, (1909).
10. 5:3401 C. Mauguin, *Ann. chim. phys.* [8], **22,** 297–369 (1911) (See *J. Chem. Soc. Abstr.*, **100,** 357 (1911)).*
11. 5:3566[6] M. O. Forster and H. Stotter, *J. Chem. Soc.* **99,** 1337–1340 (1911).
12. 6:2929[3] J. V. Braun and H. Deutsch, *Ber.* **45,** 2199–2200 (1912).
13. 7:2744[5] L. Frank, *J. Chem. Soc.* **103,** 852–861 (1913).
14. 8:681[8] L. Tschugaeff and A. Glebko, *Ber.* **46,** 2752–2762 (1913).
15. 8:917[5] T. Curtius, *J. Prakt. Chem.* **87,** 513–541 (1913).
16. 9:309[7] W. Borsche and W. Sander, *Ber.* **47,** 2815–2826 (1914).
17. 11:3262[9] T. Curtius, *J. Prakt. Chem.* **94,** 85–134 (1916).*
18. 12:698[1] H. Staudinger and R. Endle, *Ber.* **50,** 1042–1046 (1917).
19. 14:3667[9] A. Skita and H. Rolfes, *Ber.* **53B,** 1242–1255 (1920).
20. 16:415[2] L. W. Jones and C. D. Hurd, *J. Am. Chem. Soc.* **43,** 2422–2448 (1921).
21. 17:3325[2] W. H. Carothers, *J. Am. Chem. Soc.* **45,** 1734–1738 (1923).
22. 18:989[4] Society for Chemical Industry in Basle, Swiss Pats. 94,569 and 94,984 (1923) (See *J. Chem. Soc.*, **124,** Abstr. **1,** 925 (1923)).*
23. 18:2871[8] T. Curtius and R. Gund, *J. Prakt Chem.* **107,** 177–192 (1924).*
24. 19:240[5] L. W. Jones and D. H. Powers, *J. Am. Chem. Soc.* **46,** 2518–2533 (1924).
25. 19:642[5] C. D. Hurd and H. J. Brownstein, *J. Am. Chem. Soc.* **47,** 174–178 (1925).
26. 19:3083[6] K. H. Slotta and L. Lorenz, *Ber.* **58B,** 1320–1323 (1925).
27. 20:1232[9] V. T. Bickel and H. E. French, *J. Am. Chem. Soc.* **48,** 747–751 (1926).
28. 20:1616[5] K. Fries, *Ber.* **58B,** 2845–2851 (1925).
29. 20:2319[3] H. E. French and A. F. Wirtel, *J. Am. Chem. Soc.* **48,** 1736–1739 (1926).
30. 20:3010[8] H. John, *Ber.* **59B,** 1447–1452 (1926).
31. 21:78[7] H. R. Ing and R. H. F. Manske, *J. Chem. Soc.*, **1926,** 2348–2351.
32. 21:1120[6] H. Lindemann and W. Schultheis, *Ann.* **451,** 241–255 (1927).
33. 21:1632[9] K. H. Slotta and R. Tshesche, *Ber.* **60B,** 295–301 (1927).
34. 22:3345[1] K. V. Auwers, *Ber.* **61B,** 1041–1049 (1928).
35. 24:80[5] S. Kawai, *J. Chem. Soc. Japan*, **49,** 235–243 (1928).*
36. 24:1358[1] D. Vörlander and O. Ritter, *Ber.* **62B,** 2824–2831 (1929).
37. 24:2718[2] M. Montagne, *Ann. Chim* [10], **13,** 40–135 (1930).
38. 24:3217[2] T. Curtius, W. Sieber, F. Nadenheim, O. Hambsch, and W. Ritter, *J. Prakt Chem.* **125,** 152–210 (1930).*
39. 24:3232[5] T. Curtius and W. Ulmer, *J. Prakt Chem.*, **125,** 54–62 (1930).*
40. 24:3767[3] K. H. Slotta and H. Dressler, *Ber.* **63B,** 888–898 (1930).
41. 24:4256[4] S. Kawai and K. Tamura, *Proc. Imp. Acad. (Japan)* **6,** 198–200 (1930).*
42. 24:5019[3] M. Montagne and B. Casteran, *Compt. Rend* **191,** 139–141 (1930).
43. 24:5287[7] S. Kawai, *Sci. Papers Inst. Phys. Chem. Res. (Tokyo)* **13,** 260–269 (1930).*
44. 25:953[9] H. John and E. Andraschko, *J. Prakt. Chem.* **128,** 180–189 (1930).*
45. 25:2364[8] A. Dadieu, *Ber.* **64B,** 358–361 (1931).
46. 25:2686 R. L. Shriner and F. B. Cox, *J. Am. Chem. Soc.* **53,** 1601–1605 (1931).
47. 25:3616[8] C. L. Birckenbach and M. Linhard, *Ber.* **64B,** 961–968 (1931).

48. 25:4202[4] G. T. Morgan and A. E. J. Pettet, *J. Chem. Soc.*, **1931**, 1124–1126.
49. 25:4794[9] A. Dadieu, *Monatsh. Chem.* **57**, 437–468 (1938).*
50. 25:4861[7] W. H. Horne and R. L. Shriner, *J. Am. Chem. Soc.* **53**, 3186 (1931).
51. 25:4883[9] H. John and H. Lukas, *J. Prakt. Chem.* **130**, 289–292 (1931).*
52. 26:435[2] B. B. Dey and T. R. Seshadri, *J. Indian Chem. Soc.* **8**, 293–300 (1931).*
53. 26:460[5] H. John and H. Ottawa, *J. Prakt. Chem.* **131**, 301–308 (1931).*
54. 26:728[4] H. John and F. Schmit, *J. Prakt. Chem.* **132**, 15–23 (1931).*
55. 26:2430[6] C. Naegeli and P. Lendorff, *Helv. Chim. Acta* **15**, 49–59 (1932).*
56. 26:3201[1] C. W. Van Hoogstraten, *Rec. Trav. Chim.* **51**, 414–433 (1932).*
57. 26:5063[1] S. Kawai and K. Tamura, *J. Chem. Soc. Japan* **52**, 77–82 (1931).*
58. 26:5964[7] I. G. Farbenind. A. G., French Pat. 729246 (Jan. 5, 1932) (See *CA* 27:3485 for British Pat. 372,355 May 4, 1932)*
59. 27:490[1] W. J. Hoppen Brouwers, *Rec. Trav. Chim.* **51**, 951–954 (1932).*
60. 27:2382[1] H. Kopper and A. Pongratz, *Monatsh. Chem.* **62**, 78–89 (1938).*
61. 27:3459[4] N. V. Sidgwick, L. E. Sutton, and W. Thomas, *J. Chem. Soc.* **1933**, 406–412.
62. 27:3923[7] C. Naegeli and A. Tyabji, *Helv. Chim. Acta* **16**, 349–366 (1933).
63. 27:4686[9] J. Hilger, U.S. Pat. 1,916,314 (July 4, 1933).
64. 28:1696[8] A. T. Blomquist and H. B. Stevenson, *J. Am. Chem. Soc.* **56**, 146–149 (1934).
65. 28:2344[5] J. V. Braun and P. Kurtz, *Ber.* **67B**, 225–230 (1934).
66. 28:2691[3] R. L. Shriner, W. H. Horne, and R. F. B. Cox, *Org. Syn.* **14**, 72–74 (1934).
67. 28:2703[3] J. Read and R. G. Johnston, *J. Chem. Soc.* **1934**, 226–233.
68. 28:3055[9] J. S. Blair and G. E. P. Smith, Jr., *J. Am. Chem. Soc.* **56**, 907–910 (1934).
69. 28:4405[8] M. J. van Gelderen, *Rec. Trav. Chim.* **52**, 969–975 (1933).*
70. 28:4406[1] M. J. van Gelderen, *Rec. Trav. Chim.* **52**, 976–978 (1933).*
71. 28:4406[3] M. J. van Gelderen, *Rec. Trav. Chim.* **52**, 979–981 (1933).*
72. 29:765[7] G. Komppa and S. Beckmann, *Ann.* **512**, 172–185 (1934).
73. 29:778[4] H. Fischer and A. Waibel, *Ann.* **512**, 195–217 (1934).
74. 29:1399[6] D. V. N. Hardy, *J. Chem. Soc.* **1934**, 2011.
75. 29:3665[7] M. S. Newman, *J. Am. Chem. Soc.* **57**, 732–735 (1935).
76. 29:4687[6] L. Birckenbach, H. Kolb, and J. Goubeau, *Ber.* **68B**, 895–912 (1935).
77. 29:6117[4] J. R. Partington and E. G. Cowley, *Nature*, **135**, 1938 (1935).*
78. 29:7188[8] S.-C. Woo and T.-K. Liu, *J. Chem. Phys.* **3**, 544–546 (1935).
79. 30:465[1] R. Kuhn, K. Reinemund, F. Weygand, and R. Strobele, *Ber.* **68B**, 1765–1774 (1935).
80. 30:1362[1] J. Hoch, *Compt. Rend.* **201**, 733–735 (1935).
81. 30:3416[4] V. V. Vasilevskii, F. I. Bloshtein, and B. D. Kustrya, *J. Gen. Chem (USSR)* **5**, 1652–1667 (1935).*
82. 30:5084[7] E. G. Cowley and J. R. Partington, *J. Chem. Soc.* **1936**, 45–47.
83. 30:8224[9] T. B. Johnson and A. Litzinger, *J. Am. Chem. Soc.* **58**, 1936–1942 (1936).
84. 31:662[4] H. Wenker, *J. Am. Chem. Soc.* **58**, 2608 (1936).
85. 31:6676[1] I. G. Farbenind. A. G., French Pat. 809,233 (Feb. 26, 1937)*
86. 32:1686[4] K. Fries and H. Bestian, *Ann.* **533**, 72–92 (1937).*
87. 32:1695[8] H. J. Creech and W. R. Franks, *J. Am. Chem. Soc.* **60**, 127–128 (1938).
88. 32:2535[5] P. E. Spoerri and A. Erickson, *J. Am. Chem. Soc.*, **60**, 400–402 (1938).
89. 32:3387[3] H. M. Singleton and W. R. Edwards, *J. Am. Chem. Soc.* **60**, 540–544 (1938).
90. 32:7056[9] R. Greenhalgh and H. Piggott, Brit. Pat. 483,398 (Apr. 14, 1938).
91. 32:7926[2] H. A. Piggott and F. S. Statham, Brit. Pat. 485,761 (May 24, 1938).
92. 32:8973[3] M. Linhard, *Z. Anorg. Allgem. Chem.* **239**, 155–160 (1938).*

93. 33:540[3] C. Naegli, A. Tyabji, and L. Conrad, *Helv. Chim. Acta* **21**, 1127–1143 (1938).
94. 33:558[4] F. E. Ray and G. Rieveschl, Jr., *J. Am. Chem. Soc.* **60**, 2675–2677 (1938).
95. 33:1702[8] A. G. Short and J. Read, *J. Chem. Soc.* **1938**, 2016–2021.
96. 33:5774[3] P. P. T. Sah, *Rec. Trav. Chim.* **58**, 453–458 (1939).
97. 33:8189[5] Y. Asahina and H. Kawahata, *Ber.* **72B**, 1540–1548 (1939).
98. 34:1011[5] L. Fieser and H. J. Creech, *J. Am. Chem. Soc.* **61**, 3502–3506 (1939).
99. 34:1251[1] R. N. Jones, *J. Am. Chem. Soc.* **62**, 148–152 (1940).
100. 34:1583[8] P. P. T. Sah, *Rec. Trav. Chim.* **58**, 1008–1012 (1939).
101. 34:1583[9] P. P. T. Sah and T.-F. Woo, *Rec. Trav. Chim.* **58**, 1013–1017 (1939).
102. 34:1634[3] J. J. Donleavy and J. English, Jr., *J. Am. Chem. Soc.* **62**, 218–219 (1940).
103. 34:3609[2] G. S. Forbes and H. H. Anderson, *J. Am. Chem. Soc.* **62**, 761–763 (1940).
104. 34:4005[8] M. Linhard and K. Betz, *Ber.* **73B**, 177–185 (1940).
105. 34:4663[9] E. H. Eyster and R. H. Gillette, *J. Chem. Phys.* **8**, 369–377 (1940).
106. 34:5786[6] P. P. T. Sah, *Rec. Trav. Chim.* **59**, 231–237 (1940).
107. 34:5786[8] P. P. T. Sah and W.-H. Yin. *Rec. Trav. Chim.* **59**, 238–245 (1940).
108. 34:6230[1] A. J. Hill and W. M. Degnan, *J. Am. Chem. Soc.* **62**, 1595–1596 (1940).
109. 34:6661[3] H. J. Creech and R. N. Jones, *J. Am. Chem. Soc.* **62**, 1970–1975 (1940).
110. 34:7895[2] H. Goldstein and K. Stern, *Helv. Chim. Acta* **23**, 818–820 (1940).*
111. 35:11[5] E. H. Eyster, R. H. Gillette, and L. O. Brockway, *J. Am. Chem. Soc.* **62**, 3236–3243 (1940).
112. 35:85[4] J. J. Donleavy and J. English, Jr., *J. Am. Chem. Soc.* **62**, 2965–2966 (1940).
113. 35:1071[8] P. Pöhls and F. Mietzsch, U.S. Pat. 2,218,030 (Oct. 15, 1940).
114. 35:1318[2] R. N. Jones, *J. Am. Chem. Soc.* **63**, 151–155 (1941).
115. 35:2140[1] H. J. Creech, *J. Am. Chem. Soc.* **63**, 576–578 (1941).
116. 35:2858[9] T. L. Cairns, *J. Am. Chem. Soc.* **63**, 871–872 (1941).
117. 35:3970[9] D. A. Bamford and C. H. Bamford, *J. Chem. Soc.* **1941**, 30–34.
118. 35:4362[8] P. P. T. Sah and C. L. Hsu, *Rec. Trav. Chim.* **59**, 349–356 (1940).
119. 35:4363[3] P. P. T. Sah and P.-T. Young, *Rec. Trav. Chim.* **59**, 357–363 (1940).
120. 35:4363[5] P. P. T. Sah and Y-K. Wang, *Rec. Trav. Chim.* **59**, 364–368 (1940).
121. 36:6101[8] H. H. Anderson, *J. Am. Chem. Soc.* **64**, 1757–1759 (1942).
122. 37:1184[5] A. H. Coons, H. J. Creech, R. N. Jones, and E. Berliner, *J. Immunol.* **45**, 159–170 (1942).*
123. 37:1432[8] H. C. White and F. W. Bergstrom, *J. Org. Chem.* **7**, 497–507 (1942).
124. 37:4061[5] C. Mentzer, *Compt. Rend.* **213**, 581–584 (1941).*
125. 37P:4408[1] R. Greenhalgh and H. A. Piggott, U.S. Pat. 2,311,046 (Feb. 16, 1943).
126. 37:5057[6] L. C. Raiford and H. B. Freyermuth, *J. Org. Chem.* **8**, 230–238 (1943).
127. 37P:5987[8] W. E. Hanford, U.S. Pat. 2,319,057 (May 11, 1943).
128. 37:6248[1] Y.-T. Huang, K.-H. Lin, L. Li, and M.-C. Lu, *J. Chinese Chem. Soc.* **8**, 201–217 (1941).
129. 37:6576[7] A. W. Laubengayer and L. Reggel, *J. Am. Chem. Soc.* **65**, 1783–1784 (1943).
130. 39:554[6] W. Siefken and A. Doser, U.S. Pat. 2,326,501 (Aug. 10, 1944).
131. 38:693[5] G. S. Forbes and H. H. Anderson, *J. Am. Chem. Soc.* **65**, 2271–2274 (1943).
132. 38:2331[9] P. Cagniant and Buu-Hoi, *Bull. Soc. Chim. France* **10**, 349–353 (1943).
133. 38:2772[3] D. D. Coffman, U.S. Pat. 2,334,376 (Nov. 16, 1944).
134. 38:2929[3] Buu-Hoi and P. Cagniant, *Z. Physiol. Chem.* **279**, 76–86 (1943).*
135. 38:3261[7] C. Mentzer, Buu-Hoi, and P. Cagniant, *Bull Soc. Chim. France* **9**, 813–818 (1942).
136. 38:3635[3] L. H. Fraenkel-Conrat and H. S. Olcott, *J. Am. Chem. Soc.* **66**, 845 (1944).

137. 38:3954[6] A. Buhler and H. E. Fierz-David, *Helv. Chim. Acta* **26,** 2123–2136 (1943).

138. 38:3976[9] P. Cagniant, C. Mentzer, and Buu-Hoi, *Bull. Soc. Chim. France* **10,** 145–151 (1943).

139. 38:4257[7] J. B. McKinley, J. E. Nickels, and S. S. Sidhu, *Ind. Eng. Chem., Anal. Ed.* **16,** 304–308 (1944).

140. 38:4528[5] H. H. Anderson, *J. Am. Chem. Soc.* **66,** 934–935 (1944).

141. 39:4621[7] W. Kaase and E. Waltman, U.S. Pat. 2,340,757 (Feb. 1, 1944).

142. 38:5799[9] C. F. H. Allen and A. Bell, *Org. Syn.* **24,** 94–95 (1944).

143. 38:6149[5] G. S. Forbes and H. H. Anderson, *J. Am. Chem. Soc.* **66,** 1703–1706 (1944).

144. 39:1134[7] G. D. Jones, J. Zomlefer, and K. Hawkins, *J. Org. Chem.* **9,** 500–512 (1944).

145. 39:1366[3] H. H. Anderson, *J. Am. Chem. Soc.* **67,** 223–225 (1945).

146. 39:1411[8] R. H. F. Manske and H. L. Holmes, *J. Am. Chem. Soc.* **67,** 95–98 (1945).

147. 39:2764[5] J. G. Lichty and N. V. Seeger, U.S. Pat. 2,362,648 (Nov. 14, 1944).

148. 39:3555[9] M. W. Farlow, U.S. Pat. 2,374,340 (Apr. 24, 1945).

149. 39:4317[9] J. Colucci, *Can. J. Res.* **23B,** 111–112 (1945).*

150. 39:4722[6] N. H. Fell, U.S. Pat. 2,376,424 (May 22, 1945).

151. 39:5250[5] R. H. F. Manske, A. E. Ledingham, and H. L. Holmes, *Can. J. Res.* **23B,** 100–105 (1945).

152. 39:5254[5] E. Burgoine, B. Collie, and R. G. A. New, U.S. Pat. 2,379,948 (July 10, 1945).

153. 40:175[5] A. J. Hill and W. M. Degnan, U.S. Pat. 2,379,486 (July 3, 1945).

154. 40:287[1] G. S. Forbes and H. H. Anderson, *J. Am. Chem. Soc.* **67,** 1911–1914 (1945).

155. 40:1468[2] H. Heymann and L. F. Fieser, *J. Am. Chem. Soc.* **67,** 1979–1986 (1945).

156. 40:1499[5] L. C. Cheney and J. R. Piening, *J. Am. Chem. Soc.* **67,** 2252–2255 (1945).

157. 40:2071[8] R. N. Jones, *J. Am. Chem. Soc.* **67,** 2127–2150 (1945).

158. 40:2137[9] Buu-Hoi and P. Cagniant, *Compt. Rend.* **219,** 455–457 (1944).*

159. 40:2816[3] Buu-Hoi, P. Cagniant, and C. Mentzer, *Bull. Soc. Chim. France* **11,** 127–136 (1944).*

160. 40:2848[3] J. B. Dickey, J. E. Straley, and T. E. Stanin, U.S. Pat. 2,394,597 (Feb. 12, 1946).

161. 40:3410[4] R. R. McNary and L. Magid, *J. Am. Chem. Soc.* **68,** 673–674 (1946).

162. 40:5013[6] Buu-Hoi and P. Cagniant, *Rec. Trav. Chim.* **65,** 246–250 (1946).*

163. 40:5696[7] M. Montagne and T. Guilmart, *Bull. Soc. Chim France* **12,** 836–839 (1945).*

164. 40:5736[1] R. C. Elderfield, W. J. Gensler, T. H. Bembry, C. B. Kremer, J. D. Head, F. Brody, and R. Frohardt, *J. Am. Chem. Soc.* **68,** 1272–1276 (1946).

1 5. 41:386d C. E. Dunlap and S. Warren, *Cancer Res.* **6,** 454–465 (1946).

166. 41:958f L. Q. Green and H. J. Creech, *J. Am. Chem. Soc.* **68,** 2401-2402 (1946).

167. 41:992i Wingfoot Corp., British Pat. 566,532, (Jan. 3, 1945).

168. 41:1239f C. E. Schweitzer, U.S. Pat. 2,409,712 (Oct. 22, 1946).

169. 41:2747h M. W. Farlow, Brit, Pat. 567,752 (Mar. 1, 1945).

170. 41:3483f C. E. Schweitzer, U.S. Pat. 2,416,068 (Feb. 18, 1947).

171. 41:4474f M. Ishidate and T. Isshiki, *Bull. Chem. Soc. Japan* **17,** 507–509 (1942).

172. 41:5097h R. D. Haworth, A. H. Lamberton, and D. Woodcock, *J. Chem. Soc.* **1947,** 182–191.

173. 41:6533e A. E. A. Werner and J. Cray, *Sci. Proc. Roy. Dublin Soc.* **24,** 209–214 (1947).

174. 42:140c B. Witten and E. E. Reid, *J. Am. Chem. Soc.* **69,** 2470–2472 (1947).

175. 42:836a H. H. Anderson, *J. Am. Chem. Soc.* **69,** 2495–2497 (1947).

176. 42:1504h W. C. Price and A. D. Walsh, *Proc. Roy. Soc. (London) Ser. A* **191,** 22–31 (1947).

177. 42:3192b L. W. Georges and C. Hamalainen, U.S. Pat. 2,428,843 (Oct. 14, 1947).
178. 42:4544f G. S. Forbes and H. H. Anderson, *J. Am. Chem. Soc.* **70,** 1222-1223 (1948).
179. 42:4545f G. S. Forbes and H. H. Anderson, *J. Am. Chem. Soc.* **70,** 1043–1044 (1948).
180. 42:4561f P. Cagniant, *Compt. Rend.* **226,** 675–677 (1948).
181. 42:7101a F. Codignola and M. Piacenza, Ital. Pat. 415,752 (Nov. 2, 1946).*
182. 42:7257a C. Mentzer and J. Chopin, *Bull. Soc. Chim. France* **1948,** 586–590.
183. 43:168c S. Petersen and E. Muller, *Chem. Ber.* **81,** 31–38 (1948).
184. 43:1563d P. Eckert and E. Herr, *Kunstseide Zellwolle* **25,** 204–210 (1947).*
185. 43:3378b L. W. Georges and C. Hamalainen, *J. Am. Chem. Soc.* **71,** 743–744 (1949).
186. 43:4232i S. Olsen and E.-M. Enkemeyer, *Chem. Ber.* **81,** 359–361 (1948).
187. 43:4246a E. Bergmann and M. Orchin, *J. Am. Chem. Soc.* **71,** 1111–1112 (1949).
188. 43:4644i H. Bredereck and E. Reif, *Chem. Ber.* **81,** 426–438 (1948).
189. 43:5422i E. L. Kropa and A. S. Nyguist, U.S. Pat. 2,468,713 (April 26, 1949).
190. 43:5800e Laboratoires francais de chimiotherapie, British Pat. 613,111 (Nov. 23, 1948).*
191. 43:6227f E. Burgoine, B. Collie, R. G. A. New Brit. Pat. 573,017 (Nov. 2, 1945).
192. 43:6536g H. H. Anderson, *J. Am. Chem. Soc.* **71,** 1799–1801 (1949).
193. 43:6993d A. Lambert, J. D. Rose, and B. C. L. Weedon, *J. Chem. Soc.* **1949,** 42–46.
194. 44:109a W. Siefken, *Ann.* **562,** 75–136 (1949).
195. 44:115 S. Petersen, *Ann.* **562,** 205–229 (1949).
196. 44:1135i R. J. Slocombe, E. E. Hardy, and J. H. Saunders, U.S. Pat. 2,480,089 (Aug. 23, 1949).
197. 44:1136b R. J. Slocombe and E. E. Hardy, U.S. Pat. 2,480,088 (Aug. 23, 1949).
198. 44:1434c J. E. Gill, R. MacGillwray, and J. Munro, *J. Chem. Soc.* **1949** 1753–1754.
199. 44:2640i A. H. Coons and M. H. Kaplan, *J. Exptl. Med.* **91,** 1–13 (1950).
200. 44:2929b N. Sperber and R. Fricano, *J. Am. Chem. Soc.* **71,** 3352–3353 (1949).
201. 44:2974b P. Cagniant, *Bull. Soc. Chim. France* **1949,** 382–388.*
202. 44:3440h S. Masuyama, *J. Agr. Chem. Soc. Japan* **23,** 45–47 (1949).*
203. 44:3924e Y. Iwakura and K. Nagakubo, *Bull. Tokyo, Inst. Technol.* **13**(1), 25–34, 35–42 (1948).*
204. 44:4500g J. Munro, J. E. Gill, Brit. Pat. 631,007 (Oct. 25, 1949).
205. 44:5252d H. H. Anderson, *J. Am. Chem. Soc.* **72,** 196–197 (1950).
206. 44:6468f K. Zeile and M. Oetzel, *Z. Physiol. Chem.* **284,** 1–19 (1949).
207. 44:7244c R. J. Slocombe, E. E. Hardy, and J. H. Saunders, *J. Am. Chem. Soc.* **72,** 1888–1891 (1950).
208. 44:7599h L. H. Jones, J. N. Schoolery, R. G. Shulman, and D. M. Yost, *J. Chem. Phys.* **18,** 900–901 (1950).
209. 44:9930d H. Gilman, B. Hofferth, and H. W. Melvin, *J. Am. Chem. Soc.* **72,** 3045–3047 (1956).
210. 44:10713b J. W. Cook and J. S. Moffatt, *J. Chem. Soc.* **1950,** 1160–1170.
211. 45:584c M. Ishidate and T. Tani, *J. Pharm. Soc. Japan* **62,** 12–15 (1942).*
212. 45:1960h M. Harispe and J. Pernin, *Bull. Soc. Chim. France* **1950,** 660–662.*
213. 45:2711e Y. Iwakura, *Chem. High Polymers (Japan)* **4,** 91–94 (1947).*
214. 45:2910e J. B. Davenport and M. D. Sutherland, *Univ. Queensland Papers, Dept. Chem.* **1,** No. 39, 8 pp (1950).
215. 45:2917i J. H. Weisburger, E. K. Weisburger, and F. E. Ray, *J. Am. Chem. Soc.* **72,** 4253–4255 (1950).
216. 45:3409d D. X. Klein, U.S. Pat. 2,532,559 (Dec. 5, 1950).
217. 45:5117a T. Lieser and G. Nischk, *Ann.* **569,** 59–65 (1950).

218. 45:5178e A. Hrubesch, Ger. Pat. 801,993 (Feb. 1, 1951).
219. 45:5606d S. Masuyama, *J. Chem. Soc. Japan, Pure Chem. Sect.* **70,** 232–234 (1949).*
220. 45:5609b L. Herzog, M. H. Gold, and R. D. Geckler, *J. Am. Chem. Soc.* **73,** 749–755 (1951).
221. 45:6593i S. Masuyama and M. Hamada, *J. Chem. Soc. Japan, Pure Chem. Sect.* **70,** 198–200 (1949).*
222. 45:6601i E. D. Bergmann and M. Sulzbacher, *J. Org. Chem.* **16,** 84–89 (1951).
223. 45:6994e S. Masuyama, *J. Chem. Soc. Japan, Pure Chem. Sect.* **71,** 47–48 (1950).*
224. 45:7141b J. P. Mason, U.S. Pat. 2,544,709 (Mar. 13, 1951).
225. 45:7585g British Thomson-Houston Co. Ltd. British Pat. 643,941 (Sept. 27, 1950).
226. 45:9463e H. F. Piepenbrink, *Ann.* **572,** 83–95 (1951).
227. 46:907a M. W. Farlow, *Org. Syn.* **31,** 62–65 (1951).
228. 46:907c N. Sperber, R. Fricano, E. Schwenk, and D. Papa, *J. Am. Chem. Soc.* **73,** 2965 (1951).
229. 46:1520g S. Wawzonek and E. M. Smolin, *J. Org. Chem.* **16,** 746–748 (1951).
230. 46:2040b W. M. Smith, Jr., E. F. Pratt, and H. J. Creech, *J. Am. Chem. Soc.* **73,** 319–322 (1951).
231. 46:3009b R. Roesch and M. H. Gold, *J. Am. Chem. Soc.* **73,** 2959 (1951).
232. 46:4315b R. Hulme, *Acta Cryst.* **5,** 144 (1952).
233. 46:5583i A. F. McKay, *J. Org. Chem.* **16,** 1395–1404 (1951).
234. 46:6446f M. Szwarc and J. W. Taylor, *Trans. Faraday Soc.* **47,** 1293–1298 (1951).
235. 46:7117c F. C. Schaefer and E. K. Drechsel, U.S. Pat. 2,580,468 (Jan. 1, 1952).
236. 46:9122d R. J. Slocombe, E. Hardy, and J. H. Saunders, Brit. Pat. 672,685 (May 28, 1952) (see *CA* 44:1135i).
237. 46:10714c W. J. O. Thomas, *J. Chem. Soc.* **1952,** 2383–2389.
238. 47:1093g J. Read and D. S. P. Roebuck, *J. Chem. Soc.* **1952,** 812–816.
239. 47:2992h Y. Iwakura and R. Koyama, *Chem. High Polymers (Japan)* **8,** 306–309 (1951).*
240. 47:3795i S. Goldschmidt and M. Wick, *Ann.* **575,** 217–231 (1952).
241. 47:4290f M. Metayer, *Bull. Soc. Chim. France* **1951,** 802–806.
242. 47:6980b N. M. Bortnick, U.S. Pat. 2,611,782 (Sept. 23, 1952).
243. 47:8049g H. J. Creech and R. M. Peck, *J. Am. Chem. Soc.* **74,** 463–468 (1952).
244. 47:8774c A. H. Ahlbrecht and D. R. Husted, U.S. Pat. 2,617,817 (Nov. 11, 1952).
245. 47:9286d H. M. Kissman and D. S. Tarbell, *J. Am. Chem. Soc.* **74,** 4317–4319 (1952).
246. 47:11201a E. L. Anderson, J. W. Wilson, and G. E. Ullyot, *J. Am. Pharm. Assoc., Sci. Ed.* **41,** 643–650 (1952).
247. 47:11243c J. H. Werntz, U.S. Pat. 2,625,561 (Jan. 13, 1953).
248. 47:11973i H. C. Meal, *J. Am Chem. Soc.* **74,** 6121–6122 (1952).
249. 47:12259i H. H. Woeber, *J. Am. Chem. Soc.* **74,** 1354–1355 (1952).
250. 47:12302e A. Sekera, I. Jakubec, J. Kral, and C. Vrba, *Chem. Listy* **46,** 762–765 (1952).*
251. 48:712b V. P. Wystrach and J. T. Thurston, U.S. Pat. 2,626,278 (Jan. 20, 1953).
252. 48:1370f B. R. Baker, R. E. Schaub, J. P. Joseph, F. J. McEvoy, and J. H. Williams, *J. Org. Chem.* **18,** 133–137 (1953).
253. 48:2113h T. Koide, et al., Japan Pat. 2666 (July 18, 1952).*
254. 48:2477g W. H. T. Davison, *J. Chem. Soc.* **1953,** 3712–3713.
255. 48:3268h A. Burger and G. T. Fitchett, *J. Am. Chem. Soc.* **74,** 3415–3416 (1952).
256. 48:3389a Minnesota Mining and Manufg. Co., British Pat. 689,425 (May 25, 1953).
257. 48:3833f H. J. Callomon, H. W. Thompson, F. A. Anderson, and B. Bak, *J. Chem. Soc.* **1953,** 3709–3710.

258. 48:4487d N. S. Dokunikhin and I. D. Pletneva, *Zh. Obshch. Khim.* **23**, 798–800 (1953) (see *CA* 48:12731a).*

259. 48:4623g C. M. Fletcher, A. G. Lowther, and W. S. Reith, *Biochem. J. (London)* **56** 106–111 (1954).

260. 48:5212a M. G. Morningstar, C. S. Schollenberger, and G. E. Stueber, U.S. Pat. 2,642,449 (June 16, 1953).

261. 48:5570f W. J. O. Thomas, *Trans. Faraday Soc.* **49**, 855–866 (1953).

262. 48:5794b T. I. Bieber, *J. Am. Chem. Soc.* **74**, 4700–4701 (1952).

263. 48:6379h G. A. Haggis and L. N. Owen, *J. Chem. Soc.* **1953**, 389–398.

264. 48:6996e E. K. Weisburger and J. H. Weisburger, *J. Org. Chem.* **18**, 864–867 (1953).

265. 48:7018f N. S. Dokunikhin and L. A. Gaeva, *Zh. Obshch. Khim.* **23**, 606–610 (1953) (see *CA* 49:4564b).*

266. 48:7632b F C. Schaefer and E. K. Drechel, U.S. Pat. 2,640,068 (May 26, 1953).

267. 48:7633b M. D. Hurwitz and R. W. Auten, U.S. Pat. 2,640,846 (June 2, 1953).

268. 48:8465a Yu. A. Baskakov and N. N. Mel'nikov, *Zh. Obshch. Khim.* **24**, 376–379 (1954).*

269. 48:10061a T. Lieser, German Pat. 860,636 (Dec. 22, 1952).*

270. 48:11345i Y. Iwakura, Y. Matsumoto, and K. Fujikawa, *Chem. High Polymers (Japan)* **8**, 524–527 (1951).

271. 48:12491b N. W. Luft and Om P. Kharbanda, *J. Chem. Phys.* **22**, 956–957 (1954).

272. 48:12731a N. S. Dokuniklin and I. D. Pletneva, *J. Gen. Chem. USSR* **23**, 835–837 (1953) (English translation of 48:4487d).*

273. 48:13730d T. C. Aschner, U.S. Pat. 2,647,902 (Aug. 4, 1953).

274. 49:186d A. Sckera, A. Borovansky, and C. V'rba, *Chem. Listy* **47**, 591–597 (1953).*

275. 49:266a C. Grundmann and G. Ottmann, *Ann.* **582**, 163–177 (1953).

276. 49:305b Y. Sahashi, H. Akamatsu, and H. Genda, *Proc. Imperial Acad. (Tokyo)* **21**(1), 44–52 (1945).*

277. 49:852g H. D. Zook, M. Ream, and E. W. Delchamps, *J. Am. Chem. Soc.* **75**, 5590–5591 (1953).*

278. 49:1078i H. H. Zeiss and W. B. Martin, Jr., *J. Am. Chem. Soc.* **75**, 5935–5940 (1953).

279. 49:1735b R. H. Terss and W. E. McEwen, *J. Am. Chem. Soc.* **76**, 580–585 (1954).

280. 49:1823c R. Duschinsky, U.S. Pat. 2,658,067 (Nov. 3, 1953).

281. 49:2510g O. Stallmann, U.S. Pat. 2,671,082 (March 2, 1954).

282. 49:3098h N. S. Dokunikhin, L. A. Gaeva, and I. D. Pletneva, *Zh. Obshch. Khim.* **24**, 174–178 (1954) (see *CA* 49:4552e).*

283. 49:4362h W. C. von Dohlen, *Univ. Microfilms Publ.* No. 9841.

284. 49:4552e N. S. Dokunikhin, L. A. Gaeva, and I. D. Pletneva, *J. Gen. Chem. USSR* **24**, 177–180 (1954) (Eng. transl. of *CA* 49:3098h).*

285. 49:4564b N. S. Dokunikhin and L.A. Gaeva, *J. Gen. Chem. USSR* **23**, 629–633 (1953) (Eng. transl. of *CA* 48:7018f).*

286. 49:4605d P. D. Bartlett and F. D. Greene, *J. Am. Chem. Soc.* **76**, 1088–1096 (1954).

287. 49:4710d M. H. Gold, U.S. Pat. 2,680,131 (June 1, 1954).

288. 49:5526d H. Martin, U.S. Pat. 2,649,476 (Aug. 18, 1953).

289. 49:6263g E. C. Taylor, Jr. and Norman W. Kalenda, *J. Am. Chem. Soc.* **76**, 1699–1700 (1954).

290. 49:9034e C. F. Irwin, U.S. Pat. 2,683,160 (July 6, 1954).

291. 49:9571b V. N. Belov and S. D. Kustova, *Zh. Obshch. Khim* **24**, 1087–1094, (1954).*

292. 49:10257b W. L. Mosby, *J. Am. Chem. Soc.* **76**, 936–937 (1954).

293. 49:10874f H. Bredereck, E. Bader, and G. Hoschele, *Chem. Ber.* **87**, 784–788 (1954).

294. 49:11712i W. W. Thompson, U.S. Pat. 2,689,861 (Sept. 23, 1954).

295. 49:13258b F. J. Wolf, R. M. Wilson, Jr., K. Pfister III, and M. Tishler, *J. Am. Chem. Soc.* **76**, 4611–4613 (1954).

296. 49:13712c M. T. Rogers, *J. Am. Chem. Soc.* **77**, 3681–3684 (1955).

297. 49:14683i C. L. Arcus and B. S. Prydal, *J. Chem. Soc.* **1954**, 4018–4020.

298. 49:15880i E. F. Godefroi, J. S. Meek, and J. R. Clopton, *J. Am. Chem. Soc.* **76**, 5788–5789 (1954).

299. 49:15958f H. S. Bloch and D. R. Strehlau, U.S. Pat. 2,695,913 (Nov. 30, 1954).

300. 50:1371c N. M. Bortnick, U.S. Pat. 2,718,516 (Sept. 20, 1965).

301. 50:1408g W. C. Von Dohlen and G. B. Carpenter, *Acta Cryst.* **8**, 646–651 (1955).*

302. 50:2458i N. G. Gaylord and J. J. O'Brien, *Rec. Trav. Chim.* **74**, 218–219 (1955).*

303. 50:2639i J. H. Pierce, C. W. Shoppee, and G. H. R. Summers, *J. Chem. Soc.* **1955**, 679–686.

304. 50:2661f T. S. Reid, U.S. Pat. 2,706,733 (Apr. 19, 1955).

305. 50:3367c M. W. Miller, R. W. Amidon, and P. O. Tawney, *J. Am. Chem. Soc.* **77**, 2845–2848 (1955).

306. 50:4112c D. A. Barr and R. N. Haszeldine, *J. Chem. Soc.* **1955**, 1881–1889.

307. 50:4231i O. Stallmann, U.S. Pat. 2,723,265 (Nov. 8, 1955).

308. 50:5740f N. M. Bortnick, U.S. Pat. 2,713,591 (July 19, 1955).

309. 50:5765f R. Duschinsky, U.S. Pat. 2,707,186 (Apr. 26, 1955).

310. 50:7805h J. D. A. Johnson, *J. Chem. Soc.* **1955**, 3322–3324.

311. 50:7861a R. Graf, Ger. Pat. 931,225 (Aug. 4, 1955).*

312. 50:8724a Variapat, A. G. Swiss Pat. 300,283 (Oct. 1, 1954).*

313. 50:9480a Y. Hamashima, K. Hayashi, S. Takahaski, M. Kyogoku, and T. Kitasawa, *Acta Schol. Med. Univ. Kyoto* **32**, 123–137 (1955).*

314. 50:10050c M. Yoshida, *Pharm. Bull.* (*Japan*) **3**, 215–219 (1955).*

315. 50:10766f S. Melamed and B. F. Aycock, U.S. Pat. 2,727,020 (Dec. 13, 1955).

316. 50:10770 M. D. Hurwitz and R. W. Auten, U.S. Pat. 2,728,787 (Dec. 27, 1955).

317. 50:11962a V. T. Oliverio and E. Sawicki, *J. Org. Chem.* **20**, 1733–1737 (1955).

318. 50:12107a O. Stallmann, U.S. Pat. 2,729,666 (Jan. 3, 1956).

319. 50:12644f G. J. B. Crawford and J. S. Foster, *Can. J. Phys.* **34**, 653–667 (1956).

320. 50:12960e R. M. Peck, *J. Am. Chem. Soc.* **78**, 997–1001 (1956).

321. 50:13093 T. C. Allen and D. H. Chadwick, U.S. Pat. 2,733,254 (Jan. 31, 1956).

322. 50:13732b H. H. Anderson, *J. Am. Chem. Soc.* **78**, 1692–1694 (1956).

323. 50:13773g D. M. Simons and R. G. Arnold, *J. Am. Chem. Soc.* **78**, 1658–1659 (1956).

324. 50:13775b T. Mukaiyama, S. Ozaki, Y. Kobaysahi and Y. Fujita, *Bull. Chem. Soc. Japan*, 51–57 (1956).

325. 50:13831b Y. Hamashima, *Acta Schol. Med. Univ. Kyoto* **32**, 112–122 (1955) (in English).*

(557) 50:14534f C. L. Stevens, T. K. Mukherjee, and V. J. Traynelis, *J. Am. Chem. Soc.* **78**, 2264–2267 (1956).

326. 50:15566a H. C. Richards, C. W. Shoppee, J. C. P. Sly, and G. H. R. Summers, *J. Chem. Soc.* **1956**, 1054–1059.

327. 51:470a R. L. Pelley, U.S. Pat. 2,757,184 (July 31, 1956); E. Barthel, Jr., U.S. Pat. 2,757,185 (July 31, 1956).

328. 51:1817i T. Ya. Medved and M. I. Kabachnik, *Izv. Akad. Nauk SSSR, Otdel. Khim. Nauk*, **1956**, 684–691, see *CA* 51:5689a.*

329. 51:1890b D. F. Kutepov and N. S. Rozanova, *Zh. Obshch. Khim.* **26**, 1737–1740 (1956).*

330. 51:2327i T. Lieser, Ger. Pat. 888,769 (Sept. 3, 1953).
331. 51:2570d N. M. Bortnick, L. S. Luskin, M. D. Hurwitz, and A. W. Rytina, *J. Am. Chem. Soc.* **78,** 4358–4361 (1956).
332. 51:2599d C. R. Hauser and D. N. Van Eenam, *J. Am. Chem. Soc.* **78,** 5698–5699 (1956).
333. 51:2787c E. H. White and H. C. Dunathan, *J. Am. Chem. Soc.* **78,** 6055–6057 (1956).
334. 51:2814e M. B. Frankel and K. Klager, *J. Am. Chem. Soc.* **78,** 5428–5430 (1956).
335. 51:2816d R. Graf, *Chem. Ber.* **89,** 1071–1079 (1956).
336. 51:3548a D. A. Barr and R. N. Haszeldine, *J. Chem. Soc.* **1956,** 3416–3435.
337. 51:3655g Monsanto Chemical Co., British Pat. 753,741 (Aug. 1, 1956) (See U.S. Pat. 2,762,834, CA 50:13093b).
338. 51:4045d Y. Iwakura, M. Sato, T. Tamikado, and S. Mimashi, *Chem. High Polymers (Japan)* **13,** 125–131 (1956).*
339. 51:4347f M. J. S. Dewar and R. Pettit, *J. Chem. Soc.* **1956,** 2021–2025.
340. 51:4419b R. Graf, German Pat. 928,896 (June 13, 1955).
341. 51:4419d R. Graf, German Pat. 937,645 (Jan. 12, 1956).
342. 51:4431f Farbenfabriken Bayer, A. G., British Pat. 752,105 (July 4, 1956).
343. 51:4763f C. C. Unruh and A. C. Smith, Jr., U.S. Pat. 2,768,154 (Oct. 23, 1956).
344. 51:5689a T. Ya. Medved and M. I. Kabachnik, *Bull. Acad. Sci. USSR, Div. Chem. Sci.* **1956,** 695–701 (English Transl. 51:1817i).*
345. 51:5748c R. Kuhn and G. Kruger, *Chem. Ber.* **89,** 1473–1486 (1956).
346. 51:5827b Farbenfabriken Bayer, A. G., British Pat. 749,490 (May 23, 1956).
347. 51:5835h Österreichische Stickstoffwerke A. G. Austrian Pat. 187,915 (Dec. 10, 1956).*
348. 51:6528f R. Dahlbom and L.-E. Österberg, *Acta Chem. Scand.* **9,** 1553–1556 (1955) In English.*
349. 51:6583g K. G. Rutherford and M. S. Newman, *J. Am. Chem. Soc.* **79,** 213–214 (1957).
350. 51:6588f R. Pettit, *Chem. Ind. (London)* **1956,** 1306–1307.
351. 51:6590e W. V. E. Doering and L. H. Knox, *J. Am. Chem. Soc.* **79,** 352–356 (1957).
352. 51:7408f T. I. Bieber, U.S. Pat. 2,771,483 (Nov. 20, 1956).
353. 51:7420e Farbenfabriken Bayer A. G., British Pat. 752,931 (July 18, 1956).
354. 51:8141e Farbenfabriken Bayer A. G., British Pat. 748,166 (April 25, 1956).
355. 51:8669c A. Sekera, A. Borovansky, I. Jakubec, K. Palat, and C. Vrba, *Cesk. Farm.* **5,** 388–391 (1956).*
356. 51:9487f H. W. Johnson and D. E. Bublitz, *J. Am. Chem. Soc.* **79,** 753–754 (1957).
357. 51:10370d D. A. Barr and R. N. Haszeldine, *J. Chem. Soc.* **1957,** 30–39.
358. 51:10506b C. L. Arcus and B. S. Prydal, *J. Chem. Soc.* **1957,** 1091–1093.
359. 51:10571a California Research Corp., British Pat. 761,113 (Nov. 7, 1956).
360. 51:11270c D. J. Beaver, D. P. Roman, and P. J. Stoffel, *J. Am. Chem. Soc.* **79,** 1236–1245 (1957).
361. 51:11857b J. Derkosch, K. Schlögl, and H. Woidich, *Monatsh. Chem.* **88,** 35–46 (1957).
362. 51:11932i A. F. Summa and P. J. Jannke, *J. Am. Pharm. Assoc.* **46,** 363–366 (1957).
363. 51:11995a D. A. Barr and R. N. Haszeldine, *Chem. Ind. (London)* **1956,** 1050–1051.
364. 51:12837g G. J. O'Neill and F. L. M. Pattison, *J. Am. Chem. Soc.* **79,** 1956–1958 (1957).
365. 51:13762c R. Hart, *Bull. Soc. Chim. Belg.* **65,** 291–296 (1956).
366. 51:13778 J. C. Martin and P. D. Bartlett, *J. Am. Chem. Soc.* **79,** 2533–2541 (1957).
367. 51:13919i W. V. Wirth and S. E. Krahler, U.S. Pat. 2,786,864 (Mar. 26, 1957).
368. 51:14625b M. Ishidate and A. Kawada, *Pharm. Bull. (Tokyo)* **4,** 483–486 (1956).*
369. 51:15176c Rohm and Haas Co., British Patents 763,634; 763,635 and 763,636 (Dec. 12, 1956).

370. 51:15569e W. D. Smutz, U.S. Pat. 2,795,597 (June 11, 1957).

371. 51:16102b H. Hoyer, *Chem. Ber.* **89**, 2677–2681 (1956).

372. 51:16569i Farbwerke Hoechst A. G. vorm Meister Lucius and Brüning British Pat. 774,276 (May 8, 1957).

373. 51:17759h R. L. Dannley and M. Lukin, *J. Org. Chem.* **21**, 1036–1037, (1956).

374. 51:17868d E. Mataraso-Tchiroukhise and R. Quelet, *Compt. Rend.* **244**, 2518–2520 (1957).

375. 51:18016d W. Balon, U.S. Pat. 2,801,244 (July 30, 1957).

376. 51:18700b N. V. de Bataafsche Petroleum Maatschappij, British Pat. 779,806 (July 24, 1957).

377. 52:281c Y. Iwakura and I. Suzuki, *Nippon Kagaku Zasshi* **77**, 64–67 (1956).*

378. 52:397e T. L. Cairns, D. D. Coffman, and W. W. Gilbert, *J. Am. Chem. Soc.* **79**, 4405–4408 (1957).

379. 52:1082f J. A. Parker, J. J. Thomas, and C. L. Zeise, *J. Org. Chem.* **22**, 594–596 (1957).

380. 52:1095e T. Kimura, *Yakugaku Zasshi* **77**, 884–887 (1957).*

381. 52:1147e A. C. Smith, Jr. and C. C. Unruh, *J. Org. Chem.* **22**, 442–444 (1957).

382. 52:1227i C. Mentzer, Ng. Ph. Buu Hoi, and P. Cagniant, French Pat. 1,007,001 (April 30, 1952).*

383. 52:1927i R. L. Dannley, R. G. Taborsky, and M. Lukin, *J. Org. Chem.* **21**, 1318–1319 (1956).

384. 52:3735g C. A. Grob, H. Kny, and A. Gagneux, *Helv. Chim. Acta* **40**, 130–140 (1957).

385. 52:5321e S. N. Solodushenkov, V. N. Klyuev, and A. A. Spryskov, *Zh. Prikl. Khim.* **30**, 1361–1367 (1957).*

386. 52:5341d J. I. Jones and N. G. Savill, *J. Chem. Soc.* **1957**, 4392–4394.

387. 52:6248g J. Sova, A. Sekera, and C. Vrba, *Chem. Listy* **51**, 2339–2343 (1957).*

388. 52:7261e D. N. Kursanov, M. E. Vol'pin, I. S. Akhnem, and I. Ya. Kachknem, *Izv. Akad. Nauk SSSR, Otdel Khim. Nauk* **1957**, 1371–1378.*

389. 52:7724i J. R. Geigy A. G. British Pat. 786,234 (Nov. 13, 1957).

390. 52:8052c N. N. Yarovenko, S. P. Motornyi, L. I. Kirenskaya, and A. S. Vasil'eva, *Zh. Obshch. Khim.* **27**, 2243–6 (1957).*

391. 52:8188b R. Putter and D. Delfs, German Pat. 876,238 (May 11, 1953).*

392. 52:8658f B. Jolliffe and C. P. Smyth, *J. Am. Chem. Soc.* **80**, 1064–1066 (1958).

393. 52:8967a H. E. Ungnade and L. W. Kissinger, *J. Org. Chem.* **22**, 1662–1665 (1957).

394. 52:10141f H. Holtschmidt, U.S. Pat. 2,821,544 (Jan. 28, 1958).

395. 52:10188b T. Lieser, German Pat. 827,553 (Jan. 10, 1952).*

396. 52:11116b G. Steinbrunn, German Pat. 848,808 (Sept. 8, 1952).

397. 52:11937b W. Gauss, S. Petersen, and C. Hackmann, German Pat. 911,016 (May 10, 1954).*

398. 52:14581e K. Schlögel, *Monatsh. Chem.* **88**, 601–621 (1957).*

399. 52:15417e R. Berthold, *Chem. Ber.* **90**, 2743–2747 (1957).

400. 52:15566b H. Ratz, German Pat. 845,507 (July 31, 1952).*

401. 52:16212i P. Schlack, German Pat. 870,097 (Mar. 9, 1953).*

402. 52:16291a A. Modersohn, D. Delfs, and G. Schwaebel, German Pat. 848,810 (Sept. 8, 1952).

403. 52:17145b A. Sekera and C. Vrba, *Arch. Pharmacol.* **291**, 122–125 (1958).*

404. 52:17763f H. S. Block and D. R. Strehlau, U.S. Pat. 2,839,568 (June 17, 1958).

405. 52:18047i E. Soderback, *Acta Chem. Scand.* **11**, 1622-1634 (1957).*

406. 52:18272d Ng. Ph. Buu-Hoi, C. T. Long, and Ng. D. Xuong, *J. Org. Chem.* **23**, 42–45 (1958).

407. 52:18276a I. W. Logemann and D. Artini, *Chem. Ber.* **90,** 2527–2531 (1957).
408. 52:19937h H. W. Johnson and D. E. Bublitz, *J. Am. Chem. Soc.* **80,** 3150–3152 (1958).
409. 52:19938c R. Appel and H. Gerber, *Chem. Ber.* **91,** 1200–1203 (1958).
410. 53:47c R. Holland, D. W. G. Style, R. N. Dixon, and D. A. Ramsay, *Nature* **182,** 336–337 (1958).
411. 53:221g A. C. Haven, Jr., U.S. Pat. 2,835,652 (May 20, 1958).
412. 53:358d Ng. Ph. Buu-Hoi, Ng. D. Xoung, and V. T. Suu, *J. Chem. Soc.* **1958,** 2815–2821.
413. 53:1141c R. C. Bossert, *J. Org. Chem.* **23,** 906–907 (1958).
414. 53:1253i D. J. Beaver and P. J. Stoffel, U.S. Pat. 2,846,473 (Aug. 5, 1958).
415. 53:2072c J. A. Young, S. N. Tsoukalas, and R. D. Dresdner, *J. Am. Chem. Soc.* **80,** 3604–3606 (1958).
416. 53:3147e L. Jakọb, G. V. Rosenberg, O. Roser, and O. Bayer, German Pat. 862,888 (Jan. 15, 1953).*
417. 53:4153i K. Schlogl, *Monatsh. Chem.* **89,** 61–73 (1958).
418. 53:4243i A. Gemassmer, German Pat. 949,227 (Sept. 13, 1956).
419. 53:4418b L. Goodfriend and A. H. Schon, *Can. J. Biochem. Physiol.* **35,** 1177–1184 (1958).
420. 53:5112e K. Sato and O. Miyamoto, *Nippon Kagaku Zasshi* **77,** 1409–1411 (1956).*
421. 53:5839b C. E. McGinn and R. G. Spaungburgh, *Dyestuffs* **42,** 224–231 (1958).
422. 53:6759h G. L. Caldow and H. W. Thompson, *Spectrochim. Acta* **13,** 212–216, (1958).
423. 53:7063b W. Kern and W. Thoma, *Makromol. Chem.* **16,** 80–107 (1955).
424. 53:7064b J. Burkus and C. F. Eckert, *J. Am. Chem. Soc.* **80,** 5948–5950 (1958).
425. 53:7097h J. C. Kauer, U.S. Pat. 2,860,154 (Nov. 11, 1958).
426. 53:8051e W. Logemann, D. Artini, and G. Tosoline, *Chem. Ber.* **91,** 2566–2274 (1958).
427. 53:8743b W. M. Tolles and J. R. Tichy, *J. Phys. Chem.* **62,** 1606–1607 (1958).
428. 53:9069b J. F. Nobis and H. Greenberg, U.S. Pat. 2,865,940 (Dec. 23, 1958).
429. 53:9146g D. Beaver, P. C. Hamm, and P. J. Stoffel, U.S. Pat. 2,867,520 (Jan. 6, 1959).
430. 53:9172e J. Hilger and W. Altner, German Pat. 947,159 (Aug. 9, 1956).
431. 53:10126d D. J. Beaver and P. J. Stoffel, U.S. Pat. 2,870,205 (Jan. 20, 1959).
432. 53:10126e T. W. Campbell and J. Verbanc, U.S. Pat. 2,853,473 (Sept. 23, 1958).
433. 53:10173f H. Bretschneider, H. Deutscher, W. Klotzer, and M. Sander, *Monatsh. Chem.* **89,** 288–300 (1958).
434. 53:11287a R. Graf, *Chem. Ber.* **92,** 509–513 (1959).
435. 53:11330h J. C. Petropoulos, U.S. Pat. 2,855,420 (Oct. 7, 1958).
436. 53:11343a Y. Iwakura, K. Nagakubo, K. Hayashi, and M. Sakamoto, *Kogyo Kagaku Zasshi* **60,** 706–710 (1957).*
437. 53:12775h H. N. Cole, *Chem. Eng. Data Ser.* **3,** 213–215 (1958).
438. 53:13078a D. B. Denney and G. Feig, *J. Am. Chem. Soc.* **81,** 225–228 (1958).
439. 53:14042g P. E. Verkade, R. E. Strube, and P. H. Witjens, *Rec. Trav. Chim.* **77,** 1089–1095 (1958).*
440. 53:14115e W. v. E. Doering and M. J. Goldstein, *Tetrahedron* **5,** 53–69 (1959).
441. 53:14682c A. Mangini and G. Modena, *Boll. Sci. Fac. Chim. Ind. Bologna,* **16,** 18–28 (1958).*
442. 53:15080h A. Sekera, R. Pavlicek, and C. Vrba, *Bull. Soc. Chim. France* **1959,** 401–404.*
443. 53:16097b K. Schlogl and H. Seiler, *Naturwissenschaften* **45,** 337 (1958).
444. 53:16110i H. Bretschneider, W. Klotzer, W. Sachsenmaier, and M. Sander, *Monatsh. Chem.* **89,** 511–516 (1958).
445. 53:16677f D. F. Othmer and D. Zudkevitch, *Ind. Eng. Chem.* **51,** 791–796, 1022 (1959).

446. 53:16935i J. A. Young, W. S. Durrell, and R. D. Dresdner, *J. Am. Chem. Soc.* **81,** 1587–1589 (1959).
447. 53:17125d R. Pavlicek, A. Sekera, and C. Vrba, *Pharmazie* **13,** 748–752 (1958).*
448. 53:17582b M. Katz, U.S. Pat. 2,888,438 (May 26, 1959).
449. 53:18864e L. H. Jenkins and D. S. Sears, U.S. Pat. 2,873,171 (Feb. 10, 1959).
450. 53:18873a S. Komari and T. Agawa, *Technol. Rept. Osaka Univ.* **8,** 487–496 (1958).*
451. 53:19978e G. Nischk, German Pat. 1,014,986 (Sept. 5, 1957).
452. 54:358b Y. Sakata and T. Hashimoto, *Yakugaku Zasshi* **79,** 872–874 (1959).*
453. 54:1067g R. N. Dixon, *J. Chem. Phys.* **31,** 258–260 (1959).
454. 54:1539h Y. Iwakura, K. Uno, and K. Hamatani, *Nippon Kagaku Zasshi* **78,** 1416–1420 (1957).
455. 54:1555g R. Graf, German Pat. 1,000,807 (Jan. 17, 1957).
456. 54:2187g K. Poduska and J. Rudinger, *Collection Czechv. Chem. Commun.* **24,** 3449–3467 (1959).
457. 54:2269h R. Graf. German Pat. 1,010,958 (June 27, 1957).
458. 54:4434d Y. Iwakura, M. Sato, and H. Munakata, *Nippon Kagaku Zasshi* **79,** 148–152 (1958).
459. 54:5435a H. Manami and S. Nishizaki, *Kogyo Kagaku Zasshi* **61,** 214–217 (1958).
460. 54:6300i M. McCarty, Jr. and G. W. Robinson, *J. Chim. Phys.* **56,** 723–731 (1959).
461. 54:7556i K. H. Kossmann, *Can. J. Chem.* **37,** 1118 (1959).
462. 54:7667f M. J. S. Dewar and C. R. Ganellin, *J. Chem. Soc.* **1959,** 3139–3144.
463. 54:8639i K. Klager, U.S. Pat. 2,923,726 (Feb. 2, 1960).
464. 54:9413h N. A. Goldberg, V. I. Kucheryavyi, and G. N. Zinov'ev, *Zh. Prikl. Khim.* **32,** 2816–2818 (1959).
465. 54:9828d C. D. Hurd and A. G. Prapas, *J. Org. Chem.* **24,** 388–392 (1959).
466. 54:9889f H. Najer, P. Chabrier, and R. Giudicelli, *Bull. Soc. Chim. France* **1959,** 1611–1617.
467. 54:12991i H. A. Kreyssler and H. L. Needles, *J. Org. Chem.* **25,** 279–281 (1960).
468. 54:13000f B. F. Aycock and S. Melamid, German Pat. 1,028,343 (April 17, 1958).
469. 54:13036g R. Appel and A. Hauss, *Chem. Ber.* **93,** 405–411 (1960).
470. 54:13142c D. Keglevic-Brobet and A. Kornhauser, *Croat. Chem. Acta* **31,** 47–56 (1959).*
471. 54:18337a J. Goubeau and K. Paulin, *Chem. Ber.* **93,** 1111–1116 (1960).
472. 54:18618i C. S. Chadwick, P. Johnson, and E. G. Richards, *Nature* **186,** 239–240 (1960).
473. 54:19563i C. King, *J. Org. Chem.* **25,** 352–356 (1960).
474. 54:19626a R. Huisgen, E. Rauenbusch, and G. Seidel, *Chem. Ber.* **90,** 1958–1965 (1957).
475. 54:20841d I. L. Knunyants, E. G. Bykhovskaya, and V. N. Frosin, *Dokl. Akad. Nauk SSSR* **132,** 357–359 (1960).*
476. 54:22408i H. H. Anderson, *J. Am. Chem. Soc.* **81,** 4785–4787 (1959).
477. 54:23549h N. A. Gol'dberg and V. I. Kucheryavyi, *Zh. Prikl. Khim.* **33,** 1912–1913 (1960).*
478. 54:23679i N. A. Gol'dberg and G. N. Zinov'ev, *Zh. Prikl. Khim.* **33,** 1913–1915 (1960).
479. 54:24464i W. R. Boehme and J. Nickols, U.S. Pat. 2,942,026 (June 21, 1960).
480. 54:24472h Ng. Ph. Buu-Hoi and L. Petit, *J. Org. Chem.* **25,** 726–728 (1960).
481. 54:24474a I. L. Khmel'nitskaya, L. A. Epel'tsveig, and T. A. Mikhailova, *Zh. Obshch. Khim.* **30,** 581–583 (1960).*
482. 54:24776g H. W. Johnson, Jr., R. E. Lovins, and M. Reintjes, *J. Org. Chem.* **24,** 1391–1392 (1959).
483. 55:1407c J. A. Young, W. S. Durrell, and R. D. Dresdner, *J. Am. Chem. Soc.* **82,** 4553–4556 (1960).

484. 55:3427h Y. Iwakura, M. Sato, and Y. Matsuo, *Nippon Kagaku Zasshi* **80,** 502–505 (1959).
485. 55:4538b H. Hausermann, U.S. Pat. 2,929,822 (Mar. 22, 1960).
486. 55:6435i H. Bertsch, E. Ulsperger, and M. Bock, *J. Prakt. Chem.* **11,** 225–232 (1960).*
487. 55:6468c H. Najer, R. Giudicelli, and J. Menin, *Bull. Soc. Chim. France* **1960,** 1544–1548.*
488. 55:6528i H. Lettre and L. Knof, *Chem. Ber.* **93,** 2860–2864 (1960).
489. 55:7298a D. Voight, German Pat. 1,043,293 (Nov. 13, 1958).*
490. 55:7327i W. Bunge, *Angew. Chem.* **72,** 1002 (1960).
491. 55:7339f I. Rosen and J. P. Stallings, *J. Org. Chem.* **25,** 1484–1486 (1960).
492. 55:7360i D. J. Beaver and P. J. Stoffel, U.S. Pat. 2,959,612 (Nov. 8, 1960).
493. 55:7408f R. Pettit, *Tetrahedron Letters* **1960**(23), 11–13.
494. 55:7997d M. Sato, *J. Am. Chem. Soc.* **82,** 3893–3897 (1960).
495. 55:8299c K. Klager and C. R. Vanneman, U.S. Pat. 2,967,193 (Jan. 3, 1961).
496. 55:8352f F. Hostettler and E. F. Cox, Brit. Pat. 845,827 (Aug. 24, 1960).
497. 55:10054f R. N. Dixon, *Phil. Trans. Roy. Soc. London, Ser. A* **252,** 165–192 (1960).*
498. 55:11412a A. B. A. Jansen and M. Szelke, *J. Chem. Soc.* **1961,** 405–411.
499. 55:13046a R. W. Nicholls and S. L. N. G. Krishnamachari, *Can. J. Chem.* **38,** 1652–1656 (1960).
500. 55:13453i E. Kuhle and R. Wegler, German Pat. 1,073,480 (Jan. 21, 1960).
501. 55:14046g C. E. Wulfman, *J. Chem. Phys.* **33,** 1567–1576 (1960).*
502. 55:14381b Atlantic Refining Co., British Pat. 852,988 (Nov. 2, 1960).
503. 55:14393 C. L. Parris and R. M. Christenson, *J. Org. Chem.* **25,** 1888–1893 (1960).
504. 55:14979i E. Müller, O. Bayer, H. Wilms, F. V. Spulak, and M. Theis, German Pat. 1,078,115 (Mar. 24, 1960).
505. 55:16416h V. E. Shashoua, W. Sweeney, and R. F. Tietz, *J. Am. Chem. Soc.* **82,** 866–873 (1960).
506. 55:16417f J. C. Kauer and A. K. Schneider, *J. Am. Chem. Soc.* **82,** 852–853 (1960).
507. 55:16485c J. E. Franz, U.S. Pat. 2,974,164 (Mar. 7, 1961).
508. 55:16769d L. Goodfriend and A. H. Sehon, *Can. J. Biochem. Physiol.* **39,** 941–960 (1961).*
509. 55:17509e H. Hahn, German Pat. 1,083,791 (June 23, 1960)
510. 55:17539g R. Hiltmann, German Pat. 1,081,453 (May 12, 1960).
511. 55:18619g D. Peters, *J. Chem. Soc.* **1961,** 1037–1042.
512. 55:18666b H. Holtschmidt and O. Bayer, German Pat. 1,066,582 (Oct. 8, 1959).*
513. 55:19765i H. H. Anderson, *J. Org. Chem.* **26,** 276–278 (1961).
514. 55:19799d K. Klager and M. B. Frankel, U.S. Pat. 2,978,474 (Apr. 4, 1961).
515. 55:19799f K. Klager, U.S. Pat. 2,978,475 (Apr. 4, 1961).
516. 55:19799h K. Klager and C. R. Vanneman, U.S. Pat. 2,978,476 (Apr. 4, 1961).
517. 55:21006d F. Adellac, M. Lora-Tamayo, and J. L. Soto, *Anales Real Soc. Espan. Fis. Quim. (Madrid)* **56B,** 985–994 (1960).
518. 55:21139i N. S. Dokunikhin, *Org. Poluprod. i Krasiteli, Nauchn.-Issled. Inst. Org. Poluprod. i Krasitelei im. K. E. Voroshilova, Sb. Statei* **1959,** No. 1, 148–159.*
519. 55:21145b J. J. Jaruzelski and W. E. Smith, Jr., U.S. Pat. 2,980,705 (Apr. 18, 1961).
520. 55:22194c M. Kaplan, *J. Chem. Eng. Data* **6,** 272–275 (1961).
521. 55:22238e S. Hartung, E. Klauke, and H. Schwarz, German Pat. 1,092,462 (Nov. 10, 1960).
522. 55:22486i G. Poetschke, H. Uehleke, and L. Kellisch, *Z. Immunitatsforsch.* **114,** 393–405 (1957).

523. 55:23340c M. Sato, *J. Org. Chem.* **26,** 770–779 (1961).
524. 55:23393f T. Mukaiyama and H. Nohira, *J. Org. Chem.* **26,** 782–784 (1961).
525. 55:24562d R. C. Elderfield and R. S. McElhinney, *J. Org. Chem.* **26,** 1917–1923 (1961).
(556) 55:24564b Y. Tezuka, *Sen-i-Gakkaishi* **17,** 802–806 (1961).*
526. 55:24647c M. H. Benn, A. M. Creighton, L. N. Owen, and G. R. White, *J. Chem. Soc.* **1961,** 2365–2375.
527. 55:24680f W. Aumuller and R. Weyer, German Pat. 1,100,618 (Mar. 2, 1961).
528. 55:24897h G. I. Mikhailov and N. M. Kosnikova, *Metody Lyuminests Analiza Materially VIII, Soveshch. po Lyuminests., Akad. Nauk Belorussk. SSR, Inst. Fiz.* **1959,** (Pub. 1960), 140–142.*
529. 55:25795i D. E. Applequist and J. P. Kleiman, *J. Org. Chem.* **26,** 2178–2182 (1961).
530. 55:26018h G. Drefahl and S. Huneck, *Chem. Ber.* **94,** 2015–2018 (1961).
531. 55:27107a M. Zaoral and J. Rudinger, *Collection Czech. Chem. Commun.* **26,** 2316–2332 (1961).*
532. 55:27108h J. Honzl and J. Rudinger, *Collection Czech. Chem. Commun.* **26,** 2333–2344 (1961).*
533. 55:27161a R. C. Schulz and H. Hartmann, *Monatsh. Chem.* **92,** 303–309 (1961).
534. 55:27360b E. F. J. Duynstee, *Rec. Trav. Chim.* **80,** 563–571 (1961).
535. 56:315c M. Sato, *Nippon Kagaku Zasshi* **81,** 959–962 (1960).*
536. 56:1345e M. H. Gold, U.S. Pat. 3,000,921 (June 5, 1957).
537. 56:1370d W. J. Humphlett and C. V. Wilson, *J. Org. Chem.* **26,** 2507–2510 (1961).
538. 56:2327f H. Akashi, *Kogyo Kagaku Zasshi* **63,** 368–369 (1960).*
539. 56:2337f M. B. Frankel, U.S. Pat. 3,000,920 (June 9, 1955).
540. 56:2468a H. H. Anderson and A. Hendifar, *J. Org. Chem.* **26,** 3033–3034 (1961).
541. 56:4268b W. G. Fateley and F. A. Miller, *Spectrochim. Acta* **17,** 857–868 (1961).
542. 56:4595h A. Frank and K. Riedl, *Monatsh. Chem.* **92,** 725–739 (1961).
543. 56:4846d H. Bertsch, E. Ulsperger, and M. Bock, *J. Prakt. Chem.* **13,** 134–137 (1961).
544. 56:5948i G. Palazzo and G. Strani, *Gazz. Chim. Ital.* **90,** 1290–1298 (1960).*
545. 56:5952i S. L. Shapiro, V. Bandurco, and L. Freedman, *J. Org. Chem.* **26,** 3710–3712 (1961).
546. 56:7213g W. Lott, J. Bernstein, and B. Stearns, U.S. Pat. 2,999,110 (Aug. 5, 1959).
547. 56:8632b F. J. Sowa, U.S. Pat. 3,013,045 (May 8, 1958).
548 56:10105i H. Haeusermann, U.S. Pat. 2,881,186 (Apr. 7, 1959).
549. 56:10271i L. Birkofer and R. Modic, *Ann.* **628,** 162–172 (1959).
550. 56:12802c Farbenfabriken Bayer A. G., British Pat. 883, 650 (Dec. 6, 1961).*
551. 56:13688c H. Manami and S. Nishizaki, *Nippon Kagaku Zasshi* **82,** 407–408 (1961).*
552. 56:14124i T. Mukaiyama, M. Tokizawa, H. Nohira, and H. Takai, *J. Org. Chem.* **26,** 4381–4384 (1961).
553. 56:15340d M. H. Gold, M. B. Frankel, G. B. Linden, and K. Klager, *J. Org. Chem.* **27,** 334–336 (1962).
554. 56:15534e L. Birkofer, A. Ritter, and J. Schramm, *Chem. Ber.* **95,** 426–430 (1962).
555. 57:14498f Yu. A. Strepikheev, Yu. I. Baranov, and O. A. Burmistroya, *Izv. Vysshikh. Uchebn. Zavedenii, Khim. i. Khim. Tekhnol* **5**(3), 387–390 (1962).
556. 55:24564b Y. Tezuka, *Sen-i Gakkaishi* **17,** 802–806 (1961).*
557. 50:14534f C. L. Stevens, T. K. Mukherjee, and V. J. Traynelis, *J. Am. Chem. Soc.* **78,** 2264–2267 (1956).
558. 4:761 O. Wallach, *Ann.* **369,** 63–103 (1909).*

CHAPTER VIII

ISOCYANATE TOXICITY AND ANALYTICAL METHODS FOR MEASUREMENT OF HAZARDOUS CONCENTRATIONS OF ISOCYANATES IN AIR

I. INTRODUCTION

There are many common materials or "chemicals" that are toxic to some part of the human body. For most materials, it is generally recognized that there is some threshold level, above which the material becomes harmful if ingested, inhaled, or contacts a sensitive part of the body.

The toxicity of many of the chemicals used in the urethane industry have been studied extensively and reported in the literature. For a discussion of the toxicity of several of the materials dealt with in the preceding chapters, the following references should be consulted: phosgene (1–7), nitro compounds (8–14), toluene (15–18), benzene (16–22), aniline (23–27), aliphatic amines (28), glycols (29–36), and toluene diamine (37). This chapter will deal with the toxicity of some of the more commercially important isocyanates and the analytical methods for determining hazardous concentrations in the air.

Toluene diisocyanate is used in many polyurethane formulations, and due to its relatively high vapor pressure, it represents a greater hazard than some of the other components.

II. TOXICITY OF ISOCYANATES

A. Tolylenediisocyanate (TDI)

The hazard of working with TDI in polyurethane formulations was recognized early by workers in the field (38–43). Zapp (44) discussed the hazards of isocyanates in polyurethane foam production. He studied the toxicity of TDI and believed it to be representative of the other isocyanates.

The results of Zapp (44) may be summarized as follows. Isocyanates are fairly strong irritants to the skin, eyes, gastrointestinal tract, and respiratory tract. Asthmatic attacks will result in a significant proportion of men exposed to inhalation of the vapors, and skin sensitization may occur in a few exposed to vapor or liquid. Zapp suggested that the maximum concentration of TDI in the atmosphere should not exceed 0.1 ppm and that the least detectable odor of TDI was apparent at 0.4 ppm in 50% of the subjects exposed. The term "maximum allowable concentration" means the maximum average atmospheric concentration of a given component to which personnel may be exposed for an eight-hour working day.

Studies on the thermal decomposition of various plastics indicate that the pyrolysis products from urethane polymers are no more toxic or hazardous than the products from neoprene, rubber latex, or polyvinyl chloride.

Wilson and McCormick (45) and Wilson and Wilson (46) have pointed out that in general, all the isocyanates exhibit the rather unusual characteristic of a low oral toxicity with a high chronic inhalation toxicity and have the potential of producing asthma-like symptoms. These investigators reported a similar threshold limit value of 0.1 ppm and pointed out that good handling practices and excellent ventilation are necessary. This, of course, is dependent upon the type of polyurethane application.

Subsequent review of industrial experience by the American Conference of Governmental Industrial Hygienists has led to a suggested maximum hygienic threshold limit of 0.02 ppm (47). This change arose from a report to this group of a limited number of cases of respiratory difficulty where atmospheric concentrations were never found to exceed 0.05 ppm. It is believed, however, that where such cases occurred, the individuals must have been exposed to concentrations in excess of 0.1 ppm for determinate periods where these higher concentrations were not detected (47).

MacFarland and Leong (48) also studied the hazards from the thermal decomposition of polyurethanes and polyurethane-coated nylon. The pyrolysis products of both materials proved to be of about equal toxicity to rats in a single 60-min inhalation exposure to both the pyrolysis and the combustion products. A value of 2.1×10^6 mg-min/m^3 has been derived as an approximate estimate of their L(Ct)50 (mean time to death at a given concentration).

These investigators concluded that acute asphyxia from occlusion of the upper respiratory tract was the process leading to death in rats exposed to the pyrolysis products of polyurethane. In the case of nylon products, death resulted from a more slowly developing, but ultimately overwhelming, pulmonary edema. Assuming that the susceptibility of man is approximately equal or greater, these investigators estimated that 1 lb of pyrolysis products of either material distributed in 250 cu ft of air would likely prove to be lethal to man if the exposure were continued for 30 min in the absence of appreciable air change.

Duncan et al. (49), in their studies on inhalation toxicity, pathology, and mortality of TDI, subjected mice, rats, guinea pigs, and rabbits to exposures of 2, 5, or 10 ppm TDI for 4 hr and serially sacrificed them up to 4 weeks later in order to determine the sequence of pathologic changes. The reaction was generally found to be limited to the trachea and lungs and varied only in degree among the different species, the rabbits being the most susceptible in terms of pathology (14 day LC_{50} was 11 ppm). Two hours following the 10 ppm exposure the response was characterized by subepithelial edema, necrosis, ulceration of tracheal and bronchial epithelium, and mild infiltration of acute inflammatory cells. Seven days after exposure, bronchopneumonia had developed. The initial reaction to 2 ppm consisted of focal desquamation of necrotic superficial epithelium with virtually complete healing by the seventh post-exposure day. These results corroborate those of other investigators and show that an isocyanate acts as an irritant relative to the concentration of exposure, and its effect is primarily on the trachea and larger intrapulmonary air passages.

Other investigators have performed similar studies on TDI and polyurethane preparations (50–59).

Munn (60) discusses the hazards of several of the commercially important isocyanates, including toluene diisocyanate, diphenyl methane diisocyanate, hexamethylene diisocyanate, and naphthalene diisocyanate. He emphasizes that the toxicity of each should be considered individually since they have significant differences in structure. Moreover he gives considerable weight to the widely reported phenomenon that some individuals are sensitized to TDI, such that reexposure to minimal concentrations causes recurrence of severe symptoms, and ascribes this to an allergic response. He briefly mentions some unpublished work of G. Taylor (1964) in which antibodies specific for TDI were found in four "sensitized" workmen, and which appeared to confirm the allergy theory. Ehrlicher and Pilz (71–73), in an earlier publication, discounted the possibility of development of an allergy to TDI.

Scheel, Killens, and Josephson (78) investigated the effect of TDI on the amino groups in egg albumen and studied the immunologic and chemical effects of the resulting protein conjugate solutions. They also exposed animals to TDI vapor and showed that antibodies specific for TDI were developed in the blood, without increasing skin sensitivity. Several diagnostic test procedures for defining an allergic response are discussed. Their search of the literature for proven cases of sensitivity to TDI did not reveal any cases where a proven diagnosis had been made. They recommended that the term "allergy" be reserved as a specific term to describe a well-defined diagnosis.

Summary

The effect of TDI on humans may be generalized in the following statements.

(*1*) *Effects of inhalation.* Vapor inhalation results in strong respiratory irritation with retrosternal soreness and constriction of the chest. Coughing, asthma, headache, depressed feelings, and insomnia may occur.

(*2*) *Effects of Ingestion.* There are no recorded results in humans. However, animal toxicity tests show the single oral lethal dose to be a relatively high level of 4.9–6.7 g/kg (47). It is believed that swallowing of TDI probably would not be fatal in humans but would result in severe irritation and probable corrosive action on the stomach (62).

(*3*) *Effects on Skin.* Skin irritation is caused by contact with the liquid in which reddening and swelling of the exposed area occurs. Animal experiments indicate that skin sensitization may also result.

(*4*) *Effects on Eyes.* Isocyanates can be harmful to the delicate eye tissue, probably due to a dehydration reaction with the fluid.

(*5*) *Physiological Response of Man to Various Concentrations.* Response of humans indicates the least detectable odor of TDI is approximately 0.4 ppm (44). Irritation of the nose and the throat occurred at about 0.5 ppm, and appreciable odor was recognized at 0.8 ppm (55).

(*6*) *Threshold Limit Values* (*TLV*). The recommended threshold limit value for tolylene 2,4-diisocyanate has been set at 0.02 ppm by the American Conference of Governmental Industrial Hygienists since 1961 (47).

B. Methylene Diphenyl Diisocyanate (MDI)

Diisocyanates of greater molecular weight, such as MDI, which have a lower vapor pressure (see Chapter 7 for physical property data) have been reported to be relatively nontoxic. The following animal toxicity data has been reported for MDI (47).

(*1*) *Effects of Ingestion.* A single dose given orally to rats (equivalent to 31.6 g/kg of body weight) proved lethal. Smaller doses (25.1 g/kg) given daily for several days were not fatal.

(*2*) *Effects on Skin.* A single dose applied as a 25% solution in corn oil to the skin of rabbits, up to the equivalent of 10 g/kg was not fatal.

(*3*) *Effects on Eyes.* The effects upon delicate eye tissue from a 50% solution suspension in corn oil was found to be slight.

Apparently, MDI possesses toxic properties which are evidenced if the diisocyanate is able to affect a sensitive part of the body, such as the lungs, through inhalation of the material as a fine mist during spraying.

Longley (63) reported the incidence of respiratory symptoms in workers 60–120 ft from an MDI spraying area: asthmatic breathing, retrosternal soreness, constriction of the chest, coughing, depression, and headache were experienced from one to several hours after exposure by some of the subjects. After 10 days, at which time the incident was investigated, the subjects were symptom-free.

Munn (60) has concluded that the vapor pressure of MDI at ordinary temperature is so low that its vapor presents very little hazard. On the other hand, the droplets formed in a spraying operation can act as a respiratory irritant. Sensitization is rare. Other isocyanates of lesser commercial importance, such as hexamethylene diisocyanate and 1,5-naphthalene diisocyanate are also discussed by Munn.

III. ANALYTICAL METHODS

A. General Discussion

The toxicity studies discussed above pertaining to the inhalation of isocyanate vapors have created a need for analytical methods for detecting low concentrations in air. Considerable ingenuity has been shown in the development of apparatus and methods to cope with this problem. Relatively large volumes of air must be sampled in order to obtain a detectable amount of material, which must be collected in a suitable absorbing solution. Colorimetric methods are usually employed for determining the low concentration of isocyanate in the absorbing solution.

In 1956, Ehrlicher and Pilz published several papers (71–73) in journals not having a wide circulation in this country, in which the absorbing solution was prepared from equal parts of dimethylformamide and 1*N* HCl. The amine formed from the isocyanate in this solution was treated with sodium nitrite, the excess nitrite removed with sulfamic acid and then coupled with *N*-naphthylethylene diamine to produce a blue-violet color which was measured in a spectrophotometer.

The method described by Marcali (64) in 1957 is much better known in this country, and is given below under "Detailed Procedures." It differs from that of Ehrlicher and Pilz only in the composition of the absorbing solution and in the use of sodium bromide as a catalyst for the diazotization reaction. The Marcali method uses 0.4*N* hydrochloric acid and 0.4*N* acetic acid for absorbing the isocyanate from the air.

In the same paper, Marcali describes a second method (also given in detail below) for determining the urea (3,3′diisocyanato-4,4′dimethylcarbanilide) which results from the reaction of 2,4-TDI with moisture, and which he regarded as a potentially toxic substance. He assumed that it would be present as an aerosol in an atmosphere containing TDI, since the pure compound exists as a solid. Carbanilide gives a weak yellow color when absorbed in a solution containing sodium nitrite and ethyl Cellosolve; and a much more enhanced color if boric acid is also present. 2,4-TDI and 2,6-TDI also produce a yellow color in the presence of sodium nitrite and boric acid. Thus, all three components are determined by this method if present in the air stream. The concentration of the carbanilide can be estimated by deducting the concentration of the isocyanates obtained by the diazotization method from the total concentration of the three components obtained by this procedure.

A procedure attributed to K. E. Ranta, very similar to the second method of Marcali, was used by Zapp (44) in his studies of the hazards of isocyanates. In this method the isocyanates are absorbed in a solution containing sodium nitrite and ethyl Cellosolve, without the boric acid. Skonieczny (74) compared the Ranta method with the Marcali diazotization method and found that it was not satisfactory at low concentrations.

Hennschler, Assmann, and Meyer (53) studied the colors developed from the various isomers of TDI by diazotization, and the color developed in the second method described by Marcali, using paper chromatography. They could not obtain reproducible colors by the latter procedure.

Pilz (75) modified his earlier procedure by using an absorbing solution containing *N*,*N*-dimethylethylene diamine in dimethylformamide followed by diazotization and coupling with *N*-naphthylethylene diamine. In a more recent paper (76), he improved his method further by using an absorbing solution containing sulfuric acid in dimethylformamide followed by diazotization and coupling with phenol. Further concentration of the colored solution is obtained by extracting the solution with *N*-pentanol and measuring the color at 400 mμ.

Von Eicken (77) describes a method for hexamethylene diisocyanate vapors in air in which the isocyanate is absorbed in dimethyl sulfoxide containing 0.1*N* hydrochloric acid. After neutralization the solution is treated with 1-fluoro-2,4-dinitrobenzene. The condensation product is extracted with tetrachloroethane and the color measured at 366 mμ. This method is said to be sensitive to 0.002 ppm.

Robinson (66) investigated the Marcali diazotization procedure for determining TDI in the atmosphere of workrooms in which polyurethane wire-insulating varnishes are baked at temperatures up to 400°F. The pyrolysis products from such varnishes might contain benzol, toluene, formaldehyde, phenol, and cresol, all of which seriously interfered with the color development. He developed a procedure based on the reaction of TDI with *p*-dimethylamino-benzaldehyde (Ehrlich's reagent) in the presence of acetic acid. The bright yellow color, which is read in a spectrophotometer at 425 mμ, obeys Beer's law in the range of 0–50 μg of TDI per 20 ml of 0.6*N* acetic acid, and appears to be free of interference from phenol up to 40 ppm in a sample containing 0.1 ppm of TDI. The details of this procedure are also given below.

Grim and Linch (67) improved on the Marcali method by incorporating a self-powered Uni-Jet constant rate air sampler (68) and increasing the air sample to 1 cu ft (28.3 liters), collected at a rate of 0.1 cu ft/min, and reducing the volume of absorbing solution to 8.0 ml. The useful range for this procedure is 0.01–0.08 ppm of TDI. The Marcali diazotization procedure was also modified for the determination of MDI by increasing the concentration of the coupling solution (*N*-naphthylethylene diamine) to 1.0% and adjusting the acidity with sodium carbonate solution before adding the coupling solution. The color is measured at 555 mμ and concentrations of MDI of 0.01 ppm can be detected.

Luxon (70) recently reported a technique for using the *N*-1-naphthylethylenediamine method of Marcali (64) that allows the TDI in air concentration to be determined in approximately 30 min. In this technique, high efficiency sintered plate scrubbers are used. These are designed to provide an adequate depth of liquid with only 6 ml of absorbent present. This also permitted the volume of air sampled to be reduced to 18 liters. Efficiency was further increased by using a solution of 0.1% acetic acid in 2.5% hydrochloric acid since slightly lowered acidity appears to increase sensitivity.

Several comments have been made in the foregoing discussion concerning the relative hazard of TDI and MDI ascribing the difference to their different vapor pressure. In general, if two compounds have similar toxicity but different vapor pressures, the compound with the higher vapor pressure will present the greatest hazard. This volatility factor is primarily concerned with the hazards arising from inhalation of the substance causing respiratory irritation. Many of the industrial applications of isocyanates require that the components be sprayed onto a surface, thus exposing the workmen not only to fine droplets, but also to the vapors. Thus, the atmosphere may contain high concentrations of isocyanate, if it has a high vapor pressure. The increasing popularity of the high molecular weight polymeric isocyanates such as Mondur MR is due in part to its low vapor pressure. In the manufacture and handling of these polymeric isocyanates, it is sometimes desirable to demonstrate their low vapor pressure. The determination of the absolute vapor pressure requires special apparatus and careful techniques, so that the simplified method described below has proven useful in these problems. The presence of lower boiling isocyanates or solvents can readily be detected by comparing its behavior when boiling with that of a known compound or with another lot of the same material of known purity.

B. Detailed Procedures

Method 1

The diazotization procedure described by Marcali (64) is outlined in the following.

SCOPE

This method was originally designed to determine the concentration of TDI vapor in air. It can also be used for naphthylisocyanate, and with suitable modifications suggested by Grim and Linch (67) it can be used for MDI.

PRINCIPLE

The TDI vapors are absorbed in an acid solution, which converts it to the amine. The amine is diazotized and coupled with *N*-naphthylethylene diamine and the color measured in a spectrophotometer.

PROCEDURE

(*1*) *Air Sampling and Preparation of Air–Diisocyanate Mixtures*

(*a*) *Air Sampling*. Air is drawn through the all-glass bubbler and impinger assembly containing 15 ml of absorber solution, using the hand pump, at a rate of 0.1 cu ft/min for 3 min. A single bubbler will remove 95% of the TDI at levels below 2 ppm, according to Marcali.

(*b*) *Preparation of Air–Diisocyanate Mixtures*. In order to prepare samples in the range of 2–3 ppm TDI based on 0.1 cu ft of air, a known weight of TDI was vaporized into the absorber by bubbling a given volume of dry air through the isocyanate at a known temperature. During an 8-hr period, the concentration of sample in air could be duplicated with less than 1% relative deviation.

In the range of 0.1–0.5 ppm TDI, a known volume of scrupulously dried air is continuously mixed under dynamic conditions of gas flow with known TDI air compositions prepared as described above.

(*c*) *Calculations*

$$\frac{W \times 131.0 \times 1.1}{V} = \text{ppm TDI}$$

where V = total sample volume expressed in milliliters and W = weight of TDI in micrograms present in sample V.

(2) *Apparatus and Reagents*

(*a*) Any suitable, commercially obtainable spectrophotometer will suffice.

(*b*) Sodium nitrite solution: 30 g sodium nitrite and 5.0 g sodium bromide are dissolved in about 80 ml of water and diluted to 100 ml.

(*c*) Sulfamic acid solution, 10% solution by weight.

(*d*) *N*-1-Naphthylethylenediamine solution: One ml. of concentrated HCl is added to a solution of 50 mg of *N*-1-naphthylethylenediamine in 25 ml of water and diluted to 50 ml.

(*e*) Absorber medium. Aqueous solution, 0.4*N* with acetic and hydrochloric acids. To approximately 600 ml of water, 35 ml of concentrated hydrochloric acid and 22 ml of glacial acetic acid, are added and then diluted to 1 liter.

(*f*) Tolylenediisocyanate.

(*g*) Impinger, graduated, all-glass assembly (1.0 mm impinger opening)—Mine Safety Appliance Co., Catalog No. CT-42513 (65).

(*h*) Pump, hand, air sampling. Catalog No. CT-144475 or CT-14502 (65).

(3) *Calibration*

(*a*) *TDI Solution A*. Approximately 200–255 mg of TDI is weighed accurately into 660 ml of glacial acetic acid. The solution is transferred and diluted to volume in a 1-liter volumetric flask. The solution should be used within 15 min after final dilution, to prepare solution *B*.

(*b*) *TDI Solution B*. An appropriate aliquot of solution *A* containing 2200 μg of TDI is transferred to a 1-liter volumetric flask. Sufficient volume of 8.8*N* acetic acid is added so that when solution *B* is diluted to 1000 ml it will be 0.6*N* in acetic acid. The solution is then diluted to volume with water.

To a series of eight graduated absorber cylinders, 5 ml of 1.2*N* hydrochloric acid is added. Then, 10.0, 9.5, 9.0, 8.0, 7.0, 6.0, 5.0, and 0.0 ml of 0.6*N* acetic acid are added and sufficient amount of solution *B* so that the sum of the two additions equals 10.0 ml. To each cylinder 0.5 ml of 3% sodium nitrite solution is added and the same procedure as is used for air samples is followed. A calibration curve is obtained by plotting the transmittance at 550 mμ vs. μg of TDI.

(4) *Air Sample Analysis* Glass impingers are used to obtain air samples as previously described. After sampling is complete, the inner bubbler tube is loosened, and the tube is tapped against the side of the bubbler cylinder to remove the adhering absorber solution before removing completely. A separate bubbler is carried through all the steps except actual air sampling in order to act as a blank.

To each bubbler, 0.5 ml of 3% sodium nitrite solution is added and allowed to stand for about 90 sec. One ml of 10% sulfamic acid solution is added, mixed, and allowed to stand about 2 min to destroy all the excess nitrous acid present. One ml of 0.1% *N*-naphthylethylenediamine is added, mixed, and the color is allowed to develop for about 5 min. A reddish-blue color indicates the presence of TDI. The final volume is adjusted to 20 ml with water. The transmittance is read at 550 mμ using a 1 cm or 5 cm cuvet. The sample absorbance, after allowing for the absorbance of the blank, is compared with the calibration curve and concentration of TDI in ppm is calculated as above.

The precision of the method was reported to be ±0.022 ppm at the 0.26 ppm level and ±0.3 ppm at the 2.5 ppm at the 95% confidence level. All of the important variables were investigated by Marcali and the method as reported represents optimum conditions.

Marcali (64) also described a portable field kit for determining TDI based on the above procedure.

Method 2

The second or alternate procedure described by Marcali (64) is outlined in the following.

Scope

This method determines TDI together with the urea (3,3′diisocyanato-4,4′-dimethyl carbanilide) formed by the reaction of TDI with moisture. The urea concentration can be estimated by deducting the TDI content from the results of this method.

Principle

A yellow color is produced by both the above urea and TDI when added to a solution of sodium nitrite and boric acid in ethyl Cellosolve. This color is measured in a spectrophotometer at 450 mμ.

PROCEDURE

(*1*) *Air Sampling and Preparation of Air–Diisocyanate Mixtures.* As described in Method 1.

(*2*) *Apparatus and Reagents*

(*a*) Any suitable commercially obtainable spectrophotometer will suffice.

(*b*) Ethyl Cellosolve absorber medium. Five grams of sodium nitrite and 10 g of powdered boric acid are dissolved in 200 ml of water. This solution is diluted to 1 liter with ethyl Cellosolve.

(*c*) Midget portable pump with volume displacement of about 3 cu ft/hr such as the Marco Model B-2.

(*3*) *Calibration.* A stock solution of TDI in ethyl Cellosolve–nitric–absorber medium is prepared to contain 4–5 μg of TDI per ml. This solution is used to prepare standard dilutions containing between 0 and 100 μg per 25 ml of ethyl Cellosolve–nitric–boric acid absorber medium. The final solutions should be allowed to stand at room temperature for 1 hr before the transmittance of the solutions are read at 450 mμ using 2.5 cm path length calls. A calibration curve is prepared from these data taking the absorbance of a blank into account. The standard curve may also be prepared using known amounts of 3,3′-diisocyanato-4,4′-dimethylcarbanilide (if available) dissolved in absorber medium.

(*4*) *Air Sample Analysis.* Into an impinger flask, 25 ml of ethyl Cellosolve absorber medium is introduced. The impinger bubbler is assembled and connected to a pump. A 0.3–0.8 cu ft sample is collected at a rate of 0.1 cu ft in 3 min. After sampling is complete, the color is allowed to develop for an hour. The final volume is adjusted to 25 ml with absorber medium and the transmittance read at 450 mμ in 2.5 cm cells. The micrograms of TDI and 3,3′-diisocyanato-4,4′-dimethylcarbanilide are determined from a previously prepared calibration curve. The micrograms of TDI are determined separately from the *N*-1-naphthylethylenediamine procedure.

CALCULATION

$$\left(A - \frac{VB}{N}\right)\frac{0.79}{MV} = \text{ppm of 3,3'-diisocyanato-4,4'-dimethylcarbanilide}$$

where A = μg of total TDI and 3,3′-diisocyanato-4,4′-dimethylcarbanilide found by the ethyl Cellosolve method in sample volume V.

V = Sample size in cu ft used in ethyl Cellosolve method.

N = Sample size in cu ft used in *N*-1-naphthylethylenediamine method.

B = μg of TDI found in N cubic feet of sample by N-1-naphthylethylenediamine method.

M = 322 (MW of 3,3′-diisocyanato-4,4′-dimethylcarbanilide.

The precision of the method was found to be ±0.029 ppm at the 0.1 ppm carbanilide range for the 95% confidence level of duplicate determinations.

In analytical ranges below 0.03 ppm the method was found to be subject to appreciable error when TDI was present in the range of 0.25 ppm.

The calibration curves for 2,4-tolylenediisocyanate and carbanilide lie very close together as shown in Figure 8.1. This means that both materials are determined by this method, and in dilute solutions a single calibration with 3,3′-diisocyanato-4,4′-dimethylcarbanilide may be used to approximate the total concentration of 2,4-tolylenediisocyanate and carbanilide whereas the latter compound in the N-1-naphthylethylenediamine method does not produce color.

Solution concentrations, time for color development, and the effects of other variables were investigated. Materials such as acetic acid, *o*-dichlorobenzene, dimethylformamide, and xylene if present in the atmosphere, were found to interfere and hinder color development.

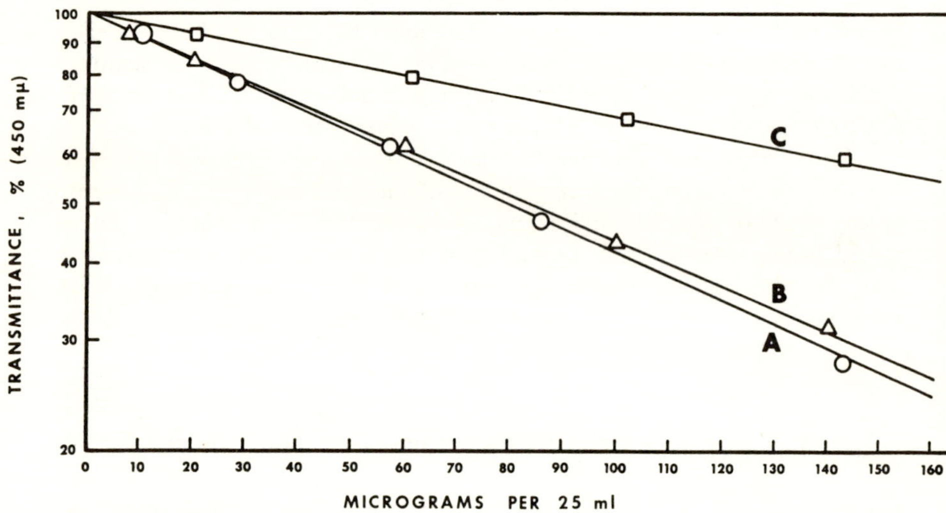

Fig. 8.1. Calibration curves for toluene-2,4-diisocyanate and 3,3-diisocyanato-4,4′-dimethylcarbanilide. (*A*) Toluene-2,4-diisocyanate, boric acid present. (*B*) 3,3-Diisocyanato-4,4′-dimethylcarbanilide, boric acid present. (*C*) 3,3′-Diisocyanato-4,4′-dimethylcarbanilide, boric acid absent. 2.5-cm cells.

Method 3

The method described by Robinson (66) is outlined below.

SCOPE

This method was designed to permit the determination of TDI in an atmosphere also containing solvent vapors and phenolic compounds resulting from the baking of polyurethane varnishes.

PRINCIPLE

TDI produces a bright yellow color by reacting with *p*-dimethylaminobenzaldehyde in the presence of acetic acid. This color is measured in a spectrophotometer at 425 mμ.

PROCEDURE

(*1*) *Reagents*

(*a*) Acetic acid, 0.6*N*.

(*b*) Dimethylaminobenzaldehyde, 1%. 0.5 g of dimethylaminobenzaldehyde is dissolved in 25 ml of glacial acetic acid and diluted with 25 ml of water. The solution is stored in a brown bottle in the refrigerator and prepared fresh daily.

(*2*) *Method.* At least 10 liters of air is collected by absorbing in a midget impinger tube containing 15 ml of 0.6*N* acetic acid at a rate of approximately 1 liter/min. The solution is transferred to a colorimeter tube and diluted to 20 ml with 0.6*N* acetic acid. After mixing, 1 ml of the dimethylaminobenzaldehyde reagent is added, after which it is mixed again and allowed to stand for 1 hr. The transmission is read at 425 mμ and the reading compared with a calibration curve obtained by introducing an appropriate concentration of TDI in 20 ml of 0.6*N* acetic acid in order to obtain the concentration of TDI in the air sample.

Method 4

Determination of the relative volatility of two liquids (79).

SCOPE

The volatility or vapor pressure of an unknown liquid may be compared with that of a known liquid by the method described below. For this purpose, the vapor pressure of the reference or known liquid should not be too far different from that of the unknown liquid.

Benzyl cinnamate having a boiling point of 350° at 760 mm pressure is useful as a reference liquid for comparison with other high boiling liquids.

This method is also useful for detecting small amounts of low boiling solvents or more volatile isocyanates in polyisocyanates such as Mondur

TABLE 8.1
Boiling Point of TDI, MDI and Benzyl Cinnamate

Pressure (mm Hg)	Boiling point (°C)		
	TDI	MDI	Benzyl cinnamate
1.0	77	170	174
5.0	107	206	206
10.0	121	224	221
40.0	—	—	256
100	180	—	281
400	220	—	327
760	247	—	350

MR. In this case a pure grade of MDI (diphenylmethane diisocyanate) may be used as the reference liquid. The boiling points of benzyl cinnamate and MDI are compared with that of TDI (toluene diisocyanate) in Table 8.1.

PRINCIPLE

The unknown and known liquids, each contained in identical flasks with accessory equipment as shown in Figure 8.2 and connected to a common vacuum system, are boiled at the same reduced pressure. The vapor temperature over the liquid having the higher vapor pressure (greater volatility) will be lower than that over the liquid having the lower vapor pressure. Likewise, if there are low boiling solvents or more volatile constituents in the unknown liquid, the vapor temperature will be lower than that of the reference liquid.

CHEMICALS

(*1*) Benzyl cinnamate (Interchemical Corp.) may be used as a reference liquid for comparison with other high boiling liquids.

(*2*) Diphenylmethane diisocyanate (MDI) may be obtained having high purity as Multrathane M.

EQUIPMENT

(*a*) Two stirrer bearings, ACE 8242.
(*b*) Two 500 ml three-neck boiling flasks, Kontes K 60605.
(*c*) Two stirrer shafts, Kontes K 78650.
(*d*) Two Teflon stirrer blades, Kontes K 78903.
(*e*) Two adapters, Kontes K 19550.

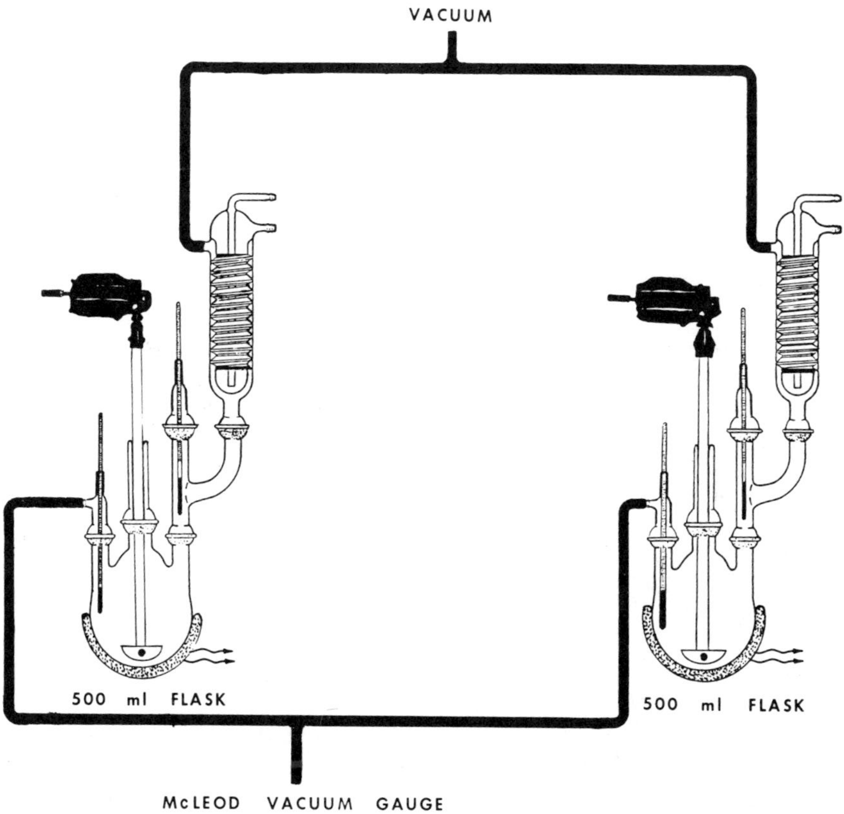

Fig. 8.2. Apparatus for the measurement of relative volatility of polyisocyanates. (79)

(*f*) Two adapters, Kontes K 16100.

(*g*) Two thermometers, −10 to 250° C, 75 mm immersion, Kontes K 87100 thermometer. These should be calibrated with the thermometers immersed in the liquid to the same depth that will occur in the test apparatus.

(*h*) Two thermometers, 12 in. 0.2°C subdivisions, 3 in. immersion, Princo, 175–210°C or 210–245° depending on the range required. These should also be calibrated in a similar manner.

(*i*) Two Friedrichs condensers, Ace 6105.

(*j*) Two Glas-Col heating mantles for 500 cc flasks.

(*k*) Two variable voltage transformer.

(*l*) McLeod gauge (range 0–10 mm Hg).

(*m*) Two variable speed stirring motors.
(*n*) Vacuum pump.
(*o*) Vacuum tubing and tees.
(*p*) Pinch clamps.
(*q*) Needle valve.
(*r*) Aluminum foil.

The above flasks and accessories have 35/25 ball joints, except for the 10/30 tapered thermometer joints.

PROCEDURE

(*1*) Assemble the equipment as shown in Figure 8.2.

(*2*) Transfer 350 ml of reference liquid to one of the flasks, and the same volume of unknown liquid to the other flask. It is desirable to keep the surface area of the two liquids exposed to heat approximately the same, to avoid unequal boiling.

(*3*) Lubricate all joints and make vacuum tight. Turn on the cooling water to the condensers, start stirring motors and vacuum pump. Apply aluminum foil to the upper surfaces of each flask, and turn on the power to the heating mantles.

(*4*) Adjust vacuum to 8–10 mm Hg. It is advisable to clamp off the rubber tubing connections to the McLeod gauge to avoid condensation in the gauge.

(*5*) Continue heating until the liquids are boiling and a gentle reflux is attained. Close the vacuum to one of the flasks with the pinch clamp, and read the pressure in the other flask. Reverse this procedure and measure the pressure in the other flask. If the pressures are not identical, adjust the heat on one of the flasks until they are equalized.

(*6*) When the two systems are refluxing at equal pressures, record the temperature of the liquid and the vapor in each system.

(*7*) To obtain the vapor pressure of an unknown liquid at other temperatures, the pressure in the flasks can be adjusted to other values and brought to equilibrium and the thermometers read under the new conditions.

RESULTS

For a pure liquid, the temperature of the vapor and the liquid will be very nearly the same, and if temperature readings at several pressures have been made, a vapor pressure–temperature plot can be made.

When comparing an unknown lot of polyisocyanate with a lot of known purity, the vapor temperature may be several degrees lower than that of the control. In general, if there is a temperature difference of 10°C, it can be concluded that the sample has been contaminated with low boiling isocyanate.

TABLE 8.2

Relative Volatility of Isocyanate Mixtures

Test	Control sample— Multrathane M MDI			Test sample— Mondur MR		
	Liquid temp. (°C)	Vapor temp. (°C)	Pressure (mm Hg)	Liquid temp. (°C)	Vapor temp. (°C)	Pressure (mm Hg)
I. Mondur MR without contamination	217	217	7.4	222	214	7.4
II. Mondur MR with 2% TDI added	211	211	6.4	210	190	6.4
III. Mondur MR with 1% phenyl isocyanate added	213	211	7.1	186	125	7.1
IV. Mondur MR with 0.5% TDI added	211	211	6.5	214	200	6.5

Discussion

Table 8.2 above shows the effect of contamination of a sample of Mondur MR with lower boiling isocyanates, as disclosed by the test procedure described above.

It can be concluded from test I that the vapor from Mondur MR contains nothing more volatile than that from the MDI, since the vapor temperatures are only 3° apart. The liquid temperature of Mondur MR is 5° higher than that for the MDI, which indicates a lower volatility in the Mondur MR.

Tests IV and II show that the vapor temperature of Mondur MR is markedly lowered (11° and 21°, respectively) by the addition of 0.5% and 2% TDI. Thus TDI increases the volatility of the polymeric isocyanate.

The presence of 1% phenyl isocyanate, having a much lower boiling point than TDI, lowers the vapor temperature of Mondur MR 85° to 90°C and lowers the liquid temperature about 36°.

The test results discussed above show that the method is capable of detecting small amounts of low boiling isocyanates, which are more volatile and which would reduce the advantages of the polymeric isocyanate for keeping the toxicity hazards at a minimum.

References

1. Hegler, C., *Deut. Med. Wochschr.*, **54,** 1551 (1928); *CA* 23:199.
2. Bloomfield, J. J., *Ind. Hyg. Foundation, Trans. 11th Ann. Meeting, Bull.*, **8,** 71 (1947); *CA* 41:3884i.

3. Konowalow, A., *Z. Ges. Schiess-Spreugstoffw.*, **22,** 152 (1927); *CA* 21,4033[4] (1927).
4. Goldston, M., Leutscher, J. A., Jr., Longcope, W. T., Ballich, N. L., Kremer, V. L., Filley, G. L., and Hopson, J. L., *J. Clin. Invest.*, **26,** 145 (1947); *CA* 43:5513e.
5. Boyd, E. M., and Perry, W. F., *J. Pharm. Pharmacol.*, **12,** 726 (1960); *CA* 55:8671c.
6. Long, J. E., and Hatch, T. F., *Am. Ind. Hyg. Assoc. J.*, **1,** 6 (1961); *CA* 55:10703a.
7. Trethewie, E. R., *Med. J. Australia*, **34,** 746 (1947); *CA* 41:7524f.
8. Capellini, A., and Zanotti, G. G., *Med. Lavoro*, **37,** 265 (1946); *CA* 41:1760h.
9. Rejsek, K., *Acta Med. Scand.*, **127,** 179 (1947); *CA* 41:4578c (1947).
10. Glahn, M., and Schack-Schou, P., *Nord. Med.*, **36,** 2135 (1947); *CA* 42:4756b.
11. Monnier, J., *Toulouse Med.*, **48,** 728 (1947); *CA* 44,10928i (1950).
12. Chary, R., *Mem. Poudres*, **37,** 351 (1955); *CA* 50,17451e (1956).
13. Yamada, Y., *Kobe J. Med. Sci.*, **4,** 227 (1958); *CA* 53:20555d.
14. Leclerc, E., and Devlaminick, F., *Bull. Centre Belge Etude Doc. Eaux (Liege)* **14,** 246 (1951); *CA* 46,10597i (1952).
15. Lurie, J. B., *S. African Med. J.*, **23,** 233 (1949); *CA* 44:232f.
16. Dmitrenko, M. T., *Farmakol. i Toksikol.*, **17**(1), 39 (1954); *CA* 48:13984b.
17. Parmeggiani, L., and Sassi, C., *Med. Lavoro*, **45,** 574 (1954); *CA* 49:5725b.
18. Hasegawa, H., and Sato, M., *Bull. Natl. Inst. Ind. Health*, 2, 1–17 (1959); *CA* 54:21492h.
19. Forssman, S., and Frykholm, O. K., *Acta Med. Scand.*, **128,** 256 (1947); *CA* 41:7581h.
20. Elmino, O., Rozera, G., and Colicchio, G., *Folia Med.*, **42,** 1228 (1959); *CA* 54:14460d.
21. Siae, H. R., *Chung Hua Wei Sheng Tsa Chih*, **6,** 327 (1958); *CA* 57:12826i (1962).
22. Elkins, H. B., *Proc. Intern. Congr. Occupational Health, 13th, New York 1960* (Pub. 1961) 981; *CA* 58:871e.
23. Navrotskii, V. K., and Orlik, I. M., *Farmakol. Toksikol.*, **10**(3), 40 (1947); *CA* 42,2680e (1948).
24. Montanaro, O., Saraco, E. G., and Paglilla, C. J., *Semana Med.*, (1952) **II,** 7; *CA* **46,** 11441g (1952).
25. Navrotskii, V. K., Lukashov, V. I., Nikolaeva, N. M., and Tiraspolskaya, M. M., *Vrachebnoe Delo*, **1958,** 59, *CA* 52:14005a.
26. Solov'eva, V. A., *Vopr. Prom. Toxsikol.*, Moscow, 1960, 29-49, *CA* 57:7561d.
27. Tkachev, P. G., *Gigiena i Sanit.*, **28**(4), 3 (1963); *CA* 59:4471a.
28. Doerr, W., *Arch. Pathol. Anat. Physiol.*, **313,** 137 (1944); *CA* 41:6983e.
29. McDonald, S. F., *Med. J. Australia*, **34,** 204 (1947); *CA* 41:7524e.
30. Karel, L., Landing, B. H., and Harvey, T. S., *J. Pharmacol. Exptl. Therap.*, **90,** 338 (1947); *CA* 42:279c.
31. Hinrichs, A., Kopf, R., and Loeser, A., *Pharmazie*, **3,** 110 (1948); *CA* 42:5567h.
32. Dotzauer, G., *Deut. Med. Wochschr.*, **73,** 22 (1948); *CA* 42:6937c.
33. Anon, *API Toxicol Rev.*, September, 1948; 3 pp, *CA* 43:5514e.
34. Scholz, J., *Klin. Wochschr.*, **28,** 69 (1950); *CA* 44:9059d.
35. Lampe, K. F., and Easterday, O. D., *J. Am. Pharm. Assoc.*, **42,** 445 (1953); *CA* 47:10129c.
36. Loeser, A., Bornmann, G., Grosskinsky, L., Hess, G., Kopf, R., Ritter, K., Schmitz, A., Sturmer, E., and Wegener, H., *Naunyn-Schmiedebergs Arch. Exptl. Pathol. Pharmakol.*, **221,** 14 (1954); *CA* 48:4698c.
37. Nakahara, Y., *Yamaguchi Igaku*, **8,** 277 (1959); *CA* 59: 2092e.
38. Reinl, W., *Farbe Lack*, **60,** 69 (1954); *CA* 48:5544g.
39. Swensson, A., Holmquist, C. E., and Lundgren, K. D., *Brit. J. Ind. Med.*, **12,** 50 (1955); *CA* 49:12015g.
40. Schürmann, D., *Deut. Med. Wochschr.*, **80,** 1661 (1955); *CA* 50:3799e.
41. Reinl, W., *Zentralbl. Arbeitsmed Arbeitsschultz*, **3,** 103 (1953).

42. Reinl, W., *Zeitschrift Bundesarbeitsblatt*, **14,** 421 (1954).
43. Fuschs, S., and Valade, P., *Arch. Maladies Profess. Med. Travail Securite Sociale*, **12,** 191 (1951); *CA* 45:6735e.
44. Zapp, J. A., Jr., *AMA Arch. Ind. Health*, **15,** 324 (April 1957); *CA* 51:9205i.
45. Wilson, R. H., and McCormick, W. E., "Health Problems in Using Isocyanates," Rubber Section, National Safety Council, April 23, 1956.
46. Wilson, R. H., and Wilson, G. L., *Toxicology of Diisocyanates, Journal of Occupational Medicine*, 1, 448-450 (1959).
47. "Toxicity and Safe Handling of Isocyanates," Mobay Chemical Company, staff report Pittsburgh, Pa., circa 1962.
48. MacFarland, H. N., and Leong, K. J., *Arch. Environ. Health*, **4,** 591 (1961); *CA* 57:6269e.
49. Duncan, B., Scheel, L. D., Fairchild, E. J., Killans, R., and Graham, S., *Am. Ind. Hyg. Assoc. J.*, **23,** 447 (1962); *CA* 58:8345g.
50. Woodbury, J. W., *Ind. Med. Surg.*, **25,** 540 (1956).
51. Johnstone, R. T., *Ind. Med. Surg.*, **26,** 33 (1957).
52. Strayer, S. C., *A.M.A. Arch. Ind. Health*, **19,** 351 (1959); *CA* 53:10594g.
53. Henschler, D., Assmann, W., and Meyer, K. O., *Arch. Toxikologie*, **19,** 364 (1962)· *CA* 59:2087h.
54. Vesely, M., and Zeman, M., *Kozarstvi*, **12,** 166 (1962); *CA* 58:10659a (1963).
55. Brugsch, H. G., and Elkins, H. B., *New Engl. J. Med.*, **268,** 353 (1963); *CA* 58:11887c.
56. Trenchard, H. J., and Harris, W. C., *Lancet* **1963-I,** 404; *CA* 58:13048b.
57. Gandevia, B., *Brit. J. Ind. Med.*, **20**(3), 204 (1963); *CA* 60:6125e.
58. Silver, H. M., *Arch. Internal Med.*, **112**(3), 401 (1963); *CA* 60:2242g.
59. Maxon, F. C., *Arch. Environ. Health*, **8**(5), 755 (1964); *CA* 61:3597d.
60. Munn, A., *Ann. Occup. Hyg.*, **8,** 163 (1965); *CA* 63:7543c.
61. *Chemicals Hazards Information Series, No. C-70*, "Isocyanate-Tolylene 2,4-Diisocyanate" April 1965, American Insurance Assoc., New York.
62. Tolylene Diisocyanate, *Chemical Safety Data Sheet SD-73*, Manufacturing Chemists' Assoc., Inc., Washington, D.C., 1959.
63. Longley, E. O., *Arch. Environ. Health*, **8,** 898 (1964); *CA* 61:7595f.
64. Marcali, K., *Anal. Chem.*, **29,** 552 (1957); *CA* 51:11177b.
65. Mine Safety Appliances Co., Pittsburgh, Pa., Catalog 7-B, pp. 3–28.
66. Robinson, D. B., *Am. Ind. Hyg. Assoc. J.*, **23,** 228 (1962); *CA* 57:12829d.
67. Grim, K. E., and Linch, A. L., *Am. Ind. Hyg. Assoc. J.*, **25**(3), 285 (1964); *CA* 61:11235d.
68. Linch, A. L., and Charsha, H. G., *Am. Ind. Hyg. Assoc. J.*, **21,** 325 (1960); *CA* 54:23143g.
69. Kohn, R. J., "Automatic "Uni-Jet" TDI Analyzer," Trade Bulletin, Union Industrial Equipment Corp., Port Chester, New York.
70. Luxon, S. G., *Ann. Occup. Hyg.*, **8,** 171 (1965); *CA* 63:12219e.
71. Ehrlicher, H., and Pilz, W., *Arbeitsschutz*, **1956,** 276 (Part I).
72. Ehrlicher, H., and Pilz, W., *Arbeitsschutz*, **1957,** 7 (Part II).
73. Ehrlicher, H., *Zbl. f. Arbeitsmed. u. Arbeitssch*, **6,** 138 (1956).
74. Skonieczny, R. F., *Am. Ind. Hyg. Assoc. J.*, **24,** 17 (1963); *CA* 59:1026b.
75. Pilz, W., *Mikrochim. Acta*, **1958,** 545; *CA* 54:19311.
76. Pilz, W., *Mikrochim. Ichnoanal Acta*, **1963,** 279; *CA* 58:14619f.
77. Von Eicken, S., *Mikrochim. Acta*, **1958,** No. 6, 731; *CA* 55:2965f.
78. Scheel, L. D., Killens, R., and Josephson, A., *Amer. Ind. Hyg. Assoc. J.*, **25,** 179 (1964); *CA* 63:6167e.
79. System Suppliers Newsletter No 6, Part II, Feb 1966, 5 pp. Mobay Chemical Company, Pittsburgh, Pa.

AUTHOR INDEX

SUBJECT INDEX